Lecture Notes in Computer Science　12602

More information about this subseries at http://www.springer.com/series/7409

Wil M. P. van der Aalst ·
Vladimir Batagelj · Dmitry I. Ignatov ·
Michael Khachay · Olessia Koltsova ·
Andrey Kutuzov · Sergei O. Kuznetsov ·
Irina A. Lomazova · Natalia Loukachevitch ·
Amedeo Napoli · Alexander Panchenko ·
Panos M. Pardalos · Marcello Pelillo ·
Andrey V. Savchenko · Elena Tutubalina (Eds.)

Analysis of Images, Social Networks and Texts

9th International Conference, AIST 2020
Skolkovo, Moscow, Russia, October 15–16, 2020
Revised Selected Papers

 Springer

Editors
Wil M. P. van der Aalst (iD)
RWTH Aachen University
Aachen, Germany

Dmitry I. Ignatov (iD)
National Research University Higher School
of Economics
Moscow, Russia

Olessia Koltsova (iD)
National Research University Higher School
of Economics
St. Petersburg, Russia

Sergei O. Kuznetsov (iD)
National Research University Higher School
of Economics
Moscow, Russia

Natalia Loukachevitch (iD)
Moscow State University
Moscow, Russia

Alexander Panchenko (iD)
Skolkovo Institute of Science
and Technology
Moscow, Russia

Marcello Pelillo (iD)
Università Ca' Foscari Venezia
Venice, Italy

Elena Tutubalina (iD)
Kazan Federal University
Kazan, Russia

Vladimir Batagelj (iD)
University of Ljubljana
Ljubljana, Slovenia

Michael Khachay (iD)
Krasovskii Institute of Mathematics
and Mechanics
Yekaterinburg, Russia

Andrey Kutuzov (iD)
University of Oslo
Oslo, Norway

Irina A. Lomazova (iD)
National Research University Higher School
of Economics
Moscow, Russia

Amedeo Napoli (iD)
LORIA
Vandœuvre lès Nancy, France

Panos M. Pardalos (iD)
University of Florida
Gainesville, FL, USA

Andrey V. Savchenko (iD)
National Research University Higher School
of Economics
Nizhny Novgorod, Russia

ISSN 0302-9743   ISSN 1611-3349 (electronic)
Lecture Notes in Computer Science
ISBN 978-3-030-72609-6   ISBN 978-3-030-72610-2 (eBook)
https://doi.org/10.1007/978-3-030-72610-2

LNCS Sublibrary: SL3 – Information Systems and Applications, incl. Internet/Web, and HCI

This Springer imprint is published by the registered company Springer Nature Switzerland AG
The registered company address is: Gewerbestrasse 11, 6330 Cham, Switzerland

Preface

This volume contains the refereed proceedings of the 9th International Conference on Analysis of Images, Social Networks, and Texts (AIST 2020)[1]. The previous conferences (during 2012–2019) attracted a significant number of data scientists – students, researchers, academics, and engineers – working on interdisciplinary data analysis of images, texts, and social networks.

The broad scope of AIST make it an event where researchers from different domains, such as image and text processing, exploiting various data analysis techniques, can meet and exchange ideas. As the test of time has shown, this leads to the cross-fertilisation of ideas between researchers relying on modern data analysis machinery.

Therefore, AIST 2020 brought together all kinds of applications of data mining and machine learning techniques. The conference allowed specialists from different fields to meet each other, present their work, and discuss both theoretical and practical aspects of their data analysis problems. Another important aim of the conference was to stimulate scientists and people from industry to benefit from knowledge exchange and identify possible grounds for fruitful collaboration.

The conference was held during October 15–16, 2020. The conference was organised in the Skolkovo Innovation Center, Russia, on the campus of the Skolkovo Institute of Science and Technology[2], but held entirely online due to the COVID-19 pandemic.

This year, the key topics of AIST were grouped into six tracks:

1. Data Analysis and Machine Learning chaired by Sergei O. Kuznetsov (HSE University, Russia) and Amedeo Napoli (Loria, France)
2. Natural Language Processing chaired by Natalia Loukachevitch (Lomonosov Moscow State University, Russia), Andrey Kutuzov (University of Oslo, Norway)
3. Social Network Analysis chaired by Vladimir Batagelj (University of Ljubljana, Slovenia) and Olessia Koltsova (HSE University, Russia)
4. Computer Vision chaired by Marcello Pelillo (University of Venice, Italy) and Andrey V. Savchenko (HSE University, Russia)
5. Theoretical Machine Learning and Optimization chaired by Panos M. Pardalos (University of Florida, USA) and Michael Khachay (IMM UB RAS and Ural Federal University, Russia)
6. Process Mining chaired by Wil M. P. van der Aalst (RWTH Aachen University, Germany) and Irina A. Lomazova (HSE University, Russia)

To facilitate easy communication and negotiation of the area chairs and our authors via only digital channels, due to the virtual character of the event, we invited additional

---

[1] https://aistconf.org.

[2] https://www.skoltech.ru/en/.

area and program co-chairs for certain tracks, respectively: Alexei Buzmakov (HSE University, Perm, Russia), Ilya Makarov (HSE University, Moscow, Russia), Anna Kalenkova (University of Melbourne, Australia), and Elena Tutubalina (Kazan Federal University, Russia).

The Programme Committee and the reviewers of the conference included 134 well-known experts in data mining and machine learning, natural language processing, image processing, social network analysis, and related areas from leading institutions of many countries including Australia, Austria, Czech Republic, France, Germany, Greece, India, Iran, Ireland, Italy, Japan, Lithuania, Norway, Qatar, Romania, Russia, Slovenia, Spain, Taiwan, Ukraine, United Kingdom, and the USA. This year, we received 115 submissions: mostly from Russia but also from Algeria, Brazil, Finland, Germany, India, Norway, Pakistan, Serbia, Spain, Ukraine, United Kingdom, and the USA.

Out of 115 submissions (not taking into account seven automatically rejected papers), only 27 full papers and four short papers were accepted as regular oral papers in the main volume. Invited talks were also included in the main volume. In order to encourage young practitioners and researchers, we included 14 full and nine short papers in this companion volume after their short presentation at the conference and four non-indexed poster abstracts as well. Thus, the acceptance rate of this volume was around 30%. Each submission was reviewed by at least three reviewers, experts in their fields, in order to supply detailed and helpful comments.

The conference featured several invited talks dedicated to current trends and challenges in the respective areas.

The invited talks from academia were on Computer Vision and Natural Language Processing, respectively:

- Marcello Pelillo (Ca' Foscari University of Venice, Italy): "Graph-Theoretic Methods in Computer Vision: Recent Advances"
- Miguel Couceiro (LORIA, Université de Lorraine, France): "Making Models Fairer Through Explanations"
- Leonard Kwuida (Bern University of Applied Sciences): "On Interpretability and Similarity in Concept Based Machine Learning"
- Santo Fortunato (Indiana University Network Science Institute, USA): "Consensus Clustering in Networks"

The invited industry speakers gave the following talks:

- Nikita Semenov (MTS AI, Russia): "Text and Speech Processing Projects at MTS AI"
- Ivan Smurov (ABBYY; Moscow Institute of Physics and Technology, Russia): "When CoNLL-2003 is not Enough: are Academic NER and RE Corpora Well-Suited to Represent Real-World Scenarios? "

An extended part of Ivan's industry talk was included in the main volume under the title "RuREBus: a Case Study of Joint Named Entity Recognition and Relation Extraction from e-Government Domain."

We would like to thank the authors for submitting their papers and the members of the Programme Committee for their efforts in providing exhaustive reviews.

According to the programme chairs, and taking into account the reviews and presentation quality, the Best Paper Awards were granted to the following papers:

- Track 1. Data Analysis and Machine Learning: "Gradient-Based Adversarial Attacks on Categorical Sequence Models via Traversing an Embedded World" by Ivan Fursov, Alexey Zaytsev, Nikita Klutchnikov, Andrey Kravchenko, and Evgeny Burnaev;
- Track 2. Natural Language Processing: "Do Topics Make a Metaphor? Topic Modeling for Metaphor Identification and Analysis in Russian" by Yulia Badryzlova, Anastasia Nikiforova, and Olga Lyashevskaya;
- Track 3. Social Network Analysis: "Detecting Automatically Managed Accounts in Online Social Networks: Graph Embedding Approach" by Ilia Karpov and Ekaterina Glazkova;
- Track 4. Computer Vision: "Deep Learning on Point Clouds for False Positive Reduction at Nodule Detection in Chest CT Scans" by Ivan Drokin and Elena Ericheva;
- Track 5. Theoretical Machine Learning and Optimization: "Fast Approximation Algorithms for Stabbing Special Families of Line Segments with Equal Disks" by Konstantin Kobylkin;
- Track 6. Process Mining: "Checking Conformance between Colored Petri Nets and Event Logs" by Julio Cesar Carrasquel, Khalil Mecheraoui, and Irina Lomazova.

We would also like to express our special gratitude to all the invited speakers and industry representatives.

We deeply thank all the partners and sponsors, especially, the hosting university and our main sponsor and the co-organiser this year, the Skolkovo Institute of Science and Technology, as well as the National Research University Higher School of Economics (including its subdivisions). Our special thanks go to Springer for their help, starting from the first conference call to the final version of the proceedings. Last but not least, we are grateful to Evgeny Burnaev and all the organisers, especially our secretary Irina Nikishina, and the volunteers, whose endless energy saved us at the most critical stages of the conference preparation.

Here, we would like to mention that the Russian word "aist" is more than just a simple abbreviation (in Cyrillic) – it means "a stork". Since it is a wonderful free bird, a symbol of happiness and peace, this stork gave us the inspiration to organise the AIST conference series. So we believe that this conference will still likewise bring inspiration to data scientists around the world!

October 2020

Wil M. P. van der Aalst
Vladimir Batagelj
Aleksey Buzmakov
Dmitry Ignatov
Anna Kalenkova
Michael Khachay
Olessia Koltsova
Andrey Kutuzov
Sergei O. Kuznetsov
Irina Lomazova
Natalia Loukachevitch
Ilya Makarov
Amedeo Napoli
Alexander Panchenko
Panos M. Pardalos
Marcello Pelillo
Andrey Savchenko
Elena Tutubalina

Organisation

The conference was organised by a joint team from Skolkovo Institute of Science and Technology (Skoltech), the divisions of the National Research University Higher School of Economics (HSE University), and Krasovskii Institute of Mathematics and Mechanics of the Russian Academy of Sciences.

Organising Institutions

- Skolkovo Institute of Science and Technology (Moscow, Russia)
- Krasovskii Institute of Mathematics and Mechanics, Ural Branch of the Russian Academy of Sciences (Yekaterinburg, Russia)
- School of Data Analysis and Artificial Intelligence, HSE University (Moscow, Russia)
- Laboratory of Algorithms and Technologies for Networks Analysis, HSE University (Nizhny Novgorod, Russia)
- International Laboratory for Applied Network Research, HSE University (Moscow, Russia)
- Laboratory for Models and Methods of Computational Pragmatics, HSE University (Moscow, Russia)
- Laboratory for Social and Cognitive Informatics, HSE University (St. Petersburg, Russia)
- Laboratory of Process-Aware Information Systems, HSE University (Moscow, Russia)
- Research Group "Machine Learning on Graphs", HSE University (Moscow, Russia)

Program Committee Chairs

Wil van der Aalst	RWTH Aachen University, Germany
Vladimir Batagelj	University of Ljubljana, Slovenia
Michael Khachay	Krasovskii Institute of Mathematics and Mechanics of Russian Academy of Sciences, Russia & Ural Federal University, Yekaterinburg, Russia
Olessia Koltsova	HSE University, Russia
Andrey Kutuzov	University of Oslo, Norway
Sergei Kuznetsov	HSE University, Moscow, Russia
Amedeo Napoli	LORIA CNRS, University of Lorraine, and Inria, Nancy, France
Irina Lomazova	HSE University, Moscow, Russia
Natalia Loukachevitch	Computing Centre of Lomonosov Moscow State University, Russia

Panos Pardalos	University of Florida, USA
Marcello Pelillo	University of Venice, Italy
Andrey Savchenko	HSE University, Nizhny Novgorod, Russia
Elena Tutubalina	Kazan Federal University, Russia

Additional Area Chairs

Aleksey Buzmakov	HSE University, Perm, Russia
Anna Kalenkova	University of Melbourne, Australia
Ilya Makarov	HSE University, Moscow, Russia

Proceedings Chair

Dmitry I. Ignatov	HSE University, Moscow, Russia

Steering Committee

Dmitry I. Ignatov	HSE University, Moscow, Russia
Michael Khachay	Krasovskii Institute of Mathematics and Mechanics of Russian Academy of Sciences, Russia & Ural Federal University, Yekaterinburg, Russia
Alexander Panchenko	Skolkovo Institute of Science and Technology, Moscow, Russia
Andrey Savchenko	HSE University, Nizhny Novgorod, Russia
Rostislav Yavorskiy	Tomsk Polytechnic University, Russia

Program Committee

Anton Alekseev	St. Petersburg Department of V. A. Steklov Institute of Mathematics of the Russian Academy of Sciences, Russia
Ilseyar Alimova	Kazan Federal University, Russia
Vladimir Arlazarov	Smart Engines Ltd. and Federal Research Centre "Computer Science and Control" of the Russian Academy of Sciences, Russia
Aleksey Artamonov	Neuromation, Russia
Ekaterina Artemova	HSE University, Moscow, Russia
Jaume Baixeries	Universitat Politécnica de Catalunya, Spain
Amir Bakarov	HSE University, Moscow, Russia
Nikita Basov	St. Petersburg State University, Russia
Vladimir Batagelj	University of Ljubljana, Slovenia
Tatiana Batura	Ershov Institute of Informatics Systems, Siberian Branch of the Russian Academy of Sciences and Novosibirsk State University, Russia
Malay Bhattacharyya	Indian Statistical Institute, India
Michael Bogatyrev	Tula State University, Russia

Elena Bolshakova	Moscow State Lomonosov University, Russia
Ivan Bondarenko	Novosibirsk State University, Russia
Evgeny Burnaev	Skolkovo Institute of Science and Technology, Russia
Aleksey Buzmakov	HSE University, Perm, Russia
Dmitry Chaly	Demidov Yaroslavl State University, Russia
Mikhail Chernoskutov	Krasovskii Institute of Mathematics and Mechanics of Russian Academy of Sciences and Ural Federal University, Russia
Alexey Chernyavskiy	Philips Innovation Labs, Russia
Massimiliano de Leoni	University of Padua, Italy
Oksana Dereza	National University of Ireland, Ireland
Boris Dobrov	Moscow State University, Russia
Ivan Drokin	BrainGarden.ai, Russia
Aleksandr Drozd	Tokyo Institure of Technology, Japan
Shiv Ram Dubey	Indian Institute of Information Technology, Sri City, India
Olga Gerasimova	HSE University, Moscow, Russia
Dmitry Granovsky	Yandex, Russia
Vera Ignatenko	HSE University, St. Petersburg, Russia
Dmitry Ignatov	HSE University, Moscow, Russia
Dmitry Ilvovsky	HSE University, Moscow, Russia
Max Ionov	Goethe University Frankfurt, Germany and Moscow State University, Russia
Vladimir Ivanov	Innopolis University, Kazan, Russia
Anna Kalenkova	University of Melbourne, Australia
Ilia Karpov	HSE University, Moscow, Russia
Egor Kashkin	Vinogradov Russian Language Institute of the Russian Academy of Sciences, Russia
Yury Kashnitsky	Elsevier, The Netherlands
Alexander Kazakov	Matrosov Institute for System Dynamics and Control Theory, Siberian Branch of the Russian Academy of Science, Russia
Michael Khachay	Krasovskii Institute of Mathematics and Mechanics of Russian Academy of Sciences, Russia
Vladimir Khandeev	Sobolev Institute of Mathematics, Siberian Branch of the Russian Academy of Sciences, Russia
Javad Khodadoust	Payame Noor University, Iran
Gregory Khvatsky	HSE University, Moscow, Russia
Donghyun Kim	Georgia State University, USA
Denis Kirjanov	HSE University, Moscow, Russia
Dmitrii Kiselev	HSE University, Moscow, Russia
Sergei Koltcov	HSE University, St. Petersburg, Russia
Olessia Koltsova	HSE University, St. Petersburg, Russia
Evgeny Komotskiy	Ural Federal University, Russia
Jan Konečný	Palacký University Olomouc, Czech Republic
Anton Konushin	Moscow State University and Samsung, Russia

Andrey Kopylov   Tula State University, Russia
Evgeny Kotelnikov   Vyatka State University, Russia
Ekaterina Krekhovets   HSE University, Nizhny Novgorod, Russia
Tomas Krilavičius   Vytautas Magnus University, Lithuania
Sofya Kulikova   HSE University, Perm, Russia
Maria Kunilovskaya   University of Wolverhampton, UK
Anvar Kurmukov   Kharkevich Institute for Information Transmission
                                     Problems of the Russian Academy of Sciences,
                                     Russia
Andrey Kutuzov   University of Oslo, Norway
Elizaveta Kuzmenko   University of Trento, Italy
Andrey Kuznetsov   Samara National Research University, Russia
Sergei O. Kuznetsov   HSE University, Moscow, Russia
Stepan Kuznetsov   Steklov Mathematical Institute of the Russian Academy
                                     of Sciences, Russia
Florence Le Ber   Université de Strasbourg, France
Alexander Lepskiy   HSE University, Moscow, Russia
Bertrand M. T. Lin   National Chiao Tung University, Taiwan
Irina Lomazova   HSE University, Moscow, Russia
Konstantin Lopukhin   Scrapinghub Inc., Ireland
Natalia Loukachevitch   Research Computing Center of Moscow State
                                     University, Russia
Ilya Makarov   HSE University, Moscow, Russia
Tatiana Makhalova   HSE University, Russia and Loria, Inria, France
Alexey Malafeev   HSE University, Nizhny Novgorod, Russia
Yury Malkov   Institute of Applied Physics of the Russian Academy
                                     of Sciences, Russia
Valentin Malykh   Institute for Systems Analysis of the Russian Academy
                                     of Sciences, Russia
Nizar Messai   Université de Tours, France
Tristan Miller   Austrian Research Institute for Artificial Intelligence,
                                     Austria
Olga Mitrofanova   St. Petersburg State University, Russia
Alexey A. Mitsyuk   HSE University, Russia
Evgeny Myasnikov   Samara National Research University, Russia
Amedeo Napoli   Loria, CNRS, Inria, and University of Lorraine, France
The Long Nguyen   Irkutsk State Technical University, Russia
Irina Nikishina   Skolkovo Institute of Science and Technology, Russia
Kirill Nikolaev   HSE University, Nizhny Novgorod, Russia
Damien Nouvel   Institut national des langues et civilisations orientales
                                     (Inalco University), France
Dimitri Nowicki   Glushkov Institute of Cybernetics of the National
                                     Academy of Sciences, Ukraine
Evgeniy M. Ozhegov   HSE University, Perm, Russia
Alexander Panchenko   Skolkovo Institute of Science and Technology, Russia
Polina Panicheva   HSE University, St. Petersburg, Russia

Panos Pardalos	University of Florida, USA
Marcello Pelillo	University of Venice, Italy
Olga Perepelkina	Lomonosov Moscow State University, Russia
Georgios Petasis	National Center for Scientific Research "Demokritos", Greece
Anna Petrovicheva	Xperience AI, Russia
Vladimir Pleshko	RCO LLC, Russia
Mikhail Posypkin	Dorodnicyn Computing Centre of the Russian Academy of Sciences, Russia
V. B. Surya Prasath	Cincinnati Children's Hospital Medical Center, USA
Ekaterina Pronoza	Saint Petersburg State University, Russia
Artem Pyatkin	Novosibirsk State University and Sobolev Institute of Mathematics, Siberian Branch of the Russian Academy of Sciences, Russia
Irina Radchenko	ITMO University, Russia
Delhibabu Radhakrishnan	Kazan Federal University, Russia and VIT University, India
Vinit Ravishankar	University of Oslo, Norway
Yuliya Rubtsova	Ershov Institute of Informatics Systems, Siberian Branch of the Russian Academy of Sciences, Russia
Alexey Ruchay	Chelyabinsk State University, Russia
Eugen Ruppert	Universität Hamburg, Germany
Christian Sacarea	Babeş-Bolyai University, Romania
Aleksei Samarin	St. Petersburg University, Russia
Andrey Savchenko	HSE University, Nizhny Novgorod, Russia
Friedhelm Schwenker	Ulm University, Germany
Oleg Seredin	Tula State University, Russia
Tatiana Shavrina	HSE University, Moscow, Russia
Andrey Shcherbakov	The University of Melbourne, Australia
Sergey Shershakov	HSE University, Moscow, Russia
Denis Sidorov	Melentiev Energy Systems Institute, Siberian Branch of the Russian Academy of Sciences, Russia
Henry Soldano	Laboratoire d'Informatique de Paris Nord, France
Alexey Sorokin	Moscow State University, Russia
Andrey Sozykin	Krasovskii Institute of Mathematics and Mechanics, Russia
Dmitry Stepanov	Program Systems Institute of Russian Academy of Sciences, Russia
Vadim Strijov	Moscow Institute of Physics and Technology, Russia
Pavel Sulimov	HSE University, Russia and BetVictor, Gibraltar
Rustam Tagiew	German Center for Rail Traffic Research at the Federal Railway Authority, Germany
Irina Temnikova	Qatar Computing Research Institute, Qatar
Mikhail Tikhomirov	Lomonosov Moscow State University, Russia
Martin Trnecka	Palacký University Olomouc, Czech Republic
Christos Tryfonopoulos	University of the Peloponnese, Greece

Evgenii Tsymbalov	Skolkovo Institute of Science and Technology, Russia
Elena Tutubalina	Kazan Federal University, Russia
Wil van der Aalst	RWTH Aachen University, Germany
Ekaterina Vylomova	The University of Melbourne, Australia
Dmitry Yashunin	Harman International, USA
Dmitry Zaytsev	HSE University, Moscow, Russia

Additional Reviewers

Vladimir Bashkin
Andrey Rivkin
Vadim Fomin
Sergey Sviridov
Artem Panin
Marketa Trneckova

Organising Committee

Evgeny Burnaev (AIST 2020 Local Organising Chair)	Skolkovo Institute of Science and Technology, Russia
Dmitry Ignatov (AIST series Head of Organisation)	HSE University, Moscow, Russia
Alexander Panchenko (AIST series Head of Organisation)	Skolkovo Institute of Science and Technology, Russia
Irina Nikishina (AIST 2020 Secretary)	Skolkovo Institute of Science and Technology, Russia
Ekaterina Artemova	HSE University, Moscow, Russia
Ilya Makarov	HSE University, Moscow, Russia

Volunteers

Daryna Dementieva	Skolkovo Institute of Science and Technology, Russia
Robiul Islam	Innopolis University, Russia
Evgenii Tsymbalov	Skolkovo Institute of Science and Technology, Russia

Abstracts of Invited Talks

Consensus Clustering in Networks

Santo Fortunato (iD)

Indiana University Bloomington, USA
santo@indiana.edu

Abstract. Algorithms for community detection are usually stochastic, leading to different partitions for different choices of random seeds. Consensus clustering is an effective technique to derive more stable and accurate partitions than the ones obtained by the direct application of the algorithm. Here we will show how this technique can be applied recursively to improve the results of clustering algorithms. The basic procedure requires the calculation of the consensus matrix, which can be quite dense if (some of) the clusters of the input partitions are large. Consequently, the complexity can get dangerously close to quadratic, which makes the technique inapplicable on large graphs. Hence we also present a fast variant of consensus clustering, which calculates the consensus matrix only on the links of the original graph and on a comparable number of additional node pairs, suitably chosen. This brings the complexity down to linear, while the performance remains comparable as the full technique. Therefore, the fast consensus clustering procedure can be applied on networks with millions of nodes and links.

Keywords: Consensus clustering · Networks · Community detection

References

1. Fortunato, S.: Community detection in graphs. Phys. Rep. **486**(3), 75–174 (2010). https://doi.org/10.1016/j.physrep.2009.11.002
2. Fortunato, S., Hric, D.: Community detection in networks: a user guide. Phys. Rep. **659**, 1–44 (2016). https://doi.org/10.1016/j.physrep.2016.09.002
3. Jeub, L.G.S., Sporns, O., Fortunato, S.: Multiresolution consensus clustering in networks. Sci. Rep. **8**(1), 3259 (2018). https://doi.org/10.1038/s41598-018-21352-7
4. Tandon, A., Albeshri, A., Thayananthan, V., Alhalabi, W., Fortunato, S.: Fast consensus clustering in complex networks. Phys. Rev. E **99**, 042301 (2019). https://doi.org/10.1103/PhysRevE.99.042301

Graph-theoretic Methods in Computer Vision: Recent Advances

Marcello Pelillo (ID)

Ca' Foscari University of Venice, Italy
`pelillo@dsi.unive.it`

Abstract. Graphs and graph-based representations have long been an important tool in computer vision and pattern recognition, especially because of their representational power and flexibility. There is now a renewed interest toward explicitly formulating computer vision problems as graph problems. This is particularly advantageous because it allows vision problems to be cast in a pure, abstract setting with solid theoretical underpinnings and also permits access to the full arsenal of graph algorithms developed in computer science and operations research. In this talk, I describe some recent developments in graph-theoretic methods which allow us to address within a unified and principled framework a number of classical computer vision problems. These include interactive image segmentation, image geo-localization, image retrieval, multi-camera tracking, and person re-identification. The concepts discussed here have intriguing connections with optimization theory, game theory and dynamical systems theory, and can be applied to weighted graphs, digraphs and hypergraphs alike.

Keywords: Computer vision · Graph theory · Weighted graphs · Hypergrpahs · Dominant sets · Image segmentation

References

1. Alemu, L.T., Pelillo, M.: Multi-feature fusion for image retrieval using constrained dominant sets. Image Vis. Comput. **94**, 103862 (2020). https://doi.org/10.1016/j.imavis.2019.103862
2. Aslan, S., Vascon, S., Pelillo, M.: Two sides of the same coin: Improved ancient coin classification using graph transduction games. Pattern Recognit. Lett. **131**, 158–165 (2020). https://doi.org/10.1016/j.patrec.2019.12.007
3. Mequanint, E.Z., Alemu, L.T., Pelillo, M.: Dominant sets for "constrained" image segmentation. IEEE Trans. Pattern Anal. Mach. Intell. **41**(10), 2438–2451 (2019). https://doi.org/10.1109/TPAMI.2018.2858243
4. Tesfaye, Y.T., Zemene, E., Prati, A., Pelillo, M., Shah, M.: Multi-target tracking in multiple non-overlapping cameras using fast-constrained dominant sets. Int. J. Comput. Vis. **127**(9), 1303–1320 (2019). https://doi.org/10.1007/s11263-019-01180-6
5. Zemene, E., Tesfaye, Y.T., Idrees, H., Prati, A., Pelillo, M., Shah, M.: Large-scale image geo-localization using dominant sets. IEEE Trans. Pattern Anal. Mach. Intell. **41**(1), 148–161 (2019). https://doi.org/10.1109/TPAMI.2017.2787132

Text and Speech Processing Projects at MTS AI

Nikita Semenov

Mobile TeleSystems, Vorontsovskaya street, Bldg. 4 109147 Moscow, Russia
nikita.semenov@mts.ru

Abstract. How to build effective systems for processing and understanding speech in a large corporation? What challenges does the researcher face? It is no secret that any solution has its own life cycle, including a solution based on natural language processing technologies. In this talk, we dwell in particular on the components of the life cycle of solutions with NLP and ASR technologies.

Keywords: Spoken language understanding · Natural language processing · Automatic speech recognition

Text and Speech Processing Projects at MTS.AI

Nikita Semenov

Contents

Computer Vision

Social Network Analysis

Data Analysis and Machine Learning

Theoretical Machine Learning and Optimization

Process Mining

Invited Papers

Making ML Models Fairer Through Explanations: The Case of LimeOut

Guilherme Alves, Vaishnavi Bhargava, Miguel Couceiro(✉),
and Amedeo Napoli

Université de Lorraine, CNRS, Inria N.G.E., LORIA, 54000 Nancy, France
{guilherme.alves-da-silva,miguel.couceiro,amedeo.napoli}@loria.fr

Abstract. Algorithmic decisions are now being used on a daily basis, and based on Machine Learning (ML) processes that may be complex and biased. This raises several concerns given the critical impact that biased decisions may have on individuals or on society as a whole. Not only unfair outcomes affect human rights, they also undermine public trust in ML and AI. In this paper we address fairness issues of ML models based on decision outcomes, and we show how the simple idea of "feature dropout" followed by an "ensemble approach" can improve model fairness. To illustrate, we will revisit the case of "LimeOut" that was proposed to tackle "process fairness", which measures a model's reliance on sensitive or discriminatory features. Given a classifier, a dataset and a set of sensitive features, LimeOut first assesses whether the classifier is fair by checking its reliance on sensitive features using "Lime explanations". If deemed unfair, LimeOut then applies feature dropout to obtain a pool of classifiers. These are then combined into an ensemble classifier that was empirically shown to be less dependent on sensitive features without compromising the classifier's accuracy. We present different experiments on multiple datasets and several state of the art classifiers, which show that LimeOut's classifiers improve (or at least maintain) not only process fairness but also other fairness metrics such as individual and group fairness, equal opportunity, and demographic parity.

Keywords: Fairness metrics · Feature importance · Feature-dropout · Ensemble classifier · LIME explanations

1 Introduction

Algorithmic decisions are now being used on a daily basis and obtained by Machine Learning (ML) processes that may be rather complex and opaque. This raises several concerns given the critical impact that such decisions may have on individuals or on society as a whole. Well known examples include the classifiers

This research was partially supported by TAILOR, a project funded by EU Horizon 2020 research and innovation programme under GA No 952215, and the Inria Project Lab "Hybrid Approaches for Interpretable AI" (HyAIAI).

© Springer Nature Switzerland AG 2021
W. M. P. van der Aalst et al. (Eds.): AIST 2020, LNCS 12602, pp. 3–18, 2021.
https://doi.org/10.1007/978-3-030-72610-2_1

which are used to predict the credit card defaulters, including multiple other datasets which may impact the government decisions. These prevalent classifiers are generally known to be biased to certain minority or vulnerable groups of society, which should rather be protected. Most of the notions of fairness thus focus on the outcomes of the decision process [15,16]. They are inspired by several anti-discrimination efforts that aim to ensure that unprivileged groups (e.g. racial minorities) should be treated fairly. Such issues can be addressed by looking into fairness individually [15] or as a group [15,16]. Actually, earlier studies [17,18] consider individual and group fairness as conflicting measures, and some studies tried to find an optimal trade-off between them. In [3] the author argues that, although apparently conflicting, they correspond to the same underlying moral concept, thus providing a broader perspective and advocating an individual treatment and assessment based on a case-by-case analysis.

The authors of [7,8] provide yet another noteworthy perspective of fairness, namely, *process fairness*. Rather than focusing on the outcome, it deals with the process leading to the outcome. In [2] we delivered a potential solution to deal with process fairness in ML classifiers. The key idea was to use an explanatory model, namely, LIME [14] to assess whether a given classifier was fair by measuring its reliance on salient or sensitive features. This component was then integrated in a human-centered workflow called *LimeOut*, that receives as input a triple (M, D, F) of a classifier M, a dataset D and a set F of sensitive features, and outputs a classifier M_{final} less dependent on sensitive features without compromising accuracy. To achieve both goals, LimeOut relies on feature dropout to produce a pool of classifiers that are then combined through an ensemble approach. Feature dropout receives a classifier and a feature a as input, and produces a classifier that does not take a into account. This preliminary study [2] showed the feasibility and the flexibility of the simple idea of feature dropout followed by an ensemble approach to improve process fairness. However, the empirical study of [2] was performed only on two families of classifiers (logistic regression and random forests) and carried out on two real-life datasets (Adult and German Credit Score). Also, it did not take into account other commonly used fairness measures. Moreover, in a recent study [6], Dimanov *et al.* question the trustfulness of certain explanation methods when assessing model fairness. In fact, they present a procedure for modifying a pre-trained model in order to manipulate the outputs of explanation methods that are based on feature importance (FI). They also observed minor changes in accuracy and that, even though the pre-trained model was deemed fair by some FI based explanation methods, it may conceal unfairness with respect to other fairness metrics.

This motivated us to revisit *LimeOut*'s framework to perform a thorough analysis that follows the tracks of [6] and extends the empirical study of [2] in several ways: (i) we experiment on many other datasets (e.g., HDMA dataset, Taiwanese Credit Card dataset, LSAC) , (ii) we make use of a larger family of ML classifiers (that include AdaBoost, Bagging, Random Forest (RF), and Logistic Regression (LR)), and (iii) we evaluate *LimeOut*'s output classifiers with respect to a wide variety fairness metrics, namely, disparate impact (DI), disparate mistreatment or equal opportunity (EO), demographic parity (DP), equal accuracy (EA), and predictive equality (PQ). As it will become clear from the empirical results, the robustness of *LimeOut*'s to different fairness view points is once again confirmed without compromising accuracy.

The paper is organised as follows. After recalling Lime explanations and various fairness measures in Subsects. 2.1 and 2.2, respectively, we briefly describe *LimeOut*'s workflow in Subsect. 2.3. We then present in Sect. 3 an extended empirical study following the tracks of [2] and the recent study [6]. First we quickly describe the datasets used (Subsect. 3.1) and the classifiers employed (Subsect. 3.2). We then present the empirical results and the various assessments with respect to the different fairness metrics considered in Subsect. 2.2. We conclude the paper in Sect. 4 with some final remarks on ongoing work and perspectives of future research.

2   Related Work

In this section, we briefly recall LIME (Subsect. 2.1), recall the different metrics used to measure model fairness (Subsect. 2.2) and revisit *LimeOut*'s framework (Subsect. 2.3).

2.1   LIME - Explanatory Method

Recall that LIME explanations [14] (Local Interpretable Model Agostic Explanations) take the form of surrogate linear models, that locally mimic the behavior of a ML model. Essentially, it tries to find the best possible linear model (i.e. explanation model) which fits the prediction of ML model of a given instance and it's neighbouring points (see below).

Let $f : \mathbb{R}^d \rightarrow \mathbb{R}$ be the function learned by a classification or regression model over training samples. LIME's workflow can be described as follows. Given an instance x and its ML prediction $f(x)$, LIME generates neighbourhood points by perturbing x and gets their corresponding predictions. These neighbouring points z are assigned weights based on their proximity to x, using the following equation:

$$\pi_x(z) = e^{\left(\frac{D(x,z)^2}{\sigma^2}\right)},$$

where $D(x, z)$ is the Euclidean distance between x and z, and σ is the hyper parameter (kernel-width). LIME then learns the weighted linear model g over the original and neighbourhood points, and their respective predictions, by solving the following optimization problem:

$$g = argmin_{g \in \mathcal{G}} \; \mathcal{L}(f, g, \pi_x(z)) + \Omega(g),$$

where $L(f, g, \pi_x(z))$ is a measure of how unfaithful g is in approximating f in the locality defined by $\pi_x(z)$. $\Omega(g)$ measures the complexity of g (regularization term). In order to ensure both interpretability and local faithfulness, LIME minimizes

$L(f, g, \pi_x(z))$ while enforcing $\Omega(g)$ to be small in order to be interpretable by humans. The obtained explanation model g is of the form

$$g(x) = \hat{\alpha}_0 + \sum_{1 \leq i \leq d'} \hat{\alpha}_i x[i],$$

where $\hat{\alpha}_i$ represents the contribution or importance of feature $x[i]$. Figure 1 presents the explanation of LIME for the classification of an instance from the Adult dataset. For instance, the value "Capital Gain" ≤ 0.0 contributes 0.29 to the class $\leq 50K$, whereas the value "Relationship"$= Husband$ contributes 0.15 to the class $> 50K$.

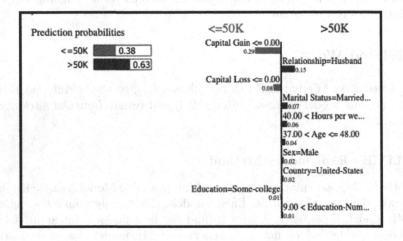

Fig. 1. LIME explanation in case of adult dataset

2.2   Model Fairness

Several metrics have been proposed in the literature in order to assess ML model's fairness. Here we recall some of the most used ones.

- **Individual Fairness**[1] [4] imposes that the instances/individuals belonging to different sensitive groups, but with similar non-sensitive attributes must receive equal decision outcomes.
- **Disparate Impact**[2] (DI) [5] is rooted in the desire for different sensitive demographic groups to experience similar rates of positive decision outcomes $(\hat{y} = pos)$. Given the ML model, $\hat{y}$ represents the predicted class. It compares two groups of the population based on a sensitive feature: the privileged $(priv)$ and the unprivileged (unp) groups. For instance, if we consider race as

---

[1] It is also referred to as *disparate treatment* or *predictive parity*.
[2] It is also referred to as *group fairness*.

sensitive feature, white people can be assigned as privileged and non-white people as unprivileged group.

$$DI = \frac{P(\hat{y} = pos | D = unp)}{P(\hat{y} = pos | D = priv)}$$

- **Equal Opportunity**[3] **(EO)** [16] proposes different sensitive groups to achieve similar rates of error in decision outcomes. It is computed as the difference in recall scores ($\frac{TP_i}{TP_i + FN_i}$, where TP_i is true positive and FN_i is false negative for a particular group i) between the unprivileged and privileged groups.

$$EO = \frac{TP_{unp}}{TP_{unp} + FN_{unp}} - \frac{TP_{priv}}{TP_{priv} + FN_{priv}}$$

- **Process Fairness**[4] [7,8] deals with the process leading to the prediction and keeps track of input features used by the decision model. In other words, the process fairness deals at the algorithmic level and ensures that the algorithm does not use any sensitive features while making a prediction.
- **Demographic Parity (DP)** [9] the difference in the predicted positive rates between the unprivileged and privileged groups.

$$DP = P(\hat{y} = pos | D = unp) - P(\hat{y} = pos | D = priv)$$

- **Equal Accuracy (EA)** [9] the difference in accuracy score ($\frac{TP_i + TN_i}{P_i + N_i}$, where TN_i is true negative of a particular group i) between unprivileged and privileged groups.

$$EA = \frac{TP_{unp} + TN_{unp}}{P_{unp} + N_{unp}} - \frac{TP_{priv} + TN_{priv}}{P_{priv} + N_{priv}}$$

- **Predictive Equality (PE)** which is defined as the difference in false positive rates ($\frac{FP_i}{FP_i + TP_i}$, where FP_i is false positive for a particular group i) between unprivileged and privileged groups. Formally,

$$PE = \frac{FP_{unp}}{FP_{unp} + TP_{unp}} - \frac{FP_{priv}}{FP_{priv} + TP_{priv}}.$$

In this paper we follow the same empirical setting of [6] and [2] and, hence, will focus mainly on disparate impact, equal opportunity, process fairness, demographic parity and equal accuracy.

2.3 *LimeOut*'s Framework

In this subsection, we briefly describe *LimeOut*'s framework, which essentially consists of two main components: LIME$_{Global}$ and ENSEMBLE$_{Out}$. Given an

---

[3] It is also referred to as *disparate mistreatment*.
[4] It is also referred to as *procedural fairness*.

input (M, D, F), where M is a classifier, D is a dataset, and F is a list of sensitive features, *LimeOut* first employs a "global variant" of LIME (LIME$_{Global}$) to assess the contribution (importance) of each feature to the classifier's outcomes. For that, LIME$_{Global}$ uses submodular pick to select instances with diverse and non-redundant explanations [14], and which are then aggregated to provide a global explanations (see [2]). The final output of LIME$_{Global}$ is thus a list of the k most important features[5].

If the k most important feature contain at least two sensitive features in F, then the model is deemed unfair (or biased), and the second component ENSEMBLE$_{Out}$ is deployed. Essentially, ENSEMBLE$_{Out}$ applies feature dropout on the sensitive features that are among the k most important features, each of which giving rise to a classifier obtained from M by removing that feature. thus resulting in a pool of classifiers. ENSEMBLE$_{Out}$ then constructs an ensemble classifier M_{final} through a linear combination of the pool's classifiers.

More precisely, if LIME$_{Global}$ outputs $a_1, a_2, \ldots, a_k$ as the k most important features, in which $a_{j_1}, a_{j_2}, \ldots, a_{j_i}$ are sensitive, then *LimeOut* trains $i + 1$ classifiers: M_t after removing a_{j_t} from the dataset, for $t = 1, \ldots, i$, and M_{i+1} after removing all sensitive features $a_{j_1}, a_{j_2}, \ldots, a_{j_i}$. The ensemble classifier M_{final} is then defined as the "average" of these $i + 1$ classifiers, i.e., by the rule: for an instance x and a class C,

$$P_{M_{final}}(x \in C) = \frac{\sum_{t=1}^{i+1} P_{M_t}(x \in C)}{i + 1}.$$

The empirical studies carried out in [2] showed that this ensemble classifier obtained by *LimeOut* is fairer with respect to process fairness than the input model M, without compromising (or even improving) M's accuracy.

3   Empirical Study

In this section, we first describe in Subsect. 3.1 the datasets that we used in our experiments, and we briefly present in Subsect. 3.2 the empirical setup. We then discuss our results from different points of view. In Subsect. 3.3 we report on the improved accuracy of *LimeOut*'s classifiers using different models and on the various datasets considered. We will then assess the fairness of *LimeOut*'s classifiers in Subsect. 3.3: first on process fairness and then on the remaining metrics of Subsect. 2.2.

3.1   Datasets

Experiments were conducted using five datasets. All datasets share common characteristics that allow us to run our experiments: a binary target feature and the presence of sensitive features. Table 1 summarizes basic information about these datasets. The details concerning each dataset are presented as follows.

---

[5] In [2] k was set to 10.

Table 1. Datasets employed in the experiments.

Dataset	# features	# sensitive	# instances
Adult	14	3	32561
German	20	3	1000
HMDA	28	3	92793
Default	23	3	30000
LSAC	11	2	26551

Adult. This dataset is available on UCI repository[6]. The target variable indicates whether a person earns more then 50k dollars per year. The goal is to predict the target feature based on census data. In this dataset, we considered as sensitive features: "Marital Status", "Race", and "Sex".

German. This is also a dataset available on UCI repository[7]. The task is to predict if an applicant has a high credit risk. In other words, if an applicant is likely to pay back his loan. We considered as sensitive features: "statussex", "telephone", and "foreign worker".

HMDA. The *Home Mortgage Disclosure Act* (HMDA)[8] aims to help identifying possible discriminatory lending practices. This public data about home mortgage contains information about the applicant (demographic information), the lender (name, regulator), the property (type of property, owner occupancy, census tract), and the loan (loan amount, type of loan, loan purpose). Here, the goal is to predict whether a loan is "high-priced", and the features that are considered sensitive are "sex", "race", and "ethnicity".

Default. This dataset is also a dataset available on the UCI repository[9]. The goal is to predict the probability of default payments using data from Taiwanese credit card users, e.g., credit limit, gender, education, marital status, history of payment, bill and payment amounts. We consider as sensitive features in this dataset: "sex" and "marriage".

LSAC. The *Law School Admissions Council* (LSAC)[10] dataset contains information about approx. 27K students through law school, graduation, and sittings for bar exams. This information was collected from 1991 through 1997, and it describes students' gender, race, year of birth (DOB_yr), full-time status,

---

[6] http://archive.ics.uci.edu/ml/datasets/Adult.

[7] https://archive.ics.uci.edu/ml/datasets/statlog+(german+credit+data).

[8] https://www.consumerfinance.gov/data-research/hmda/.

[9] https://archive.ics.uci.edu/ml/datasets/default+of+credit+card+clients.

[10] http://www.seaphe.org/databases.php.

family income, Law School Admission Test score (`lsat`), and academic perfor-
mace (undegraduate GPA (`ugpa`), standardized overall GPA (`zgpa`), standardized 1st year GPA (`zfygpa`), weighted index using 60% of LSAT and 40% of
ugpa (`weighted_lsat_ugpa`)). Here, the goal is to predict whether a law stu-
dent passes in the bar exam. In this dataset, features that could be considered
sensitive are "race" and "sex".

3.2 Empirical Setup

To perform our experiments[11], we split each dataset into 70% training set and
30% testing. As the datasets are imbalanced, we used Synthetic Minority Oversampling Technique (SMOTE[12]) over training data to generate the samples synthetically. We trained original and ensemble models on the balanced (augmented)
datasets using Scikit-learn implementations [12] of the following five algorithms:
AdaBoost (ADA), *Bagging*, *Random Forest* (RF), and *Logistic Regression* (LR).
For ADA, Bagging, RF, and LR we kept the default parameters of Scikit-learn
documentation[13].

3.3 Accuracy Assessment

Table 2 shows the average accuracy obtained in all experiments. We repeated the
same experiment 10 times. For each dataset, we indicate the average accuracy
of the original model ("Original") and the average accuracy of the *LimeOut*
ensemble model (line "*LimeOut*"). Our analysis is based on the comparison
between the accuracy of the original and the ensemble models. Since we drop
sensitive features, it is expected that the accuracy of model decreases. However,
it is evident that *LimeOut* ensemble models maintain the level of accuracy, even
though sensitive features were dropped out.

We notice a slight improvement in the accuracy of the ensemble models
when we use Bagging over German, Adult and Default datasets. Although in
some cases we notice a difference between original and ensemble models, in all
scenarios the difference is statistically negligible.

3.4 Fairness Assessment

We now assess model fairness with respect to two points of view, namely, in
terms of *process fairness* and in terms of various *fairness metrics*.

---

[11] The gitlab repository of LimeOut can be found here:
https://gitlab.inria.fr/orpailleur/limeout.
[12] https://machinelearningmastery.com/threshold-moving-for-imbalanced-
classification/.
[13] We used version 0.23.1 of Scikit-learn.

Table 2. Average accuracy assessment, where *LimeOut* stands for the ensemble model built by our proposed framework. Numbers in parentheses indicate standard deviation. No accuracy values are reported on the HMDA dataset for logistic regression, and on the Default dataset for random forest and logistic regression, since in each of these cases the original model was deemed fair.

		ADA	Bagging	RF	LR
German	Original	0.757 (0.015)	0.743 (0.019)	**0.772** (0.016)	0.769 (0.021)
	LimeOut	0.765 (0.014)	0.755 (0.021)	0.769 (0.016)	0.770 (0.021)
Adult	Original	**0.855** (0.003)	0.841 (0.002)	0.808 (0.007)	0.845 (0.004)
	LimeOut	0.856 (0.003)	0.849 (0.002)	0.808 (0.004)	0.849 (0.004)
HMDA	Original	0.879 (0.001)	**0.883** (0.001)	0.882 (0.001)	0.878 (0.001)
	LimeOut	0.880 (0.001)	0.884 (0.000)	0.884 (0.000)	–
LSAC	Original	0.857 (0.003)	**0.861** (0.002)	0.852 (0.002)	0.820 (0.006)
	LimeOut	0.859 (0.002)	0.866 (0.002)	0.859 (0.002)	0.822 (0.005)
Default	Original	**0.817** (0.003)	0.804 (0.003)	0.807 (0.003)	0.779 (0.004)
	LimeOut	0.817 (0.003)	0.812 (0.002)	–	–

Process Fairness. In this section we analyze the impact of feature dropout and the dependence on sensitive features. We employ LIME_{Global} to compute feature contributions and build the list of the most important features. Instead of providing the lists of feature contributions for all combinations of datasets and classifiers, for each dataset, we select the classifier that provides the highest accuracy, as we did in Subsect. 3.3.

We thus look at the explanations obtained from LIME_{Global} for these selected combinations. Tables 3, 4, 5, 6 and 7 present the list of most important features for these datasets. In all cases, we can notice that *LimeOut* decreases the dependence on sensitive features. In other words, the ensemble models provided by our framework have less sensitive features in the list of most important features. Also, LIME explanations show that the remaining sensitive features (the ones that appeared in the list of the ensemble model) contributed less to the global prediction compared to the original model.

For all datasets we used $k = 10$, except for the HMDA dataset. Indeed, in the latter case we took $k = 15$ (Table 5). This is due to the fact that all models were considered fair by *LimeOut* if only the first 10 important features were taken into account. We thus decided to investigate whether considering more features would show a different result, as it turned out to be the case when applying Bagging on HMDA.

Fairness Metrics In this section, we assess fairness using the fairness metrics introduced in Sect. 2. We compute fairness metrics using IBM AI Fairness 360 Toolkit[14] [1]. Our goal is to have a different perspective on the fairness of

---

[14] https://github.com/Trusted-AI/AIF360.

LimeOut ensemble models since we only assessed fairness by using LIME explanations. In this analysis, we compare the original and ensemble models for each combination of classifier and sensitive feature.

Table 3. LIME explanations in the form of pairs feature/contribution for the original AdaBoost model and the ensemble variant (*LimeOut*'s output) on the on Adult dataset.

Original		Ensemble	
Feature	Contrib.	Feature	Contrib.
CapitalGain	−18.449067	CapitalGain	−19.147673
CapitalLoss	−4.922207	CapitalLoss	−9.682837
Hoursperweek	3.297749	Hoursperweek	1.173417
Workclass	−0.997601	fnlwgt	0.974685
fnlwgt	−0.890244	Workclass	−0.423646
MaritalStatus	0.873829	Education-Num	−0.259837
Sex	0.694676	**Sex**	−0.244728
Education-Num	−0.603877	Country	−0.162728
Relationship	0.277705	Education	0.127105
Occupation	0.173059	Age	0.124858

Table 4. LIME explanations of RF on German dataset.

Original		Ensemble	
Feature	Contrib.	Feature	Contrib.
Foreignworker	2.664899	Otherinstallmentplans	−1.487604
Otherinstallmentplans	−1.354191	Housing	−1.089726
Housing	−1.144371	Savings	0.679195
Savings	0.984104	Duration	−0.483643
Property	−0.648104	**foreignworker**	0.448643
Purpose	−0.415498	Property	−0.386355
existingchecking	0.371415	Credithistory	0.258375
Telephone	0.311451	Job	−0.252046
Credithistory	0.263366	Existingchecking	−0.21358
Duration	−0.223288	Residencesince	−0.138818

Figures 2, 3 and 4 show values for all fairness metrics in each graphic. Red points indicate the values for *LimeOut* ensemble models while blue points indicate values for original models. The dashed line is the reference for a fair model (optimal value), i.e., 0 for all metrics except DI where the optimal is 1.

Results for the German dataset are depicted in Fig. 2. It is evident that *LimeOut* produces ensemble models that are fairer according to metrics DP and

Table 5. LIME explanation of Bagging on HMDA dataset.

Original		Ensemble	
Feature	Contrib.	Feature	Contrib.
derived_loan_product_type	4.798847	derived_loan_product_type	6.457707
balloon_payment_desc	4.624029	balloon_payment_desc	5.054243
intro_rate_period	4.183828	intro_rate_period	4.638744
loan_to_value_ratio	2.824717	balloon_payment	1.512304
balloon_payment	2.005847	prepayment_penalty_term	−1.267424
prepayment_penalty_term	0.683618	interest_only_payment	0.777766
reverse_mortgage	−0.659169	loan_to_value_ratio	0.704758
applicant_age_above_62	0.532331	negative_amortization_desc	0.61936
derived_ethnicity	−0.409255	reverse_mortgage_desc	0.508204
co_applicant_age_above_62	−0.333838	interest_only_payment_desc	−0.393068
property_value	−0.326801	applicant_credit_score_type_desc	−0.379852
derived_race	−0.318802	negative_amortization	−0.353717
applicant_age	−0.304565	applicant_age_above_62	0.349847
loan_term	0.270951	property_value	-0.316311
negative_amortization	−0.229379	applicant_credit_score_type	−0.192114

Table 6. LIME explanations of AdaBoost on Default dataset.

Original		Ensemble	
Feature	Contrib.	Feature	Contrib.
PAY_0	0.014194	PAY_2	−0.024354
MARRIAGE	−0.013986	PAY_0	0.008862
PAY_2	−0.013513	PAY_5	0.008729
PAY_6	−0.011724	PAY_AMT6	−0.00566
PAY_AMT1	0.011664	LIMIT_BAL	−0.003584
PAY_AMT6	0.008088	BILL_AMT2	0.00329
PAY_AMT2	0.007735	PAY_6	−0.00307
PAY_3	0.00735	AGE	−0.002058
EDUCATION	0.0032	PAY_AMT1	0.001592
SEX	0.000732	PAY_3	−0.001492

EQ. Red points are closer to zero compared to blue points, which means that *LimeOut* ensemble models are fairer than pre-trained models. We can also notice general improvement on DI. However, we observe that the only problematic sensitive feature is "foreignworker", where no improvement is observed. For all other sensitive features, we observe an improved fairness behaviour. In a few cases, the differences are negligible, which indicates that *LimeOut* either improves or at least maintains the fairness metrics.

Table 7. LIME explanations of Bagging on LSAC dataset.

Original			Ensemble	
Feature	Contrib.		Feature	Contrib.
isPartTime	−12.588169		isPartTime	−9.294158
race	−3.943962		cluster_tier	−3.464014
cluster_tier	−1.873394		zgpa	2.835836
DOB_yr	−1.235803		family_income	−1.292526
zgpa	−0.71457		DOB_yr	−0.923861
zfygpa	0.314865		**race**	−0.895484
ugpa	0.123805		zfygpa	0.238397
family_income	−0.08999		weighted_lsat_ugpa	0.060846
lsat	−0.07596		ugpa	−0.055593
Sex	−0.068117		**Sex**	−0.041478

Figure 3 shows the results on fairness metrics for the Adult dataset. In this dataset, *LimeOut* ensemble models keep values of all metrics in almost scenarios. We only see a deterioration of fairness when we compute EQ for Logistic

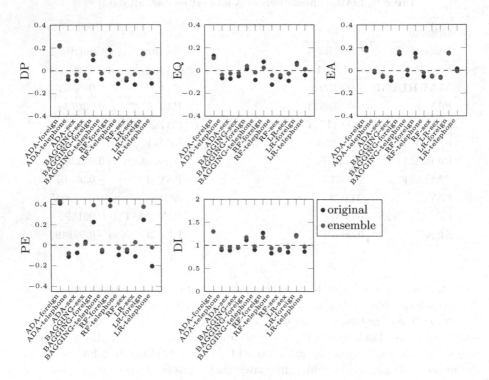

Fig. 2. Fairness metrics for German credit score dataset

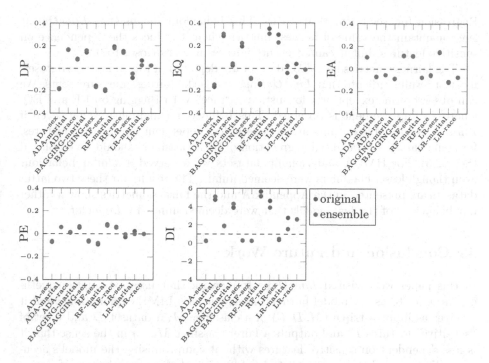

Fig. 3. Fairness metrics for adult dataset.

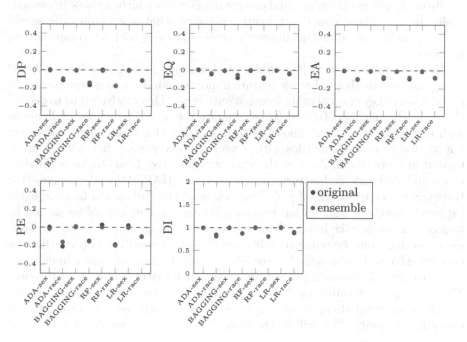

Fig. 4. Fairness metrics for LSAC dataset.

Regression focuses on marital status. This behaviour means that *LimeOut* at least maintain the value of fairness metrics when it reduces the dependence on sensitive features, but it cannot ensure fairness metrics closer to 0.

The fairness metrics for LSAC dataset are depicted in Fig. 4. For this dataset, most of results indicate that *LimeOut* maintains the fairness measurements. We can observe some exceptions, for instance, "race" with Bagging on PE and EQ, where an improvement is observed. This behaviour can indicate that, even if *LimeOut*'s ensemble outputs are in general less dependent on sensitive features, for some datasets a weighted aggregation of pool classifiers should be employed (Sect. 2.3). For HMDA and Default datasets we observed a similar behaviour even though lesser classifiers were deemed unfair. The results for these two latter datasets are presented in the Appendix A and the fairness metrics show a rather fair behaviour of the few models that were deemed unfair by *LimeOut*.

4 Conclusion and Future Work

In this paper we revisited *LimeOut*'s framework that uses explanation methods in order to assess model fairness. *LimeOut* uses LIME explanations, and it receives as input a triple (M, D, F) of a classifier M, a dataset D and a set of "sensitive" features F, and outputs a fairer classifier M_{final} in the sense that it is less dependent on sensitive features without compromising the model's accuracy. We extended the empirical study of [2] by including experiments of a wide family of classifiers on various and diverse datasets on which fairness issues naturally appear. These new experiments reattested what was empirically shown in [2], namely, that *LimeOut* improves process fairness without compromising accuracy.

However, the authors of [6] raised several concerns in such an approach based on explanation methods that use feature importance indices to determine model fairness since they conceal other forms of unfairness. This motivated us to deepen the thorough analysis of *LimeOut* to evaluate the model outcomes of *LimeOut* with respect to several well known fairness metrics. Our results show consistent improvements in most metrics with a very few exceptions that will be investigated in more detail. Also, we have already adapted *LimeOut* to other data types and different explanatory models such as SHAP [11] and Anchors [13]. However, the construction of global explanations like [10] should be thoroughly explored. Also, the aggregation rule to produce classifier ensembles should be improved in order take into account classifier weighting, as well as other classifiers resulting from the removal of different subsets of sensitive features (here we only considered the removal of one or all features). Finally, we took a human and context-centered approach for identifying sensitive features in a given use-case. There is hope to automating this task while taking into account domain knowledge and using statistical dataset characteristics and utility-based approaches to quantify sensitivity. This will be the topic of a follow up contribution.

A   Appendix

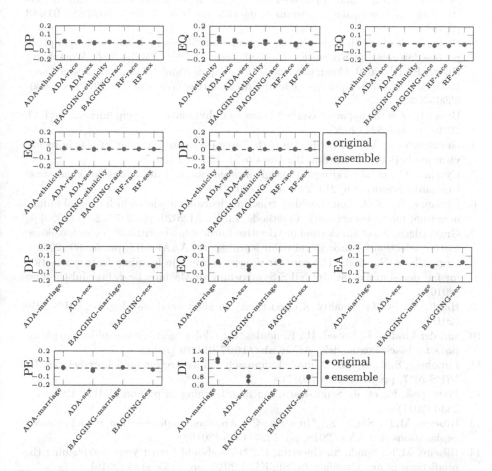

Fig. 5. Fairness metrics for the HMDA dataset (first and second lines) and the Default dataset (third and fourth lines). For both datasets, lesser original models were deemed unfair, namely, ADA, Bagging and RF on HMDA, and ADA and Bagging on Default. Even though these models were deemed unfair by *LimeOut*, most of the fairness metrics actually indicate a rather fair behaviour by the original and *LimeOut*'s ensemble models.

References

1. Bellamy, R.K.E., et al.: AI Fairness 360: An extensible toolkit for detecting, understanding, and mitigating unwanted algorithmic bias. ArXiv **abs/1810.01943** (2018)
2. Bhargava, V., Couceiro, M., Napoli, A.: LimeOut: an ensemble approach to improve process fairness. In: Koprinska, I., et al. (eds.) ECML PKDD 2020 Workshops. ECML PKDD 2020. Communications in Computer and Information Science, vol. 1323, pp. 475–491. Springer, Cham (2020). https://doi.org/10.1007/978-3-030-65965-3_32
3. Binns, R.: On the apparent conflict between individual and group fairness. In: FAT 2020, pp. 514–524 (2020)
4. Chouldechova, A.: Fair prediction with disparate impact: a study of bias in recidivism prediction instruments. Big Data **5**(2), 153–163 (2017)
5. Cynthia, D., et al.: Fairness through awareness. In: Innovations in Theoretical Computer Science, pp. 214–226. ACM (2012)
6. Dimanov, B., et al.: You shouldn't trust me: learning models which conceal unfairness from multiple explanation methods. In: ECAI 2020, pp. 2473–2480 (2020)
7. Grgić-Hlača, N., et al.: Beyond distributive fairness in algorithmic decision making: feature selection for procedurally fair learning. In: AAAI 2018, pp. 51–60 (2018)
8. Grgic-Hlaca, N., et al.: The case for process fairness in learning: feature selection for fair decision making. In: NIPS Symposium on Machine Learning and the Law (2016)
9. Hardt, M., et al.: Equality of opportunity in supervised learning. In: NIPS 2016 (2016)
10. van der Linden, I., Haned, H., Kanoulas, E.: Global aggregations of localexplanations for black box models. ArXiv **abs/1907.03039** (2019)
11. Lundberg, S.M., Lee, S.: A unified approach to interpreting model predictions. In: NIPS 2017, pp. 4765–4774 (2017)
12. Pedregosa, F., et al.: Scikit-learn: machine learning in python. JMLR **12**, 2825–2830 (2011)
13. Ribeiro, M.T., Singh, S., Guestrin, C.: Anchors: high-precision model-agnostic explanations. In: AAAI 2018, pp. 1527–1535 (2018)
14. Ribeiro, M.T., Singh, S., Guestrin, C.: "why should i trust you?": explaining the predictions of any classifier. In: SIGKDD 2016, pp. 1135–1144 (2016)
15. Speicher, T., et al.: A unified approach to quantifying algorithmic unfairness: measuring individual & group unfairness via inequality indices. In: SIGKDD 2018, pp. 2239–2248 (2018)
16. Zafar, M.B., et al.: Fairness beyond disparate treatment & disparate impact: learning classification without disparate mistreatment. In: WWW 2017, pp. 1171–1180 (2017)
17. Zafar, M.B.E.: Fairness constraints: mechanisms for fair classification. In: AISTATS 2017, pp. 962–970 (2017)
18. Zemel, R.E.: Learning fair representations. In: ICML 2013, pp. 325–333 (2013)

RuREBus: A Case Study of Joint Named Entity Recognition and Relation Extraction from E-Government Domain

Vitaly Ivanin[1,2], Ekaterina Artemova[3], Tatiana Batura[4,7],
Vladimir Ivanov[5,7], Veronika Sarkisyan[3], Elena Tutubalina[6,7],
and Ivan Smurov[1,2(✉)]

[1] ABBYY, Milpitas, USA
ivan.smurov@abbyy.com
[2] Moscow Institute of Physics and Technology, Dolgoprudny, Russia
[3] National Research University Higher School of Economics, Moscow, Russia
[4] Novosibirsk State University, Novosibirsk, Russia
[5] Innopolis University, Innopolis, Russia
[6] Kazan Federal University, Kazan, Russia
[7] Lomonosov Moscow State University, Moscow, Russia

Abstract. We show-case an application of information extraction methods, such as named entity recognition (NER) and relation extraction (RE) to a novel corpus, consisting of documents, issued by a state agency. The main challenges of this corpus are: 1) the annotation scheme differs greatly from the one used for the general domain corpora, and 2) the documents are written in a language other than English. Unlike expectations, the state-of-the-art transformer-based models show modest performance for both tasks, either when approached sequentially, or in an end-to-end fashion. Our experiments have demonstrated that fine-tuning on a large unlabeled corpora does not automatically yield significant improvement and thus we may conclude that more sophisticated strategies of leveraging unlabelled texts are demanded. In this paper, we describe the whole developed pipeline, starting from text annotation, baseline development, and designing a shared task in hopes of improving the baseline. Eventually, we realize that the current NER and RE technologies are far from being mature and do not overcome so far challenges like ours.

Keywords: Information extraction · Named entity recognition · Relation extraction

1 Introduction

Information extraction tasks, named entity recognition (NER) and relation extraction (RE), have been studied extensively. NER and RE are sometimes

---

The extended notes for invited talk "When CoNLL-2003 is not Enough: are Academic NER and RE Corpora Well-Suited to Represent Real-World Scenarios?" delivered by Ivan Smurov.

W. M. P. van der Aalst et al. (Eds.): AIST 2020, LNCS 12602, pp. 19–27, 2021.
https://doi.org/10.1007/978-3-030-72610-2_2

thought of as easy and almost solved problems. However, outside of the ideal-
istic academic setup, many complications may arise. The most used datasets,
leveraged to compare new methods and establish state-of-the-art (SoTA) results
are CoNLL03 [27], TACRED [32], SemEval-2010 Task 8 [10], CoNLL04 [4], ACE
2005 [28], OntoNotes [11]. However, as the choice of open sources for dataset
construction appears to be quite limited, these datasets are usually assembled
from news articles. What is more, the annotation scheme typically is driven
by academic interest, rather than practical considerations. Real-life applications
though may vary a lot and target domains other than news. Such applications
cover Legal Tech (previous studies had focused on extraction of organization
and person names [8], while more recent studies look beyond classical NER
types [3,18,19,26], medical domain [12,29] list more than forty corpora) and
noisy user texts [25], and may require a domain-specific and application-driven
annotation scheme. It might prove to be difficult in practice to adopt exciting
approaches to NER and RE to other domains, as straightforward domain adap-
tion techniques do not lead to the desired quality. A question of how big the gap
between academic benchmarks and real-life applications is rarely explored.

Adaptation to languages other than English complicates the usage of the
SoTA methods. If no corpora, similar to the one developed for English, in terms
of size, domain, and annotation quality, is available, it is almost impossible to
draw a fair comparison. For example, for the Russian language, used in this
paper, the only corpora for joint NER and RE are FactRuEval [23], significantly
smaller than OntoNotes or TACRED, and RuRED [9], which partially replicates
TACRED annotation. As no identical setup for evaluation of NER and RE methods is available for different languages, it may be difficult to investigate whether
the same methods deliver comparable results for different languages. Transfer
learning [31] is a promising paradigm that helps to re-use cross-lingual models,
trained for English, for other languages. However, early attempts show that the
application of transfer learning techniques turns out to be rather challenging.
For example, so far, neither NER nor RE tasks benefit from transfer learning
approaches, when applied to TACRED and RuRED.

In this paper, we explore a typical industrial case: prototyping NER and RE
models in a specific application domain based on existing SoTA approaches. We
describe a problematic real-life setup, which requires both 1) adaptation to a
new domain and an unconventional annotation scheme and 2) processing text,
written in a language other than English. Our results show a significant decrease
in quality when compared to SoTA academic results. We aim to bring more
attention to the challenges of NER and RE tasks and show that existing methods
so far can be treated as off the shelf solutions only in a limited scope. The task
under consideration comes from e-Goverment domain: we investigate the corpus
of strategic planning documents[1], which are annually issued by the Ministry of
Economic Development of the Russian Federation. The entities considered relate
to different types of state assets and enterprises. At the same time, the relations

---

[1] The corpus is open and available online on the Ministry of Economic Development
of the Russian Federation website.

express various aspects of strategic planning, i.e., goal setting and forecasting. As the current approach to strategic planning is in desperate need of innovative organizational development, the NER and RE methods should be at the forefront of automation efforts. Extracted entities and relations between them allow for faster retrieval, ontology-based analysis, and compliance testing.

The remainder is organized as follows. Section 2 introduces the corpus and the annotation scheme. Section 3 presents with the methods used for NER and RE as well as with a shared task, which was held in hopes of improving baseline solution quality. Section 4 concludes by discussing the results and outlining the directions for future work.

2   Corpus Annotation

We develop guidelines for entity and relation identification in order to maintain uniformity of annotation in our corpus.

Table 1. Entities description and examples (translated to English)

Entity	Description	Examples
MET (metric)	Indicator or object on which the comparison operation is defined	Unemployment rate, total length of roads, average life expectancy
ECO (economics)	Economic entity or infrastructure object	Private business, PJSC Sberbank, hospital complex
BIN (binary)	Binary characteristics or single action	Modernization, development, invest
CMP (comparative)	Comparative characteristic	Reduction of level, positive dynamics, increase of
QUA (qualitative)	Quality characteristic	Ineffective, fault tolerant, stable
ACT (activity)	Activities, events or measures taken by the authorities	Restoration work, educational project "Silver University" , drug prevention
INST (institutions)	institutions, structures and organizations	Cultural center, region administration , youth employment center
SOC (social)	Social object	Leisure activities, historical heritage, population of the country

We define eight types of entities described in Table 1 and nine types of relations that describe actions taken in the past and present time and also forecasts. We distinguish them in terms of tonality, whether the actions or state of affairs are positive, negative, or neutral. Another two relations are GOL, used for abstract goals, and TSK used for specific tasks. Such relations tightly correspond to the domain: strategic planning is based on setting goals and targets due to past, current, and predicted state of affairs.

All annotations were obtained using a Brat Rapid Annotation Tool (BRAT) [24]. Annotation instructions are available at the GitHub repository[2]. Each document in the corpus was annotated by two annotators independently, while a moderator resolved disagreements. To speed up and facilitate the annotation process, we used active learning techniques [22]. We applied widely used architecture, namely char-CNN-BiLSTM-CRF described in [17] and [20] and used pre-trained FastText embeddings [2] from RusVectores [16]. For RE, we employed morphological, syntactical, and semantic features obtained from Compreno [1, 33] and some hand-made features, such as capitalization templates and dependency tree distance between relation members.

The resulting corpus **contains** 394,966 tokens, 120,989 entities and 12,648 relations. Annotation consistency is evaluated by measuring annotators **agreement** on documents, which were marked up twice, leading to Cohen's kappa equal to 0.698.

3 Baselines and Evaluation

As a baseline for the NER task, we employed standard BERT-based [7] architecture, fine-tuned for the Russian language, namely RuBERT [15] with an MLP on top of it. Although being close to SoTA on most academic corpora, this model yielded a rather disappointing strict token-based f1-score of **0.53**.

To explore our corpus and double-check ourselves, we decided to conduct an external evaluation of both NER and RE tasks. As for our internal evaluation, we have chosen to continue working with NER (leaving the RE models exploration to the external evaluation). External evaluation, organized in the form of RuREBus-2020 Shared Task, is vital in broadening the scope of tested models and providing additional validation for the scores obtained on the corpus. To provide an additional grounding for our results, we were also able to draw some comparisons between our setting and some other non-classical sequence-labelling tasks.

3.1 External Evaluation

We provide a full account on the RuREBus Shared Task results in [13]. Here we publish only the ones most useful for further analysis of the corpus.

Unsurprisingly the most fruitful approach in the NER task was based on contextualised word embeddings in particular on BERT. While some participants attempted to use some additional layers such as BLSTMs and CRFs on top of contextualized word embeddings, the two systems with highest scores both employed standard, but powerful MLP on top of BERT model. The scores

---

obtained by the two systems are **0.561** and **0.547**[3]. There is no significant difference between both models other than the version of BERT the participant used (multilingual uncased base BERT in the first case and RuBERT for the second one). We should also note that both systems only fitted BERTs on the train set and did not employ the finetuning on the 299M token unmarked corpus provided.

RE task yielded diverse models. Two top models, while once again both used BERT, had different architectures. One of the systems employed R-BERT [30] based solution and was able to obtain 0.441 on RE task (given gold standard NERs). Another system used a SpanBERT [14] inspired model. While two systems have substantial differences, scores obtained by them are roughly comparable (**0.441** for R-BERT and **0.394** for SpanBERT).

3.2 Internal Evaluation

As a part of our internal evaluation, we have decided to fine-tune the language model of the contextual encoders for the NER task on the 299M token unmarked corpus.

In this ongoing work so far, we have been able to obtain some rather unexpected results. Fine-tuning our BERT-based baseline model did not leave to a significant improvement the performance and scored **0.54** on the NER task. We also tried fine-tuning multilingual BERT, but it scored only **0.44** on the NER task. In contrast fine-tuning ELMo [21] yielded the absolute best score obtained in both internal and external evaluation: **0.57**. We should also note that the ELMo model used for fine-tuning was pre-trained on English. Thus it had essentially non-random weights only on "middle layers" (as both embeddings and softmax were pre-trained on different vocabulary). We intend to explore this unexpected result further.

3.3 Evaluation Analysis

During the internal and external evaluation, several SoTA-like models were tested, scoring **0.53–0.57** for the NER task and **0.39–0.44** for the RE task. While these results can be improved, we can interpret them as a sort of industrial baseline for the corpus. Such results can be obtained by a specialist rigorously following academic publications, but not conducting large-scale research independently.

One can easily notice the contrast between these scores and the results obtained on most often cited academic corpora such as CoNLL-2003 and SemEval2010 Task 8. In our opinion, this can be explained by domain-specific content of the corpus and by the nature of entities and relations (that are often longer and have less well-defined boundaries than "standard" entities).

The last assumption can be illustrated by the fact that there is a direct correlation between the average length of the entity and the difference between token-based f-measure for entities and char-based f-measure, see Table 2.

---

[3] Obtained after the shared task deadline.

Table 2. Differences in char-based f-measure and span-based

Metrics	ACT	BIN	CMP	ECO	INST	MET	QUA	SOC
Average span-based f1	0.23	0.55	0.79	0.43	0.4	0.47	0.53	0.36
Average f1 diff	0.28	0.03	0.00	0.23	0.21	0.27	0.00	0.19
Mean no. chars	34	12	10	24	27	31	12	21
Mean no. tokens	4.74	1.05	1.16	2.78	3.69	4.23	1.14	2.77

We can draw a direct comparison between RuREBus corpus and SemEval-2020 Task 11 corpus for propaganda detection [6]. While these two corpora have completely different domains and are in different languages, both involve span extraction of long entities with sometimes less-than-clear borders and yield comparable results (0.57 f-measure for RuREBus, 0.52 for SemEval-2020 Task 11). While not all entities in industrial settings are of this type, some are, and thus, RuREBus can be treated as "worst-case business scenario".

4   Conclusion

In this paper, we deal with a real-world situation when one applies SoTA methods for NER and RE tasks. To this end, we have retrieved a large domain-specific text collection and manually annotated a small fraction of it with a 'non-standard' annotations (RuREBus corpus). The BERT-based baseline, as well as other independently developed and tested models, have shown low results (f1-score **0.53–0.57** for the NER task and **0.39–0.44** for RE task). This negative result helps to learn about the extent of the gap between the academic evaluations of SoTA models and the results of the same models in practical applications. Our result is consistent with another study (in a different domain) of information extraction models (SemEval-2020 Task 11).

Indeed, our ad-hoc approach can be criticized for many reasons (e.g., for the lack of deep analysis of errors, for the lack of diverse methods, or for the presence of 'non-standard' types in annotation schema). However, we argue that in industrial cases, many parameters may be less controllable than the *in-vitro* setting, which leads to more laborious tasks. Thus, the RuREBus corpus can be considered as a typical "worst-case business scenario" for NER and RE tasks. Future work direction include investigating domain adaptation and fine-tuning strategies and leveraging semi-supervided methods, such as cross-view training [5] to make reasonable use of unlabelled texts.

Acknowledgements. Work on maintenance of the annotation system, discussions of results, and manuscript preparation was carried out by Elena Tutubalina, Vladimir Ivanov, Tatiana Batura and supported by the Russian Science Foundation grant no. 20-11-20166. Ekaterina Artemova and Veronika Sarkisyan worked on text annotation, discussions of results, and manuscript. Their work was supported by the framework of the HSE University Basic Research Program and Russian Academic Excellence Project "5–100".

References

1. Anisimovich, K., Druzhkin, K., Minlos, F., Petrova, M., Selegey, V., Zuev, K.: Syntactic and semantic parser based on abbyy compreno linguistic technologies. In: Computational Linguistics and Intellectual Technologies: Proceedings of the International Conference "Dialog" [Komp'iuternaia Lingvistika i Intellektual'nye Tehnologii: Trudy Mezhdunarodnoj Konferentsii "Dialog"], Bekasovo, Russiavol. 2, pp. 90–103 (2012)
2. Bojanowski, P., Grave, E., Joulin, A., Mikolov, T.: Enriching word vectors with subword information. Trans. Assoc. Comput. Linguist. **5**, 135–146 (2017)
3. Cardellino, C., Teruel, M., Alemany, L.A., Villata, S.: A low-cost, high-coverage legal named entity recognizer, classifier and linker. In: Proceedings of the 16th Edition of the International Conference on Artificial Intelligence and Law, pp. 9–18 (2017)
4. Carreras, X., Màrquez, L.: Introduction to the CoNLL-2004 shared task: Semantic role labeling. In: Proceedings of the Eighth Conference on Computational Natural Language Learning (CoNLL-2004) at HLT-NAACL 2004, pp. 89–97. Association for Computational Linguistics, Boston (2004). https://www.aclweb.org/anthology/W04-2412
5. Clark, K., Luong, M.T., Manning, C.D., Le, Q.: Semi-supervised sequence modeling with cross-view training. In: Proceedings of the 2018 Conference on Empirical Methods in Natural Language Processing, pp. 1914–1925 (2018)
6. Da San Martino, G., Barrón-Cedeño, A., Wachsmuth, H., Petrov, R., Nakov, P.: SemEval-2020 task 11: detection of propaganda techniques in news articles. In: Proceedings of the 14th International Workshop on Semantic Evaluation. SemEval 2020, Barcelona, Spain (2020)
7. Devlin, J., Chang, M.W., Lee, K., Toutanova, K.: Bert: pre-training of deep bidirectional transformers for language understanding. In: Proceedings of the 2019 Conference of the North American Chapter of the Association for Computational Linguistics: Human Language Technologies, vol. 1 (Long and Short Papers), pp. 4171–4186 (2019)
8. Dozier, C., Kondadadi, R., Light, M., Vachher, A., Veeramachaneni, S., Wudali, R.: Named entity recognition and resolution in legal text. In: Francesconi, E., Montemagni, S., Peters, W., Tiscornia, D. (eds.) Semantic Processing of Legal Texts. LNCS (LNAI), vol. 6036, pp. 27–43. Springer, Heidelberg (2010). https://doi.org/10.1007/978-3-642-12837-0_2
9. Gordeev, D., Davletov, A., Rey, A., Akzhigitova, G., Geymbukh, G.: Relation extraction dataset for the russian language. In: Computational Linguistics and Intellectual Technologies: Proceedings of the International Conference "Dialog" [Komp'iuternaia Lingvistika i Intellektual'nye Tehnologii: Trudy Mezhdunarodnoj Konferentsii "Dialog"] Moscow, Russia (2020)
10. Hendrickx, I., et al.: SemEval-2010 task 8: multi-way classification of semantic relations between pairs of nominals. In: Proceedings of the 5th International Workshop on Semantic Evaluation, pp. 33–38. Association for Computational Linguistics, Uppsala (2010). https://www.aclweb.org/anthology/S10-1006
11. Hovy, E., Marcus, M., Palmer, M., Ramshaw, L., Weischedel, R.: Ontonotes: the 90% solution. In: Proceedings of the Human Language Technology Conference of the NAACL, Companion, NAACL-Short 2006, vol. Short Papers, p. 57–60. Association for Computational Linguistics, USA (2006)

12. Huang, C.C., Lu, Z.: Community challenges in biomedical text mining over 10 years: success, failure and the future. Briefings Bioinf. **17**(1), 132–144 (2015)
13. Ivanin, V., et al.: Rurebus-2020 shared task: Russian relation extraction for business. In: Computational Linguistics and Intellectual Technologies: Proceedings of the International Conference "Dialog" [Komp'iuternaia Lingvistika i Intellektual'nye Tehnologii: Trudy Mezhdunarodnoj Konferentsii "Dialog"], Moscow, Russia (2020)
14. Joshi, M., Chen, D., Liu, Y., Weld, D.S., Zettlemoyer, L., Levy, O.: Spanbert: improving pre-training by representing and predicting spans. Trans. Assoc. Comput. Linguist. **8**, 64–77 (2020)
15. Kuratov, Y., Arkhipov, M.: Adaptation of deep bidirectional multilingual transformers for russian language. In: Computational Linguistics and Intellectual Technologies: Proceedings of the International Conference "Dialog" [Komp'iuternaia Lingvistika i Intellektual'nye Tehnologii: Trudy Mezhdunarodnoj Konferentsii "Dialog"], pp. 333–339 (2019)
16. Kutuzov, A., Kuzmenko, E.: WebVectors: a toolkit for building web interfaces for vector semantic models. In: Ignatov, D.I., et al. (eds.) AIST 2016. CCIS, vol. 661, pp. 155–161. Springer, Cham (2017). https://doi.org/10.1007/978-3-319-52920-2_15
17. Lample, G., Ballesteros, M., Subramanian, S., Kawakami, K., Dyer, C.: Neural architectures for named entity recognition, pp. 260–270 (2016)
18. Leitner, E., Rehm, G., Moreno-Schneider, J.: Fine-grained named entity recognition in legal documents. In: Acosta, M., Cudré-Mauroux, P., Maleshkova, M., Pellegrini, T., Sack, H., Sure-Vetter, Y. (eds.) SEMANTiCS 2019. LNCS, vol. 11702, pp. 272–287. Springer, Cham (2019). https://doi.org/10.1007/978-3-030-33220-4_20
19. Leitner, E., Rehm, G., Moreno-Schneider, J.: A dataset of german legal documents for named entity recognition. arXiv preprint arXiv:2003.13016 (2020)
20. Ma, X., Hovy, E.: End-to-end sequence labeling via bi-directional LSTM-CNNS-CRF. In: Proceedings of the 54th Annual Meeting of the Association for Computational Linguistics, vol. 1: Long Papers, pp. 1064–1074 (2016)
21. Peters, M.E., et al.: Deep contextualized word representations. In: Proceedings of NAACL-HLT, pp. 2227–2237 (2018)
22. Shen, Y., Yun, H., Lipton, Z.C., Kronrod, Y., Anandkumar, A.: Deep active learning for named entity recognition. In: Proceedings of the 2nd Workshop on Representation Learning for NLP, pp. 252–256 (2017)
23. Starostin, A., et al.: Factrueval 2016: Evaluation of named entity recognition and fact extraction systems for Russian. In: Computational Linguistics and Intellectual Technologies: Proceedings of the International Conference "Dialog" [Komp'iuternaia Lingvistika i Intellektual'nye Tehnologii: Trudy Mezhdunarodnoj Konferentsii "Dialog"] pp. 702–720 (2016)
24. Stenetorp, P., Pyysalo, S., Topić, G., Ohta, T., Ananiadou, S., Tsujii, J.: Brat: a web-based tool for NLP-assisted text annotation. In: Proceedings of the Demonstrations at the 13th Conference of the European Chapter of the Association for Computational Linguistics, pp. 102–107. Association for Computational Linguistics (2012)
25. Strauss, B., Toma, B., Ritter, A., De Marneffe, M.C., Xu, W.: Results of the wnut16 named entity recognition shared task. In: Proceedings of the 2nd Workshop on Noisy User-generated Text (WNUT), pp. 138–144 (2016)

26. Teruel, M., Cardellino, C., Cardellino, F., Alemany, L.A., Villata, S.: Legal text processing within the mirel project. In: Proceedings of the Eleventh International Conference on Language Resources and Evaluation (LREC 2018) (2018)
27. Tjong Kim Sang, E.F., De Meulder, F.: Introduction to the conll-2003 shared task: Language-independent named entity recognition. In: Proceedings of the Seventh Conference on Natural Language Learning at HLT-NAACL 2003, CONLL 2003, vol. 4, p. 142–147. Association for Computational Linguistics (2003)
28. Walker, C., Strassel, S., Medero, J., Maeda, K.: ACE 2005 Multilingual Training Corpus. LDC2006T06. Philadelphia: Linguistic Data Consortium (2006)
29. Weber, L., Münchmeyer, J., Rocktäschel, T., Habibi, M., Leser, U.: Huner: improving biomedical NER with pretraining. Bioinformatics **36**(1), 295–302 (2020)
30. Wu, S., He, Y.: Enriching pre-trained language model with entity information for relation classification. In: Proceedings of the 28th ACM International Conference on Information and Knowledge Management, pp. 2361–2364 (2019)
31. Yang, Z., Salakhutdinov, R., Cohen, W.W.: Transfer learning for sequence tagging with hierarchical reccurent networks. arXiv preprint arXiv:1703.06345 (2017)
32. Zhang, Y., Zhong, V., Chen, D., Angeli, G., Manning, C.D.: Position-aware attention and supervised data improve slot filling. In: Proceedings of the 2017 Conference on Empirical Methods in Natural Language Processing (EMNLP 2017), pp. 35–45 (2017). https://nlp.stanford.edu/pubs/zhang2017tacred.pdf
33. Zuev, K.A., Indenbom, M.E., Judina, M.V.: Statistical machine translation with linguistic language model. In: Computational Linguistics and Intellectual Technologies: Proceedings of the International Conference "Dialog" [Komp'iuternaia Lingvistika i Intellektual'nye Tehnologii: Trudy Mezhdunarodnoj Konferentsii "Dialog"], Bekasovo, Russia, vol. 2, pp. 164–172 (2013)

On Interpretability and Similarity in Concept-Based Machine Learning

Léonard Kwuida[1(✉)] and Dmitry I. Ignatov[2,3]

[1] Bern University of Applied Sciences, Bern, Switzerland
leonard.kwuida@bfh.ch
[2] National Research University Higher School of Economics,
Moscow, Russian Federation
dignatov@hse.ru
[3] St. Petersburg Department of Steklov Mathematical Institute of Russian Academy
of Sciences, St. Petersburg, Russia

Abstract. Machine Learning (ML) provides important techniques for classification and predictions. Most of these are black-box models for users and do not provide decision-makers with an explanation. For the sake of transparency or more validity of decisions, the need to develop explainable/interpretable ML-methods is gaining more and more importance. Certain questions need to be addressed:

– How does an ML procedure derive the class for a particular entity?
– Why does a particular clustering emerge from a particular unsupervised ML procedure?
– What can we do if the number of attributes is very large?
– What are the possible reasons for the mistakes for concrete cases and models?

For binary attributes, Formal Concept Analysis (FCA) offers techniques in terms of intents of formal concepts, and thus provides plausible reasons for model prediction. However, from the interpretable machine learning viewpoint, we still need to provide decision-makers with the importance of individual attributes to the classification of a particular object, which may facilitate explanations by experts in various domains with high-cost errors like medicine or finance.

We discuss how notions from cooperative game theory can be used to assess the contribution of individual attributes in classification and clustering processes in concept-based machine learning. To address the 3rd question, we present some ideas on how to reduce the number of attributes using similarities in large contexts.

Keywords: Interpretable Machine Learning · Concept learning · Formal concepts · Shapley values · Explainable AI

1   Introduction

In the notes of this invited talk, we would like to give the reader a short introduction to Interpretable Machine Learning (IML) from the perspective of Formal

W. M. P. van der Aalst et al. (Eds.): AIST 2020, LNCS 12602, pp. 28–54, 2021.
https://doi.org/10.1007/978-3-030-72610-2_3

Concept Analysis (FCA), which can be considered as a mathematical framework for concept learning, Frequent Itemset Mining (FIM) and Association Rule Mining (ARM).

Among the variety of concept learning methods, we selected the rule-based JSM-method named after J.S. Mill in its FCA formulation. Another possible candidate is Version Spaces. To stress the difference between concept learning paradigm and formal concept we used concept-based learning term in case of usage of FCA as a mathematical tool and language.

We assume, that interpretation by means of game-theoretic attribute ranking is also important in an unsupervised setting as well, and demonstrate its usage via attribution of stability indices of formal concepts (concept stability is also known as the robustness of closed itemset in the FIM community).

Being a convenient language for JSM-method (hypotheses learning) and Frequent Itemset Mining, its direct application to large datasets is possible only under a reasonable assumption on the number of attributes or data sparseness. Direct computation of the Shapley value for a given attribute also requires enumeration of almost all attribute subsets in the intent of a particular object or concept. One of the possibilities to cope with the data volume is approximate computations, while another one lies in the reduction of the number of attributes or their grouping by similarity.

The paper is organised as follows. Section 2 observes several closely related studies and useful sources on FCA and its applications. Section 3 is devoted to concept-based learning where formal intents are used as classification hypotheses and specially tailored Shapley value helps to figure out contributions of attributes in those hypotheses when a particular (e.g., unseen) object is examined. Section 4 shows that the Shapley value approach can be used for attribution to stability (or robustness) of formal concepts, thus we are able to rank single attributes of formal intents (closed itemsets) in an unsupervised setting. Section 5 sheds light on the prospects of usage attribute-based similarity of concepts and attribute reduction for possibly large datasets (formal contexts). Section 6 concludes the paper.

2   Related Work

Formal Concept Analysis is an applied branch of modern Lattice Theory suitable for knowledge representation and data analysis in various domains [15]. We refer the reader to a modern textbook on FCA with a focus on attribute exploration and knowledge extraction [14], surveys on FCA models and techniques for knowledge processing and representation [36,53] as well as on their applications [52]. Some of the examples in subsequent sections are also taken from a tutorial on FCA and its applications [18].

Since we deal with interpretable machine learning, we first need to establish basic machine learning terminology in FCA terms. In the basic case, our data are Boolean object-attribute matrices or formal contexts, which are not necessarily labeled w.r.t. a certain target attribute. Objects can be grouped into clusters (concept extents) by their common attributes, while attributes compose a cluster (concept intent) if they belong to a certain subset of objects. The pairs of subsets of objects and attributes form the so-called formal concepts, i.e. maximal submatrices (w.r.t. of rows and attribute permutations) of an input context full of ones in its Boolean representation. Those concepts form hierarchies or concept lattices (Galois lattices), which provide convenient means of visualization and navigation and enables usage of suprema and infima for incomparable concepts.

The connection between well-known concept learning techniques (for example, Version Spaces, and decision tree induction) from machine learning and FCA was well established in [12,31]. Thus Version Spaces studied by T. Mitchell [49] also provides hierarchical means for hypotheses learning and elimination, where hypotheses are also represented as conjunctions of attributes describing the target concept. Moreover, concept lattices can be used for searching for globally optimal decision trees in the domains where we should not care about the trade-off between time spent for the training phase and reached accuracy (e.g., medical diagnostics) but should rather focus on all valid paths in the global search space [4,25].

In case we deal with unsupervised learning, concept lattices can be considered as a variant of hierarchical clustering where one has the advantage to use multiple inheritance in both bottom-up and top-down directions [6,7,64,66]. Another fruitful property of formal concepts allows one not only to receive a cluster of objects without any clue why they are similar but to reveal objects' similarity in terms of their common attributes. This property allows considering a formal concept as bicluster [19,27,48], i.e. a biset of two clusters of objects and attributes, respectively.

Another connection between FCA and Frequent Itemset Mining is known for years [45,51]. In the latter discipline, transactions of attributes are mined to find items frequently bought together [1]. The so-called closed itemsets are used to cope with a huge number of frequent itemsets for large input transaction bases (or contexts), and their definition coincides with the definition of concept intents (under the choice of constraint on the concept extent size or itemset support). Moreover, attribute dependencies in the form of implications and partial implications [47] are known as association rules, which appeared later in data mining as well [1][1].

This is not a coincidence that we discuss data mining, while stressed interpretability and machine learning in the title. Historically, data mining was formulated as a step of the Knowledge Discovery in Databases process that is "the nontrivial process of identifying valid, novel, potentially useful, and ultimately understandable patterns in data." [10]. While understandable patterns are a must for data mining, in machine learning and AI in general, this property should be instantiated as something extra, which is demanded by analysts to ease decision making as the adjectives explainable (AI) and interpretable (ML) suggest [50].

---

[1] One of the earlier precursors of association rules can be also found in [17] under the name of "almost true implications".

To have a first but quite comprehensive reading on interpretable ML we suggest a freely available book [50], where the author states that "Interpretable Machine Learning refers to methods and models that make the behaviour and predictions of machine learning systems understandable to humans".

The definition of interpretability may vary from the degree to which a human can understand the cause of a decision to the degree to which a human can consistently predict the model's result.

The taxonomy of IML methods has several aspects. For example, models can be roughly divided into *intrinsic* and *post hoc* ones. The former include simpler models like short rules or sparse linear models, while among the latter black-box techniques with post hoc processing after their training can be found. Some researchers consistently show that in case of the necessity to have interpretable models, one should not use post hoc techniques for black-box models but trust naturally interpretable models [57]. Another aspect is the universality of the method, the two extremes are *model-specific* (the method is valid for only one type of models) and or *model-agnostic* (all models can be interpreted with the method). There is one more important aspect, whether the method is suitable for the explanation of the model's predictions for a concrete object (*local method*) or it provides an interpretable picture for the entire model (*global method*). Recent views on state-of-the-art techniques and practices can be found in [8,26].

FCA provides interpretable patterns a priori since it deals with such understandable patterns as sets of attributes to describe both classes (by means of classification rules or implications) and clusters (e.g., concept intents). However, FCA theory does not suggest the (numeric) importance of separate attributes. Here, a popular approach based on Shapley value from Cooperative Game Theory [59] recently adopted by the IML community may help [24,46,63].

The main idea of Shapley value based approaches in ML for ranking separate attributes is based on the following consideration: each attribute is considered as a player in a specific game-related to classification or regression problem and attributes are able to form (winning) coalitions. The importance of such a player (attribute) is computed over all possible coalitions by a combinatorial formula taking into account the number of winning coalitions where without this attribute the winning state is not reachable.

One of the recent popular implementations is SHAP library [46], which however cannot be directly applied to our concept-based learning cases: JSM-hypotheses and stability indices. The former technique assumes that unseen objects can be left without classification or classified contradictory when for an examined object there is no hypothesis for any class or there are at least two hypotheses from different classes [11,38]. This might be an especially important property for such domains as medicine and finance where wrong decisions may lead to regrettable outcomes. We can figure out what are the attributes of the contradictory hypotheses we have but which attributes have the largest positive or negative impact on the classification is still unclear without external means. The latter case of stability indices, which were originally proposed for ranking JSM-hypotheses by their robustness to the deletion of object subsets from the input contexts (similarly to cross-validation) [33,37], is considered in an unsupervised setting. Here, supervised interpretable techniques like SHAP are not directly applicable. To fill the gap we formulated two corresponding games with specific valuation functions used in the Shapley value computations.

Mapping of the two proposed approaches onto the taxonomy of IML methods says that in the case of JSM-hypotheses it is an intrinsic model, but applying Shapley values on top of it is post hoc. At the same time, this concrete variant is rather model-specific since it requires customisation. This one is local since it explains the classification of a single object. As for attribution of concept stability, this one is definitely post hoc, model-specific, and if each pattern (concept) is considered separately this one is rather local but since the whole set of stable concepts can be attributed it might be considered as a global one as well.

It is important to note that one of the stability indices was rediscovered in the Data Mining community and known under the name of the robustness of closed itemsets [34,65] (where each transaction/object is kept with probability $\alpha = 0.5$). So, the proposed approach also allows attribution of closed itemsets.

Classification and direct computation of Shapley values afterwards might be unfeasible for large sets of attributes [8]. So, we may think of approximate ways to compute Shapley values [63] or pay attention to attribute selection, clarification, and reduction known in the FCA community. We would like to draw the reader's attention to scale coarsening as feature selection tools [13] and a comparative survey on FCA-based attribute reduction techniques [28,29]. However, we prefer to concentrate on attribute aggregation by similarity[2] as an attribute reduction technique which will not allow us to leave out semantically meaningful attributes even if they are highly-correlated and redundant in terms of extra complexity paid for their processing otherwise.

The last note on related works, which is unavoidable when we talk about IML, is the relation to Deep Learning (DL) where black-box models predominate [60]. According to the textbook [16], "Deep Learning is a form of machine learning that enables computer to learn from experience and understand the world in terms of a hierarchy of concepts." The authors also admit that there is no need for a human computer operator to formally specify all the knowledge that the computer needs and obtained hierarchy of concepts allows the computer to learn complicated concepts by building them out of simpler ones. The price of making those concepts intelligible for the computer but not necessary for a human is paid by specially devised IML techniques in addition to DL models.

Since FCA operates with concept hierarchies and is extensively used in human-centric applications [52], the question "What can FCA do for DL?" is open. For example, in [58] closure operators on finite sets of attributes were encoded by a three-layered feed-forward neural network, while in [35] the authors were performing neural architecture search based on concept lattices to avoid overfitting and increase the model interpretability.

---

[2] Similarity between concepts is discussed in [9].

3 Supervised Learning: From Hypotheses to Attribute Importance

In this section, we discuss how interpretable concept-based learning for JSM-method can be achieved with Shapley Values following our previous study on the problem [20]. Let us start with a piece of history of inductive reasoning. In XIX century, John Stuart Mill proposed several schemes of inductive reasoning. Let us consider, for example, the Method of Agreement [23]: "If two or more instances of the phenomenon under investigation have only one circumstance in common, ... [it] is the cause (or effect) of the given phenomenon."

The JSM-method (after J.S. Mill) of hypotheses generation proposed by Viktor K. Finn in the late 1970s is an attempt to describe induction in purely deductive form [11]. This new formulation was introduced in terms of many-valued many-sorted extension of the First Order Predicate Logic [32].

This formal logical treatment allowed usage of the JSM-method as a machine learning technique [37]. While further algebraic redefinitions of the logical predicates to express similarity of objects as an algebraic operation allowed the formulation of JSM-method as a classification technique in terms of formal concepts [32,39].

3.1 JSM-hypotheses in FCA

In FCA, a formal concept consists of an extent and an intent. The intent is formed by all attributes that describe the concept, and the extent contains all objects belonging to the concept. In FCA, the JSM-method is known as rule-based learning from positive and negative examples with rules in the form "concept intent → class".

Let a formal context $\mathbb{K} := (G, M, I)$ be our universe, where the binary relation $I \subseteq G \times M$ describes if an object $g \in G$ has an attribute $m \in M$. For $A \subseteq G$ and $B \subseteq M$ the derivation (or Galois) operators are defined by:

$$A' = \{ m \in M \mid \forall a \in A \; a I m \} \text{ and } B' = \{ g \in G \mid \forall b \in B \; g I b \}.$$

A (formal) concept is a pair (A, B) with $A \subseteq G$, $B \subseteq M$ such that $A' = B$ and $B' = A$. We call B its intent and A its extent. An implication of the form $H \rightarrow m$ holds if all objects having the attributes in H also have the attribute m, i.e. $H' \subseteq m'$.

The set of all concepts of a given context $\mathbb{K}$ is denoted by $\mathfrak{B}(G, M, I)$; the concepts are ordered by the "to be a more general concept" relation as follows: $(A, B) \geq (C, D) \iff C \subseteq A$ (equivalently $B \subseteq D$).

The set of all formal concepts $\mathfrak{B}(G, M, I)$ together with the introduced relation form the *concept lattice*, which line diagram is useful for visual representation and navigation through the concept sets.

Let $w \notin M$ be a *target attribute*, then w partitions G into three subsets:

- *positive examples*: $G_+ \subseteq G$ of objects known to satisfy w,
- *negative examples*: $G_- \subseteq G$ of objects known not to have w,
- *undetermined examples*: $G_\tau \subseteq G$ of objects for which it remains unknown whether they have the target attribute or do not have it.

This partition gives rise to three subcontexts $\mathbb{K}_\varepsilon := (G_\varepsilon, M, I_\varepsilon)$ with $\varepsilon \in \{-, +, \tau\}$.

- The *positive context* $\mathbb{K}_+$ and the *negative context* $\mathbb{K}_-$ form the training set called by *learning context*:

$$\mathbb{K}_\pm = (G_+ \cup G_-, M \cup \{w\}, I_+ \cup I_- \cup G_+ \times \{w\}).$$

- The subcontext $\mathbb{K}_\tau$ is called the *undetermined context* and is used to predict the class of not yet classified objects.

The whole *classification context* is the context

$$\mathbb{K}_c = (G_+ \cup G_- \cup G_\tau, M \cup \{w\}, I_+ \cup I_- \cup I_\tau \cup G_+ \times \{w\}).$$

The derivation operators in the subcontexts $\mathbb{K}_\varepsilon$ are denoted by $(\cdot)^+$ $(\cdot)^-$, and $(\cdot)^\tau$, respectively. The goal is to classify the objects in G_τ with respect to w.

To do so let us form the positive and negative hypotheses as follows. A *positive hypothesis* $H \subseteq M$ ($H \neq \emptyset$) is a intent of $\mathbb{K}_+$ that is not contained in the intent of a negative example; i.e. $H^{++} = H$ and $H' \subseteq G_+ \cup G_\tau$ ($H \rightarrow w$). A *negative hypothesis* $H \subseteq M$ ($H \neq \emptyset$) is an intent of $\mathbb{K}_-$ that is not contained in the intent of a positive example; i.e. $H^{--} = H$ and $H' \subseteq G_- \cup G_\tau$ ($H \rightarrow \overline{w}$).

An intent of $\mathbb{K}_+$ that is contained in the intent of a negative example is called a *falsified (+)-generalisation*. A *falsified (−)-generalisation* is defined in a similar way.

To illustrate these notions we use the credit scoring context in Table 1 [22]. Note that we use *nominal scaling* to transform many-valued context to one-valued context [15] with the following attributes, Ma, F (for two genders), Y, MI, O (for young, middle, and old values of the two-valued attribute Age , resp.), HE, Sp, SE (for higher, special, and secondary education, resp.), Hi, L, A (for high, low, and average salary, resp.), and w and $\overline{w}$ for the two-valued attribute Target.

For example, the intent of the red node labelled by the attribute A in the left line diagram (Fig. 1), is $\{A, Mi, F, HE\}$, and this is not contained in the intent of any node labelled by the objects g_5, g_6, g_7, and g_8. So we believe in the rule $H \rightarrow w$. Note that the colours of the nodes in Fig. 1 represent different types of hypotheses: the red ones correspond to minimal hypotheses (cf. the definition below), the see green nodes correspond to negative hypotheses, while light grey nodes correspond to non-minimal positive and negative hypotheses for the left and the right line diagrams, respectively.

The undetermined examples g_τ from G_τ are classified according to the following rules:

Table 1. Many-valued classification context for credit scoring

G/M	Gender	Age	Education	Salary	Target
1	Ma	Young	Higher	High	+
2	F	Middle	Special	High	+
3	F	Middle	Higher	Average	+
4	Ma	Old	Higher	High	+
5	Ma	Young	Higher	Low	−
6	F	Middle	Secondary	Average	−
7	F	Old	Special	Average	−
8	Ma	Old	Secondary	Low	−
9	F	Young	Special	High	τ
10	F	Old	Higher	Average	τ
11	Ma	Middle	Secondary	Low	τ
12	Ma	Old	Secondary	High	τ

- If $g_\tau^\mathcal{T}$ contains a positive, but no negative hypothesis, then g_τ is *classified positively*.
- If $g_\tau^\mathcal{T}$ contains a negative, but no positive hypothesis, then g_τ is *classified negatively*.
- If $g_\tau^\mathcal{T}$ contains both negative and positive hypotheses, or if $g_\tau^\mathcal{T}$ does not contain any hypothesis, then this object classification is *contradictory* or *undetermined*, respectively.

To perform classification by the aforementioned rules, it is enough to have only *minimal hypotheses* (w.r.t. $\subseteq$) of both signs.

Let $\mathcal{H}_+$ (resp. $\mathcal{H}_-$) be the set of minimal positive (resp. minimal negative) hypotheses. Then,

$$\mathcal{H}_+ = \big\{\{F, Mi, HE, A\}, \{HS\}\big\} \text{ and } \mathcal{H}_- = \big\{\{F, O, Sp, A\}, \{Se\}, \{Ma, L\}\big\}.$$

We proceed to classify the four undetermined objects below.

- $g_9' = \{F, Y, Sp, HS\}$ contains the positive hypothesis $\{HS\}$, and no negative hypothesis. Thus, g_9 is classified positively.
- $g_{10}' = \{F, O, HE, A\}$ does not contain neither positive nor negative hypotheses. Hence, g_{10} remains undetermined.
- $g_{11}' = \{Ma, Mi, Se, L\}$ contains two negative hypotheses: $\{Se\}$ and $\{Ma, L\}$, and no positive hypothesis. Therefore, g_{11} is classified negatively.
- $g_{12}' = \{Ma, O, Se, HS\}$ contains the negative hypothesis $\{Se\}$ and the positive hypothesis $\{HS\}$, which implies that g_{12} remains undetermined.

Even though we have a clear explanation of why a certain object belongs to one of the classes in terms of contained positive and negative hypotheses, the

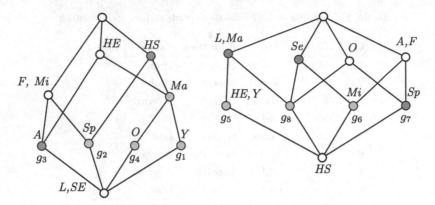

Fig. 1. The line diagrams of the lattice of positive context (left) and the lattice of negative context (right).

following question arises: Do all attributes play the same role in the classification of certain examples? If the answer is no, then one more question appears: How can we rank attributes with respect to their importance in classifying examples, for example, g_{11} with attributes Ma, Mi, Se, and L? Game Theory offers several indices for such comparison: e.g., the Shapley value and the Banzhaf index. For the present contribution, we concentrate on the use of Shapley values.

3.2 Shapley Values and JSM-hypotheses

To answer the question "What are the most important attributes for classification of a particular object?" in our case, we follow to basic recipe studied in [46, 50, 63].

To compute the Shapley value for an example x and an attribute m, one needs to define $f_x(S)$, the expected value of the model prediction conditioned on a subset S of the input attributes.

$$\phi_m = \sum_{S \subseteq M \backslash \{m\}} \frac{|S|!(|M| - |S| - 1)!}{|M|!} \left(f_x(S \cup \{m\}) - f_x(S) \right), \tag{1}$$

where M is the set of all input attributes and S a certain coalition of players, i.e. set of attributes.

Let $\mathbb{K}_c = (G_+ \cup G_- \cup G_\tau, M \cup \{w\}, I_+ \cup I_- \cup I_\tau \cup G_+ \times \{w\})$ be our classification context, and $\mathcal{H}_+$ (resp. $\mathcal{H}_-$) the set of minimal positive (resp. negative) hypotheses of $\mathbb{K}_c$.

Since we deal with hypotheses (i.e. sets of attributes) rather than compute the expected value of the model's prediction, we can define a valuation function v directly. For $g \in G$, the *Shapley value of an attribute* $m \in g'$:

$$\varphi_m(g) = \sum_{S \subseteq g' \backslash \{m\}} \frac{|S|!(|g'| - |S| - 1)!}{|g'|!} \left(v(S \cup \{m\}) - v(S) \right), \tag{2}$$

where

$$v(S) = \begin{cases} 1, & \exists H_+ \in \mathcal{H}_+ : H_+ \subseteq S \text{ and } \forall H_- \in \mathcal{H}_- : H_- \not\subseteq S, \\ -1, & \exists H_- \in \mathcal{H}_- : H_- \subseteq S \text{ and } \forall H_+ \in \mathcal{H}_+ : H_+ \not\subseteq S \\ 0, & \text{otherwise} \end{cases}$$

The Shapley value $\varphi_m(g)$ is set to 0 for every $m \in M \setminus g'$. The Shapley vector for a given object g is denoted by $\Phi(g)$. To differentiate between the value in cases when $m \in M \setminus g'$ and $m \in g'$, we will use decimal separator as follows, 0 and 0.0, respectively.

For the credit scoring context, the minimal positive and the negative hypotheses are

$$\mathcal{H}_+ = \{\{F, Mi, HE, A\}, \{HS\}\}; \quad \mathcal{H}_- = \{\{F, O, Sp, A\}, \{Se\}, \{M, L\}\}.$$

The Shapley values for JSM-hypotheses have been computed with our freely available Python scripts[3] for the objects in G_τ:

- $g_9' = \{F, Y, Sp, HS\} \supseteq \{HS\}$, and g_9 is classified positively. $\varphi_{HS}(g_9) = 1$ and and its Shapley vector is $\Phi(g_9) = (0, 0.0, 0.0, 0, 0, 0, 0.0, 0, 1.0, 0, 0)$.
- $g_{10}' = \{F, O, HE, A\}$ and g_{10} remains undetermined. Its Shapley vector is $\Phi(g_{10}) = (0, 0.0, 0, 0, 0.0, 0.0, 0, 0, 0, 0.0, 0)$.
- $g_{11}' = \{Ma, Mi, Se, L\} \supseteq \{Se\}$, $\{Ma, L\}$. Its Shapley vector is $\Phi(g_{11}) = (-1/6, 0, 0, 0.0, 0, 0, 0, -2/3, 0, 0, -1/6)$.
- $g_{12}' = \{Ma, O, Se, HS\} \supseteq \{HS\}$, $\{Se\}$. $\varphi_{Se}(g_{12}) = -1$, $\varphi_{HS}(g_{12}) = 1$. Its Shapley vector is $\Phi(g_{12}) = (0.0, 0, 0, 0, 0.0, 0, 0, -1.0, 1.0, 0, 0)$.

Let us examine example g_{11}. Its attribute Mi has zero importance according to the Shapley value approach since it is not in any contained hypothesis used for the negative classification. The most important attribute is Se, which is alone two times more important than the attributes Ma and L together. It is so, since the attribute Se, which is the single attribute of the negative hypothesis $\{Se\}$, forms more winning coalitions $S \cup \{Se\}$ with $v(S \cup \{Se\}) - v(S) = 1$ than Ma and L, i.e. six vs. two. Thus, $\{Se\} \uparrow \setminus \{Ma, L\} \uparrow = \{\{Se\}, \{Ma, Se\}, \{Mi, Se\}, \{Se, L\}, \{Mi, Se, L\}, \{Ma, Mi, Se\}\}$[4] are such winning coalitions for Se, while $\{Ma, L\}$, $\{Ma, Mi, L\}$, are those for Ma and L.

The following properties hold:

Theorem 1 ([20]). *The Shapley value, $\varphi_m(g)$, of an attribute m for the JSM-classification of an object g, fulfils the following properties:*

1. $\sum_{m \in g'} \varphi_m(g) = 1$ *if g is classified positively;*
2. $\sum_{m \in g'} \varphi_m(g) = -1$ *if g is classified negatively.*

---

[3] https://github.com/dimachine/Shap4JSM.
[4] $S \uparrow$ is the up-set of S in the Boolean lattice $(\mathcal{P}\{Ma, Mi, Se, L\}, \subseteq)$.

3. $\sum_{m \in g'} \varphi_m(g) = 0$ *if g is classified contradictory or undetermined.*

The last theorem expresses the so-called *efficiency property* or axiom [59], where it is stated that the sum of Shapley values of all players in a game is equal to the total pay-off of their coalition, i.e. $v(g')$ in our case.

It is easy to check $\varphi_m(g) = 0$ for every $m \in g'$ that does not belong to at least one positive or negative hypothesis contained in g'. Moreover, in this case for any $S \subseteq g' \setminus \{m\}$ it also follows $v(S) = v(S \cup \{m\})$ and these attributes are called *null* or *dummy players* [59].

We also performed experiments on the Zoo dataset[5], which includes 101 examples (animals) and their 17 attributes along with the target attribute (7 classes of animals). The attributes are binary except for the number of legs, which can be scaled nominally and treated as categorical.

We consider a binary classification problem where *birds* is our positive class, while all the rest form the negative class.

There are 19 positive examples (birds) and 80 negative examples since we left out two examples for our testing set, namely, chicken and warm. The hypotheses are $\mathcal{H}_+ = \{\{feathers, eggs, backbone, breathes, legs_2, tail\}\}$ and

$$\mathcal{H}_- = \{\{venomous\}, \{eggs, aquatic, predator, legs_5\}, \{legs_0\}, \{eggs, legs_6\},$$
$$\{predator, legs_8\}, \{hair, breathes\}, \{milk, backbone, breathes\}, \{legs_4\},$$
$$\{toothed, backbone\}\}.$$

The intent $aardvark' = \{hair, milk, predator, toothed, backbone, breathes, legs_4, catsize\}$ contains four negative hypotheses and no positive one.

The Shapley vector for the **aardvark** example is

$$(-0.1, 0, 0, -0.0167, 0, 0, 0.0, -0.1, -0.133, -0.133, 0, 0, -0.517, 0, 0, 0, 0, 0, 0, 0, 0.0).$$

Backbone, breathes, and **four legs** are the most important attributes with values −0.517, −0.133, and −0.133, respectively, while **catsize** is not important in terms of Shapley value (Fig. 2).

A useful interpretation of classification results could be an explanation for true positive or true negative cases. However, in the case of our test set both examples, **chicken** and **warm**, are classified correctly as bird and non-bird, respectively. Let us have a look at their Shapley vectors. Our test objects have the following intents

$$chicken' = \{feathers, eggs, airborne, backbone, breathes, legs_2, tail, domestic\}$$

and

$$warm' = \{eggs, breathes, legs_0\}.$$

---

[5] https://archive.ics.uci.edu/ml/datasets/zoo.

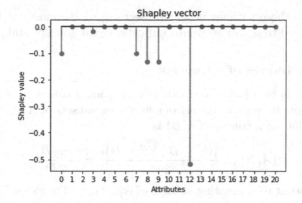

Fig. 2. The Shapley vector diagram for the **aardvark** example

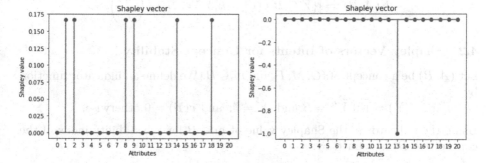

Fig. 3. The Shapley vector diagram for the **chicken** (left) and **warm** (right) examples

Thus, for the **chicken** example all six attributes that belong to the single positive hypothesis have equal Shapley values, i.e. 1/6. The attributes **airborne** and **domestic** have zero importance (Fig. 3, left). The **warm** example has only one attribute with non-zero importance, i.e. the absence of legs with importance −1 (Fig. 3, right) . It is so since the only negative hypothesis, $\{legs_0\}$, is contained in the object intent.

4   Unsupervised Learning: Contribution to Stability and Robustness

(Intensional) stability indices were introduced to rank the concepts (intents) by their robustness under objects deletion and provide evidence of the non-random nature of concepts [56]. The extensional stability index is defined as the proportion of intent subsets generating this intent; it shows the robustness of the concept extent under attributes deletion [56]. Our goal here is to find out whether all attributes play the same role in the stability indices. To measure the importance of an attribute for a concept intent, we compare generators with this attribute to those without it. In this section, we demonstrate how Shapley values can be used to assess attribute importance for concept stability.

4.1   Stability Indices of a Concept

Let $\mathbb{K} := (G, M, I)$ be a formal context. For any closed subset X of attributes or objects, we denote by $\text{gen}(X)$ the set of generating subsets of X. The *extensional stability index* [56] of a concept (A, B) is

$$\sigma_e(A, B) := \frac{|\{Y \subseteq B \mid Y'' = B\}|}{2^{|B|}} = \frac{|\text{gen}(B)|}{2^{|B|}}.$$

We can also restrict to generating subsets of equal size. The extensional stability index of the k-th level of (A, B) is

$$J_k(A, B) := |\{Y \subseteq B \mid |Y| = k, Y'' = B\}| \Big/ \binom{|B|}{k}.$$

4.2   Shapley Vectors of Intents for Concept Stability

Let (A, B) be a concept of (G, M, I) and $m \in B$. We define an indicator function by

$$v(Y) = 1 \text{ if } Y'' = B \text{ and } Y \neq \emptyset, \text{ and } v(Y) = 0 \text{ otherwise.}$$

Using the indicator v, the Shapley value of $m \in B$ for the stability index of the concept (A, B) is defined by:

$$\varphi_m(A, B) := \frac{1}{|B|} \sum_{Y \subseteq B \setminus \{m\}} \frac{1}{\binom{|B|-1}{|Y|}} \Big(v(Y \cup \{m\}) - v(Y) \Big). \tag{3}$$

The Shapley vector of (A, B) is then $(\varphi_m(A, B))_{m \in B}$. An equivalent formulation is given using upper sets of minimal generators [21]. In fact, for $m \in X_m \in \text{mingen}(B)$ and $m \notin X_{\overline{m}} \in \text{mingen}(B)$, we have

$$\varphi_m(A, B) = \frac{1}{|B|} \sum_{D \sqcup \{m\} \in \bigcup X_m \uparrow \setminus \bigcup X_{\overline{m}} \uparrow} \frac{1}{\binom{|B|-1}{|D|}},$$

where $\sqcup$ denotes the disjoint union, X_m and $X_{\overline{m}}$ the minimal generators of B with and without m, respectively (Fig. 4).

To compute φ_m, additional simplifications are useful:

Theorem 2 ([21]). *Let (A, B) be a concept and $m \in B$.*

(i) $\varphi_m(A, B) = \sum_{k=1}^{|B|} \frac{J_k(A,B)}{k} - \sum_{D \subseteq B \setminus \{m\}} \frac{1}{|D| \binom{|B|-1}{|D|}} v(D).$

(ii) If $m \in X_m \in \text{mingen}(B)$ and $Y \subseteq B \setminus \{m\}$ with $(A, B) \prec (Y', Y)$ then

$$\varphi_m(A, B) = \frac{1}{|B|} \sum_{D \in \bigcup [X_m \setminus \{m\}, Y]} \frac{1}{\binom{|B|-1}{|D|}}.$$

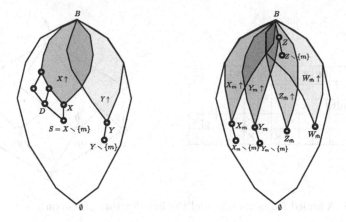

Fig. 4. Computing Shapley vectors for concept stability

(iii) If $m \in X \in \mathrm{mingen}(B)$ and $|\mathrm{mingen}(B)| = 1$, then

$$\varphi_m(A, B) = \sum_{k=1}^{|B|} \frac{J_k(A, B)}{k} = \frac{1}{|X|}. \tag{4}$$

To illustrate the importance of attributes in concept stability, we consider the the fruits context [31], where we extract the subcontext with the first four objects (Table 2).

Table 2. A many-valued context of fruits

G\M	Color	Firm	Smooth	Form
1 Apple	Yellow	No	Yes	Round
2 Grapefruit	Yellow	No	No	Round
3 Kiwi	Green	No	No	Oval
4 Plum	Blue	No	Yes	Oval

After scaling we get the binary context and its concept lattice diagram (Fig. 5).

For each concept, the stability index σ_e and its Shapley vector ϕ are computed (Table 3).

For the Zoo dataset we obtain 357 concepts in total. The top-3 most stable are c_1, c_2, c_3 with extent stability indices: $\sigma_e(G, \emptyset) = 1$, $\sigma_e(\emptyset, M) = 0.997$, $\sigma_e(A, A') = 0.625$, respectively, where
$A' = \{feathers, eggs, backbone, breathes, legs_2, tail\}$ and
$A = \{11, 16, 20, 21, 23, 33, 37, 41, 43, 56, 57, 58, 59, 71, 78, 79, 83, 87, 95, 100\}$.

Fruits	w	y	g	b	f	$\bar{f}$	s	$\bar{s}$	r	$\bar{r}$
1 apple		×				×	×		×	
2 grapefruit		×				×		×	×	
3 kiwi			×			×		×		×
4 plum				×		×	×			×

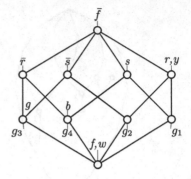

Fig. 5. A scaled fruits context and the line diagram of its concept lattice

Table 3. The concepts of fruits context and their stability indices along with Shapley vectors

Concepts	σ_e	Φ
$(\{4\}, \{b, \bar{f}, s, \bar{r}\})$	0.625	(2/3, 0.0, 1/6, 1/6)
$(\{3\}, \{g, \bar{f}, \bar{s}, \bar{r}\})$	0.625	(2/3, 0.0, 1/6, 1/6)
$(\{3, 4\}, \{\bar{f}, \bar{r}\})$	0.5	(0.0, 1.0)
$(\{2\}, \{y, \bar{f}, \bar{s}, r\})$	0.375	(1/6, 0.0, 2/3, 1/6)
$(\{2, 3\}, \{\bar{f}, \bar{s}\})$	0.5	(0.0, 1.0)
$(\{1\}, \{y, \bar{f}, s, r\})$	0.375	(1/6, 0.0, 2/3, 1/6)
$(\{1, 4\}, \{\bar{f}, s\})$	0.5	(0.0, 1.0)
$(\{1, 2\}, \{y, \bar{f}, r\})$	0.75	(0.5, 0.0, 0.5)
$(\{1, 2, 3, 4\}, \{\bar{f}\})$	1	(0.0)

$$\sigma_e(\emptyset, \{w, y, g, b, f, \bar{f}, s, \bar{s}, r, \bar{r}\}) = 0.955$$
$$\Phi = (0.256, 0.069, 0.093, 0.093, 0.260, 0.0, 0.052, 0.052, 0.069, 0.052)$$

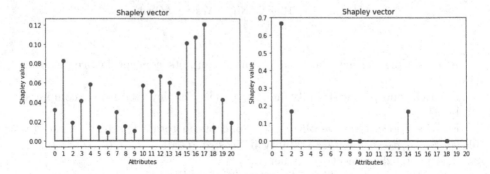

Fig. 6. The Shapley vector for concept $c_2 = (\emptyset, M)$ (left) and c_3 (right)

The most important attributes are **six legs**, **eight legs**, **five legs**, **feathers**, and **four legs** for c_2 (Fig. 6, left), and **feathers**, **eggs**, and **two legs** for c_3 (Fig. 6, right), w.r.t. to the Shapley vectors.

The demo is available on GitHub[6]. Shapley values provide a tool for assessing the attribute importance of stable concepts. Comparison with other (not only Game-theoretic) techniques for local interpretability is desirable. We believe that the attribute importance can be lifted at the context level, via an aggregation, and by then offer a possibility for attribute reduction, similar to the principal component analysis (PCA) method.

5 Attribute Similarity and Reduction

Computation of attribute importance could lead to ranking the attributes of the context, and by then classifying the attributes with respect to their global importance, similar to principal component analysis. Therefore cutting off at a certain threshold could lead to attribute reduction in the context. Other methods leading to attributes reduction are based on their granularity, an ontology or an is-a taxonomy, by using coarser attributes. Less coarse attributes are then put together by going up in the taxonomy and are considered to be similar. In the present section, we briefly discuss the effect of putting attributes together on the resulting concept lattice. Doing this leads to the reduction of the number of attributes, but not always in the reduction of the number of concepts.

Before considering such compound attributes, we would like to draw the readers' attention to types of data weeding that often overlooked outside of the FCA community [28,29,55], namely, clarification and reduction.

5.1 Clarification and Reduction

A context (G, M, I) is called *clarified* [15], if for any objects $g, h \in G$ from $g' = h'$ it always follows that $g = h$ and, similarly, $m' = n'$ implies $m = n$ for all $m, n \in M$. A clarification consists in removing duplicated lines and columns from the context. This context manipulation does not alter the structure of the concept lattice, though objects with the same intents and attributes with the same extents are merged, respectively.

The structure of the concept lattice remains unchanged in case of removal of *reducible attributes* and *reducible objects* [15]; An attribute m is reducible if it is a combination of other attributes, i.e. $m' = Y'$ for some $Y \subseteq M$ with $m \notin Y$. Similarly, an object g is reducible if $g' = X'$ for some $X \subseteq G$ with $g \notin X$. For example, full rows ($g' = M$) and full columns ($m' = G$) are always reducible.

However, if our aim is a subsequent interpretation of patterns, we may wish to keep attributes (e.g. in aggregated form), rather than leaving them out before knowing their importance.

---

[6] https://github.com/dimachine/ShapStab/.

5.2   Generalised Attributes

As we know, FCA is used for conceptual clustering and helps discover patterns in terms of clusters and rules. However, the number of patterns can explode with the size of an input context. Since the main goal is to maintain a friendly overview of the discovered patterns, several approaches have been investigated to reduce the number of attributes without loss of much information [28,55]. One of these suggestions consists in using is-a taxonomies. Given a taxonomy on attributes, how can we use it to discover *generalised* patterns in the form of clusters and rules? If there is no taxonomy, can we (interactively) design one? We will discuss different scenarios of grouping attributes or objects, and the need of designing similarity measures for these purposes in the FCA setting.

To the best of our knowledge the problem of mining generalised association rules was first introduced around 1995 in [61,62], and rephrased as follows: Given a large database of transactions, where each transaction consists of a set of items, and a taxonomy (is-a hierarchy) on the items, the goal is to find associations between items at any level of the taxonomy. For example, with a taxonomy that says that `jackets` is-a `outerwear` and `outerwear` is-a `clothes`, we may infer a rule that "people who buy `outerwear` tend to buy `shoes`". This rule may hold even if rules that "people who buy `jackets` tend to buy `shoes`", and "people who buy `clothes` tend to buy `shoes`" do not hold (See Fig. 7).

(a) Database $\mathcal{D}$

Transaction	Items bought
100	Shirt
200	Jacket, Hiking Boots
300	Ski Pants, Hiking Boots
400	Shoes
500	Shoes
600	Jacket

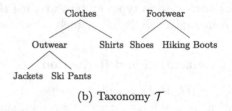

(b) Taxonomy $\mathcal{T}$

(c) Frequent itemsets

Itemset	Support
Jacket	2
Outwear	3
Clothes	4
Shoes	2
Hiking Boots	2
Footwear	4
Outwear, Hiking Boots	2
Clothes, Hiking Boots	2
Outwear, Footwear	2
Clothes, Footwear	2

(d) Association rules

Rule	Support	Confidence
Outwear → Hiking Boots	1/3	2/3
Outwear → Footwear	1/3	2/3
Hiking Boots ↛ Outwear	1/3	1
Hiking Boots ↛ Clothes	1/3	1

Fig. 7. A database of transactions, taxonomies and extracted rules [61,62]

A generalised association rule is a (partial) implication $X \to Y$, where X, Y are disjoint itemsets and no item in Y is a generalisation of any item in X [61,62]. We adopt the following notation: $\mathcal{I} = \{i_1, i_2, \cdots, i_m\}$ is a set of items and $\mathcal{D} = \{t_1, t_2, \cdots, t_n\}$ a set of transactions. Each transaction $t \in \mathcal{D}$ is a subset of items $\mathcal{I}$. Let $\mathcal{T}$ be a set of taxonomies (i.e. directed acyclic graph on items and generalised items). We denote by $(\mathcal{T}, \leq)$ its transitive closure. The elements of $\mathcal{T}$ are called "general items". A transaction t *supports* an item x (resp. a general item y) if x is in t (resp. y is a generalisation of an item x in t). A set of transactions T *supports* an itemset $X \subseteq \mathcal{I}$ if T supports every item in X.

In FCA setting, we build a generalised context $(\mathcal{D}, \mathcal{I} \cup \mathcal{T}, I)$, where the set of objects, $\mathcal{D}$, is the set of transactions (strictly speaking transaction-ID), and the set of attributes, $M = \mathcal{I} \cup \mathcal{T}$, contains all items $(\mathcal{I})$ and general items $(\mathcal{T})$. The incidence relation $I \subseteq \mathcal{D} \times M$ is defined by

$$tIm \iff \begin{cases} m \in \mathcal{I} \text{ and } m \in t \\ m \in \mathcal{T} \text{ and } \exists n \in \mathcal{I}, n \in t \text{ and } n \leq m. \end{cases}$$

Below is the context associated to the example on Fig. 7.

	Shirt	Jacket	Hiking Boots	Ski Pants	Shoes	Outerwear	Clothes	Footwear
100	×						×	
200		×	×			×	×	×
300			×	×		×	×	×
400					×			×
500					×			×
600		×				×	×	

The basic interestingness measures for a generalised rule $X \to Y$ are support and confidence (see association rules in Fig. 7(d)). Its *support* $supp(X \to Y)$ is defined as $\frac{|(X \cup Y)'|}{|\mathcal{D}|}$, while its *confidence* $conf(X \to Y)$ is $\frac{|(X \cup Y)'|}{|X'|}$.

For some applications, it would make sense to work only with the subcontext $(\mathcal{D}, \mathcal{T}, I \cap \mathcal{D} \times \mathcal{T})$ instead of $(\mathcal{D}, \mathcal{I} \cup \mathcal{T}, I)$, for example if the goal is to reduce the number of attributes, concepts or rules. Sometimes, there is no taxon available to suggest that considered attributes should be put together. However, we can extend the used taxonomy, i.e. put some attributes together in a proper taxon, and decide when an object satisfies the grouped attributes.

5.3   Generalising Scenarios

Let $\mathbb{K} := (G, M, I)$ be a context. The attributes of $\mathbb{K}$ can be grouped to form another set of attributes, namely S, whose elements are called **generalised** attributes. For example, in basket market analysis, items (products) can be generalised into product lines and then product categories, and even customers may be generalised to groups according to specific features (e.g., income, education). This replaces (G, M, I) with a context (G, S, J) where S can be seen as an index set such that $\{m_s \mid s \in S\}$ covers M. How to define the incidence relation J, is domain dependent. Let us consider several cases below [42–44]:

($\exists$) $gJs : \iff \exists m \in s, \, g\,I\,m$. When companies are described by the locations of their branches then cities can be grouped to regions or states. A company g operates in a state s if g has a branch in a city m which is in s.

($\forall$) $gJs : \iff \forall m \in s, \, g\,I\,m$. For exams with several components (e.g. written, oral, and thesis), we might require students to pass all components in order to succeed.

($\alpha\%$) $gJs : \iff \frac{|\{m \in s \mid g\,I\,m\}|}{|s|} \geq \alpha_s$ with α_s a threshold. In the case of exams discussed above, we could require students to pass just some parts, defined by a threshold.

Similarly, objects can also be put together to get "generalised objects". In [54] the author described *general on objects* as classes of individual objects that are considered to be extents of concepts of a formal context. In that paper, different contexts with general objects are defined and their conceptual structure and relation to other contexts is analysed with FCA methods. Generalisation on both objects and attributes can be carried out with the combinations below, with $A \subseteq G$ and $B \subseteq M$:

1. AJB iff $\exists a \in A, \, \exists b \in B$ such that $a\,I\,b$ (i.e. some objects from A are in relation with some attributes in B);
2. AJB iff $\forall a \in A, \, \forall b \in B \; a\,I\,b$ (i.e. each object in A has all attributes in B);
3. AJB iff $\forall a \in A, \, \exists b \in B$ such that $a\,I\,b$ (i.e. each object in A has at least one attribute in B);
4. AJB iff $\exists b \in B$ such that $\forall a \in A \; a\,I\,b$ (i.e. an attribute in B is satisfied by all objects of A);
5. AJB iff $\forall b \in B, \, \exists a \in A$ such that $a\,I\,b$ (i.e. each property in B is satisfied by an object of A);
6. AJB iff $\exists a \in A$ such that $\forall b \in B \; a\,I\,b$ (i.e. an object in A has all attributes in B);
7. AJB iff $\dfrac{\left|\{a \in A \mid \frac{|\{b \in B \mid a\,I\,b\}|}{|B|} \geq \beta_B\}\right|}{|A|} \geq \alpha_A$ (i.e. at least $\alpha_A\%$ of objects in A have each at least $\beta_B\%$ of the attributes in B);
8. AJB iff $\dfrac{\left|\left\{b \in B \mid \frac{|\{a \in A \mid a\,I\,b\}|}{|A|} \geq \alpha_A\right\}\right|}{|B|} \geq \beta_B$ (i.e. at least $\beta_B\%$ of attributes in B belong altogether to at least $\alpha_A\%$ of objects in the group A);
9. AJB iff $\frac{|A \times B \cap I|}{|A \times B|} \geq \alpha$ (i.e. the density of the rectangle $A \times B$ is at least α).

5.4   Generalisation and Extracted Patterns

After analysing several generalisation cases, including simultaneous generalisations on both objects and attributes as above, the next step is to look at the extracted patterns. From contexts, knowledge is usually extracted in terms of clusters and rules. When dealing with generalised attributes or objects, we coin the term "generalised" to all patterns extracted. An immediate task is to compare knowledge gained after generalising with those from the initial context.

New and interesting rules as seen in Fig. 7 can be discovered [61,62]. Experiments have shown that the number of extracted patterns quite often decreases. Formal investigations are been carried out to compare these numbers. For $\forall$-generalisations, the number of concepts does not increase [42]. But for $\exists$-generalisations, the size can actually increase [40–44].

In [3] the authors propose a method to control the structure of concept lattices derived from Boolean data by specifying granularity levels of attributes. Here a taxonomy is already available, given by the granularity of the attributes. They suggest that granularity levels should be chosen by a user based on his expertise and experimentation with the data. If the resulting formal concepts are too specific and there is a large number of them, the user can choose to use a coarser level of granularity. The resulting formal concepts are then less specific and can be seen as resulting from a zoom-out. Similarly, one may perform a zoom-in to obtain finer, more specific formal concepts. Through all these precautions, the number of concepts can still increase when attributes are coarser: "The issue of when attribute coarsening results in an increase in the number of formal concepts needs a further examination, as well as the possibility of informing automatically a user who is selecting a new level of granularity that the new level results in an increase in the number of concepts" [3].

In [41] a more succinct analysis of $\exists$-generalisations presents a family of contexts where generalising two attributes results in an exponential increase in the number of concepts. An example of such context is given in the Table 4 (left).

Table 4. A formal context (left) and its $\exists$-generalisation that puts m_1 and m_2 together. The number of concepts increases from 48 to 64, i.e. by 16.

	1	2	3	4	5	6	m_1	m_2
1		×	×	×	×	×	×	
2	×		×	×	×	×	×	×
3	×	×		×	×	×	×	×
4	×	×	×		×	×	×	×
5	×	×	×	×		×	×	×
6	×	×	×	×	×			×
g_1	×	×	×	×	×	×		

$\Longrightarrow$

	1	2	3	4	5	6	m_{12}
1		×	×	×	×	×	×
2	×		×	×	×	×	×
3	×	×		×	×	×	×
4	×	×	×		×	×	×
5	×	×	×	×		×	×
6	×	×	×	×	×		×
g_1	×	×	×	×	×	×	

Putting together some attributes does not always reduce the number of extracted patterns. It's therefore interesting to get measures that suggest which attributes can be put together, in the absence of a taxonomy. The goal would be to not increase the size of extracted patterns.

5.5   Similarity and Existential Generalisations

This section presents investigations on the use of certain similarity measures in generalising attributes. A *similarity measure* on a set M of attributes is a function $S : M \times M \rightarrow \mathbb{R}$ such that for all m_1, m_2 in M,

(i) $S(m_1, m_2) \geq 0$,   *positivity*
(ii) $S(m_1, m_2) = S(m_2, m_1)$ *symmetry*
(iii) $S(m_1, m_1) \geq S(m_1, m_2)$ *maximality*

We say that S is *compatible* with generalising attributes if whenever m_1, m_2 are more similar than m_3, m_4, then putting m_1, m_2 together should not lead to more concepts than putting m_3, m_4 together does. To give the formula for some known similarity measures that could be of interest in FCA setting (Table 5), we adopt the following notation for m_1, m_2 attributes in $\mathbb{K}$:

$$a = |m_1' \cap m_2'|, \quad d = |m_1' \Delta m_2'|, \quad b = |m_1' \setminus m_2'|, \quad c = |m_2' \setminus m_1'|.$$

Table 5. Some similarity measures relevant in FCA

Name	Formula	Name	Formula
Jaccard (Jc)	$\dfrac{a}{a+b+c}$	Sneath/Sokal (SS$_1$)	$\dfrac{2(a+d)}{2(a+d)+b+c}$
Dice (Di)	$\dfrac{2a}{2a+b+c}$	Sneath/Sokal (SS$_2$)	$\dfrac{0.5a}{0.5a+b+c}$
Sorensen (So)	$\dfrac{4a}{4a+b+c}$	Sokal/Michener (SM)	$\dfrac{a+d}{a+d+b+c}$
Anderberg (An)	$\dfrac{8a}{8a+b+c}$	Rogers/Tanimoto (RT)	$\dfrac{0.5(a+d)}{0.5(a+d)+b+c}$
Orchiai (Or)	$\dfrac{a}{\sqrt{(a+b)(a+c)}}$	Russels/Rao (RR)	$\dfrac{a}{a+d+b+c}$
Kulczynski (Ku)	$\dfrac{0.5a}{a+b} + \dfrac{0.5a}{a+c}$	Yule/Kendall (YK)	$\dfrac{ad}{ad+bc}$

For the context left in Table 4, we have computed $S(m_1, x)$, $x = 1, \ldots, 6, m_2$ (Table 6). Although m_1 is more similar to m_2 than any attribute $i < 6$, putting m_1 and m_2 together increases the number of concepts. Note that putting m_1 and 6 together is equivalent to removing m_1 from the context, and thus, reduces the number of concepts.

Let $\mathbb{K}$ be a context (G, M, I) with $a, b \in M$ and $\mathbb{K}_{00}$ be its subcontext without a, b. Below, $\mathrm{Ext}(\mathbb{K}_{00})$ means all the extents of concepts of the context $\mathbb{K}_{00}$. In order to describe the increase in the number of concepts after putting a, b together, we set

$$\mathcal{H}(a) := \{A \cap a' \mid A \in \mathrm{Ext}(\mathbb{K}_{00}) \text{ and } A \cap a' \notin \mathrm{Ext}(\mathbb{K}_{00})\}$$
$$\mathcal{H}(b) := \{A \cap b' \mid A \in \mathrm{Ext}(\mathbb{K}_{00}) \text{ and } A \cap b' \notin \mathrm{Ext}(\mathbb{K}_{00})\}$$
$$\mathcal{H}(a \cup b) := \{A \cap (a' \cup b') \mid A \in \mathrm{Ext}(\mathbb{K}_{00}), \ A \cap (a' \cup b') \notin \mathrm{Ext}(\mathbb{K}_{00})\}$$
$$\mathcal{H}(a \cap b) := \{A \cap (a' \cap b') \mid A \in \mathrm{Ext}(\mathbb{K}_{00}), \ A \cap (a' \cap b') \notin \mathrm{Ext}(\mathbb{K}_{00})\}.$$

Table 6. The values of considered similarity measures $S(m, i)$

	Jc	Di	So	An	SS_2	Ku	Or	SM	RT	SS_1	RR
$i \in S_5$	0.57	0.80	0.89	0.94	0.50	0.80	0.80	0.71	0.56	0.83	0.57
$i = 6$	0.83	0.91	0.95	0.97	0.71	0.92	0.91	0.75	0.75	0.92	0.71
$i = m_2$	0.67	0.80	0.89	0.94	0.50	0.80	0.80	0.71	0.56	0.83	0.57

The following proposition shows that the increase can be exponential.

Theorem 3 ([41]). *Let (G, M, I) be an attribute reduced context and a, b be two attributes such that their generalisation $s = a \cup b$ increases the size of the concept lattice. Then $|\mathfrak{B}(G, M, I)| = |\mathfrak{B}(G, M \setminus \{a, b\}, I \subseteq G \times M \setminus \{a, b\})| + |\mathcal{H}(a, b)|$, with*

$$|\mathcal{H}(a, b)| = |\mathcal{H}(a) \cup \mathcal{H}(b) \cup \mathcal{H}(a \cap b)| \leq 2^{|a'| + |b'|} - 2^{|a'|} - 2^{|b'|} + 1.$$

This upper bound can be reached.

The difference $|\mathcal{H}(a, b)|$ is then used to define a compatible similarity measure. We set

$$\psi(a, b) := |\mathcal{H}(a \cup b)| - |\mathcal{H}(a, b)|, \quad \delta(a, b) := \begin{cases} 1 & \text{if } \psi(a, b) \leq 0 \\ 0 & \text{else} \end{cases}, \text{ and define}$$

$$S(a, b) := \frac{1 + \delta(a, b)}{2} - \frac{|\psi(a, b)|}{2n_0} \text{ with } n_0 = \max\{\psi(x, y) \mid x, y \in M\}. \text{ Then}$$

Theorem 4 ([30]). *S is a similarity measure compatible with the generalisation.*

$$S(a, b) \geq \frac{1}{2} \iff \psi(a, b) \leq 0$$

6 Conclusion

The first two parts contain a concise summary of the usage of Shapley values from Cooperative Game Theory for interpretable concept-based learning in the FCA playground with its connection to Data Mining formulations. We omitted results related to algorithms and their computational complexity since they deserve a separate detailed treatment.

The lessons drawn from the ranking attributes in JSM classification hypotheses and those in the intents of stable concepts show that valuation functions should be customised and are not necessarily zero-one-valued. This is an argument towards that of Shapley values approach requires specification depending on the model (or type of patterns) and thus only conditionally is model-agnostic. The other lesson is about the usage of Shapley values for pattern attribution concerning their contribution interestingness measures like stability or robustness.

The third part is devoted to attribute aggregation by similarity, which may help to apply interpretable techniques to larger sets of attributes or bring additional aspects to interpretability with the help of domain taxonomies. The desirable property of similarity measures to provide compatible generalisation helps to reduce the number of output concepts or JSM-hypotheses as well. The connection between attribute similarity measures and Shapley interaction values [46], when the interaction of two or more attributes on the model prediction is studied, is also of interest.

In addition to algorithmic issues, we would like to mention two more directions of future studies. The first one lies in the interpretability by means of Boolean matrix factorisation (decomposition), which was used for dimensionality reduction with explainable Boolean factors (formal concepts) [5] or interpretable "taste communities" identification in collaborative filtering [22]. In this case, we are transitioned from the importance of attributes to attribution of factors. The second one is a closely related aspect to interpretability called fairness [2], where, for example, certain attributes of individuals should not influence much to the model prediction (disability, ethnicity, gender, etc.).

Acknowledgements. The study was implemented in the framework of the Basic Research Program at the National Research University Higher School of Economics and funded by the Russian Academic Excellence Project '5–100'. The second author was also supported by Russian Science Foundation under grant 17-11-01276 at St. Petersburg Department of Steklov Mathematical Institute of Russian Academy of Sciences, Russia. The second author would like to thank Fuad Aleskerov, Alexei Zakharov, and Shlomo Weber for the inspirational lectures on Collective Choice and Voting Theory.

References

1. Agrawal, R., Imielinski, T., Swami, A.N.: Mining association rules between sets of items in large databases. In: Buneman, P., Jajodia, S. (eds.) Proceedings of the 1993 ACM SIGMOD International Conference on Management of Data, Washington, DC, USA, 26–28 May 1993, pp. 207–216. ACM Press (1993)
2. Alves, G., Bhargava, V., Couceiro, M., Napoli, A.: Making ML models fairer through explanations: the case of limeout. CoRR abs/2011.00603 (2020)
3. Belohlávek, R., Baets, B.D., Konecny, J.: Granularity of attributes in formal concept analysis. Inf. Sci. **260**, 149–170 (2014)
4. Belohlávek, R., Baets, B.D., Outrata, J., Vychodil, V.: Inducing decision trees via concept lattices. Int. J. Gen. Syst. **38**(4), 455–467 (2009)
5. Belohlávek, R., Vychodil, V.: Discovery of optimal factors in binary data via a novel method of matrix decomposition. J. Comput. Syst. Sci. **76**(1), 3–20 (2010)
6. Bocharov, A., Gnatyshak, D., Ignatov, D.I., Mirkin, B.G., Shestakov, A.: A lattice-based consensus clustering algorithm. In: Huchard, M., Kuznetsov, S.O. (eds.) Proceedings of the Thirteenth International Conference on Concept Lattices and Their Applications, Moscow, Russia, CEUR Workshop Proceedings, 18–22 July 2016, vol. 1624, pp. 45–56. CEUR-WS.org (2016)
7. Carpineto, C., Romano, G.: A lattice conceptual clustering system and its application to browsing retrieval. Mach. Learn. **24**(2), 95–122 (1996)

8. Caruana, R., Lundberg, S., Ribeiro, M.T., Nori, H., Jenkins, S.: Intelligible and explainable machine learning: best practices and practical challenges. In: Gupta, R., Liu, Y., Tang, J., Prakash, B.A. (eds.) KDD 2020: The 26th ACM SIGKDD Conference on Knowledge Discovery and Data Mining, Virtual Event, CA, USA, 23–27 August 2020, pp. 3511–3512. ACM (2020)
9. Eklund, P.W., Ducrou, J., Dau, F.: Concept similarity and related categories in information retrieval using Formal Concept Analysis. Int. J. Gen. Syst. **41**(8), 826–846 (2012)
10. Fayyad, U.M., Piatetsky-Shapiro, G., Smyth, P.: From data mining to knowledge discovery in databases. AI Mag. **17**(3), 37–54 (1996)
11. Finn, V.: On machine-oriented formalization of plausible reasoning in F. Bacon-J.S. Mill Style. Semiotika i Informatika **20**, 35–101 (1983). (in Russian)
12. Ganter, B., Kuznetsov, S.O.: Hypotheses and version spaces. In: Ganter, B., de Moor, A., Lex, W. (eds.) ICCS-ConceptStruct 2003. LNCS (LNAI), vol. 2746, pp. 83–95. Springer, Heidelberg (2003). https://doi.org/10.1007/978-3-540-45091-7_6
13. Ganter, B., Kuznetsov, S.O.: Scale coarsening as feature selection. In: Medina, R., Obiedkov, S. (eds.) ICFCA 2008. LNCS (LNAI), vol. 4933, pp. 217–228. Springer, Heidelberg (2008). https://doi.org/10.1007/978-3-540-78137-0_16
14. Ganter, B., Obiedkov, S.A.: Conceptual Exploration. Springer, Heidelberg (2016). https://doi.org/10.1007/978-3-662-49291-8
15. Ganter, B., Wille, R.: Formal Concept Analysis - Mathematical Foundations. Springer, Heidelberg (1999). https://doi.org/10.1007/978-3-642-59830-2
16. Goodfellow, I.J., Bengio, Y., Courville, A.C.: Deep Learning. Adaptive Computation and Machine Learning. MIT Press, Cambridge (2016)
17. Hájek, P., Havel, I., Chytil, M.: The GUHA method of automatic hypotheses determination. Computing **1**(4), 293–308 (1966)
18. Ignatov, D.I.: Introduction to formal concept analysis and its applications in information retrieval and related fields. In: Braslavski, P., Karpov, N., Worring, M., Volkovich, Y., Ignatov, D.I. (eds.) RuSSIR 2014. CCIS, vol. 505, pp. 42–141. Springer, Cham (2015). https://doi.org/10.1007/978-3-319-25485-2_3
19. Ignatov, D.I., Kuznetsov, S.O., Poelmans, J.: Concept-based biclustering for internet advertisement. In: 12th IEEE International Conference on Data Mining Workshops, ICDM Workshops, Brussels, Belgium, 10 December 2012, pp. 123–130 (2012)
20. Ignatov, D.I., Kwuida, L.: Interpretable concept-based classification with shapley values. In: Alam, M., Braun, T., Yun, B. (eds.) ICCS 2020. LNCS (LNAI), vol. 12277, pp. 90–102. Springer, Cham (2020). https://doi.org/10.1007/978-3-030-57855-8_7
21. Ignatov, D.I., Kwuida, L.: Shapley and banzhaf vectors of a formal concept. In: Valverde-Albacete, F.J., Trnecka, M. (eds.) Proceedings of the Fifthteenth International Conference on Concept Lattices and Their Applications, Tallinn, Estonia, CEUR Workshop Proceedings, June 29–July 1, 2020, vol. 2668, pp. 259–271. CEUR-WS.org (2020)
22. Ignatov, D.I., Nenova, E., Konstantinova, N., Konstantinov, A.V.: Boolean matrix factorisation for collaborative filtering: An FCA-based approach. In: Agre, G., Hitzler, P., Krisnadhi, A.A., Kuznetsov, S.O. (eds.) AIMSA 2014. LNCS (LNAI), vol. 8722, pp. 47–58. Springer, Cham (2014). https://doi.org/10.1007/978-3-319-10554-3_5
23. John, S.: Mill, A System of Logic, Ratiocinative and Inductive, Being a Connected View of the Principles of Evidence and the Methods of Scientific Investigation. Green, and Co., Longmans, London (1843)

24. Kadyrov, T., Ignatov, D.I.: Attribution of customers' actions based on machine learning approach. In: Proceedings of the Fifth Workshop on Experimental Economics and Machine Learning co-located with the Seventh International Conference on Applied Research in Economics (iCare7), Perm, Russia, 26 September 2019, vol-2479, pp. 77–88. CEUR-ws (2019)
25. Kashnitsky, Y., Kuznetsov, S.O.: Global optimization in learning with important data: an FCA-based approach. In: Huchard, M., Kuznetsov, S.O. (eds.) Proceedings of the Thirteenth International Conference on Concept Lattices and Their Applications, Moscow, Russia, CEUR Workshop Proceedings, 18–22 July 2016, vol. 1624, pp. 189–201. CEUR-WS.org (2016)
26. Kaur, H., Nori, H., Jenkins, S., Caruana, R., Wallach, H.M., Vaughan, J.W.: Interpreting interpretability: understanding data scientists' use of interpretability tools for machine learning. In: Bernhaupt, R., et al. (eds.) CHI 2020: CHI Conference on Human Factors in Computing Systems, Honolulu, HI, USA, 25–30 April, 2020, pp. 1–14. ACM (2020)
27. Kaytoue, M., Kuznetsov, S.O., Macko, J., Napoli, A.: Biclustering meets triadic concept analysis. Ann. Math. Artif. Intell. **70**(1–2), 55–79 (2014)
28. Konecny, J.: On attribute reduction in concept lattices: methods based on discernibility matrix are outperformed by basic clarification and reduction. Inf. Sci. **415**, 199–212 (2017)
29. Konecny, J., Krajca, P.: On attribute reduction in concept lattices: experimental evaluation shows discernibility matrix based methods inefficient. Inf. Sci. **467**, 431–445 (2018)
30. Kuitché, R.S., Temgoua, R.E.A., Kwuida, L.: A similarity measure to generalize attributes. In: Ignatov, D.I., Nourine, L. (eds.) Proceedings of the Fourteenth International Conference on Concept Lattices and Their Applications, CLA 2018, Olomouc, Czech Republic, CEUR Workshop Proceedings, 12–14 June 2018, vol. 2123, pp. 141–152. CEUR-WS.org (2018)
31. Kuznetsov, S.O.: Machine learning and formal concept analysis. ICFCA **2004**, 287–312 (2004)
32. Kuznetsov, S.O.: Galois connections in data analysis: contributions from the soviet era and modern Russian research. In: Ganter, B., Stumme, G., Wille, R. (eds.) Formal Concept Analysis. LNCS (LNAI), vol. 3626, pp. 196–225. Springer, Heidelberg (2005). https://doi.org/10.1007/11528784_11
33. Kuznetsov, S.O.: On stability of a formal concept. Ann. Math. Artif. Intell. **49**(1–4), 101–115 (2007)
34. Kuznetsov, S.O., Makhalova, T.P.: On interestingness measures of formal concepts. Inf. Sci. **442–443**, 202–219 (2018)
35. Kuznetsov, S.O., Makhazhanov, N., Ushakov, M.: On neural network architecture based on concept lattices. In: Kryszkiewicz, M., Appice, A., Slezak, D., Rybinski, H., Skowron, A., Ras, Z.W. (eds.) ISMIS 2017. LNCS (LNAI), vol. 10352, pp. 653–663. Springer, Cham (2017). https://doi.org/10.1007/978-3-319-60438-1_64
36. Kuznetsov, S.O., Poelmans, J.: Knowledge representation and processing with formal concept analysis. Wiley Interdiscip. Rev. Data Min. Knowl. Disc. **3**(3), 200–215 (2013)
37. Kuznetsov, S.: Jsm-method as a machine learning method. Method. Itogi Nauki i Tekhniki ser. Informatika **15**, 17–53 (1991). (in Russian)
38. Kuznetsov, S.: Stability as an estimate of the degree of substantiation of hypotheses derived on the basis of operational similarity. Nauchn. Tekh. Inf. Ser. **2**(12), 217–29 (1991). (in Russian)

39. Kuznetsov, S.: Mathematical aspects of concept analysis. J. Math. Sci. **80**(2), 1654–1698 (1996)
40. Kwuida, L., Kuitché, R., Temgoua, R.: On the size of ∃-generalized concepts. ArXiv:1709.08060 (2017)
41. Kwuida, L., Kuitché, R.S., Temgoua, R.E.A.: On the size of ∃-generalized concept lattices. Discret. Appl. Math. **273**, 205–216 (2020)
42. Kwuida, L., Missaoui, R., Balamane, A., Vaillancourt, J.: Generalized pattern extraction from concept lattices. Ann. Math. Artif. Intell. **72**(1–2), 151–168 (2014)
43. Kwuida, L., Missaoui, R., Boumedjout, L., Vaillancourt, J.: Mining generalized patterns from large databases using ontologies (2009). ArXiv:0905.4713
44. Kwuida, L., Missaoui, R., Vaillancourt, J.: Using taxonomies on objects and attributes to discover generalized patterns. In: Szathmary, L., Priss, U. (eds.) Proceedings of The Ninth International Conference on Concept Lattices and Their Applications, Fuengirola (Málaga), CEUR Workshop Proceedings, Spain, 11–14 October 2012, vol. 972, pp. 327–338. CEUR-WS.org (2012)
45. Lakhal, L., Stumme, G.: Efficient mining of association rules based on formal concept analysis. In: Ganter, B., Stumme, G., Wille, R. (eds.) Formal Concept Analysis. LNCS (LNAI), vol. 3626, pp. 180–195. Springer, Heidelberg (2005). https://doi.org/10.1007/11528784_10
46. Lundberg, S.M., Lee, S.I.: A unified approach to interpreting model predictions. In: Guyon, I. et al. (eds.) Advances in Neural Information Processing Systems, vol. 30, pp. 4765–4774. Curran Associates, Inc. (2017)
47. Luxenburger, M.: Implications partielles dans un contexte. Mathématiques et Sci. Humaines. **113**, 35–55 (1991)
48. Mirkin, B.: Mathematical Classification and Clustering. Kluwer Academic Publishers, Amsterdam (1996)
49. Mitchell, T.M.: Version spaces: a candidate elimination approach to rule learning. In: Reddy, R. (ed.) Proceedings of the 5th International Joint Conference on Artificial Intelligence 1977, pp. 305–310. William Kaufmann (1977)
50. Molnar, C.: Interpretable Machine Learning (2019). https://christophm.github.io/interpretable-ml-book/
51. Pasquier, N., Bastide, Y., Taouil, R., Lakhal, L.: Efficient mining of association rules using closed itemset lattices. Inf. Syst. **24**(1), 25–46 (1999)
52. Poelmans, J., Ignatov, D.I., Kuznetsov, S.O., Dedene, G.: Formal concept analysis in knowledge processing: a survey on applications. Expert Syst. Appl. **40**(16), 6538–6560 (2013)
53. Poelmans, J., Kuznetsov, S.O., Ignatov, D.I., Dedene, G.: Formal concept analysis in knowledge processing: a survey on models and techniques. Expert Syst. Appl. **40**(16), 6601–6623 (2013)
54. Prediger, S.: Formal concept analysis for general objects. Discret. Appl. Math. **127**(2), 337–355 (2003)
55. Priss, U., Old, L.J.: Data weeding techniques applied to Roget's thesaurus. In: Wolff, K.E., Palchunov, D.E., Zagoruiko, N.G., Andelfinger, U. (eds.) KONT/KPP -2007. LNCS (LNAI), vol. 6581, pp. 150–163. Springer, Heidelberg (2011). https://doi.org/10.1007/978-3-642-22140-8_10
56. Roth, C., Obiedkov, S., Kourie, D.: Towards concise representation for taxonomies of epistemic communities. In: Yahia, S.B., Nguifo, E.M., Belohlavek, R. (eds.) CLA 2006. LNCS (LNAI), vol. 4923, pp. 240–255. Springer, Heidelberg (2008). https://doi.org/10.1007/978-3-540-78921-5_17

57. Rudin, C.: Stop explaining black box machine learning models for high stakes decisions and use interpretable models instead. Nat. Mach. Intell **1**(5), 206–215 (2019)
58. Rudolph, S.: Using FCA for encoding closure operators into neural networks. In: Priss, U., Polovina, S., Hill, R. (eds.) ICCS-ConceptStruct 2007. LNCS (LNAI), vol. 4604, pp. 321–332. Springer, Heidelberg (2007). https://doi.org/10.1007/978-3-540-73681-3_24
59. Shapley, L.S.: A value for n-person games. Contrib. Theory Games **2**(28), 307–317 (1953)
60. Shrikumar, A., Greenside, P., Kundaje, A.: Learning important features through propagating activation differences. In: Precup, D., Teh, Y.W. (eds.) Proceedings of the 34th International Conference on Machine Learning. Proceedings of Machine Learning Research, vol. 70, pp. 3145–3153. PMLR, International Convention Centre, Sydney (2017)
61. Srikant, R., Agrawal, R.: Mining generalized association rules. In: Dayal, U., Gray, P.M.D., Nishio, S. (eds.) VLDB 95, Proceedings of 21th International Conference on Very Large Data Bases, Zurich, Switzerland, 11–15 September 1995, pp. 407–419. Morgan Kaufmann (1995)
62. Srikant, R., Agrawal, R.: Mining generalized association rules. Future Gener. Comput. Syst. **13**(2–3), 161–180 (1997)
63. Strumbelj, E., Kononenko, I.: Explaining prediction models and individual predictions with feature contributions. Knowl. Inf. Syst. **41**(3), 647–665 (2014)
64. Stumme, G., Taouil, R., Bastide, Y., Lakhal, L.: Conceptual clustering with iceberg concept lattices. In: Proceedings of GI-Fachgruppentreffen Maschinelles Lernen, vol. 1 (2001)
65. Tatti, N., Moerchen, F.: Finding robust itemsets under subsampling. ICDM **2011**, 705–714 (2011)
66. Valtchev, P., Missaoui, R.: Similarity-based clustering versus galois lattice building: strengths and weaknesses. In: Huchard, M., Godin, R., Napoli, A. (eds.) Contributions of the ECOOP 2000 Workshop, "Objects and Classification: a Natural Convergence", European Conference on Object-Oriented Programming (2000), vol. Research Report LIRMM, no. 00095, p. w13 (2000)

Natural Language Processing

DaNetQA: A Yes/No Question Answering Dataset for the Russian Language

Taisia Glushkova[1], Alexey Machnev[1], Alena Fenogenova[2],
Tatiana Shavrina[2], Ekaterina Artemova[1](✉), and Dmitry I. Ignatov[1]

[1] National Research University Higher School of Economics, Moscow, Russia
toglushkova@edu.hse.ru, {amachnev,elartemova,dignatov}@hse.ru
[2] Sberbank, Moscow, Russia
{Fenogenova.A.S,Shavrina.T.O}@sberbank.ru
https://cs.hse.ru/en/ai/computational-pragmatics/

Abstract. DaNetQA, a new question-answering corpus, follows
BoolQ [2] design: it comprises natural yes/no questions. Each question
is paired with a paragraph from Wikipedia and an answer, derived from
the paragraph. The task is to take both the question and a paragraph as
input and come up with a yes/no answer, i.e. to produce a binary output.
In this paper, we present a reproducible approach to DaNetQA creation
and investigate transfer learning methods for task and language transferring. For task transferring we leverage three similar sentence modelling
tasks: 1) a corpus of paraphrases, Paraphraser, 2) an NLI task, for which
we use the Russian part of XNLI, 3) another question answering task,
SberQUAD. For language transferring we use English to Russian translation together with multilingual language fine-tuning.

Keywords: Question answering · Transfer learning · Transformers

1 Introduction

The creation of new datasets, aimed at new, challenging tasks, describing complex phenomena, related to various aspects of language understanding and usage,
is core to the current view of modern language technologies. However, the majority of the datasets, created and published at the best venues, target the English-
language tasks and cultural aspects, related to English-speaking society.

In response to the bias towards English, new datasets and benchmarks are
developed, that comprise multiple languages. One of the well-known examples
of such multilingual datasets is XNLI [4], which is a natural language inference
dataset. Although this dataset is developed for 15 languages, including low-
resource ones, the approach to its creation still excessively utilizes English data:
the dataset entries are manually translated from English to other languages,

---

T. Glushkova and A. Machnev—The first two authors have equal contribution.

© Springer Nature Switzerland AG 2021
W. M. P. van der Aalst et al. (Eds.): AIST 2020, LNCS 12602, pp. 57–68, 2021.
https://doi.org/10.1007/978-3-030-72610-2_4

without any specific adjustments. Although the translation-based approach is quick and dirty and allows us to overcome the lack of dataset for any language, other concerns arise. From a general point of view, we understand, that translating is different from natural everyday language usage [8]. Thus translated datasets may have different statistics and word usage in comparison to text, composed from scratch. This may affect the quality of the models, trained further on the translated datasets, when applied to real-life data.

Another approach to dataset creation involves collecting datasets following the guidelines and annotation schemes, designed for original datasets in English. RuRED [7] is a recent example of such a dataset: it is created following TACRED [25] annotation scheme and collection pipeline. This approach however is criticized for the obvious lack of novelty.

Nevertheless, in this paper we stick to the second approach to the dataset creation and more or less follow the pipeline, developed for BoolQ [2], to create a new dataset for binary questions in Russian, which we refer to as DaNetQA. We only deviate from the BoolQ pipeline, if we do not have access to proprietary data sources and instead use crowdsourcing. The motivation to re-create BoolQ lies, first, in the lack of question-answering dataset for Russian, and second, in the fact, that binary question answering appears to be a more challenging task, when compared to SQuAD-like and natural language inference tasks [23]. Thus we hope that DaNetQA may become of great use both for chat-bot technologies, which massively use question answering data, and for a thorough evaluation of deep contextual encoders, such as BERT [5] or XLM-R [3].

As the annotations for DaNetQA are crowd-sourced and require payment, we explore different strategies that can help to increase the quality without the need to annotate larger amounts of data. This leads us to two strategies of transfer learning: transferring from tasks, which have a similar setting, and transferring from English, keeping in mind, that DaNetQA recreates BoolQ, but in a different language. Our main contributions are the following:

1. We create and intent to publish in open access a new middle-scale dataset for the Russian language, DaNetQA, which comprises yes/no questions, paired with paragraphs, providing enough information to answer the questions (Subsect. 2.1) and report its statistics (Subsect. 2.2);
2. We establish a simple baseline and a more challenging deep baseline for DaNetQA (Subsect. 2.2);
3. We explore the applicability of transfer learning techniques, of which some overcome the established baseline (Sect. 3).

The paper is organised as follows. Section 2 describes the data collection process and provides the reader with their basic statistics. In Sect. 3, all the conducted experiments are described. Section 4 discusses the obtained results. In Sect. 5, related work is summarised. Sect. 6 concludes the paper.

2   Dataset

2.1   Collection

Our approach to DaNetQA dataset creation is inspired by the work of [2], where the pipeline from NQ [16] is used as a base.

Questions are generated on the crowdsourcing platform Yandex.Toloka[1], which is a good source for Slavic languages data generation compared to other language groups due to the origin of the platform. Questions are created by crowd workers following an instruction, that suggests phrases that can be used to start a phrase – a broad list of templates (more than 50 examples).

Generated yes/no questions are then treated as queries in order to retrieve relevant Wikipedia pages with the use of Google API. Questions are only kept if 3 Wikipedia pages could be returned, in which case the question and the text of an article are passed forward to querying a pre-trained BERT-based model for SQuAD to extract relevant paragraphs.

Finally, crowd workers label question and paragraph pairs with "yes" or "no" answers, and the questions that could not be answered based on the information form the paragraph are marked as "not answerable". Annotating data in such a manner is quite expensive since not only crowd workers read through and label thousands of pairs manually, but there is also used a high overlap of votes to ensure the high quality of the dataset. In addition, Google API also requires some payments to process high numbers of queries.

The final labels for each pair are picked based on the majority of votes. In case of uncertainty when there is no label picked by the majority, the pair is being checked and labeled manually by the authors.

Aside from the overlap used on the final step, we use a number of gold-standard control questions that are randomly mixed into the tasks to make sure that crowd workers do the annotation responsibly. Also, each annotator goes through a small set of learning tasks before starting labelling the actual pairs.

2.2   Statistics

In the following section, we analyze our corpus to better understand the nature of the collected questions and paragraphs and compute all kinds of descriptive statistics. Statistics presented below, include the minimum, average, and maximum length of questions and paragraphs in tokens (Table 1) and the distribution of text lengths in the entire dataset as a whole. The total number of yes/no questions in Dev/Test/Train sets can be found in Table 2.

We also provide a brief overview of categories covered by our question and paragraph pairs (Table 3) and the t-SNE visualization of an LDA model for 15 topics, trained on concatenated questions and paragraphs (Fig. 1). Each topic is labelled with top-3 tokens that represent a topic summary.

Since the crowd workers were provided (but not restricted) with question starter templates, in order to generate yes/no questions, we decided to check

---

[1] https://toloka.yandex.ru.

Table 1. Descriptive statistics of collected question/paragraph pairs (in tokens).

	Count	Mean	Min	25%	50%	75%	Max
Questions	2691	5	2	5	5	6	14
Paragraphs	2691	90	37	72	88	106	206

Table 2. The number of yes/no questions in Dev, Test and Train sets. Symbol # stands for the total number of yes (or no) questions (columns 3 and 5) and % stands for the ratio of yes (or no) questions in the dataset (columns 4 and 6).

	Size	#Yes	Yes%	#No	No%
Dev set	821	672	81.8	149	18.1
Test set	805	620	77	185	22.9
Train set	1065	805	75.5	260	24.4

Table 3. Question categorization of DaNetQA. Top-10 question topics.

Category	Example	Percent	Yes%
Война (War/Military)	Был ли взят калинин немцами?	10.51	54.7
История (History)	Были ли фамилии у крепостных крестьян?	9.55	87.1
Космос (Space/Galaxy)	Есть ли жизнь на других планетах солнечной системы?	9.10	79.5
СССР (USSR)	Была ли конституция в СССР?	7.61	83.9
Здоровье (Health)	Передаётся ли чумка от кошки к человеку?	7.17	84.9
Семья (Family)	Был ли у гагарина брат?	6.76	81.3
Искусство (Art/Literature)	Существует ли жанр эклектика в живописи?	6.54	75.5
Евросоюз (EU)	Входит ли чехия в евросоюз?	6.28	86.9
География (Geography)	Была ли албания в составе югославии?	6.05	76.6
Связи (Communications)	Был ли лермонтов знаком с пушкиным?	5.42	82.8

the frequency of the used bigrams that the generated questions begin with (see Table 4).

Note that this list of starter phrases is consistent with the one in the [2] pipeline. Indeed, in the paper, authors select questions that begin with a certain set of words ("did", "do", "does", "is", "are", "was", "were", "have" , "has", "can", "could", "will", "would"). Our list and the list used in the creation of BoolQ are somewhat similar, if we omit the particle "ли", which is added in Russian when creating interrogative sentences. The majority of the questions in the corpus are starting with these words.

3   Experiments

The DaNetQA task is formulated as a binary classification task. The input to the model consists of the question and paragraph pair. The task is to return either 0 or 1, such that the positive answer indicates that the answer to the question, based on the information from the paragraph is "yes". Otherwise, 0 stands for

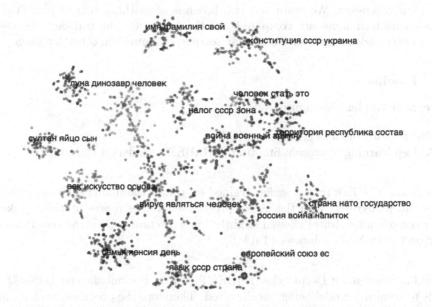

Fig. 1. The t-SNE visualization of an LDA model for 15 topics, trained on concatenated questions and paragraphs.

Table 4. Top-15 frequent bigrams, that start the questions.

Indicator words	Frequency
был ли (was (*3rd pers., sing., masc.*) there)	575
есть ли (is (*3rd pers., sing., masc.*) there)	362
была ли (was (*3rd pers., sing., fem.*) there)	317
были ли (was (*3rd pers., plur*) there)	302
входит ли (is included (*3rd pers., sing.*))	201
едят ли (do [they] eat (*3rd pers., plur*))	130
правда ли (is [it] true)	130
разрешено ли (is [it] allowed (*3rd pers., sing., neutr.*))	74
передаётся ли (is [it] transmitted (*3rd pers., sing.*))	70
состоит ли (does [it] consist (*3rd pers., sing.*) in)	68
бывает ли (is (*3rd pers., sing., fem.*) there)	55
вреден ли (is [it] harmful (*3rd pers., sing., masc.*))	55
вредна ли (is [it] harmful (*3rd pers., sing., fem.*))	50
существует ли (does [it] exist (*3rd pers., sing.*))	41
а была (was (*3rd pers., sing., fem.*) there)	30

the negative answer. We evaluated two baseline algorithms, to which multiple transferring techniques are compared. As the source for the transfer, we used either other tasks, which have a similar setting, or sources in other languages.

3.1   Baseline

We exploit two baseline approaches:

1. FastText for binary classification [11];
2. A deep learning approach: fine-tuning RuBERT model on DaNetQA.

FastText. The FastText classifier trained on the concatenated vectors of questions and paragraphs with the use of pre-trained vector representations (taken from the official website) showed slightly lower accuracy, than the one trained only on the DaNetQA dataset (Table 5).

BERT. Fine-tuning DeepPavlov RuBERT [15] on the imbalanced DaNetQA, leads to resulting model being uncalibrated. Therefore the predicted probability for the positive class is higher in most cases. To tackle this problem, we aim to find an optimal predicted probability threshold for validation dataset. This is done in two steps:

1. For each threshold t in some discrete subset of range $[0, 1]$, "yes" answer $precision(t)$ and $recall(t)$ calculated;
2. $F_1(t) = \dfrac{2 \cdot precision(t) \cdot recall(t)}{precision(t) + recall(t)}$ is calculated;
3. Threshold t for the largest $F_1(t)$ is selected.

We train a model for 5 epochs, saving checkpoint each 40 training steps, and then select the best checkpoint according to the accuracy value. For this task, we set the learning rate to 3e–5, and linearly decrease it up to zero to the end of the training.

Table 5. Accuracy for binary classification with FastText.

	FastText	Pre-trained FastText	RuBERT
ACC	0.81366	0.80745	0.7975

FastText can be seen as a strong baseline, as it manages to achieve results comparable to the larger RuBERT model. Further, we will refer to the RuBERT baseline to compare with transfer learning techniques.

3.2 Task Transferring

To transfer learning from other tasks, we make two steps:

1. Fine-tuning the pre-trained transformer model on a similar task;
2. Fine-tuning the model on DaNetQA.

The tasks, used for the first step, are Paraphraser, the Russian part of XNLI, and Task A of the SberQUAD. For each task, we initialize model weights with DeepPavlov RuBERT. All three datasets are used independently.

Paraphraser [17] is a dataset for the paraphrasing task: it consists of sentence pairs, each of which is labeled as paraphrase, not paraphrase or maybe paraphrase. This task is close to DaNetQA as the model is required to detect linkage between sentences. We transform this task into a binary classification problem, where the label is true only for definitely paraphrase.

XNLI [4] is the dataset for the language inference task, translated to 15 languages, including Russian. The task is to predict if one sentence is entailment or contradiction of another, or two sentences are neutral. Each pair has initial annotation and five other crowd workers annotations. For our task, we kept only the Russian part of the dataset and only sentences where the majority label is the same as initial. As for Paraphraser, we transformed the task into a binary classification problem, and the only entailment is treated as the positive class.

SberQUAD task A is the part of the Sberbank question-answer challenge. In task A, a model needs to predict if a given text contains an answer to a given question.

Since each task in the first step is, as DaNetQA, a binary classification problem, we use the same approach with training and selecting the best model, as for fine-tuning BERT on DaNetQA in the baseline approach. The best model selection and evaluation of further DaNetQA fine-tuning is also the same as in the baseline solution.

The model is fine-tuned on the first-step task with initial learning rate 3e–5 and then fine-tuned on DaNetQA with the same initial learning rate, linearly decreasing to zero in each case. During training, a checkpoint is saved each 40 steps for Paraphraser, 75 steps for XNLI, and 3000 steps for SberQUAD, then the best one is selected by accuracy value on the pre-training dataset. We perform 5 different pre-training runs on each dataset, then fine-tune each pre-trained model 5 different times on DaNetQA. We show mean and standard deviation values over different fine-tuning runs after the best pre-training run, selected by mean accuracy value.

3.3 Language Transferring

Similarly to task transferring, for language transferring we fine-tune the transformer on a similar dataset and then further fine-tune it on DaNetQA. The difference is that the dataset in the first-step task is not in Russian, so we should train the model on its machine translation or use a multilingual pre-trained model. We tested the following configurations:

– Fine-tune RuBERT on translated BoolQ and test on DaNetQA;
– Fine-tune RuBERT on translated BoolQ and then fine-tune and test on DaNetQA;
– Fine-tune XLM-R on BoolQ and then fine-tune on DaNetAQ.

The hyperparameters and algorithm of the best model selection and evaluation are the same as in the baseline approach for fine-tuning RuBERT on DaNetQA, apart from the initial learning rate for pre-training and fine-tuning XLM-R model is 1e–5.

Intuitively, when pre-trained on BoolQ, which inspired DaNetQA, the model should have the relevant knowledge to solve our task. As the BoolQ dataset is in English, we should use some kind of multilingual technique to enable transfer learning. The core methods in this case are:

1. To use machine translation models to translate either BoolQ to Russian, or DaNetQA to English, and fine-tune a monolingual model on BoolQ and further use it for DaNetQA;
2. To use a multi-lingual model and BoolQ for pre-training in a straightforward way.

To translate BoolQ into Russian we used the Helsinki-NLP/opus-mt-en-ru machine translation model from HuggingFace Transformers[2] library.

XLM-R [3] is a transformer-based model, trained on different languages. This model should capture semantic information in higher layers, so fine-tuning on English data could be good initialization to process further DaNetQA.

4 Results

To evaluate the transfer learning approaches and compare against baselines, we calculated accuracy and $F1$ scores for "Yes" answers. However, since "Yes" is the most common answer, we should pay more attention to "No" answers, which are less frequent. Therefore we report precision-recall AUC for "No" answers. To understand how sensitive is the model we calculated ROC-AUC. The result values of each configuration are described in Table 6.

Three Russian datasets in the task transferring approach help the model to achieve higher quality than without fine-tuning. Nevertheless, the highest improvement is achieved when transferring from SberQUAD Task A in line with similar experiments of [2]. As the tasks differ from Yes/No question answering, results are still lower than those, achieved by the language transferring approach.

Our results show, that fine-tuning the XLM-R model on BoolQ (both translated and not) and using the model for further evaluation on DaNetQA, fails in comparison to other techniques. A possible explanation for this is the difference in the topics, covered by both datasets. However, if the model is further fine-tuned on DaNetQA, the superior results are achieved due to the cross-lingual nature of the model.

---

[2] https://huggingface.co/transformers/.

Table 6. Results of the task solving approaches.

Transfer dataset	Model	Fine-tune on DaNetQA	ACC	ROC - AUC	"No" precision -recall AUC
–	RuBERT	Yes	82.46 ± 0.48	75.92 ± 0.38	57.16 ± 2.38
–	XLM-R	Yes	81.44 ± 1.28	67.76 ± 2.53	49.41 ± 3.43
Paraphraser	RuBERT	Yes	82.48 ± 0.53	75.02 ± 0.48	55.22 ± 1.97
XNLI (ru)	RuBERT	Yes	81.89 ± 0.76	78.56 ± 1.99	57.15 ± 2.47
SberQUAD Task A	RuBERT	Yes	82.63 ± 0.88	**80.74 ± 1.16**	**63.02 ± 1.54**
BoolQ translated	RuBERT	No	74.04 ± 1.88	62.74 ± 2.61	32.64 ± 2.86
	RuBERT	Yes	82.56 ± 0.29	**80.81 ± 1.50**	62.49 ± 1.84
BoolQ	XLM-R	No	78.68 ± 0.97	66.53 ± 3.21	42.25 ± 4.65
	XLM-R	Yes	**83.20 ± 0.30**	79.55 ± 1.63	61.84 ± 1.00

5 Related Work

Binary question answering is a significant part of machine reading comprehension problem. Binary questions are present in the following datasets in English: CoQA [18], QuAC [1], HotPotQA [24] and ShARC [20]. Some of these datasets, in particular, HotPotQA require multi-hop reasoning, for which answering binary questions is crucial. However BoolQ [2] to the best of our knowledge is the only dataset, devoted to binary questions exclusively.

Transfer learning is one of the leading paradigms in natural language processing. It leverages pre-training on large-scale datasets as a preliminary step before fine-tuning a contextualized encoder for the task under consideration. Applications of transfer learning range from part-of-speech tagging [13] to machine translation [10]. As the language modelling can be seen as independent tasks, some researchers see pre-training with language modelling objective as a part of the transfer learning paradigm [6,9]. Pre-training on natural language understanding tasks, in particular, on sentence modelling tasks, help not only to improve the quality of the task under consideration [2,12,21], but also to derive semantically meaningful sentence embeddings that can be compared using cosine-similarity [19].

New datasets for the Russian language are rarely published. Although Russian is one of the most widely spoken languages, the amount of datasets, suitable for NLP studies is quite limited. Dialogue, AIST and AINL conferences are the main venues for the Russian datasets to appear. Among recent datasets are RuRED [7] and SentiRusColl [14]. One more example of the previous studies with a large collection of question queries in Russian and the associated results can be found in [22].

6 Conclusion

In this paper, a new question answering dataset, DaNetQA, is presented. It comprises binary yes/no questions, paired with paragraphs, which should be used to answer the questions. The overall collection procedure follows the design of the BoolQ dataset, which is a magnitude larger in size than DaNetQA, partially due to the use of proprietary sources. We establish a straightforward baseline, exploiting FastText and RuBERT models and experiment with multiple transfer learning settings. Our results show, that on the one hand, the English dataset can be leveraged to improve the results for the Russian one. However, we can not confirm, that we can re-use BoolQ for training the model while keeping DaNetQA for evaluation only. This brings us to the following conclusion: although the re-creation of English datasets in other languages may seem like a redundant and secondary activity, the current state of the cross-lingual models does not allow for perfect language transfer. It is not enough to train the model on the English data. It seems impossible to gain high-quality results if the model is not trained in the target language. This highlights the need for future development: the development of more advanced cross-lingual contextualized encoders as well as more sophisticated datasets to evaluate cross-lingual tasks. As for DaNetQA development, we plan to enlarge the dataset with more question-paragraph pairs and to extend the dataset with an unanswerable question, affecting though the task setting.

Acknowledgements. This paper was prepared in the Laboratory for Model and Methods of Computational Pragmatics within the framework of the HSE University Basic Research Program and funded by the Russian Academic Excellence Project '5–100'.

References

1. Choi, E., et al.: Quac: question answering in context. In: Proceedings of the 2018 Conference on Empirical Methods in Natural Language Processing, pp. 2174–2184 (2018)
2. Clark, C., Lee, K., Chang, M.W., Kwiatkowski, T., Collins, M., Toutanova, K.: Boolq: exploring the surprising difficulty of natural yes/no questions. arXiv preprint arXiv:1905.10044 (2019)
3. Conneau, A., et al.: Unsupervised cross-lingual representation learning at scale. arXiv preprint arXiv:1911.02116 (2019)
4. Conneau, A., et al.: Xnli: evaluating cross-lingual sentence representations. In: Proceedings of the 2018 Conference on Empirical Methods in Natural Language Processing, pp. 2475–2485 (2018)
5. Devlin, J., Chang, M.W., Lee, K., Toutanova, K.: Bert: pre-training of deep bidirectional transformers for language understanding. In: Proceedings of the 2019 Conference of the North American Chapter of the Association for Computational Linguistics: Human Language Technologies, vol. 1 (Long and Short Papers), pp. 4171–4186 (2019)
6. Golovanov, S., Kurbanov, R., Nikolenko, S., Truskovskyi, K., Tselousov, A., Wolf, T.: Large-scale transfer learning for natural language generation. In: Proceedings of the 57th Annual Meeting of the Association for Computational Linguistics, pp. 6053–6058 (2019)
7. Gordeev, D., Davletov, A., Rey, A., Akzhigitova, G., Geymbukh, G.: Relation extraction dataset for the russian. In: Proceedings of Dialogue (2020)

8. Hickey, L.: The pragmatics of translation, vol. 12. Multilingual matters (1998)
9. Howard, J., Ruder, S.: Universal language model fine-tuning for text classification. In: Proceedings of the 56th Annual Meeting of the Association for Computational Linguistics, vol. 1: Long Papers, pp. 328–339 (2018)
10. Ji, B., Zhang, Z., Duan, X., Zhang, M., Chen, B., Luo, W.: Cross-lingual pre-training based transfer for zero-shot neural machine translation. Proceedings of the AAAI Conference on Artificial Intelligence, vol. 34, pp. 115–122 (2020)
11. Joulin, A., Grave, É., Bojanowski, P., Mikolov, T.: Bag of tricks for efficient text classification. In: Proceedings of the 15th Conference of the European Chapter of the Association for Computational Linguistics, vol. 2, Short Papers, pp. 427–431 (2017)
12. Kamath, S., Grau, B., Ma, Y.: How to Pre-train your model? comparison of different pre-training models for biomedical question answering. In: Cellier, P., Driessens, K. (eds.) ECML PKDD 2019. CCIS, vol. 1168, pp. 646–660. Springer, Cham (2020). https://doi.org/10.1007/978-3-030-43887-6_58
13. Kim, J.K., Kim, Y.B., Sarikaya, R., Fosler-Lussier, E.: Cross-lingual transfer learning for pos tagging without cross-lingual resources. In: Proceedings of the 2017 Conference on Empirical Methods in Natural Language Processing, pp. 2832–2838 (2017)
14. Kotelnikova, A., Kotelnikov, E.: SentiRusColl: Russian collocation lexicon for sentiment analysis. In: Ustalov, D., Filchenkov, A., Pivovarova, L. (eds.) AINL 2019. CCIS, vol. 1119, pp. 18–32. Springer, Cham (2019). https://doi.org/10.1007/978-3-030-34518-1_2
15. Kuratov, Y., Arkhipov, M.: Adaptation of deep bidirectional multilingual transformers for Russian language. arXiv preprint arXiv:1905.07213 (2019)
16. Kwiatkowski, T., et al.: Natural questions: a benchmark for question answering research. Trans. Assoc. Comput. Linguist. **7**, 453–466 (2019)
17. Pronoza, E., Yagunova, E., Pronoza, A.: Construction of a Russian paraphrase corpus: unsupervised paraphrase extraction. In: Braslavski, P., et al. (eds.) RuSSIR 2015. CCIS, vol. 573, pp. 146–157. Springer, Cham (2016). https://doi.org/10.1007/978-3-319-41718-9_8
18. Reddy, S., Chen, D., Manning, C.D.: Coqa: a conversational question answering challenge. Trans. Assoc. Comput. Linguist. **7**, 249–266 (2019)
19. Reimers, N., Gurevych, I.: Sentence-bert: sentence embeddings using siamese bert-networks. In: Proceedings of the 2019 Conference on Empirical Methods in Natural Language Processing and the 9th International Joint Conference on Natural Language Processing (EMNLP-IJCNLP), pp. 3973–3983 (2019)
20. Saeidi, M., et al.: Interpretation of natural language rules in conversational machine reading. In: Proceedings of the 2018 Conference on Empirical Methods in Natural Language Processing, pp. 2087–2097 (2018)
21. Shang, M., Fu, Z., Yin, H., Tang, B., Zhao, D., Yan, R.: Find a reasonable ending for stories: Does logic relation help the story cloze test? Proceedings of the AAAI Conference on Artificial Intelligence, vol. 33, pp. 10031–10032 (2019)
22. Völske, M., Braslavski, P., Hagen, M., Lezina, G., Stein, B.: What users ask a search engine: Analyzing one billion russian question queries. In: Proceedings of the 24th ACM International Conference on Information and Knowledge Management, CIKM 2015, Melbourne, VIC, Australia, 19–23 October 2015, pp. 1571–1580 (2015). https://doi.org/10.1145/2806416.2806457
23. Wang, A., et al.: Superglue: a stickier benchmark for general-purpose language understanding systems. In: Advances in Neural Information Processing Systems, pp. 3266–3280 (2019)

24. Yang, Z., et al.: Hotpotqa: a dataset for diverse, explainable multi-hop question answering. In: EMNLP (2018)
25. Zhang, Y., Zhong, V., Chen, D., Angeli, G., Manning, C.D.: Position-aware attention and supervised data improve slot filling. In: Proceedings of the 2017 Conference on Empirical Methods in Natural Language Processing (EMNLP 2017), pp. 35–45 (2017). https://nlp.stanford.edu/pubs/zhang2017tacred.pdf

Do Topics Make a Metaphor? Topic Modeling for Metaphor Identification and Analysis in Russian

Yulia Badryzlova[1]([⊠]) [iD], Anastasia Nikiforova[1] [iD], and Olga Lyashevskaya[1,2] [iD]

[1] National Research University Higher School of Economics, Moscow, Russia
[2] Vinogradov Russian Language Institute RAS, Moscow, Russia

Abstract. The paper examines the efficiency of topic models as features for computational identification and conceptual analysis of linguistic metaphor on Russian data. We train topic models using three algorithms (LDA and ARTM – sparse and dense) and evaluate their quality. We compute topic vectors for sentences of a metaphor-annotated Russian corpus and train several classifiers to identify metaphor with these vectors. We compare the performance of the topic modeling classifiers with other state-of-the-art features (lexical, morphosyntactic, semantic coherence, and concreteness-abstractness) and their different combinations to see how topics contribute to metaphor identification. We show that some of the topics are more frequent in metaphoric contexts while others are more characteristic of non-metaphoric sentences, thus constituting topic predictors of metaphoricity, and discuss whether these predictors align with the conceptual mappings attested in literature. We also compare the topical heterogeneity of metaphoric and non-metaphoric contexts in order to test the hypothesis that metaphoric discourse should display greater topical variability due to the presence of Source and Target domains.

Keywords: Metaphor identification · Topic modelling · LDA · ARTM · Topical predictors of metaphoricity · Topical profiles · Topical heterogeneity

1 Introduction

1.1 The Task of Computational Metaphor Identification

Contemporary cognitive theory states that human reasoning is intrinsically metaphorical and imaginative, based on various kinds of prototypes, framings, and metaphors [1, 2]. Our abstract conceptual representations are grounded in sensorimotor systems, and conceptual metaphor connects these two realms by mapping the domain of familiar, concrete and distinct experiences (the Source Domain) onto the domain of predominantly abstract and complex concepts (the Target Domain), thus enabling us to conceptualize the rich fabric of the reality that surrounds us. The Source-Target mappings are systematic, i.e. they reproduce themselves across similar situations; some of them are claimed to be universal, while others may be culture-specific.

© Springer Nature Switzerland AG 2021
W. M. P. van der Aalst et al. (Eds.): AIST 2020, LNCS 12602, pp. 69–81, 2021.
https://doi.org/10.1007/978-3-030-72610-2_5

Conceptual metaphors manifest themselves in language and discourse as linguistic metaphors, that is, the lexical units and constructions which express the Target, the Source, and the relations between them. An example of a conceptual metaphoric mapping is CORRUPTION IS A DISEASE which may be linguistically conveyed, for example, by the following English sentences (with T and S indicating the Source and the Target terms, respectively): "Corrupt *(T)* officials are infecting *(S)* our government at every level." or "Our government is afflicted *(S)* with the cancer *(S)* of corruption *(T)*." [3].

Evidence from psycholinguistic research demonstrates that metaphor guides reasoning and decision-making in societal, economic, health-related, educational, and environmental issues [see, for example, 4–7]. As deeply as conceptual metaphor is engrained in the mind, as much linguistic metaphor is integrated into the language and its usage, forming an organic part of them. According to various estimates, up to nearly one third of words in a corpus may be used metaphorically [8, 9]. Such pervasiveness of metaphor in language and thought – as well as the ambiguity it creates – make metaphor a challenge to various NLP applications, such as machine translation, information retrieval and extraction, question answering, opinion mining, etc. The interest of the NLP community to computational metaphor research expressed itself in the series of dedicated workshops in 2013–2016 [10–13] and the two metaphor detection shared tasks in 2018 and 2020 [14, 15].

How can the underlying conceptual properties of metaphoric utterances be captured in order to train machine learning algorithms to tell them apart from non-metaphoric ones? Different types of features have been explored in the state-of-the-art research:

- Lexical features [16];
- Morphological and syntactic features [17, 18];
- Distributional semantic features [19, 20];
- Features from lexical thesauri and ontologies: e.g., WordNet [21], FrameNet [22], VerbNet [23], ConceptNet [18], and the SUMO ontology [24, 25];
- Psycholinguistic features: concreteness and abstractness, imageability, affect, and force [26–29];
- Topic modelling [16, 30, 31] – the feature which is explored in the present paper.

1.2 Topic Modelling in Metaphor Identification: Previous Work

Application of topic modelling to identification of metaphor relies on the assumption that metaphoric contexts should contain terms from both the Source and the Target domains, whereas non-metaphoric sentences should be more homogeneous in terms of topical composition; the topics are regarded as proxies for the conceptual domains.

Heintz et al. [30] use topic models to identify linguistic metaphors belonging to the Target domain of Governance in English and Spanish. They train LDA models on the full text of Wikipedia in these languages and automatically align the topics with the manually collected lists of seed words representing the Target and the Source domains. A sentence is judged to contain a linguistic metaphor on the account of the strength of association between topics and the sentence, between the annotated words and the topics, and between the topics and their aligned concepts. The authors carry out two evaluations of their system. In the first, the predictions of the algorithm on the English data are compared to the judgements of two annotators, with the reported F1-scores of 0.66 and 0.5, respectively; however, the agreement between the annotators (κ) was rather low (0.48). In the second evaluation, the annotation task was crowdsourced, and the metaphoricity of a sentence was defined as the fraction of the subjects who annotated it as being metaphoric. Sixty-five per cent of the English sentences that were judged metaphoric by the algorithm had human-generated metaphoricity scores greater than 0.25, and 73% greater than 0.2; on the Spanish data, the respective results were 60 and 73%.

Ghavidel et al. [31] train an LDA model to detect linguistic metaphors in Persian. They generate a topic vector of each sentence in the corpus, and run the rule-based classifier to check whether there is any word which does not belong to the overall topic of the sentence. If the topic of a word is recognized as deviant, the sentence is marked as metaphoric, and non-metaphoric if otherwise. The system is reported to yield the F1-score of 0.68 when evaluated on a random sample of 100 sentences.

Klebanov et al. [16] use LDA topic modelling in combination with other features (lexical unigrams, part of speech tags, and concreteness indexes) to identify metaphor on the word level (i.e. to tag each content word in running text as either metaphoric or non-metaphoric). The F1-score ranges between 0.21 to 0.67 depending on the dataset and the genre. The authors investigate the relative contribution of each feature and report that topics is the second most effective feature (after lexical unigrams).

Besides, topic models, along with the other types of features, were suggested for use to the participants of the First and the Second shared tasks on metaphor detection [14, 15].

In this paper, we apply topic modelling to sentence-level metaphor identification in Russian on a representative metaphor-annotated corpus [32]. We also compare the performance of the topic models in metaphor classification to other state-of-the-art features, and estimate the contribution of topics to the most efficient classifier. Moreover, we take an in-depth look into the topic models of metaphoric and non-metaphoric discourse in order to identify the topical cues of metaphoricity. We also examine the topical heterogeneity of metaphoric and non-metaphoric contexts in order to explore the hypothesis that metaphoric contexts should feature a greater variety of topics due to the presence of two conceptual domains (the Source and the Target).

To the best of our knowledge, this is the first research to apply topic modelling to the problem of metaphor identification in Russian.

2 Topic Modelling for Metaphor Identification in Russian

2.1 Training the Topic Models: LDA and ARTM

For our experiments, we train two types of topic models: LDA and ARTM.

LDA (latent Dirichlet allocation) [33] is the topic modelling method which is most widely used in NLP tasks. In LDA, the parameters Φ (the matrix of term probabilities for the topics) and Θ (the matrix of topic probabilities for the documents) are constrained by an assumption that vectors φt and θd are drawn from Dirichlet distributions with hyperparameters $\beta = (\beta w)w \in W$ and $\alpha = (\alpha t)t \in T$ respectively (where T is a set of topics, W is a set of all terms in a collection of texts).

Two major problems arise when training topic models with LDA – noise from stop-words and other high-frequency words, and assigning words to multiple topics, which negatively affects the overall interpretability of the topics. This issue is addressed by Additive Regularization of Topic Models (ARTM) [34]; in this study, we used the following regularizers available in the BigARTM library[1]:

1. The smoothing/sparsing regularization of terms over topics, where the smoothing regularizer sends high-frequency words into dedicated background topics, and the sparsing regularizer highlights the lexical nuclei of domain-specific topics covering a relatively small proportion of the vocabulary;
2. The smoothing/sparsing regularization of topics over documents, in which the smoothing regularizer indicates the background words in each document of the collection, while the sparsing regularizer pinpoints the domain-specific words in each document.

As a result of such regularization, zero probability is assigned to words that do not describe domain-specific topics, as well as to high-frequency and general vocabulary; each term is assigned to a relatively small number of topics, so that the resulting topics become more interpretable. The smoothing and the sparsing regularizations of matrices Φ and Θ are presented in the equation:

$$R(\Phi, \Theta) = \sum_{t \in T} \sum_{w \in W} \beta_{wt} ln\phi_{wt} + \sum_{d \in D} \sum_{t \in T} \alpha_{td} ln\theta_{td},$$

where D is a collection (set) of texts, $\beta_0 > 0$, $\alpha_0 > 0$ are regularization coefficients, and β_{wt}, α_{td} are user-defined hyperparameters, so that.

- $\beta_{wt} > 0, \alpha_{td} > 0$ results in smoothing,
- $\beta_{wt} < 0, \alpha_{td} < 0$ results in sparsing
- $\beta_{wt} > -1, \alpha_{td} > -1$ results in an LDA model.

For our study we trained two types of ARTM models: in the sparse models, the Θ matrix was regularized using the sparsing coefficient $\tau = -0.1$; in the dense models, the smoothing coefficient $\tau = 0.1$ was applied. In both types of models, the Φ matrix was regularized using the sparsing coefficient $\tau = 0.25$ and the topic decorrelation coefficient $\tau - 10^4$ (so that words with high frequency throughout the collection received lower weights in each document).

All the models (LDA, sparse ARTM, and dense ARTM) were trained on $\approx$ 600,000 randomly sampled entries from Russian Wikipedia (the dump of 1 March 2020[2]). The data was cleaned with the corpuscula[3] tool, and lemmatized and POS-tagged using the pymorphy2 parser[4]; bigram collocations (e.g. чемпионат_мир 'world_cup') were identified using gensim[5]. The Wikipedia corpus was chosen on the assumption that it is

---

[1] https://bigartm.readthedocs.io/en/stable/intro.html.
[2] https://dumps.wikimedia.org/ruwiki.
[3] https://github.com/fostroll/corpuscula.
[4] https://pymorphy2.readthedocs.io/en/latest/.
[5] https://radimrehurek.com/gensim/models/phrases.html.

likely to represent a large variety of common topics. The Wiki data was vectorized using count vectorization; the topic models were incorporated with BERT word embeddings [35] by concatenating topic vectors with averaged BERT vectors.

All the resources related to this project (the preprocessed Wikipedia dump, the trained topic models, the Russian metaphor-annotated corpus, and the scripts) are available in a github repository[6].

2.2 Experimental Setup

The metaphor identification experiment was run on the Russian corpus of metaphor-annotated sentences [32]. The corpus consists of 7,020 sentences; each of them contains one of the 20 polysemous target verbs (e.g. *бомбардировать* 'to bombard', *нападать* 'to attack', *утюжить* 'to iron (about clothes)', *взвешивать* 'to weigh', etc.) which is used either metaphorically or non-metaphorically. The number of sentences per target verb ranges from 225 to 693; each of these subsets is balanced by class. Below are examples of metaphoric and non-metaphoric sentences with the target verb *взрывать* 'to explode (smth)'; the first metaphoric sentence contains an unconventional metaphor, while the second metaphoric sentence demonstrates a conventionalized metaphor:

– Example 1: (Metaphoric) *Ксенофобия – это то, что, возможно, станет бомбой замедленного действия, которая < взорвет > наше общество.* 'Xenophobia is what may become a ticking bomb which will < explode > our society.'
– Example 2: (Metaphoric) *Для нее было необходимо < взорвать > ситуацию любым способом...* 'It was necessary for her to < explode > the situation by any means.'
– Example 3: (Non-metaphoric) *Главнокомандующий князь Горчиков приказал < взорвать > уцелевшие укрепления и оставить город.* 'The commander-in-chief, Prince Gorchikov, gave orders to < explode > the remaining fortifications and to flee the town.'

The metaphor identification task was formulated as sentence-level binary classification. We experimented with several conventional ML algorithms (logistic regression, SVM, Naïve Bayes, Random Forest, etc., including a simple neural network – multilayer perceptron); no deep learning methods (such as LSTM or CNN) were applied to the task, firstly, due to the relatively small size of the experimental corpus and, secondly, due to the fact that in topic modelling documents are represented as bags of words. For each of the three types of topic models (LDA, ARTM dense, and ARTM sparse), we took vectors varying between 30 and 130 topics in size. The experiments were run using 5-fold cross-validation.

2.3 Results

The best classification results (in terms of accuracy) – 0.7 – were yielded by the logistic regression (LogReg) and the multilayer perceptron (NN) models with 40, 50, 80, and 90

---

[6] https://github.com/steysie/topic-modelling-metaphor.

topics vectors, as summarized in Table 1 (models with 60 and 70 topics are not shown since they produced slightly lower results). It can be seen that somewhat higher results are obtained with the non-regularized LDA-based models.

At the same time, most of the highest results in terms of F1-score are delivered by the SVM classifier on the ARTM sparse and the ARTM dense models.

Table 1. Classification results with topic modelling.

Number of topics	Classifier	LDA				ARTM sparse				ARTM dense			
		Acc	Prec	Rec	F1	Acc	Prec	Rec	F1	Acc	Prec	Rec	F1
40	LogReg	**0.70**	0.70	0.72	0.71	0.69	0.69	0.72	0.70	0.69	0.69	0.71	0.70
	SVM	0.67	0.64	0.80	0.71	0.67	0.63	0.81	0.71	0.67	0.63	0.82	0.71
	NN	**0.70**	0.70	0.71	0.71	0.69	0.68	0.71	0.70	0.69	0.69	0.71	0.70
50	LogReg	**0.70**	0.69	0.70	0.70	0.69	0.68	0.71	0.69	0.69	0.69	0.71	0.70
	SVM	0.67	0.63	0.79	0.70	0.66	0.63	0.80	0.70	0.66	0.63	0.80	0.70
	NN	0.69	0.68	0.72	0.70	0.69	0.68	0.70	0.69	0.69	0.68	0.72	0.70
80	LogReg	**0.70**	0.72	0.67	0.69	0.69	0.68	0.70	0.69	0.69	0.69	0.70	0.69
	SVM	0.67	0.66	0.73	0.69	0.65	0.61	0.86	0.71	0.67	0.62	0.83	0.71
	NN	**0.70**	0.72	0.67	0.69	0.68	0.68	0.71	0.69	0.69	0.68	0.71	0.70
90	LogReg	**0.70**	0.70	0.67	0.69	0.68	0.67	0.69	0.68	0.68	0.68	0.70	0.69
	SVM	0.67	0.65	0.77	0.70	0.65	0.61	0.84	0.70	0.66	0.61	0.84	0.71
	NN	0.69	0.69	0.69	0.69	0.68	0.67	0.70	0.69	0.68	0.67	0.72	0.69

To compare the results of the topic-based classifiers to other state-of-the-art features, we replicated the features proposed by Badryzlova [36]: lexical (LEX), morphosyntactic (POS), Concreteness-Abstractness (CONC), and semantic coherence (SEM) features. In order to assess the contribution of the topic-based classifier to metaphor identification, we conducted an ablation experiment in which the performance of each feature, as well as their combinations, was evaluated with the topic-based feature (+TM) and without it (- TM) – see Table 2 (there we show the results for the LDA model with 80 features).

When comparing the performance of the topic-based classifier to the other uni-feature classifiers, we see that the accuracy of TM surpasses the result of the classifier informed with morphosyntactic features (POS); TM is slightly outperformed by the classifiers operating on Concreteness-Abstractness (CONC) and semantic coherence indexes (SEM). In comparison to the lexical classifier, the topic-based classifier falls behind by a tangible margin – similarly to the other three types of features.

When analyzing the contribution of the topic-based model to the other uni-feature models, we observe that addition of TM improves the performance of LEX, CONC, and

Table 2. Feature ablation experiment (accuracy). Asterisk denotes statistically significant differences between combinations with and without the topic-based model.

Feature/classifier	SVM-TM	LogReg-TM	NN-TM	SVM+TM	LogReg+TM	NN+TM
LEX	0.8164	0.8173	0.8179	0.8318	0.8287	0.8301
POS	0.6757	0.6749	0.6702	0.6032	0.5668*	0.5958*
CONC	0.7173	0.7158	0.7178	0.7595*	0.7473*	0.7603*
SEM	0.7195	0.7310	0.7359	0.7319	0.7430	0.7372
TM	0.6715	0.7033	0.7018	---	---	---
LEX+SEM	0.8074	0.8484	0.8340	0.8094	0.8517	0.8382
LEX+POS	0.8204	0.8201	0.8204	0.8353	0.8294	0.8320
LEX+CONC	0.8327	0.8327	0.8323	0.8384	0.8331	0.8359
LEX+POS+CONC	0.8352	0.8350	0.8337	0.8418	0.8339	0.8377
LEX+POS+SEM+CONC	0.8176	0.8544	0.8542	0.8121	0.8537	0.8377

SEM; however, only the CONC + TM increase proves statistically significant[7]. At the same time, addition of TM to the POS model considerably worsens the result.

Adding topicality to multi-feature models increases the efficiency of classification in at least one of the classifiers in almost all combinations of features (the exception is the last, most complex model); however, this increase of accuracy is rather narrow and does not prove to be statistically significant.

Overall, the highest results are attained with combinations of three to five features, one of which is lexical (LEX). The importance of this feature for metaphor classification is consistent with previous findings [16] and is closely examined by Badryzlova [36]. Lexical cues seem to be the most potent predictors of metaphoricity; therefore, adding other features does not dramatically affect the performance of the classifier. Five of the features implemented in present study bear on the lexico-distributional properties of words: LEX, SEM, CONC, and TM – and thus they complement each other in this regard. In contrast, the POS feature is based on patterns of words' morphosyntactic combinability which are highly idiosyncratic and thus less generalizable and reliable in metaphor prediction [36]. The substantial drop in classification accuracy in the POS+TM model most likely occurs because the POS predictor, rather weak as it is, is collapsed with the topic-based model, which is intended to capture a different type of distributions, and, moreover, is not the strongest predictor in itself.

---

[7] Wilcoxon signed-rank test [37] (SciPy implementation) is used in this study to evaluate the statistical significance of results.

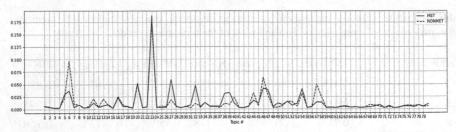

Fig. 1. Distribution of topics in metaphoric (MET) and non-metaphoric (NONMET) contexts (LDA, 80 topics).

3 Analysis of Topic Distribution in Metaphoric and Non-metaphoric Contexts

In order to test whether metaphoric and non-metaphoric contexts contain different sets of topics, we applied the Kolmogorov-Smirnov statistic [38] which tests the hypothesis that two sets belong to the same distribution. On all our matrices of topics the *p-value* proved above the significance level – therefore, we cannot claim that the distributions of topics in metaphoric vs. Non-metaphoric contexts are statistically different. However, this does not mean that the topics are distributed uniformly across these two types of discourse. Analysis of distribution revealed that there are topics that are indicative of either metaphoric or non-metaphoric utterances.

Figure 1 shows the distribution of topics in the metaphoric (MET) and the non-metaphoric (NONMET) subcorpora, as generated by the LDA model with vector dimensionality of 80 topics. It is easy to notice that topics 27, 32, 38, and 39 prevail in metaphoric contexts, while topics 6, 11, 44, 46, 57, and 58 are more salient in non-metaphoric sentences. Remarkably, topics 16, 20, and 23 are equally frequent in both subcorpora.

Analyses of the topic matrixes generated with the LDA, the ARTM dense and the ARTM sparse models indicated the following topical cues of metaphoricity (the names of the topics were assigned manually): Literature and Writing, Economy, Judicial System, Corporate Management, and Railway. While the metaphoricity of the first four topics is quite expected, explained by the high frequency of analogies and comparisons in their metaphoric contexts, the metaphoricity of the Railway topic arises from the conventionalized indirect meanings of some of the target verbs, for example, *пилить* 'to travel a long distance' (lit. 'to saw'):

– Example 4: (Metaphoric) *Поезд подошел и оказалось, что до нашего вагона еще < пилить > и < пилить >*. The train pulled in, and we discovered that we had to < do a great deal of sawing > (lit. 'to walk a long distance') to reach our carriage

The topics that prevail in the non-metaphoric subcorpus are: Biology, Language, Cars, Chemistry, Aviation, Peoples and Traditions, and Religion, e.g.:

- Example 5: (Non-metaphoric) *В верхней части карты Таро находится божественная фигура, обычно представленная крылатым ангелом, [который] смотрит из облаков и < трубит > в трубу.* The upper part of the Tarot card depicts a divine figure which is usually represented by a winged angel who is looking down from the clouds, < trumpeting >.

The topics that have high frequency in both metaphoric and non-metaphoric contexts are: Military and Warfare, Cinema, TV Series and Computer Games, and Architecture and Construction. The following sentences demonstrate examples of metaphoric and non-metaphoric occurrences of the Military and Warfare topic:

- Example 6: (Metaphoric) *Когда немцы с земли и воздуха < утюжили > снарядами и бомбами наши армейские позиции, только воля божья спасла их на дне окопа и в землянке.* When Germans were < ironing > (lit. 'bombing out') our army's positions with shells and bombs, it was but for the grace of God that they survived at the bottom of a trench and in a dugout.
- Example 7: (Non-metaphoric) *Враги рыли под землей галереи, чтобы, заложив мины, < взорвать > русские укрепления.* The enemies were digging underground galleries in order to plant mines and < explode > the Russian fortifications.

The identified topical cues seem to reflect certain broadly defined realms of reality rather than the more fine-grained conceptual structures suggested by the cognitive metaphor theory and attested in empirical linguistic research [e.g. 3]. Thus, the present implementation of topic modeling for metaphor analysis falls short of capturing the expected conceptual mappings. Yet, it demonstrates that differences exist in the topical profiles of metaphoric and non-metaphoric discourse, calling for further investigation. Besides, it should be borne in mind that the inventory of topical predictors of metaphoricity/non-metaphoricity in the present study is by no means exhaustive: it is limited by the scope and the size of the experimental corpus, and is likely to alter with expansion of the corpus.

4 Heterogeneity of Topic Distribution in Metaphoric and Non-metaphoric Discourse

According to the conceptual metaphor theory, metaphoric contexts may represent at least two topics associated with the Target and Source Domains (see Politics and Military/Warfare in Example (1) above) while non-metaphoric contexts can be limited to one topic space (see Military/Warfare in Example (3)). Therefore, we can expect more salient topics per sentence in the MET class than in the NONMET class. Besides that, the Source Domain can be mapped to different Target Domains in different sentences, which assumes the topic space to be potentially more variable in MET than in NONMET.

We used several probability thresholds to empirically define the number of salient topics per sentence for the LDA matrix with k = 80 topics. The average number of topics is significantly larger in the MET class as compared to the NONMET class for thresholds below 0.1 (t-test at the threshold 0.05: $t = 5.718$, $p = 1.122e-08$). As for the individual verbs comprising the metaphor-annotated corpus (see Sect. 2.2), this trend holds for 11–15 out of the 20 verbs. However, the verb *уколоть* 'prick' follows a different pattern, with metaphoric contexts having in general fewer topics per sentence as compared to non-metaphoric ones. We can explain this by the specifics of the Wikipedia-based topic modeling as both everyday physical events (Source Domain) and emotional reactions (Target Domain) are underrepresented in the training Wiki data and therefore in the topic clusters. This is in line with another observation that the verbs of everyday activity such as *утюжить* 'to iron (about clothes)' and *причесать* 'comb' form a subgroup that shows fewer topics per sentence in non-metaphoric discourse than other verbs.

We run latent profile analysis [39] to visualize most common topical profiles in each verb in MET and NONMET. We conclude that there is not enough evidence to prove the heterogeneity hypothesis from the point of view of the variability of topics in metaphoric and non-metaphoric discourse since verbs are inconsistent in their behaviour in the current settings. All this suggests that other pre-trained topic models, with a greater number of domains covered, could be used to further test the hypothesis of topic heterogeneity. For example, topic models trained on a corpus of fiction could be expected to reveal the currently underrepresented topics (such as everyday activity or emotional reactions) and, besides, to capture the topics formed by indirect, figurative usages of words.

5   Conclusions

We applied topic-based features to the task of sentence-level metaphor identification in Russian. In doing so, we compared three types of topic models – a conventional LDA model and two models with additive regularization – ARTM dense and ARTM sparse. When taken alone, the topic-based classifier yields the accuracy of 0.7; in comparison to other state-of-the-art features, topic-based classifier performs on the par with Concreteness-Abstractness and semantic coherence indexes, yet it underperforms in comparison to the lexical baseline. Combining the topic-based model with the other features resulted in statistically significant improvement only in the combination with the Concreteness-Abstractness model; integrating the topic-based model into the morphosyntactic one led to a sharp decrease in performance, which is likely due to the weak predictive power of both features and the differences in patterns (morphosyntactic vs. Lexico-distributional combinability) captured by them.

However, application of topic modelling to metaphor analysis allowed us to test two hypotheses about the conceptual nature of metaphor suggested in linguistic literature and practice of metaphor studies.

Firstly, we analyzed the topical profiles (i.e. the distribution of topics) in metaphoric and non-metaphoric discourse, and identified the topical cues, that is, the topics that are indicative of metaphoric and non-metaphoric contexts. These cues do not resemble the Source and Target domains attested in linguistic studies; yet, the existence of these cues suggests a promising direction for further research.

The second hypothesis concerned topic heterogeneity of metaphoric and non-metaphoric discourse. According to the conceptual metaphor theory, metaphoric contexts should be more topically heterogeneous (due to the presence of two conceptual domains, the Source and the Target) than non-metaphoric ones. We found some evidence that metaphoric uses are associated with a larger number of topics than those identified in non-metaphoric uses. However, larger studies are needed to support our findings; for example, applying topic models trained on corpora other than Wikipedia (e.g. fiction) might be able to capture the topics that are currently underrepresented in our models.

References

1. Lakoff, G., Johnson, M.: Philosophy in the Flesh. Basic books, New York (1999)
2. Lakoff, G., Johnson, M.: Metaphors We Live by. University of Chicago Press, Chicago (1980)
3. Honga, J., Stickles, E., Dodge, E.: The MetaNet metaphor repository: formalized representation and analysis of conceptual metaphor networks. In: 12th International Cognitive Linguistics Conference. Edmonton, Canada (2013)
4. Flusberg, S.J., Matlock, T., Thibodeau, P.H.: Metaphors for the war (or race) against climate change. Environ. Commun. **11**, 769–783 (2017)
5. Hauser, D.J., Schwarz, N.: The war on prevention: bellicose cancer metaphors hurt (some) prevention intentions. Pers. Soc. Psychol. Bull. **41**, 66–77 (2015)
6. Landau, M.J., Oyserman, D., Keefer, L.A., Smith, G.C.: The college journey and academic engagement: how metaphor use enhances identity-based motivation. J. Pers. Soc. Psychol. **106**, 679 (2014)
7. Thibodeau, P.H., Boroditsky, L.: Metaphors we think with: the role of metaphor in reasoning. PLoS ONE **6**, e16782 (2011). https://doi.org/10.1371/journal.pone.0016782
8. Shutova, E., Teufel, S.: Metaphor corpus annotated for source-target domain mappings. In: LREC, p. 2 (2010)
9. Steen, G., Herrmann, B., Kaal, A., Krennmayr, T., Pasma, T.: A Method for Linguistic Metaphor Identification: From MIP to MIPVU. John Benjamins Publishing Company, Amsterdam (2010)
10. Klebanov, B.B., Shutova, E., Lichtenstein, P.: Proceedings of the second workshop on metaphor in NLP. In: Proceedings of the Second Workshop on Metaphor in NLP. Association for Computational Linguistics, Baltimore, MD (2014)
11. Klebanov, B.B., Shutova, E., Lichtenstein, P.: Proceedings of the fourth workshop on metaphor in NLP. In: Proceedings of the Fourth Workshop on Metaphor in NLP. Association for Computational Linguistics, San Diego, California (2016)
12. Shutova, E., Klebanov, B.B., Tetreault, J., Kozareva, Z.: Proceedings of the first workshop on metaphor in NLP. In: Proceedings of the First Workshop on Metaphor in NLP. Association for Computational Linguistics, Atlanta, Georgia (2013)
13. Shutova, E., Klebanov, B.B., Lichtenstein, P. (eds): Proceedings of the third workshop on metaphor in NLP. Association for Computational Linguistics, Denver, Colorado (2015)
14. Leong, C.W.B., Klebanov, B.B., Shutova, E.: A report on the 2018 VUA metaphor detection shared task. In: Proceedings of the Workshop on Figurative Language Processing. pp. 56–66 (2018)
15. Leong, C.W., Klebanov, B.B., Hamill, C., Stemle, E., Ubale, R., Chen, X.: A report on the 2020 VUA and TOEFL metaphor detection shared task. In: Proceedings of the Second Workshop on Figurative Language Processing, pp. 18–29. Association for Computational Linguistics, Online (2020)
16. Klebanov, B.B., Leong, B., Heilman, M., Flor, M.: Different texts, same metaphors: unigrams and beyond. In: Proceedings of the Second Workshop on Metaphor in NLP, pp. 11–17 (2014)

17. Hovy, D., et al.: Identifying metaphorical word use with tree kernels. In: Proceedings of the First Workshop on Metaphor in NLP, pp. 52–57. Citeseer (2013)

18. Ovchinnikova, E., Israel, R., Wertheim, S., Zaytsev, V., Montazeri, N., Hobbs, J.: Abductive inference for interpretation of metaphors. In: Proceedings of the Second Workshop on Metaphor in NLP, pp. 33–41 (2014)

19. Shutova, E., Kiela, D., Maillard, J.: Black holes and white rabbits: Metaphor identification with visual features. In: Proceedings of the 2016 Conference of the North American Chapter of the Association for Computational Linguistics: Human Language Technologies, pp. 160–170 (2016)

20. Panicheva, P.: Analiz parametrov semantičeskoj svjaznosti s pomošč'ju distributivny xsemantičeskix modelej (na materiale russkogo jazyka) [Analysis of parameters of semantic coherence by means of distributional semantic models (on Russian data)] (2019)

21. Gandy, L., et al.: Automatic identification of conceptual metaphors with limited knowledge. In: AAAI (2013)

22. Gedigian, M., Bryant, J., Narayanan, S., Ciric, B.: Catching metaphors. In: Proceedings of the Third Workshop on Scalable Natural Language Understanding, pp. 41–48. Association for Computational Linguistics (2006)

23. Klebanov, B.B., Leong, C.W., Gutierrez, E.D., Shutova, E., Flor, M.: Semantic classifications for detection of verb metaphors. In: Proceedings of the 54th Annual Meeting of the Association for Computational Linguistics, vol. 2: Short Papers, pp. 101–106 (2016)

24. Dunn, J.: Evaluating the premises and results of four metaphor identification systems. In: Gelbukh, A. (ed.) CICLing 2013. LNCS, vol. 7816, pp. 471–486. Springer, Heidelberg (2013). https://doi.org/10.1007/978-3-642-37247-6_38

25. Dunn, J.: What metaphor identification systems can tell us about metaphor-in-language. In: Proceedings of the First Workshop on Metaphor in NLP, p. 10 (2013)

26. Neuman, Y., et al.: Metaphor identification in large texts corpora. PLoS ONE **8**, e62343 (2013). https://doi.org/10.1371/journal.pone.0062343

27. Strzalkowski, T., et al: Robust extraction of metaphors from novel data. In: Proceedings of the First Workshop on Metaphor in NLP, pp. 67–76 (2013)

28. Turney, P.D., Neuman, Y., Assaf, D., Cohen, Y.: Literal and metaphorical sense identification through concrete and abstract context. In: Proceedings of the Conference on Empirical Methods in Natural Language Processing, pp. 680–690. Association for Computational Linguistics (2011)

29. Klebanov, B.B., Leong, C.W., Flor, M.: Supervised word-level metaphor detection: experiments with concreteness and reweighting of examples. In: Proceedings of the Third Workshop on Metaphor in NLP, pp. 11–20. Association for Computational Linguistics, Denver (2015). https://doi.org/10.3115/v1/W15-1402

30. Heintz, I., et al.: Automatic extraction of linguistic metaphors with LDA topic modeling. In: Proceedings of the First Workshop on Metaphor in NLP, pp. 58–66 (2013)

31. Abdi Ghavidel, H., Khosravizadeh, P., Rahimi, A.: Impact of topic modeling on rule-based persian metaphor classification and its frequency estimation. Int. J. Inf. Commun. Technol. Res. **7**, 33–40 (2015)

32. Badryzlova, Y., Panicheva, P.: A multi-feature classifier for verbal metaphor identification in Russian texts. In: Ustalov, D., Filchenkov, A., Pivovarova, L., Žižka, J. (eds.) AINL 2018. CCIS, vol. 930, pp. 23–34. Springer, Cham (2018). https://doi.org/10.1007/978-3-030-012 04-5_3

33. Blei, D.M., Ng, A.Y., Jordan, M.I.: Latent Dirichlet allocation. J. Mach. Learn. Res. **3**, 993–1022 (2003)

34. Vorontsov, K., Potapenko, A.: Tutorial on probabilistic topic modeling: additive regularization for stochastic matrix factorization. In: Ignatov, D.I., Khachay, M.Y., Panchenko, A., Konstantinova, N., Yavorskiy, R.E. (eds.) AIST 2014. CCIS, vol. 436, pp. 29–46. Springer, Cham (2014). https://doi.org/10.1007/978-3-319-12580-0_3
35. Wolf, T., et al.: HuggingFace's Transformers: State-of-the-art Natural Language Processing. ArXiv. arXiv–1910 (2019).
36. Badryzlova, Y.: Automated metaphor identification in Russian texts (2019)
37. Wilcoxon, F.: Individual comparisons by ranking methods. In: Kotz, S., Johnson, N.L. (eds.) Breakthroughs in statistics, pp. 196–202. Springer New York, New York, NY (1992). https://doi.org/10.1007/978-1-4612-4380-9_16
38. Massey, Jr. F.J.: The Kolmogorov-Smirnov test for goodness of fit. J. Am. Stat. Assoc. **46**, 68–78 (1951)
39. Scrucca, L., Fop, M., Murphy, T.B., Raftery, A.E.: mclust 5: clustering, classification and density estimation using Gaussian finite mixture models. R J. **8**, 289 (2016)

Metagraph Based Approach for Neural Text Question Generation

Marina Belyanova, Sergey Chernobrovkin, Igor Latkin,
and Yuriy Gapanyuk(✉)

Bauman Moscow State Technical University, Moscow, Russia
gapyu@bmstu.ru

Abstract. The paper considers the task of generating questions by given text. The generation of high-quality questions allows us to solve many problems, for example, in the field of teaching – for the automatic creation of tests for training materials or the enrichment of interaction techniques for question-answering systems. Leading research in this area shows that models based on the Seq2Seq architecture achieve the best quality. However, such models do not use the hidden structure of the text, which is essential for generating semantically correct questions. New works on this topic use additional data in the form of the graphs, representing the dependencies of the words in a sentence. In this article, an approach that uses a metagraph model of text as the initial structure for storing and enriching data with additional information and semantic relationships is considered. After generating a metagraph model of the text, the metagraph is decomposed into a multipartite graph, which allows its usage in existing models for generating text questions without losing information about the additional hierarchical and semantical dependencies of the text.

Keywords: Question generation · Graph neural networks · Metagraph · Metavertex

1 Introduction

Asking questions is an essential function of the human mind that affects the results of the learning process. In the book [1], it is stated that "questions are 'workhorses' in building foundational knowledge." In order to ask the question right, it is necessary to know the context of the question: the information field, in which the question appears to confirm or get the new information. Thus, in the case of question generation from the text, the context may be presented by the text itself or by the different ontological characteristics of the words and phrases in it.

The task of automatic question generation based on the text can be used for the testing of the read text understanding by students, for the optimization of user query prediction in search engines. Also, this task can be helpful for the chatbot systems to start, continue, and enrich dialogues.

W. M. P. van der Aalst et al. (Eds.): AIST 2020, LNCS 12602, pp. 82–95, 2021.
https://doi.org/10.1007/978-3-030-72610-2_6

The motivation for the idea of the testing materials generation task is to reduce the amount of manual work and time that is spent on making the tests for the text understanding check. On the other hand, there is also the motivation to enrich the dialog systems' chatting techniques to provide better user experience for future communication with artificial intelligence. There's also an interesting idea about how the pre-generated question from the text may predict user search query. This particular work's motivation is to check whether the metagraph-based approach can provide comparable or better results for the task of question generation compared with the previous research. In this article, we propose:

- The architecture of the question generation system, which is based on the Hybrid Intelligent Information System (HIIS) approach.
- Application of the metagraph approach for the text model description.
- Evaluation of the metagraph text model on the problem of text question generation.

2 The Review of Recent Approaches to the Question Generation from the Text

The question generation approaches from the text are usually divided into two categories: the logical rules approach and the machine learning approach. In the first category, the question is generated from the text, basing on the manually provided rules that map declarative sentences to the question. That requires in-depth knowledge of the text language structure. For the second category, the learning on the massive amount of texts with the pre-written questions is required, which provides a flexible generalization of question generation rules.

As an example of first category task solving, in the research [2], the mapping from the text to the question was held with the set of the rules: using the syntax parser, the words in the text were tagged with parts of speech and, depending on which predefined part of speech structure was found in the sentence, it was rebuilt to get the questions. The authors used an array of a massive amount of regular expressions to define the rules of the rebuilding of the text.

To reduce the necessity to construct the inference rules, the strategy called "overgenerate-and-rank" was used in [3]. First of all, the text was summarized into several simple facts. Then these facts were transformed to the question and ranked by the model, trained on manually written text.

In work [4], the question generation is made on the ontological structure, for example, the phylogenetic tree of species, presented as a graph, where the vertices are the categories titles and the connections are the relations between categories. Using the "expanding" of the categories titles and their relationships with the manually written rules for question generation, authors received the questions that allow them to check structural knowledge of the topic.

As for the second category, nowadays, the Seq2Seq (sequence-to-sequence) approach, which allows us to generate the sequence based on another sequence, is becoming more and more popular. It is inspired by the task of machine translation.

Just as in the machine translation, the most common metric for the quality of the model is BLEU [5] – the intersection of n-grams between the machine-generated text and the manually provided text.

$$BLEU = e^{min(0, \frac{n-L}{n})} \prod_{i=1}^{N} P(i))^{\frac{1}{N}},$$

where n – the number of words in the reference text, L – the number of words in the output of the model, $P(i)$ – the amount of matching i-grams for the generated and reference text, N – the maximum n for n-gram, which is commonly equal to 4.

In the article [6], the approach to generate questions based on the text, and an answer with an additional reinforcement learning is described. As an input, the sentence and the text of the answer are used, encoded in the graph form with the vertices, explaining the words in the text, and the edges, representing the relations between words and sentences. The output of the model is the text of the question. The input of the sentence is encoded by the bidirectional graph neural network (BiGNN) [7]. This approach allows authors to encounter not only forward but also backward dependencies in the sentence.

As the training and testing set, the SQuAD Benchmark dataset (split-1 and split-2) was used. It is based on the manually tagged questions and answers of the Wikipedia articles.

The approach called "semantic hypergraphs" is proposed in the paper [8]. By this approach, authors wanted to solve two problems: first, to provide the visualization of the text, that allows saving recursive connections between sentence parts; second, to provide the mapping, that allows storing connections that would disappear if the syntactic tree structure is used for the text presentation. New attributes could increase the quality of models that are trained using them, although authors do not provide the results of model training using their data structure.

In this paper, as the graph structure of the text given to the neural network model, the metagraph structure is proposed. Metagraphs allow building ontological structures to extract additional information about the entities from the text and enrich the sentence structure with extra features.

3   The Architecture of the Question Generation System

The architecture of the proposed question generation system is based on the HIIS (Hybrid Intelligent Information System) approach [10]. The generalized HIIS architecture includes the following components: the environment; the subconsciousness module (MS); the consciousness module (MC); the boundary model of consciousness and subconsciousness.

The MS is related to the environment in which a HIIS operates. Because the environment can be represented as a set of continuous signals, the data processing techniques of the MS are mostly based on neural networks, fuzzy logic, and combined neuro-fuzzy methods.

The MC is based on conventional data and knowledge processing, which may be based on traditional programming, workflow, or rule-based approaches.

The boundary model of consciousness and subconsciousness is intended for deep integration of modules of consciousness and subconsciousness and represents an interface between these modules with the function of data storage. The data is a complex ontology that is used by both the consciousness and subconsciousness modules. The main task of the subconsciousness module is to recognize elements of ontology from the environment. If we consider the consciousness module as a kind of expert system, then the recognized elements of the ontology can be considered as elements of the operating memory of the expert system that trigger the corresponding rules. Depending on the goals of the system, rules can generate output information for the user or signals for the subconscious module that have the desired effect on the environment.

The proposed HIIS concept is considered as a generalized approach that should be adapted to create information systems in specific subject areas. The architecture of the proposed system based on the HIIS concept is presented in Fig. 1.

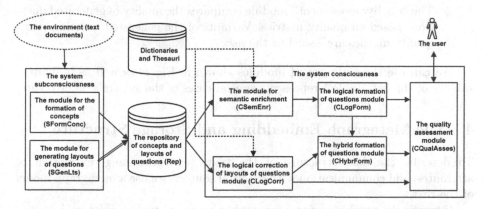

Fig. 1. The architecture of the proposed question generation system.

- The environment in the proposed system is text documents used to generate questions.
- The subconsciousness of the system includes a "module for the formation of concepts" and a "module for generating layouts of questions:"
  1. The "module for the formation of concepts" implements the selection of concepts and the relationships between them from the source text.
  2. The "module for generating layouts of questions" uses machine learning techniques for question generation. This module can simultaneously call several methods for generating question layouts and create several question variants. Therefore, this module is connected to the storage using a double arrow.

– "The repository of concepts and layouts of questions" acts as a boundary model of consciousness and subconsciousness. On the one hand, this storage contains concepts selected from the text and links between them. On the other hand, generated question layouts are placed in this storage. The storage is implemented on the basis of a metagraph data model, which allows semantic enrichment of concepts, as well as linking concepts with fragments of generated question layouts.

– The consciousness of the system includes the described below modules:

  1. The "module for semantic enrichment" is used to expand concepts useful for the formation of questions with information based on dictionaries and thesauri.
  2. The "logical formation of questions module" is used to form questions based on enriched concepts.
  3. The "logical correction of layouts of questions" module is used to correct the generated question layouts based on dictionaries and thesauri: combating typos, eliminating word inconsistencies in the text.
  4. The "hybrid formation of questions" module is used to generate questions both on the basis of enriched concepts and on the basis of question layouts.
  5. The "quality assessment" module compares the quality of generated questions based on quality metrics. Variants of the generated questions with quality metrics are issued to the user.

We can use the entire set of modules shown in Fig. 1, as well as individual subsets of this set. Table 1 represents special cases of the system's architecture.

4   The Metagraph Embedding and Storage Structure

To describe the data structure of the text, text processing, extraction of attributes, and communication with external sources, we use a metagraph model of the text.

Increasingly, graph structures are used to represent textual information since they allow us to reflect the dependencies between parts of the text, make links to external data, and therefore improve the quality of the entire algorithm. However, the description of the structure of text and knowledge in the general case in the form of a graph can be limited, since it does not allow natural hierarchies of entities to be constructed. For such purposes, we use a metagraph model for text representation.

A metagraph model was proposed in [11] and adapted for use in intelligent information systems in [10]: The metagraph MG consists of metavertices MV that contain vertices V that store data and update attributes during processing:

$$MG = \langle MV, ME, V, E \rangle$$

$$MV_i = \langle V \rangle, MV_i \in MV, V_i = \{attr_i\},$$

$$e_i = \langle v_s, v_e, eo, \{atr_k\} \rangle, e_i \in E, eo = true|false,$$

Table 1. The special cases of the architecture of the system.

No.	Modules	Description
I	SFormConc-CLogForm	The classical architecture of concept-based question generation includes a chain of two modules: "The module for the formation of concepts" and "The logical formation of questions module." In this case, concepts are extracted from the text, and questions are generated based on rules. The main disadvantage of this option is the great complexity of developing a system of rules.
II	SGenLts-CQualAsses	The classical architecture of generating questions based on machine learning methods also includes a chain of two modules: "The module for generating layouts of questions" and "The quality assessment module." In this case, machine learning methods are used to generate questions, and then a quality assessment is performed. The main disadvantages of this option are inherent in almost all machine learning methods: lack of explanation (the model cannot explain why it generated such a question), the inability to correct the algorithm for generating questions using rules.
III	SFormConc-Rep-CSemEnr-CLogForm	In this case, semantic enrichment is applied to the recognized concepts. The generated question can include synonyms, hyponyms, or hyperonyms (based on a particular example, a question related to a more general concept is formed). This option retains the disadvantages of the option I.
IV	SGenLts-Rep-CLogCorr-CQualAsses	In this case, the logical models using the rules and untrained algorithms based on dictionaries and thesauri are applied to the original question layouts (which were generated in the subconsciousness of the system using machine learning methods). For example, to combat typos, an algorithm based on the Levenshtein distance can be used, and rules can be used to eliminate word inconsistencies in the text.
V	All modules	A complete workflow, including all modules shown in Fig. 1. In this case, the hybrid question-building module uses the results of the work of the logical-question-building and logical-correction modules. A unified metagraph data model facilitates data integration when implementing a hybrid approach. The quality assessment module identifies the best question model based on the data generated by previous modules

$$me_i = \langle v_s, v_e, attr_k, MG_j \rangle, me_i \in ME,$$

where MG - data metagraph, v_s, v_e are vertices connected by the edge, $\{attr_k\}$ is the set of attributes, MG_j is a fragment of the metagraph, that is embedded in the metaedge, e_i – edge of the metagraph, v_s and v_e – vertices, connected by the edge, eo – represents the direction of the edge.

A metagraph has an important feature that allows us to combine many vertices in a hierarchy by including them in one metavertex. Simultaneously, the same vertex can be included in several metavertices, which can determine different semantic properties of such inclusion. The same applies to the metagraph edges – they can connect vertices belonging to different "hierarchy levels" –

metavertices. A common text processing pipeline usually contains the following steps:

- Segmentation – definition of "levels" of text (paragraphs, sentences, words).
- Preprocessing – clearing text from unnecessary data, such as stop words.
- Morphological analysis – definition of morphological attributes of words.
- Syntax parsing – defining the syntactic structure of sentences.
- Semantic analysis, knowledge extraction, text enrichment, etc. – a processing stage that is specific to the application.

While processing the text, we use the following formal description of the metagraph model that meets the basic requirements for the presentation of data at each of the steps described above:

$$MG_{text} = \langle MV_{base}, L \rangle,$$

$$MV_{base} = \langle MV_{inner}, MV_{leaf} \rangle, MV_{leaf} \supset MV_{inner},$$

$$L = \{paragraph, sentence, word, ...\},$$

where L – given set of levels, into which the text is divided, MG_{text} is a metagraph of text, MV_{base} is a top-level metavertex, MV_{inner} – metavertices included in the text on the lower level, MV_{leaf} – low-level metavertices. This structure describes the division of the text into its component parts. To carry out morphological analysis and save its results, we use the following leaft metavertex expansion:

$$v_j = \langle l, text, processed, mv_{morph} \rangle, v_j \in MV_{leaf},$$

$$l \in L, \{text, processed\} \in \{atr_k\},$$

$$mv_{morph} = \langle pos, g, n, lemma, stem, ..., morph \rangle,$$

$$pos, g, n, lemma, stem, ...morph \in \{atr_k^{mv_{morph}}\}$$

where mv_{morph} is a meta-vertex of morphological characteristics, pos - part of speech, g - gender, n - number, $lemma, stem, ...morph$ – any other morphological attributes, $\{atr_k\}$ – set of meta-vertex attributes, $\{atr_k^{mv_{morph}}\}$ – set of metavertex attributes of morphological characteristics, $text$ – text inside of the selected vertex, $processed$ – pre-processed text.

To preserve the syntactic structure of sentences, the following model is used:

$$v_i = \langle l, MV_{inner}^i, MV_{leaf}^i, Syntax \rangle, v_i \in MV_{inner}, l \in L,$$

$$v_k = \langle l, Syntax \rangle, v_k \in MV_{inner}, l \in L, if\ l = sentence,$$

$$Syntax = \langle MV_{syntax}, E_{dep} \rangle,$$

$$v_{si} = \langle tag, atr_k^{vs} \rangle, v_{si} \in MV_{syntax}, tag \in \{NP, VP, PP, ...\},$$

$$e_{dep_i} = \langle mv_j, mv_k, \{atr_p^{edep}\} \rangle, e_{dep_i} \in E_{dep}$$

where *Syntax* is a fragment of a metagraph, that displays parsing tree, *tag* – pos-tag of a word, atr_k^{vs} – set of the syntactic attributes of a word, atr_p^{edep} – set of syntactic link attributes.

This model stores the entire structure of the text in a single format, allowing us to use data at any stage of text processing. For example, the following different type of sentence parsing structures can be stored in a single metagraph model without changing its original structure, as represented in Fig. 2.

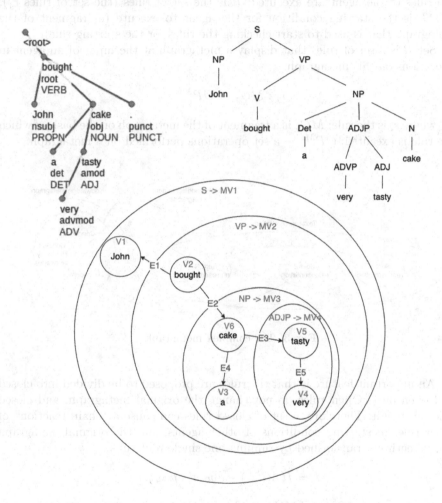

Fig. 2. The metagraph representation of the syntax tree.

Also, the metagraph allows us to preserve more complex hierarchies and relationships between heterogeneous data, as shown in Fig. 3

To form the model, we use the agent approach described in [12]. Since the text in the model is represented by a metagraph, any changes, additions, and calculations on this metagraph (and therefore throughout the text/texts corpus) can be the result of executing the metagraph agents $AG = \{ag_i\}$, where AG is the set of agents; ag_i is an agent.

One type of agent is a metagraph agent:

$$ag^M = \langle MG_D, R, AG^{ST} \rangle, R = \{r_j\},$$

where ag^M is a metagraph agent, MG_D – metagraph, on the basis of which the rules of the agent are executed; R is the set of rules (the set of rules r_j), $AG^S T$ is the starting condition for the agent to execute (a fragment of the metagraph that is used to start checking the rules, or the starting rule).

Set R is a set of rules that display a metagraph at the input of an agent in operations on this metagraph:

$$r_i : MG_j \rightarrow OP^{MG},$$

where r_i is the rule; MG_j is a fragment of the metagraph on the basis of which the rule is executed; OP^{MG} – a set operations performed on a metagraph.

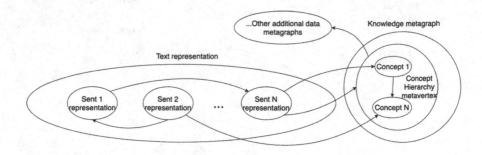

Fig. 3. The complex metagraph.

An important feature is that the rules are proposed to be divided into closed and open ones. Open rules do not change the original metagraph, and closed rules change it. Therefore, a set of closed rules can cause a "chain reaction" of other rules or starting conditions of other agents, and the original metagraph can be entirely transformed by running one single root rule.

$$T = f(AG^{ST}) - agent's\ trigger,$$

$$AG_p^{ST} - \{MV_i, atr_{ik}\} - tracked\ parameters.$$

Agent trigger – the process of checking the start condition of an agent. Represents as a function that accepts a metagraph for rules and returns an agent launch flag.

Agent tracked parameters – a set of pointers to metavertices and their attributes, which serve as input parameters for an agent trigger. They can be used for more granular configuration of agent launch and optimizations.

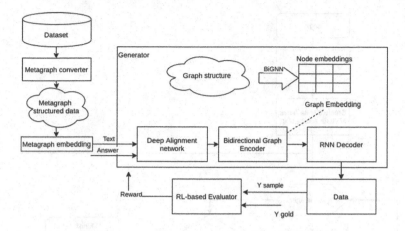

Fig. 4. The algorithm for the text metagraph model generation.

According to this definition, we can describe a general text processing algorithm for a metagraph model generation represented in Fig. 4.

Such an algorithm for the metagraph model generation makes it possible to use a declarative description of the text processing process, which provides several advantages, for example, the ability to conveniently parallelize agent execution.

The metagraph model of text representation allows us to save data and text attributes, store heterogeneous text structures, hierarchies of word dependencies, enrich the text using links to external knowledge graphs, also represented as fragments of a metagraph. This allows us to significantly expand the set of source data for use in various data processing systems and machine learning tasks.

After the metagraph model generation, the metagraph must be decomposed into a multipartite graph, for example, using the method described in [13]. Next, a flat graph is embedded in a vector representation and is used in the target model.

5   Evaluation

For evaluation, the model from [6] was used, with an additional metagraph module from the section above, that encoded input data into the metagraph form before training. Afterward, the resulting metagraph was embedded back to fit into the neural network. It was chosen to remove the BERT encoder from the training model due to memory consumption and a sufficient training time increase. However, we decided to keep the Deep Alignment network step.

The structure of the neural network system is shown in Fig. 5. There the metagraph converter is the realization of "The module for semantic enrichment," and the generator is the realization of "The quality assessment module" from

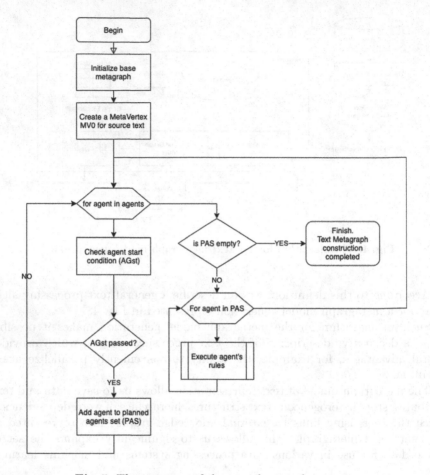

Fig. 5. The structure of the neural network system.

Fig. 1. In Table 2 the comparison between the proposed model with metagraph encoding and several other models on the same task is shown. The results of other models were taken from the [6] paper.

The model was fine-tuned using a grid search for the RL-ratio and the dropout parameter. The random seed kept constant, but due to the duration of the training not all the models with different parameters were trained to the end. The decision, whether to continue training or not, was made manually depending on what was the BLEU-4 parameter on the 15–20 epoch.

If there was no option to train the model and measure train or preprocessing time, the time is marked with "–". Python 3 and Pytorch framework were used for the training and evaluation. Spacy was used for POS and NER tagging; additional tagging was made with Python 3. All trained models were trained on NVIDIA GeForce GTX 1050 graphic card with CUDA 10.2 using the SQuAD dataset split-2 from the paper [9]. It is based on the manually tagged questions and answers of the Wikipedia articles. In this research, the questions and answers are encoded into the syntax trees with additional POS-tagging and custom NER tags with the metavertex structure.

Table 2. Model evaluation results.

	Train BLEU-4	Test BLEU-4	Train time	Data pre-processing time
$G2S_{sta} + BERT + RL$	–	17.94	–	–
Proposed metagraph model	15.32	16.86	11 h 16 m 41 s	13 m 30 s
$G2S_{sta}w/oBiGGNN, w/Seq2Seq$	15.34	16.72	12 h 54 m 50 s	0
$G2S_{sta}w/GGNN - forward$	–	16.53	–	–

As shown in Table 2, the model trained on the data from the proposed metagraph model provides slightly better results on the test evaluation, but the advantage can be considered modest, and the model needs further enhancement. The advantage of the provided model is that it learns faster, even if encountering the additional data preprocessing time, taken by the conversion to the metagraph, and then to the graph embedding.

In order to check whether the metagraph approach can be used for the task of question generation, an additional survey was made. (https://forms.gle/wBESKFcrtsD3Lkoa8) It consisted of ten texts with two questions (one question for each trained model from 2), that were selected using the random generator from the generation results, and third variant "None/Both." The survey was taken by 22 respondents, who marked their English knowledge as level B1 and higher. This survey showed that the respondents preferred six questions out of ten generated by the proposed metagraph model. If the choice of 1 generated question is equivalent to 1 point, our model has 84 points out of 155. The advantage is small, yet it encourages further model improvements.

Several samples, showing particular qualities of generated questions from the trained models, are provided in Table 3. Both models that were trained end-to-end in this research act equally unsatisfying for the long and complex questions, perhaps due to the exposure bias, which is a common problem for seq2seq models. However, many of the examples are human-readable. The metagraph connections between nodes allow the proposed model to learn better, how words, that are not evidently connected in the sentence, but connected through metaedges, behave, that eventually allows to generate questions with the words, better connected with the implicit context of the sentence. The specific context that could be critical for the right generation is sometimes missed or overlapped by features that are more significant for the model.

Table 3. Examples of questions generated by models.

Passage: there are 5 polytechnics in singapore.

Gold: how many polytechnics are there in singapore?

Proposed metagraph model: how many polytechnics are there in singapore?

G2S$_{sta}$w/oBiGGNN, w/Seq2Seq: how many polytechnics are in singapore?

Passage: the hudson river separates the city from the u.s. state of new jersey

Gold: the hudson river serves as a dividing line between new york and what state?

Proposed metagraph model: the hudson separates the city from what state?

G2S$_{sta}$w/oBiGGNN, w/Seq2Seq: from what state is the hudson river?

Passage: as of november 2008, 67 % of registered voters in the city are democrats

Gold: in november 2008, how many new yorkers were registered as democrats?

Proposed metagraph model: as of november 2008, what percentage of registered voters are democrats?

G2S$_{sta}$w/oBiGGNN, w/Seq2Seq: in november 2008, what percentage of the city are democrats in the city?

---

6   Conclusions

The task of automatic generation of questions on the text can be used to automate checking the quality of understanding of the text read by students, as well as in chatbot systems.

There are two main approaches to the generation of questions in the text: the logical approach and the approach based on machine learning. Each approach has its own advantages and disadvantages.

The proposed approach based on the HIIS allows experimenting with different versions of the architecture of the system for generating questions based on texts. These can be variants implemented both on the rules and on the basis of machine learning methods, as well as a hybrid variant that includes both approaches. In this paper, the ability to use the metagraph approach for the task of question generation was proved.

The subject of further research consists of broader usage of metagraph features, optimizing the preprocessing part, and constructing a metagraph framework for a more general task of natural language processing.

References

1. Walsh, J.A., Sattes, B.D.: Quality questioning: Research-based practice to engage every learner. Corwin Press (2016)
2. Mitkov, R.,Ha, L.: Computer-aided generation of multiple-choice tests. In: Proceedings of the HLT-NAACL 03 Workshop on Building Educational Applications using Natural Language Processing - Volume 2 (HLT-NAACL-EDUC '03), pp. 17–22. Association for Computational Linguistics, USA (2003). https://doi.org/10.3115/1118894.1118897

3. Heilman, M., Smith, N.A.: Question generation via overgenerating transformations and ranking. Technical report no. CMU-LTI-09-013. Carnegie-Mellon University, Pittsburgh (2009)
4. Stasaski, K., Hearst, M.A.: Multiple choice question generation utilizing an ontology. In: Proceedings of the 12th Workshop on Innovative Use of NLP for Building Educational Applications, pp. 303–312. Association for Computational Linguistics, Copenhagen, Denmark (2017)
5. Papineni, K., Roukos, S., Ward, T., Zhu W.-J.: Bleu: a method for automatic evaluation of machine translation. In: Proceedings of the 40th Annual Meeting of the Association for Computational Linguistics, pp. 311–318. Association for Computational Linguistics, Philadelphia, Pennsylvania, USA (2002)
6. Chen, Y., Wu, L., Zaki M.J.: Reinforcement learning based graph-to-sequence model for natural question generation. arXiv preprint arXiv:1908.04942 (2019)
7. Scarselli, F., Gori, M., Tsoi, A.C., Hagenbuchner, M., Monfardini, G.: The graph neural network model. IEEE Trans. Neural Networks **20**(1), 61–80 (2009)
8. Menezes, T., Roth, C.: Semantic Hypergraphs. arXiv preprint arXiv:1908.10784 (2019)
9. Zhou, Q., Yang, N., Wei, F., Tan, C., Bao, H., Zhou, M.: Neural question generation from text: a preliminary study. In: Huang, X., Jiang, J., Zhao, D., Feng, Y., Hong, Y. (eds.) Natural Language Processing and Chinese Computing, NLPCC 2017. LNCS, vol. 10619, pp. 662–671. Springer, Cham (2018). https://doi.org/10.1007/978-3-319-73618-1_56
10. Chernenkiy, V., Gapanyuk, Yu., Terekhov, V., Revunkov, G., Kaganov, Y.: The hybrid intelligent information system approach as the basis for cognitive architecture. Procedia Comput. Sci. **145**, 143–152 (2018)
11. Basu, A., Blanning, R.W.: Metagraphs and their Applications. Springer, Boston (2007)
12. Chernenkiy, V.M., Gapanyuk, Yu.E., Nardid, A.N., Gushcha, A.V., Fedorenko, Yu.S.: The hybrid multidimensional-ontological data model based on metagraph approach. In: Petrenko, A., Voronkov, A. (eds.) Perspectives of System Informatics, PSI 2017. LNCS, vol. 10742, pp. 72–87. Springer, Cham (2018) https://doi.org/10.1007/978-3-319-74313-4_6
13. Chernenkiy, V.M., Gapanyuk, Yu.E., Kaganov, Yu.T., Dunin, I.V., Lyaskovsky, M.A., Larionov, V.S.: Storing metagraph model in relational, document-oriented, and graph databases. In: Selected Papers of the XX International Conference on Data Analytics and Management in Data Intensive Domains (DAMDID/RCDL 2018), pp. 82–89. Moscow, Russia (2018)

Abstractive Summarization of Russian News Learning on Quality Media

Daniil Chernyshev[(✉)] and Boris Dobrov

Lomonosov Moscow State University, Moscow, Russia

Abstract. Summarization is becoming a demanded task in the modern world of ever-increasing document flow. This task allows to compress existing text while maintaining all salient information. However, building a neural summarization model requires training data which is scarce in some languages. In this work, we consider the problem of abstract summarization of news texts in Russian. We propose a new method for obtaining training data that uses the news leads of high-quality media that publishes news in accordance with the classical model. We prove dataset eligibility for training by building an abstractive summarization framework based on pre-trained language models and comparing summarization results with extractive baselines.

Keywords: Abstractive summarization · News summarization · BERT

1 Introduction

Summarization is representing the meaning of the analyzed text in the form of a short abstract. It is one of the most popular tasks in the modern world of ever-increasing document flow. Extractive summary is formed from fragments of the analyzed text. For abstractive summary, words and phrases that were not in the source text can be used. Both extractive and abstractive automatic summarization approaches pose serious challenges, however, only recently the former started receiving solutions with quality comparable to a human-written summary [14]. State-of-the-art architectures employ pre-trained language models which increase comprehension power and process text without complex preprocessing.

In this paper, we propose an algorithm for automatic synthesizing of a dataset for news article summarization. According to the inverted pyramid model[1], the beginning of the "correct" news text ("lead") should reflect the main content of the news. We consider the first paragraph of news published in high-quality media as a source of abstractive summary for other media. We demonstrate the eligibility of the resulting dataset for abstractive summarization experiments by comparing with existing counterparts and building an encoder-decoder framework similar to the state-of-the-art summarization model [9] that outperforms the baseline and produce coherent paraphrases.

---

[1] https://en.wikipedia.org/wiki/Inverted_pyramid_(journalism).

© Springer Nature Switzerland AG 2021
W. M. P. van der Aalst et al. (Eds.): AIST 2020, LNCS 12602, pp. 96–104, 2021.
https://doi.org/10.1007/978-3-030-72610-2_7

2 Related Work

The emergence of pre-trained language models has opened new ways of improving performance in various natural language processing tasks. By pre-training on tasks dedicated to learning contextual representations, these models extend their comprehension power. For instance, commonly used Bidirectional Encoder Representations from Transformers (BERT) [3] is pre-trained with a masked language modeling and a "next sentence prediction" task.

Abstractive summarization task can be formulated as the extraction of salient concepts from the text and then organizing them into a coherent summary. Among the first successful works in neural abstractive summarization were the work of Rush et al. [15] who employed sequence to sequence model. Their approach was later augmented with recurrent decoders [2]. However, the method was suited only for the problem of generating news article headings. Nallapati et al. [11] extended the model to a multi-sentence summary construction task by adapting CNN/Daily Mail dataset and introducing a hierarchical network. See et al. [14] proposed a pointer generator network to deal with the out-of-vocabulary issue, and coverage mechanism to reduce token repetition. Paulus et al. [13] applied a deep reinforcement learning model to the task and designed a special algorithm to alleviate text degeneration of long summaries. Celikymaz et al. [1] enhanced the original pointer generator network approach with multiple deep communicating agents encoder and decoder with a hierarchical attention mechanism. Gehrmann et al. [4] adopted a bottom-up attention approach to improve the detection of salient tokens. Narayan et al. [12] proposed a new model for generating extremely compressed summaries, based on convolutional neural networks and additionally conditioned on topic distributions. Liu et al. [9] adapted BERT encoder for summarization task and proposed a method for transferring extractive summarization experience to the abstractive summarization model. Zhang et al. [18] designed a two-stage transformer-based decoder that refines the resulting summary with BERT's language knowledge.

3 Dataset

For English news summarization task several datasets were proposed. CNN/Daily Mail [7] dataset was originally proposed for question answering task and contains articles and associated facts. New York Times dataset shares a similar structure however according to Narayan et al. [12], it has less bias towards extractive methods. The main problem with these datasets is that fact set requires additional processing to construct a coherent summary. Xsum [12] has the most abstractive summaries but is intended for extreme summarization which is limited only to one sentence and thus may not contain all important information.

The largest news dataset published to date, Newsroom [5], contains 1.3 million articles with human-written summaries. The dataset was obtained by content scraping of article pages from selected publishers and choosing only articles with summaries with low text overlapping.

To our best knowledge, there are only two published Russian headline generation datasets Lenta.Ru-News[2] and Rossiya Segodnya[3] which does not contain multi-sentence summaries. Gusev [6] published a Russian dataset similar to Xsum but due to lower document quantity it requires extension for more stable model training. It would be convenient to translate the Newsroom dataset using machine translation methods, however, this may corrupt the contents by introducing translation artifacts. Instead, we ease the task by overlooking the extraction of all salient facts and proceeding just to paraphrasing with compression. Inspired by Grusky et al. [5] approach, we design a method to automatically construct a dataset for such task from raw internet resources.

We expanded Lenta.Ru-News dataset by collecting articles from "Коммерсантъ"[4] dating from 2016 to 2019. These publishers do not provide a summary as a separate text, but often incorporate a semblance in the main body as the first paragraph. A human-written summary may be found in metadata, however, for most articles, it takes the form of concatenated text parts of the first paragraph. Simple metadata extraction and filtering high-quality abstractive summaries would yield only a small fraction of data, insufficient for complex model training. To achieve universality and quantity we utilize the first paragraph as pseudo-summary. This approach will require excluding the first paragraph from the source text to prevent data leakage. However, there is no guarantee that the information presented in the excluded part will be reflected in the rest of the text.

To tackle this issue, we exploit the fact that publishers may cover the same story. By finding related articles we can construct pseudo-summary-source pairs by taking the first paragraph from one source and full text from another.

We use "Lenta.ru"[5] articles as a source since this publisher tends to cover most of the stories and usually presents the material earlier than others. "Коммерсантъ" is not a random choice for a target summaries either; this publisher has a special emphasis on business news, so it is expected to provide articles with more analytical background and, thus, more informative sentences. The difference in nature between these two publishers ensures the abstractiveness of paraphrase and strict role assignment provides a consistent style difference which allows to define the task as style transfer with compression.

Pairing is achieved with the following algorithm. For each possible article pair with the same publishing date we calculate TF-IDF cosine similarity. Then for each article-source (text source) of publisher B list top k articles-candidates (summary source) from publisher A according to similarity score (it is assumed that $|A| \leq |B|$). For these candidates in the top list we calculate context similarity by calculating BERT embeddings for each sentence in articles and then building cosine similarity matrix M between sentences of candidate and source.

---

[2] https://github.com/yutkin/Lenta.Ru-News-Dataset.

[3] https://github.com/RossiyaSegodnya/ria_news_dataset.

[4] https://www.kommersant.ru/.

[5] https://lenta.ru/.

Table 1. Dataset statistics.

	Russian news	CNN	Daily mail	NY times
Mean article length (words)	220.0	760.50	653.33	800.04
Mean summary length (words)	52.6	45.70	54.65	45.54
Lead-3 ROUGE-1	34.69	29.15	40.68	31.85
Lead-3 ROUGE-2	12.21	11.13	18.36	15.86
Lead-3 ROUGE-L	27.69	25.95	37.25	23.75

Sentence embeddings are obtained by average pooling second-to-last hidden layer of BERT of all of the tokens in the sentence. The context similarity between candidate c and target s is defined as the minimum of column-wise and row-wise maximums of matrix M:

$$\text{context}(c, s) = \min\{\max_i M_i;\ \max_i M_i^T\} \tag{1}$$

Finally, we assign candidate articles to the source which maximizes overall context similarity:

$$\sum_{\substack{i \in |A| \\ j \in |B|}} \text{context}(c_i, s_j) \to \max \tag{2}$$

$$\text{pair}(c_i) = s_i;\quad pair(c_i) \neq pair(c_j),\ i \neq j$$

The convergence of the algorithm depends on the selected number of top articles k. In experiments, we found that good estimation is

$$k = \max\{|A|, |B|\} \cdot 0.15 \tag{3}$$

However, the resulting set of article pairs requires additional refining as it will contain low similarity (or exact) pairs. Low quality pairs are the result of duplicate articles within source set (extensions of previous articles) or lack of event coverage in candidate set. To ensure text-summary connection (possibility of summary extraction) we use ROUGE-N (percentage of overlapping N-grams) and ROUGE-L (share of longest common subsequence). Low similarity pairs are filtered out by simply setting threshold $t_1 = 0.25$ for ROUGE-1 score (F1 measure) between paraphrase and source. And to remove pairs with high text overlapping we set a limit $t_2 = 0.35$ for ROUGE-L.

Our news dataset consists of 26555 article-paraphrase pairs. The data is divided into training (80%), development (10%) and test (10%). Statistics are represented in Table 1. Interestingly, despite the same origin of article and paraphrase (both are parts of main article body), the dataset Lead-3 ROUGE scores are comparable to abstractive summarization counterparts with exclusive human-written summaries. Since summaries tend to repeat some parts of source article this demonstrates acceptable level of data leakage as well as indicates the similarity of tasks.

4   Model

We use a standard encoder-decoder framework for abstractive summarization [14]. The encoder captures salient text information and encodes it in vector form. Many approaches employ separate attention mechanism to determine word (token) importance, however, modern Transformer-based [16] language models such as BERT [3] incorporate it in the encoding mechanism. In addition, BERT is specifically pre-trained for text comprehension which makes it a preferable choice for encoding of salient information.

The decoder takes encoded vector and attempts to decode it in "the right way". "The right way" depends on problem formulation or, to be more precise, target structure. If the problem is formulated as the extraction of salient sentences, then the decoder tries to reconstruct these sentences by extracting information from input. And if it is style transfer, the decoder builds a new version of the text with respect to salient word distribution. Text compression is not a separate task but in fact, a simplification that removes the minimum length constraint and allows the decoder to produce sentences of any length. In training phase decoder learns to produce position-vise token distribution so length constraint is imposed by target sentence length distribution. The coherence of produced sentences is also determined by target token distribution as it means the dependency of current position token distribution from previous. And paraphrasing is a difference between source and target global token distribution.

As for encoder, we use BERT and decoder consists of 8-layer Transformer. There are other decoder architectures that have proved to be efficient, however, our goal is to demonstrate the similarity of style transfer with compression and abstractive summarization tasks, so we aim for simplicity.

The BERT is prepared in accordance with Liu et al. [9] approach. However, we do not pre-train BERT on extractive summarization task. Taking into account the nature of the training dataset it is evident that pre-training on extractive summarization would lead to favoring sentences at the beginning of the source, lowering the comprehension power of the abstractive summarization model. Alternatively, we use a special version of RuBERT [8] pre-trained on Russian news article language[6].

Since the encoder is pre-trained and the decoder must be trained from scratch the training could become unstable. For example, the encoder might overfit the data before the decoder finished the fine-tuning, or vice-versa. To alleviate this issue, we adopt a scheduled learning rate mechanism [16].

Both encoder and decoder use Adam optimizer with $\beta_1 = 0.9$ and $\beta_2 = 0.999$, but the learning rate is set according to the formula:

$$lr = \hat{lr} \cdot \min\{step^{-0.5}, step \cdot \text{warmup}^{-1.5}\} \tag{4}$$

where $\hat{lr} = 2e^{-3}$ and warmup = 40000 for encoder, $\hat{lr} = 2e^{-1}$ and warmup = 20000 for decoder. This way by the end of the encoder warmup stage the decoder will accumulate enough gradients to become stable.

---

[6] https://github.com/dciresearch/RuBERT-News.

Table 2. ROUGE F1 results on test set of Russian news dataset.

Model	ROUGE-1	ROUGE-2	ROUGE-L
Baseline			
Oracle	47.92	29.62	44.99
Lead-3	33.28	13.01	27.30
Pointer generator	31.63	13.82	31.62
TextRank	27.95	10.95	27.16
Ours			
RuBERTAbs (Standard)	30.18	12.45	27.17
RuBERTAbs (News)	36.52	15.80	33.59

5 Implementation

For model implementation we used PyTorch. We applied dropout with probability 0.1 and label smoothing with smoothing factor 0.1. The Transformer decoder has 768 hidden units and the hidden size for all feed-forward layers is 2048. The models were trained for 150000 steps on a single Tesla P100. The choice of training steps was determined by model performance on validation set, where it was concluded that further training resulted in text degeneration and overfitting.

For paraphrase decoding we used beam search with beam size 5 and length normalization [17] with $\alpha = 0.7$. To avoid token repetition, we adopt a trigram blocking mechanism [13]. In consequence to BERT's WordPiece tokenizer we do not need any copy mechanism to deal with out-of-vocabulary words as it gives the ability to generate substitution according to context.

6 Results

For dataset we evaluated unigram and bigram overlap (ROUGE-1 and ROUGE-2) and longest common subsequence (ROUGE-L) without stemming. These metrics were used in all previous works and proved to be informative in terms of similarity to a human-written summary.

We show our results in Table 2. The first section of the table covers the baseline methods: TextRank [10], Pointer generator [14], Lead-3 and Oracle. Lead-3 is just the first 3 sentences of the source text. The Oracle is the best possible result that could be obtained with extractive summarization methods. To construct an oracle summary, we use a greedy algorithm to select a set of source sentences that maximizes the ROUGE-2 score for the target summary. This set could be considered as an upper bound for any summarization task.

The second section demonstrates the results of two variants of our model: RuBERT based abstractive summarization model (RuBERTAbs) with standard RuBERT (Standard) and a special version for news language (News). As it can be seen, the standard variant falls short of even Lead-3 baseline while the variant pre-trained on news articles non-Oracle baselines in terms of all ROUGE scores. That suggests that pretraining BERT encoder on tasks with the same language (set of possible words) as the main one is an efficient way to boost the performance of the full model. The difference between Lead-3 and RuBERT (News) may be considered not substantial, however, the results of the best models on CNN/Daily Mail dataset show a similar margin [9].

Table 3. ROUGE-1 F1 comparison Prediction with Gold and Lead-3

Gold vs. Lead-3	Prediction vs. Lead-3	Prediction vs. Gold
33.28	40.96	36.52

We investigated some properties of the resulting solution (Prediction) compared with the leads of the analyzed news: Lead-3 (news lead of Lenta.ru) and Gold (news lead of "Коммерсантъ"). It turns out to be closer to each of the Lead-3 and Gold than they are among themselves (Table 3).

Table 4. ROUGE-1 Recall results on test set of Russian news dataset.

Model	Lead-3	Gold	Prediction
Lead-3 ∧ Gold	37.08	30.18	60.13

Prediction is much closer to the intersection of Lead-3 and Gold than each of the news leads (Table 4). This indirectly indicates the automatic selection in the abstractive summary more important details of news content and ignoring the secondary. Table 5 provides a translated example of the resulting annotation.

7   Conclusion

In this paper, we demonstrated the application of pre-trained language models for abstractive summarization of Russian news. We proposed a method for building a learning dataset for the task and developed a fine-tuning process for proper training. The experimental results across the dataset show that the model outperforms simple extractive and abstractive baselines and indicate the importance of BERT pretraining on task's language. While we were preparing this article, we discovered a parallel work [6] that also considered the problem of abstractive summarization and proposed a new dataset with human-written summaries for Russian language. In further work, we plan to use this data to improve our approach and adopt methods with additional conditions for a summary generation.

Acknowledgements. We thank Mikhail Tikhomirov (Research Computing Center of Lomonosov Moscow State University) for providing pre-trained models for our research.

9   Appendix

Table 5. Translated example of output summaries on Russian news dataset.

Original text				
British boxer Tyson Fury announced his return to the professional ring. The athlete announced this on his Twitter account. "Breaking news! The big comeback will take place on May 13th. We are working on finding an opponent. Follow the news," Fury wrote in his microblog. On December 3, Fury announced his return to boxing after recovering from drug addiction...				
Model	Summary	R1	R2	RL
Lead-3	British boxer Tyson Fury announced his return to the professional ring. The athlete announced this on his Twitter account. "Breaking news!	43.47	24.99	43.47
Pointer generator	British boxer Tyson Fury has negotiated with the footballer's rival WBO and WBA World Pion, Fury said. We are working on finding an opponent on Twitter	51.06	32.65	55.31
RuBERTAbs (News)	British boxer Tyson Fury announced his return to the United States. "We want to fight on May 13," he tweeted	55.55	37.29	51.85
Oracle	British boxer Tyson Fury announced his return to the professional ring. The athlete announced this on his Twitter account. The big comeback will take place on May 13th. We are working on finding an opponent. Follow the news, "Fury wrote in his microblog.	65.51	39.39	65.51
Gold	British boxer Tyson Fury said he will return to the ring on May 13. "We are working on finding an opponent," he wrote on Twitter			

References

1. Celikyilmaz, A., Bosselut, A., He, X., Choi, Y.: Deep communicating agents for abstractive summarization. In: Proceedings of the 2018 Conference of the North American Chapter of the Association for Computational Linguistics: Human Language Technologies, Volume 1 (Long Papers), pp. 1662–1675 (2018). https://doi.org/10.18653/v1/N18-1150

2. Chopra, S., Auli, M., Rush, A.M.: Abstractive sentence summarization with attentive recurrent neural networks. In: Proceedings of the 2016 Conference of the North American Chapter of the Association for Computational Linguistics: Human Language Technologies, pp. 93–98 (2016). https://doi.org/10.18653/v1/N16-1012

3. Devlin, J., Chang, M.W., Lee, K., Toutanova, K.: BERT: Pre-training of deep bidirectional transformers for language understanding. In: Proceedings of the 2019 Conference of the North American Chapter of the Association for Computational Linguistics: Human Language Technologies, Volume 1 (Long and Short Papers), pp. 4171–4186 (2019). https://doi.org/10.18653/v1/N19-1423

4. Gehrmann, S., Deng, Y., Rush, A.: Bottom-up abstractive summarization. In: Proceedings of the 2018 Conference on Empirical Methods in Natural Language Processing, pp. 4098–4109 (2018). https://doi.org/10.18653/v1/D18-1443

5. Grusky, M., Naaman, M., Artzi, Y.: Newsroom: A dataset of 1.3 million summaries with diverse extractive strategies. In: Proceedings of the 2018 Conference of the North American Chapter of the Association for Computational Linguistics: Human Language Technologies, Volume 1 (Long Papers), pp. 708–719 (2018). https://doi.org/10.18653/v1/N18-1065
6. Gusev, I.: Dataset for automatic summarization of russian news (2020), arXiv:2006.11063
7. Hermann, K., Kovciský, T., Grefenstette, E., Espeholt, L., Kay, W., Suleyman, M., Blunsom, P.: Teaching machines to read and comprehend. In: NIPS'15, p. 14 (2015)
8. Kuratov, Y., Arkhipov, M.: Adaptation of deep bidirectional multilingual transformers for russian language (2019), arXiv:1905.07213
9. Liu, Y., Lapata, M.: Text summarization with pretrained encoders, pp. 3721–3731 (2019). https://doi.org/10.18653/v1/D19-1387
10. Mihalcea, R., Tarau, P.: TextRank: bringing order into text. In: Proceedings of the 2004 Conference on Empirical Methods in Natural Language Processing, pp. 404–411 (2004)
11. Nallapati, R., Zhou, B., dos Santos, C., Gulcehre, Ç., Xiang, B.: Abstractive text summarization using sequence-to-sequence RNNs and beyond. In: Proceedings of The 20th SIGNLL Conference on Computational Natural Language Learning, pp. 280–290 (2016). https://doi.org/10.18653/v1/K16-1028
12. Narayan, S., Cohen, S.B., Lapata, M.: Don't give me the details, just the summary! topic-aware convolutional neural networks for extreme summarization. In: Proceedings of the 2018 Conference on Empirical Methods in Natural Language Processing. pp. 1797–1807 (2018). https://doi.org/10.18653/v1/D18-1206
13. Paulus, R., Xiong, C., Socher, R.: A deep reinforced model for abstractive summarization (2017)
14. See, A., Liu, P., Manning, C.: Get to the point: Summarization with pointer-generator networks, pp. 1073–1083 (2017). https://doi.org/10.18653/v1/P17-1099
15. Sutskever, I., Vinyals, O., Le, Q.: Sequence to sequence learning with neural networks. In: NIPS'14, p. 10 (2014)
16. Vaswani, A., et al.: Attention is all you need. In: NIPS'17 (2017)
17. Wu, Y., et al.: Google's neural machine translation system: Bridging the gap between human and machine translation (2016), arXiv:1609.08144
18. Zhang, H., Cai, J., Xu, J., Wang, J.: Pretraining-based natural language generation for text summarization. In: Proceedings of the 23rd Conference on Computational Natural Language Learning (CoNLL) pp. 789–797 (2019). https://doi.org/10.18653/v1/K19-1074

RST Discourse Parser for Russian: An Experimental Study of Deep Learning Models

Elena Chistova[1](✉), Artem Shelmanov[2], Dina Pisarevskaya[1], Maria Kobozeva[1],
Vadim Isakov[1], Alexander Panchenko[2], Svetlana Toldova[3], and Ivan Smirnov[1]

[1] FRC CSC RAS, Moscow, Russia
{chistova,kobozeva,ivs}@isa.ru
[2] Skolkovo Institute of Science and Technology, Moscow, Russia
{a.shelmanov,a.panchenko}@skoltech.ru
[3] NRU Higher School of Economics, Moscow, Russia
stoldova@hse.ru

Abstract. This work presents the first fully-fledged discourse parser for Russian based on the Rhetorical Structure Theory of Mann and Thompson (1988). For the segmentation, discourse tree construction, and discourse relation classification we employ deep learning models. With the help of multiple word embedding techniques, the new state of the art for discourse segmentation of Russian texts is achieved. We found that the neural classifiers using contextual word representations outperform previously proposed feature-based models for discourse relation classification. By ensembling both methods, we are able to further improve the performance of the discourse relation classification achieving the new state of the art for Russian.

Keywords: Rhetorical structure theory · Discourse parsing · Deep learning · Pre-trained language models

1 Introduction

Many natural language processing applications require understanding of a text structure beyond the individual sentences. It has been shown that methods for machine translation [9], deception detection [24], summarization [11], sentiment analysis [2], and other tasks can take advantage of a text discourse structure that provides relationships between text segments that go beyond sentence boundaries. One of the widely-used underlying linguistic models for discourse is Rhetorical Structure Theory (RST) [18]. According to this theory, a discourse structure of a text can be represented as a discourse tree (DT). DT is a constituency tree, whose leaf nodes called elementary discourse units (EDUs) are textual segments corresponding to simple sentences and clauses, and the intermediate nodes called discourse units (DUs) are the higher-order combinations of the underlying EDUs or DUs marked with discourse relationship types (see Fig. 1). The

© Springer Nature Switzerland AG 2021
W. M. P. van der Aalst et al. (Eds.): AIST 2020, LNCS 12602, pp. 105–119, 2021.
https://doi.org/10.1007/978-3-030-72610-2_8

inventory of discourse relationships usually includes "Cause", "Condition", "Elaboration", "Concession", "Sequence", "Contrast", etc. The practical usefulness of RST discourse parsing was shown in various NLP applications, such as text summarization, automatic essay scoring, and sentiment analysis [2,4,19].

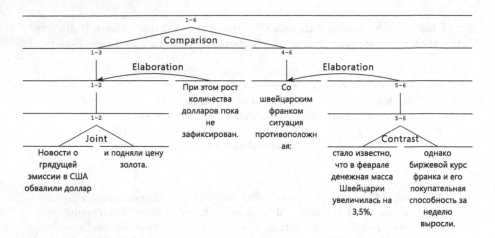

Fig. 1. An example of a discourse tree

There is a long history of research publications on automatic text-level rhetorical parsing for English [8,10,16,28] inter alia. Recently, the Russian-language corpus annotated with rhetorical structures (Ru-RSTreebank) [23] was released, which has unlocked the possibility of research on discourse parsing also for Russian. With the help of this corpus, some attempts to creating a model for discourse relation classification have been performed using feature engineering and classical machine learning techniques [5]. We expand this effort by training deep learning models on this corpus and assembling a fully-fledged pipeline for automatic discourse parsing of Russian-language texts, which includes segmentation, DT construction, and discourse relation classification. The developed parser takes advantage of the available deep pre-trained ELMo [22] language model for Russian. We evaluate the components of the parser and compare the novel relation classification model with the models developed in the previous work. By ensembling deep neural models with feature-based models, we achieve the new state of the art for the Russian language in discourse relation classification and discourse segmentation tasks, as well as establish the first results for the rhetorical tree construction. Summarizing, the contributions of this paper are the following:

- We present the first fully-fledged discourse parser for Russian.
- We establish the first results for rhetorical tree construction for Russian.
- We achieve state-of-the-art results for Russian in discourse segmentation and relation classification.

The rest of the paper is structured as follows. Section 2 reviews the recent related work on discourse parsing for English and Russian. In Sect. 3, we discuss the deep neural models for segmentation, discourse tree construction, and rhetorical relation classification. We also discuss the novel feature-based model for rhetorical relation classification. Results of the experiments are presented in Sect. 4. In Sect. 5, we perform the qualitative analysis of the results. Finally, Sect. 6 concludes and outlines future work.

2   Related Work

RST parsers have been created for several languages, including English [29,32], Chinese [12], Basque [1], Spanish, Portuguese, German, and Dutch [3,21]. The architecture of most of the existing RST-style parsers contains two modules: for discourse segmentation and for DT construction.

The first evaluation of discourse segmentation for Russian is conducted by participants of the DISRPT 2019 Shared Task [33], which was aimed at analysis of discourse unit segmentation and connective detection for 10 languages. The best score for the Russian dataset, as well as for the English RST-DT corpus (Rhetorical Structure Theory Discourse Treebank) [4], is achieved by the Tony segmenter [21] employing BiLSTM models with multilingual BERT embeddings. Recently, it was outperformed by the model presented in [7], which employs contextual embeddings for joint prediction of part-of-speech tags and dependency features for each token in a sentence. This model achieves 96.7% F-score on the English RST-DT corpus improving upon the Tony model by 0.7%.

There are two state-of-the-art methods for text-level discourse parsing for English that take advantage of transition-based bottom-up parsing algorithms [29,32]. In [29], the parsing is performed in two stages. In the first stage, a transition-based parser constructs an unlabeled tree. In the second stage, relations in the tree are labeled with SVM classifiers. The authors train three separate classifiers: sentence-level, paragraph-level, and text-level; each of them uses different feature sets. Yu et al. [32] additionally propose to use as features hidden states from a neural syntactic dependency parser.

Although the aforementioned systems provide state-of-the-art results, they subject to low computational performance. Recently, various top-down parsing approaches were proposed for simultaneous discourse parsing and elementary unit segmentation that address the computational performance issue. In [16], the segmenter and the sentence-level parser are trained jointly as parts of the unified encoder-decoder architecture, achieving superior results in the parsing performance, as well as in the parsing speed compared to previous end-to-end sentence-level bottom-up discourse parsers, namely SPADE [26] and DCRF [13]. Kobayashi et al. [15] has recently proposed a top-down method that takes into account granularity levels of spans, namely document, paragraph, and sentence. The authors show that the granularity levels are important features for discourse parsing and achieve 60% micro F score on the RST-DT corpus.

The first attempts to discourse relation classification for Russian are described in [5]. The authors apply the classical machine learning models to relation and nuclearity classification. By grounding on the previous works for other languages, they propose various lexical, quantitative, and morphological features for these tasks. In this work, we construct the fully-fledged discourse parser for Russian that features not only discourse relation classification, but also unlabeled tree construction and discourse segmentation. We implement deep learning models for these tasks and evaluate all components on the RuRSTreebank corpus. For relation classification, we expand the original feature set used by Chistova et al. [6] and compare the deep learning models with the feature-based models.

3   Methods

Given input plain text, first of all, we perform its tokenization and sentence splitting with the UDPipe [27] pre-trained models for Russian. We correct the output of the sentence segmenter with paragraph boundaries. Then, we perform discourse segmentation that splits the sentences into EDUs (simple sentences and clauses). The EDUs are assembled into a DT with a bottom-up algorithm along with a structure classifier. Finally, the tree constituents are labeled with a discourse relation classifier.

3.1   Discourse Segmentation Model

Following [30] and [21], we train a BiLSTM-CRF sequence labeling model on the contextual word embeddings and character-level convolution filters. As a contextual embedder, we use an ELMo model from RusVectores[1] pre-trained on Russian National Corpus and Russian Wikipedia, and a multilingual BERT model. We do not fine-tune pre-trained models during training. For each token in a sentence, the segmentation model predicts one of two labels: "Begin" or "Inside", which further allows us to decode segments. The BiLSTM-CRF model consists of a BiLSTM encoder containing a single layer with a CRF layer on top. Hidden layers in the BiLSTM networks contain 100 units. To reduce overfitting, the dropout and batch normalization were employed.

3.2   Discourse Tree Construction Algorithm

For discourse tree construction, we adopt the greedy algorithm presented in [10]. It employs a cascade of models for structure and label prediction. The structure classifier determines whether two units can be organized into a higher-level unit. The label classifier assigns to the created higher-level unit a relation and a nuclearity type.

In our pipeline, the tree construction algorithm is applied to each granularity level sequentially. At the sentence level, the input discourse units are elementary

---

[1] http://rusvectores.org/en/.

discourse units predicted by the segmentation model. At the paragraph and document granularity level, the input discourse units are those that are predicted on the previous levels. On each level, higher-level units are built while there is a pair of discourse units to merge and the predicted merge score for this pair is greater than the confidence threshold of the current level. The description of this pipeline is presented in Algorithm 1.

Scores for merging of discourse units are obtained using the structure classifier. The structure classifier is a binary classifier. Negative instances for its training and evaluation are adjacent text spans unconnected in gold standard trees. The confidence threshold for structure prediction is a hyperparameter adjusted between 0.0 and 1.0 for each granularity level individually.

Algorithm 1: Greedy tree parsing algorithm

Input: List of discourse units $[e_1, e_2, ..., e_n]$
Output: Discourse trees
$Trees \leftarrow [e_1, e_2, ..., e_n]$
$Scores \leftarrow \emptyset$
for $i \leftarrow 1$ *to* $n - 1$ **do**
| $Scores[i] = getScore(e_i, e_{i+1})$
end
while $|Trees| > 1$ *and* $any(Scores) > confidenceThreshold$ **do**
| $j = argmax(Scores)$
| $NewDU = mergeNodes(j, j + 1)$
| $NewDU.relation = getRelation(NewDU)$
| Replace $Trees[j]$ and $Trees[j + 1]$ with $NewDU$
| **if** $j \neq 0$ **then**
| | $Scores[j - 1] = getScore(L[j - 1], NewDU)$
| **end**
| **if** $j \neq length(Scores)$ **then**
| | $Scores[j] = getScore([NewDU, L[j + 1])$
| **end**
end
return $Trees$

The parser first predicts a labeled tree for each separate paragraph, constructing intra-sentence trees in the first place. Then it builds relations between paragraph-level trees on the document level. The choice of such a strategy is justified by the fact that each document in the corpus is presented not as a single RST tree but as a set of trees of various lengths. We train one document-level structure classifier and use different confidence thresholds for structure prediction on paragraph and document levels. Tree construction on the sentence level relies on the separate sentence-level structure classifier. The confidence threshold for intra-paragraph parsing is significantly lower than the document-level threshold, but it is still positive, as each separate paragraph in the gold-standard corpus do not always represent a separate subtree. For the same reason, the intra-sentential threshold is lower than the intra-paragraph threshold, but it is also positive.

3.3   Models for Unlabeled Tree Construction and Discourse Relation Classification

For the classification subtasks, we use the BiMPM [31] multi-perspective symmetric matching model, originally proposed for sentence matching and used to extract implicit interactions between text spans [17]. In the BiMPM model, two textual inputs are encoded in the BiLSTM encoder and the generated vectors are matched in both directions on each time step. Matching results are then processed by another BiLSTM layer, and both outputs of the aggregation layer are concatenated into a matching vector used for the final classification. As an input of the BiMPM model, we use a concatenation of ELMo from RusVectores and character embeddings. To exhibit the effectiveness of the neural relation and structure classifiers, the results are compared against a baseline, which is the feature-based method proposed in [5].

1. **Relation classifier.** Predicts the relation type along with the nuclearity, e.g. Cause-effect_NS and Cause-effect_SN are separate classes with different nuclear orientations.
   Baseline. The baseline method for relation classification is an ensemble of a CatBoost model with selected features and a linear SVM using various lexical (explicit discourse markers), quantitative (number of occurrences of markers and stop-words, TF-IDF vectors), morphological, and semantic features (distances and correlations between various feature vectors extracted from both DUs).
   Baseline+AF. The performance of a baseline model is improved with an accurate feature selection. In particular, we reduce the list of discourse markers relying on feature importances given by a CatBoost model on crossvalidation. Furthermore, additional features are developed, namely, an indication of the DUs presence in the same sentence, another value for the presence in the same paragraph and sentiment scores of the left and right DUs obtained with the model presented in [25].
2. **Structure classifier.** The structure classifier is used for scoring discourse units for combining them into a higher-level DU. It takes as an input a concatenation of ELMo and character embeddings along with binary granularity features indicating whether both text spans are located in the same sentence and in the same paragraph.
   Baseline. As a baseline for structure classification, we use a linear SVM model and the feature set with additional features developed for the baseline relation classification.

4   Experiments

4.1   Dataset

Ru-RSTreebank is the first open discourse corpus for Russian within the framework of the Rhetorical Structure Theory[2]. It contains three parts: news and

---

[2] https://rstreebank.ru/dataset.

popular science texts (129 texts); scientific papers about Linguistics and Computer Science (99 texts); blog texts about different topics, i.e. traveling, sports and health, psychology, IT and tech, politics (104 texts). There are about 408,000 tokens in total. The corpus was manually annotated by 11 experts. For all texts, titles, subtitles, reference lists, and other metainformation were excluded from labeling. Image captions in blogs were annotated in different ways, depending on deictic indications linking them to other text fragments. In some cases, markers for images (IMG) are considered as EDU parts.

4.2   Pre-processing

For the model training and evaluation, scientific papers were excluded from the corpus due to their specific structure. Science themed part of the corpus was annotated during the first stage of the corpus development and followed the earliest annotation instructions that may conflict with more modern ones. Hence, the pipeline is evaluated on 233 texts including news, popular science articles, and blog texts.

The RuRSTreebank corpus pre-processing includes structure and relations readjustment. Non-binary relations, such as Joint and Sequence, are converted into cascades of binary relations. It was noted that punctuation marks, such as dots and commas, are in some cases shifted from the end of the previous EDU to the beginning of the next EDU, which is corrected during corpus pre-processing. Some relations indicated in [23] as combined or excluded, for example, Antithesis or Evaluation, are automatically pre-processed following the aforementioned paper. Examples with relations of the underrepresented classes (less than 90 training instances) are added to the most represented non-causal classes of the same nuclearity orientation, e.g. Elaboration_NS for Solutionhood_NS and Preparation_SN for Elaboration_SN. Considering nuclearity, we have a total set of 22 relations.

Following [21], URLs and special symbols are replaced with service tokens.

4.3   Results and Discussion

The performance of the pipeline components on the development and test sets is presented in Table 1. The structure classifier is used in the final pipeline for span prediction, and the relation classifier is used for both label and nuclearity prediction.

The baseline for discourse segmentation is a single-layer BiLSTM model on top of multilingual BERT, which is used for cross-lingual segmentation in [21]. The results show that the model with the CRF layer that uses language-specific contextual embeddings outperforms the baseline by 0.86%.

The neural model for structure classification performs comparably to the feature-based baseline model in terms of F-score, but it has a higher recall score.

The test scores for label classification for the baseline feature-based method, baseline method with additional features (AF), BiMPM classifier, and an ensemble of the baseline and the BiMPM classifier are presented in Fig. 2. It can be

Table 1. Parser components evaluation

Component	Method	Features	Dev			Test		
			P	R	F1	P	R	F1
Segmentation	Baseline	BERT-M	91.79	85.40	88.48	**89.66**	85.56	87.56
	BiLSTM+CRF	BERT-M	90.75	**88.72**	89.72	87.80	**88.99**	88.39
		ELMo	**92.10**	87.53	**89.76**	89.09	87.86	**88.42**
Structure classification	Baseline	Baseline+AF	57.42	75.42	65.20	**58.42**	76.38	66.21
	BiMPM	ELMo	54.95	**82.95**	66.11	54.54	82.82	65.77
	Baseline+BiMPM	Baseline+AF, ELMo	**57.62**	82.56	**67.87**	57.66	**83.06**	**68.07**
Relation classification	Baseline	Baseline	45.54	42.07	42.70	42.48	41.32	40.63
		Baseline+AF	48.76	44.15	45.45	46.54	44.18	44.19
	BiMPM	ELMo	50.85	46.80	46.76	47.35	45.40	44.64
	Baseline+BiMPM	Baseline+AF, ELMo	**54.28**	**48.59**	**49.17**	**49.89**	**47.73**	**47.50**

Table 2. Performance on manual EDU segmentation

	Micro			Macro			Macro F1	
	P	R	F1	P	R	F1	Blogs	News
Span	38.04	38.59	38.31	39.61	44.67	41.98	40.35	43.26
Nuclearity	30.96	31.40	31.18	31.72	35.28	33.40	32.08	34.43
Relation	23.87	24.21	24.04	24.78	27.69	26.16	24.53	27.43
Full	23.58	23.91	23.75	24.44	27.25	25.77	23.96	27.18

observed that the neural classifier outperforms baselines on 12 of 22 rhetorical relations. However, the baseline model better predicts the most represented classes, namely Joint (21.9% of instances) and Elaboration (20.4% of instances), as it tends to the overgeneralization. Ensembling of the baseline method and the BiMPM model improves the average F-score by 2.86%.

Table 2 contains the evaluation results for tree parsing on manual discourse segmentation using the components that achieve best results in Table 1. Following Morey et al. [20], we present both micro-averaged scores computed globally over the predicted and manually annotated spans from all documents and macro-averaged scores obtained by averaging scores for each document in the corpus. The results are calculated according the standard Parseval trees evaluation between binarized gold-standard and predicted trees [20]. We also present the macro F1 score for individual domains: "News" and "Blogs". The results show that the parser has slightly higher performance on the "News" subset.

5   Qualitative Analysis

Examples of segmentation mistakes are given in Table 3 in Appendix. The first issue is related to the fact that some EDUs in the corpus contain multiple sentences or multiple paragraphs (1). It often occurs in the "Blogs" subsection with

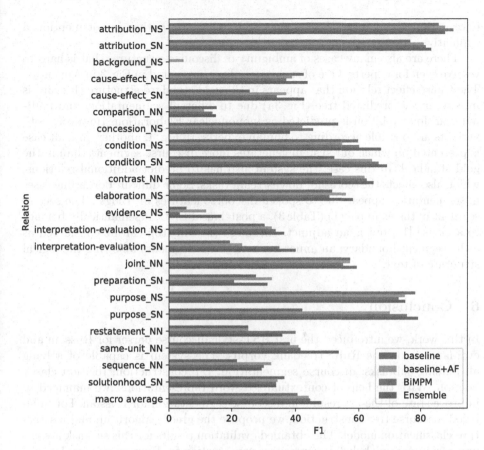

Fig. 2. Relation classification performance for each class

image-replacement tags and image captions, as well as with lists manually annotated as a single EDU, because such text fragments are usually treated by the
system as separate paragraphs. This problem often seriously affects the performance of all downstream components. The parsing algorithm cannot deal with
such cases since it performs segmentation and tree building on a sentence level,
paragraph and document levels consequentially. The second issue is related to
discrepancies in corpus annotation. In the example (2), two separate discourse
segments predicted by the model are annotated as one in the gold standard.
However, other syntactically and semantically similar cases in the corpus are
annotated as two segments linked by the Elaboration relation. Such discrepancies make it difficult to tune thresholds for various discourse levels, which results
in underprediction (3) and overprediction (4) of discourse segments even within
the same document. Another frequent case of discrepancy is the annotation of
the Attribution relation (2). Annotators sometimes mark the statement and the

information about its author as one EDU when the author is not mentioned explicitly in text.

There are also many cases of ambiguity of discourse structures that is hard to resolve even for experts. One of such examples is presented in Fig. 3 in Appendix. The Cause-effect relation that appears in the gold-standard structure (Fig. 3b) is missing in the predicted tree (Fig. 3a) due to different segmentation, and multinuclear Joint relation is predicted as mononuclear Elaboration. However, both variants are possible according to human experts. In Fig. 4, another difficult case is presented, in which our system generates more discourse segments than in the gold standard. In this case, the system also has to predict additional relations, which also affects the scores for downstream tasks. Some difficult borderline cases in segmentation appear due to spoken discourse features in "Blogs". The second segment in the example (1) (Table 3), a postclausal EDU, it semantically belongs to a clause [14] but is an adjunct. Although the full stop can be considered as a the segment boundary, an annotator united the segments following the logical structure of text.

6   Conclusion

In this work, we introduced the first RST-style discourse parser for Russian and evaluated it on the RuRSTreebank corpus. The system is capable of solving all parsing subtasks: discourse segmentation, DT construction, relation classification. With the help of contextualized word representations, we managed to improve state-of-the-art results in discourse segmentation for Russian. For unlabeled discourse tree construction, we propose the greedy algorithm and a structure classification model. The obtained evaluation results for this subtask are the first results for unlabeled discourse tree construction for Russian and can be used as a baseline for the subsequent works. We expand the feature set of the relation classification model presented in the previous work achieving improved results. We further improve results by introducing a deep learning-based model and by ensembling the deep model with the feature-based model. The obtained evaluation results are the new state of the art for discourse relation classification for Russian language. We release the source code of the models and experiments[3]. We hope that this work will foster further research in the area of discourse parsing for Russian language.

Acknowledgements. This work was partially supported by the Ministry of Science and Higher Education of the Russian Federation, project No. 075-15-2020-799.

---

[3] http://nlp.isa.ru/discourse_parser.

Appendix

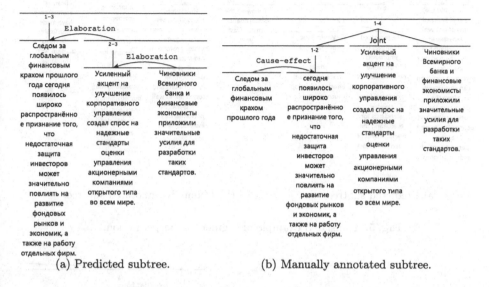

(a) Predicted subtree.   (b) Manually annotated subtree.

Fig. 3. Example of segment underprediction (translation is in Fig. 5).

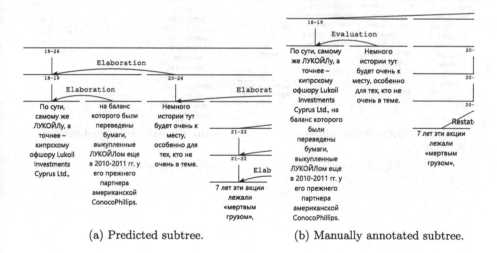

(a) Predicted subtree.   (b) Manually annotated subtree.

Fig. 4. Example of segment overprediction (translation is in Fig. 6).

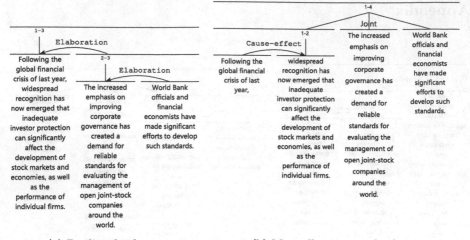

(a) Predicted subtree.   (b) Manually annotated subtree.

Fig. 5. Translated example of segment underprediction.

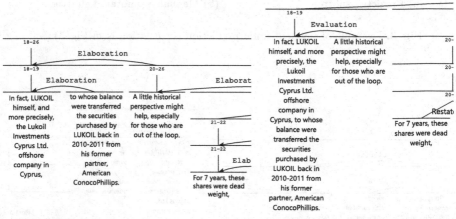

(a) Predicted subtree.   (b) Manually annotated subtree.

Fig. 6. Translated example of segment overprediction.

Table 3. Incorrect segmentation examples (translated examples are in the Appendix Table 4).

Predicted EDUs	Gold EDUs
(1) Осталось только копченый сыр порезать. \<SEP> В суп, ага. \<SEP> IMG	Осталось только копченый сыр порезать. В суп, ага.\nIMG
(2) Несмотря на то, что российский медиарынок на порядок отстаёт от американского, \<SEP> издатели отечественной прессы единодушны: \<SEP>планшетный компьютер [...] даёт отрасли новую надежду на выживание.	Несмотря на то, что российский медиарынок на порядок отстаёт от американского, \<SEP> издатели отечественной прессы единодушны: планшетный компьютер [...] даёт отрасли новую надежду на выживание.
(3) Что происходит, когда резервов не хватает, \<SEP> нам хорошо известно.	Что происходит, \<SEP> когда резервов не хватает, \<SEP> нам хорошо известно.
(4) Хотя бы до осени, \<SEP> когда прояснится ситуация с долгами и нефтяными ценами.	Хотя бы до осени, когда прояснится ситуация с долгами и нефтяными ценами.

Table 4. Translated incorrect segmentation examples.

Predicted EDUs	Gold EDUs
(1) It remains only to cut the smoked cheese. \<SEP> Yep, in the soup. \<SEP> IMG	It remains only to cut the smoked cheese. Yep, in the soup.\nIMG
(2) Despite the fact that the Russian media market is an order of magnitude behind the American one, \<SEP> the publishers of the domestic press are unanimous: \<SEP> the tablet [...] gives the industry new hope.	Despite the fact that the Russian media market is an order of magnitude behind the American one, \<SEP> the publishers of the domestic press are unanimous: the tablet [...] gives the industry new hope.
(3) What happens when reserves are scarce, \<SEP> we are well aware.	What happens \<SEP> when reserves are scarce, \<SEP> we are well aware.
(4) At least until the fall, \<SEP> when the situation with debts and oil prices becomes clearer.	At least until the fall, when the situation with debts and oil prices becomes clearer

References

1. Atutxa, A., Bengoetxea, K., de Ilarraza, A.D., Iruskieta, M.: Towards a top-down approach for an automatic discourse analysis for Basque: segmentation and central unit detection tool. PLOS ONE **14**(9), 1–25 (2019)
2. Bhatia, P., Ji, Y., Eisenstein, J.: Better document-level sentiment analysis from RST discourse parsing. In: Proceedings of the 2015 Conference on Empirical Methods in Natural Language Processing, pp. 2212–2218. Association for Computational Linguistics (2015)

3. Braud, C., Coavoux, M., Søgaard, A.: Cross-lingual RST discourse parsing. In: Proceedings of the 15th Conference of the European Chapter of the Association for Computational Linguistics: Volume 1, Long Papers, pp. 292–304 (2017)

4. Carlson, L., Marcu, D., Okurowski, M.E.: Building a discourse-tagged corpus in the framework of rhetorical structure theory. In: van Kuppevelt, J., Smith, R.W. (eds.) Current and New Directions in Discourse and Dialogue. Text, Speech and Language Technology, vol. 22, pp. 85–112. Springer, Dordrecht (2003). https://doi.org/10.1007/978-94-010-0019-2_5

5. Chistova, E., Kobozeva, M., Pisarevskaya, D., Shelmanov, A., Smirnov, I., Toldova, S.: Towards the data-driven system for rhetorical parsing of Russian texts. In: Proceedings of the Workshop on Discourse Relation Parsing and Treebanking 2019, pp. 82–87. Association for Computational Linguistics (2019)

6. Chistova, E., Shelmanov, A., Kobozeva, M., Pisarevskaya, D., Smirnov, I., Toldova, S.: Classification models for RST discourse parsing of texts in Russian. In: Computational Linguistics and Intellectual Technologies. Papers from the 2019 Annual International Conference "Dialogue", vol. 18, pp. 163–176 (2019)

7. Desai, T., Dakle, P.P., Moldovan, D.: Joint learning of syntactic features helps discourse segmentation. In: Proceedings of The 12th Language Resources and Evaluation Conference, pp. 1073–1080 (2020)

8. Feng, V.W., Hirst, G.: A linear-time bottom-up discourse parser with constraints and post-editing. In: Proceedings of the 52nd Annual Meeting of the Association for Computational Linguistics (Volume 1: Long Papers), pp. 511–521 (2014)

9. Guzmán, F., Joty, S., Màrquez, L., Nakov, P.: Using discourse structure improves machine translation evaluation. In: Proceedings of the 52nd Annual Meeting of the Association for Computational Linguistics, pp. 687–698 (2014)

10. Hernault, H., Prendinger, H., Ishizuka, M., et al.: HILDA: a discourse parser using support vector machine classification. Dialogue Discourse 1(3), 1–33 (2010)

11. Hirao, T., Yoshida, Y., Nishino, M., Yasuda, N., Nagata, M.: Single-document summarization as a tree knapsack problem. In: Proceedings of the 2013 Conference on Empirical Methods in Natural Language Processing, pp. 1515–1520 (2013)

12. Hung, S.S., Huang, H.H., Chen, H.H.: A complete shift-reduce Chinese discourse parser with robust dynamic oracle. In: Proceedings of the 58th Annual Meeting of the Association for Computational Linguistics, pp. 133–138 (2020)

13. Joty, S., Carenini, G., Ng, R.: A novel discriminative framework for sentence-level discourse analysis. In: Proceedings of the 2012 Joint Conference on Empirical Methods in Natural Language Processing and Computational Natural Language Learning, pp. 904–915 (2012)

14. Kibrik, A.: The problem of non-discreteness and spoken discourse structure. Comput. Linguist. intellect. Technol. 1(14(21)), 225–233 (2015)

15. Kobayashi, N., Hirao, T., Kamigaito, H., Okumura, M., Nagata, M.: Top-down RST parsing utilizing granularity levels in documents. In: AAAI, pp. 8099–8106 (2020)

16. Liu, L., Lin, X., Joty, S., Han, S., Bing, L.: Hierarchical pointer net parsing. In: Proceedings of the 2019 Conference on Empirical Methods in Natural Language Processing and the 9th International Joint Conference on Natural Language Processing (EMNLP-IJCNLP), pp. 1006–1016 (2019)

17. Liu, X., Ou, J., Song, Y., Jiang, X.: On the importance of word and sentence representation learning in implicit discourse relation classification. In: Proceedings of the 29th International Joint Conference on Artificial Intelligence, IJCAI-20, pp. 3830–3836. International Joint Conferences on Artificial Intelligence Organization (2020)

18. Mann, W.C., Thompson, S.A.: Rhetorical structure theory: toward a functional theory of text organization. Text Interdisc. J. Study Discourse **8**(3), 243–281 (1988)
19. Marcu, D.: The Theory and Practice of Discourse Parsing and Summarization. MIT Press, Cambridge (2000)
20. Morey, M., Muller, P., Asher, N.: How much progress have we made on RST discourse parsing? A replication study of recent results on the RST-DT. In: Proceedings of the 2017 Conference on Empirical Methods in Natural Language Processing, pp. 1319–1324 (2017)
21. Muller, P., Braud, C., Morey, M.: ToNy: Contextual embeddings for accurate multilingual discourse segmentation of full documents. In: NAACL HLT 2019 (2019)
22. Peters, M.E., et al.: Deep contextualized word representations. In: Proceedings of NAACL-HLT, pp. 2227–2237 (2018)
23. Pisarevskaya, D., et al.: Towards building a discourse-annotated corpus of Russian. In: Computational Linguistics and Intellectual Technologies: Proceedings of the International Conference Dialogue, p. 23 (2017)
24. Pisarevskaya, D., Galitsky, B.: An anatomy of a lie: discourse patterns in ultimate deception dataset. In: Proceedings of the International Conference on Computational Linguistics and Intellectual Technologies "Dialogue 2019" (2019)
25. Rogers, A., Romanov, A., Rumshisky, A., Volkova, S., Gronas, M., Gribov, A.: RuSentiment: an enriched sentiment analysis dataset for social media in Russian. In: Proceedings of the 27th International Conference on Computational Linguistics, pp. 755–763. Association for Computational Linguistics (2018)
26. Soricut, R., Marcu, D.: Sentence level discourse parsing using syntactic and lexical information. In: Proceedings of the 2003 Conference of the North American Chapter of the Association for Computational Linguistics on Human Language Technology, vol. 1, pp. 149–156 (2003)
27. Straka, M., Straková, J.: Tokenizing, POS tagging, lemmatizing and parsing UD 2.0 with UDPipe. In: Proceedings of the CoNLL 2017 Shared Task: Multilingual Parsing from Raw Text to Universal Dependencies, pp. 88–99 (2017)
28. Subba, R., Di Eugenio, B.: An effective discourse parser that uses rich linguistic information. In: Proceedings of Human Language Technologies: the 2009 Annual Conference of the North American Chapter of the Association for Computational Linguistics, pp. 566–574 (2009)
29. Wang, Y., Li, S., Wang, H.: A two-stage parsing method for text-level discourse analysis. In: Proceedings of the 55th Annual Meeting of the Association for Computational Linguistics (Volume 2: Short Papers), pp. 184–188 (2017)
30. Wang, Y., Li, S., Yang, J.: Toward fast and accurate neural discourse segmentation. In: Proceedings of the 2018 Conference on Empirical Methods in Natural Language Processing, pp. 962–967 (2018)
31. Wang, Z., Hamza, W., Florian, R.: Bilateral multi-perspective matching for natural language sentences. In: Proceedings of the 26th International Joint Conference on Artificial Intelligence, pp. 4144–4150 (2017)
32. Yu, N., Zhang, M., Fu, G.: Transition-based neural RST parsing with implicit syntax features. In: Proceedings of the 27th International Conference on Computational Linguistics, pp. 559–570 (2018)
33. Zeldes, A., Das, D., Maziero, E.G., Antonio, J., Iruskieta, M.: The DISRPT 2019 shared task on elementary discourse unit segmentation and connective detection. In: Proceedings of the Workshop on Discourse Relation Parsing and Treebanking 2019, pp. 97–104. Association for Computational Linguistics (2019)

A Comparative Study of Feature Types for Age-Based Text Classification

Anna Glazkova[1]([✉])[iD], Yury Egorov[1][iD], and Maksim Glazkov[2][iD]

[1] University of Tyumen, ul. Volodarskogo 6, 625003 Tyumen, Russia
`a.v.glazkova@utmn.ru`
[2] Organization of cognitive associative systems LLC,
ul. Gertsena 64, 625000 Tyumen, Russia

Abstract. The ability to automatically determine the age audience of a novel provides many opportunities for the development of information retrieval tools. Firstly, developers of book recommendation systems and electronic libraries may be interested in filtering texts by the age of the most likely readers. Further, parents may want to select literature for children. Finally, it will be useful for writers and publishers to determine which features influence whether the texts are suitable for children. In this article, we compare the empirical effectiveness of various types of linguistic features for the task of age-based classification of fiction texts. For this purpose, we collected a text corpus of book previews labeled with one of two categories – children's or adult. We evaluated the following types of features: readability indices, sentiment, lexical, grammatical and general features, and publishing attributes. The results obtained show that the features describing the text at the document level can significantly increase the quality of machine learning models.

Keywords: Text classification · Fiction · Corpus · Age audience · Content rating · Text difficulty · RuBERT · Neural network · Natural language processing · Machine learning

1 Introduction

Nowadays, there are quite a lot of approaches to text classification according to document subjects, genre, author or according to other attributes. However, modern challenges in the field of natural language processing (NLP) and information retrieval (IR) increasingly require classification based on more complex characteristics. For example, it may be necessary to determine whether the text contains elements of propaganda or whether it has similar plot characteristics to other texts. One of such urgent and complex classification tasks is the division of literary texts into suitable for children and for adults. Age-based classification tools could find wide practical application. For instance, they would be useful in the personal selection of fiction or in filtering content not intended for children.

Supported by the grant of the President of the Russian Federation no. MK-637.2020.9.

W. M. P. van der Aalst et al. (Eds.): AIST 2020, LNCS 12602, pp. 120–134, 2021.
https://doi.org/10.1007/978-3-030-72610-2_9

Despite the fact that many scholars considered the issue of text difficulty estimation, formally, text difficulty does not indicate the age of the intended reader. The question of whether the features describing text difficulty are suitable for age-based classification needs to be investigated. In addition, the severity of the features depends on the text genre. It is necessary to find out how these or those features are presented in the literary text and whether they contain information about the age audience of the text.

In this paper, we systematically evaluated different feature types on age-based classification task. In addition to popular text difficulty features, we consider special publishing attributes intrinsic fiction books, such as age rating score and abstract features. We collected the corpus of Russian fiction texts and applied two commonly used machine learning models, these are random forest (RF) and linear support vector classifier (LSVC). For comparison, we evaluated a transformer model based on RuBERT and a convolutional neural network (CNN) trained on Word2Vec embeddings. Finally, we evaluated feedforward neural network (FNN) trained on RuBERT text embeddings and age rating scores.

The LSVC model using a combination of a baseline and publishing attributes showed the best result of 95.77% (F1-score). RuBERT achives 90.16%. The FNN model combining RuBERT embeddings and age ratings showed 94.78%. The results show that the features describing the text at the document level gives an advantage in case of long texts. Moreover, publishing attributes provide valuable information for the age-based classifier. We also found that some features used to determine text difficulty positively affect the quality of age-based classification.

The paper is organised as follows. In Sect. 2 we present a brief review of related works. Section 3 describes feature types evaluated in the paper. Section 4 contains the description of our dataset. Section 5 presents the structure of the models and the evaluation results. Finally, Sect. 6 is a conclusion.

2   Related Works

In the modern world, the constant growth of information resources gives rise to the need for filtering and ranking texts. One of the significant characteristics of a text is its complexity. The question of determining text difficulty naturally looks related to the task of age-based classification.

The task of estimating texts by complexity is not new. It appeared at the beginning of the last century in the context of evaluating the readability of educational texts. Further, during the XX century, researchers have proposed a number of tests to determine readability based on the quantitative characteristics of texts. Readability tests usually use quantitative text features, such as counting syllables, words, and sentences. There are several common readability texts for text difficulty estimation. For instance, these are: the Flesch–Kincaid readability test, the Coleman–Liau index, the automated readability index (ARI), the SMOG grade, the Dale-Chall formula [6,10].

– The Flesch–Kincaid readability test is based on the idea that the shorter the sentences and words, the simpler the text. The specific mathematical formula is:

$$R_F = 206.835 - 1.015 \cdot ASL - 84.6 \cdot ASW, \tag{1}$$

ASL – average sentence length, ASW – average number of syllables per word (i.e., the number of syllables divided by the number of words).

– The Coleman–Liau index uses letters instead of syllables. The formula takes into account the average number of letters per word and the average number of words per sentence.

$$R_C = 0.0588 \cdot L - 0.296 \cdot S - 15.8, \tag{2}$$

where L – average number of letters per 100 words, S – average number of sentences per 100 words.

– The ARI formula takes into account the number of letters. In the past, this allowed the use of this index to measure the complexity of texts in real time in electric typewriters.

$$R_A = 4.71 \cdot \frac{characters}{words} + 0.5 \cdot \frac{words}{sentences} - 21.43, \tag{3}$$

where $characters$ – number of letters and numbers, $words$ – number of words, $sentences$ – number of sentences.

– The main idea of the SMOG grade is that the complexity of the text is most affected by complex words. Complex words are words with many syllables (more than 3). The more syllables the more complicated the word.

$$R_S = 1.043 \cdot \sqrt{polysyllable \cdot \frac{30}{sentences}} + 3.1291, \tag{4}$$

where $polysyllable$ – number of polysyllable words, $sentences$ – number of sentences.

– The Dale-Chall formula uses a count of "hard" words. These "hard" words are words that do not appear on a specially designed list of common words familiar to most 4th-grade students.

$$R_D = 0.1579 \cdot \frac{difficult}{words} \cdot 100 + 0.0496 \cdot \frac{words}{sentences}, \tag{5}$$

where $difficult$ – number of difficult words, $words$ – number of words, $sentences$ – number of sentences.

   In addition to the above, there are many other readability tests that are also actively used, e.g. the Fry Graph readability formula, the Spache index, the Linsear Write formula and others. The values obtained from readability tests are called readability indices. The Readability Index characterizes the difficulty of perceiving a text or the expected level of education that is required to understand it.

The readability formulas listed above are metrics for English texts. At the same time, the quantitative characteristics of other languages can differ significantly. For instance, Russian sentences are on average shorter than English, and words are longer. Therefore, the readability formulas nced to be processed for use in other languages. Up to now, several studies have suggested the adaptation of readability tests for Russian. For example, I. Oborneva [27] proposed the coefficients for the Flesch–Kincaid formula for Russian texts. The project [41] offers the adaptation of several readability formulas. M. Solnyshkina et al. presented a new approach to reading difficulty prediction in Russian texts [37,38].

Readability is however only one aspect of age-based classification. Scholars have proposed more complex techniques for text complexity estimation using features of different nature. Thus, Yu. Tomina [42] considered the lexical and syntactic features of the text complexity level. A. Laposhina et al. [22] evaluated a wide range of different types of features, such as readability, semantic, lexical, grammatical and others. M. Shafaei et al. [33] estimated age suitability rating of movie dialogs using genre and sentiment features. L. Flekova et al. [16] proposed an approach to describing the story complexity for literary text. Y. Bertills [4] wrote about the features of literary characters and named entities in books for children. Finally, in our previous research, we evaluated the informativeness of some quantitative and categorical features for age-based text classification [12]. The modern methodology for text difficulty estimation is based in most cases on machine learning approaches. Thus, R. Balyan et al. [3] showed that applying machine learning methods increased accuracy by more than 10% as compared to classic readability metrics (e.g., Flesch–Kincaid formula). To date, a number of studies confirmed the effectiveness of various machine learning techniques for text difficulty estimation, such as support vector machine (SVM) [36,39], random forest [26], and neural networks [2,7,35].

Another aspect of assessing the age category of text readers is the safety of the information it contains. Currently, in many countries, publishers are required to label books (including fiction) and other informational sources [9,11,14,18,30] according to their age rating. For these purposes, there are special laws that rank information in terms of the potential harm it can bring. So, in Russia there is a Russian Age Rating System (RARS).

The RARS was introduced in 2012 when the Federal law of Russian Federation no. 436-FZ of 2010-12-23 "On Protection of Children from Information Harmful to Their Health and Development" was passed [15]. The law prohibits the distribution of "harmful" information that depicts violence, unlawful activities, substance abuse, or self-harm. The RARS includes 5 categories, such as for children under the age of six (0+), for children over the age of six (6+), for children over the age of twelve (12+), for children over the age of sixteen (16+), and prohibited for children (18+). As a rule, an age rating is assigned to a book by editors or experts. As far as we know, there are currently no published research of how age rating correlates with other attributes of text, such as readability.

The reviewed studies and sources clearly indicate that age-based classification of fiction texts includes several aspects. First, the research topic is related to works on text difficulty evaluation. Text difficulty is characterized by different features, these are lexical, semantic, grammatical and other types. However, the measure of the difficulty of the text does not guarantee that this text is targeted to a particular age audience. It is required to evaluate the effectiveness of the existing text difficulty features for age-based classification. In addition, it would be interesting to evaluate the role of publishing attributes (for example, age rating labels) as classification features. Finally, the studies presented thus far provide evidence that machine learning approaches show the highest results in the task of estimating texts by difficulty. Based on this, it is reasonable to evaluate the text features for age-based classification using machine learning methods.

3   Feature Types

According to the related works, we consider the following types of classification features.

1. **General features.** This type includes features that reflect the quantitative characteristics of the text:
   - the average and median length of words (avg_words_len, med_words_len);
   - the average and median length of sentences (avg_sent_len, med_sent_len), e.g. average or median number of symbols in each sentence;
   - the average number of syllables (avg_count_syl);
   - the percentage of long words with more than 4 syllables ($many\_syllables$);
   - the Type-Token Ration, TTR (ttr) [40]. The main idea of the metric is that if the text is more complex, the author uses a more varied vocabulary so there's a larger number of unique words. So, the TTR's value is calculated as the number of unique words divided by the number of words. As a result, the higher the TTR, the higher the variety of words;
   - the TTR for nouns (ttr_n), adjectives (ttr_a), and verbs (ttr_v). The values of TTR calculated separately for parts of speech;
   - the NAV metric (nav). The NAV metric is a TTR-based ratio of (TTR A + TTR N)/TTR V proposed in [37].
2. **Readability features.** We used the readability formulas with the coefficients for the Russian language proposed by the project [41]. In this study, we evaluated five types of readability indices using the following metrics: the Flesch–Kincaid readability test ($index\_fk$); the Coleman–Liau index ($index\_cl$); the ARI index ($index\_ari$); the SMOG grade ($index\_SMOG$); the Dale-Chall formula ($index\_dc$).
3. **Lexical features.** In this category, we included features constructed by the evaluation of the text in accordance with frequency dictionaries. As frequency dictionaries, we used the lists of Russian frequency words presented in [25,34]:
   - the percentage of words included in the list of 5000 most frequent Russian words (5000_proc);
   - the average frequency of the words included in the 5000 most frequent words (5000_freq);

- the average frequency of words per 1 million occurrences (ipm, *words_fr*);
- the average frequency of nouns, verbs, adjectives, adverbs and proper names per 1 million occurrences (*s_fr*, *v_fr*, *adj_fr*, *adv_fr*, *prop_fr*);
- the average number of topic segments of the corpus[1] where the word was encountered (out of 100 possible, *words_r*);
- the average number of the corresponding topic segments for nouns, verbs, adjectives, adverbs and proper names (*s_r*, *v_r*, *adj_r*, *adv_r*, *prop_r*);
- the average value of Juilland's usage coefficients (*words_d*). This Juilland's usage coefficient measures the dispersion of the word's subfrequencies over n equally-sized subcategories of the corpus [17];
- the average value of Juilland's usage coefficients for nouns, verbs, adjectives, adverbs and proper names (*s_d*, *v_d*, *adj_d*, *adv_d*, *prop_d*);
- the number of documents in the corpora in which a word occurs (averaged over the text, *words_doc*);
- the average number of documents in the corpora in which a word occurs (for nouns, verbs, adjectives, adverbs and proper names, *s_fr*, *v_fr*, *adj_fr*, *adv_fr*, *prop_fr*).

4. **Grammatical features.** We evaluated the percentage of nouns, verbs, and adjectives (*count_n*, *count_v*, *count_a*).

5. **Sentiment features.** These features obtained with Russian Sentiment Lexicon [24]. We separately evaluated the percentage of positive and negative words for each of the topic categories, these are opinion, feeling (private state), or fact (sentiment connotation) (*neg_opinion*, *neg_feeling*, *neg_fact*, *pos_opinion*, *pos_feeling*, *pos_fact*).

6. **Publishing features.** Here we have included features based on publishing attributes, i.e. on the book characteristics assigned by an editor or publisher, such as age rating according to the RARS (*age_rating*) and TF-IDF scores for book abstracts.

4 Dataset

For feature evaluation, we collected a dataset of fiction books published in Russian. Due to copyright restrictions, the full texts of the books are not publicly available. Therefore, we used a collection of previews presented in online libraries in the public domain. Typically, the preview is 5–10% of the total book volume.

The corpus consists of 5592 texts of children's and adult book previews. We have divided the texts into two parts. The first part included 4492 texts. It was used to train the models. The remaining 1000 texts were served as an independent text sample. The main characteristics of the data is presented in Table 1. Table 2 shows short text examples of adults and children's categories.

---

[1] The frequency dictionary was created on the basis of the modern subcorpus of the Main Corpus and the Oral Corpus of the Russian National Corpus (1950–2007) [32] with a total volume of 92 million tokens [25].

Table 1. Characteristics of the corpus.

Characteristic	Training sample		Test sample	
	Children's	Adult	Children's	Adult
Number of text	2108	2384	500	500
Avg number of symbols	3134.38	3326.11	3048.69	3319.86
Avg number of tokens	488.55	499.52	479.3	498.16
Avg number of sentences	37.35	35.2	36.05	36.49

Table 2. Example short fragments.

Category	Age Rating	Genre	Fragment
Adults	16+	Modern romance novels	A tall young man dressed in jeans, an inconspicuous jacket and a baseball cap pulled down with a visor over his eyes, approached the entrance of a seventeen-story apartment building and stood as if waiting for someone, and when a mother with a stroller appeared at the door, he quickly jumped inside - he did not know the code. I walked up to the fifth floor, putting on thin gloves on the go, looked around, and then deftly opened the door of one of the apartments.[a]
Adults	12+	Histori- cal adventures	What a wonderful autumn it was in Southern Poland that year! Almost without rain and cold winds, tenderly warm, quiet, crimson-gold. Fabulous autumn - in such an autumn it is good, having climbed into the spurs of the Beskydy, from dawn to noon to wander along the slopes of hills overgrown with beech and hazel, and to your fill, drunk to breathe in the cool and crystal clear mountain air.[b]
Children's	6+	Child- ren's adventures	In a big, big city, where there are many, many houses, many, many cars and even more people, and the crows cannot be counted at all, there lived a ginger cat on a short street consisting of only two courtyards. His name was Ostrich.[c]
Children's	12+	Child- ren's fantastic tales	The hands of the clock were approaching half past seven, but the setting sun, reluctantly sliding behind the houses, continued to burn the city with rays, and the approaching twilight did not promise the long-awaited coolness.[d]

[a] Fragment from the book "The Men We Choose" by Evgenia Perova (translated from Russian). [b] "Sold Poland" by Alexander Usovsky (translated from Russian). [c] "Greetings from cutlets" by Evgenia Malinkina (translated from Russian). [d] "Vlad and the Secret Ghost" by Sasha Gotti (translated from Russian).

Table 3 shows the most informative quantitative features with their means and standard deviation values. The informativeness is measured using the method of cumulative frequencies [1,44]. The main idea of this method consists in dividing the range of feature values for each class into n intervals. The cumulative frequency

Table 3. Top-10 of the most informative quantitative features (according to the method of cumulative frequencies).

Feature	Mean (adult)	std (adult)	Mean (children's)	std (children's)
avg_sent_len	105.65	54.51	88.69	30.64
med_sent_len	97.9	59.06	79.69	34
index_dc	7.85	2.64	6.36	2.09
adj_doc	3484.28	1074.03	3669.48	1173.78
index_ari	9.29	3.53	7.4	2.93
adj_fr	135.3	53.56	146.96	60.39
s_doc	3705.73	838.45	3370.52	766.39
index_fk	9.21	3.66	7.36	2.95
v_doc	6255.7	2243.76	6481.53	1937.8
adv_fr	239.17	84.26	277.43	90.93

of characteristic values is calculated for each interval. The informativeness indicator is defined as the maximum absolute value of the difference in the accumulated frequencies for the corresponding intervals in the classes.

Figure 1 presents the distribution of the age rating labels (age rating is a categorical feature) in the classes of the training data. It is interesting to note that some of the books from the children's class are labelled with the 18+ age category. Notable examples of this type of books are love stories for teens.

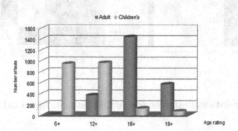

Fig. 1. The distribution of age rating categories.

5   Experiments

This section describes our feature evaluation experiments. We built two types of baseline models and sequentially enriched them with different types of features. Further, we compared the results obtained with our models with the results of CNN, RuBERT, and FNN trained on RuBERT embeddings and age rating scores. Our dataset and models are available at [5].

5.1   Baselines

We built two classifiers for model evaluation. The first one was a Random Forest Classifier trained on bootstrap samples. The number of trees in the forest was equal to 100 and the Gini impurity was implemented to measure the quality of a split. The second model was a Linear Support Vector Classifier with the "l2" penalty and the squared hinge loss function. Both models were implemented using Scikit-learn [29] and Python 3.6.

5.2   Preprocessing

To preprocess our data, we used min-max normalization. Moreover, it is obvious that some of the features are correlated. For instance, most readability indices show a cross-correlation greater than 0.8. Another example of correlated feature pairs is average and median length of sentences or TTR values for all words and for particular parts of speech (Fig. 2). To reduce the influence of feature correlation on the LSVC model, we applied linear dimensionality reduction using Singular Value Decomposition of the data with the minimum number of principal components such that 95% of the variance is retained.

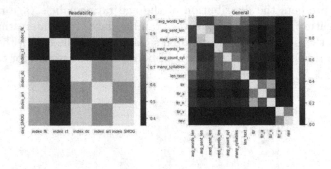

Fig. 2. Correlation matrices for readability and general features.

5.3   Experiments and Results

We used models trained of TF-IDF vectors as baselines. Further, we systematically evaluated each type of features.

To begin this process, we connected the TF-IDF vector of the text with the corresponding vector of features of a certain type. Book previews are rather long texts. Since the RuBERT model that participated in the comparison can only process a sequence of limited length, we used the same fragments of 256 tokens to train both neural networks and to construct TF-IDF vectors[2]. TF-IDF vectors

---

[2] The maximum sequence length for BERT is 512 tokens. However, due to the rather large volume of the corpus, we were also limited in computational resources.

were built over the top 2000 words ordered by term frequency across the corpus. At the same time, the values of additional features were calculated for the full preview texts.

To build TF-IDF vectors and CNN, the texts were pre-processed. The preprocessing included the following steps: special character removal, lowercase translation, lemmatization, stop-word removal. Text preprocessing was implemented using NLTK [23] and Pymorphy2 [19].

Table 4 shows the results obtained for each type of features (e.g. RF baseline + readability, LSVC baseline + readability) and the results of the models trained only on additional features without TF-IDF vectors (e.g. RF (readability), LSVC (readability)). For publishing attributes, we evaluated two separate types of models. The first type used book abstracts as supplementary information. In other words, we added the texts of abstract to the book preview and built new TF-IDF vectors. The second type used baseline TF-IDF vectors with an additional feature of age rating. Finally, we evaluated three types of combined models, such as using all considered features, only all additional features, and all features with the exception of editorial attributes. The results obtained were compared with the results of three neural models:

- RuBERT [20], based on BERT architecture [8]. BERT showed state-of-the-art results on a wide range of NLP tasks. RuBERT was trained on the Russian part of Wikipedia and news data. The model was implemented using PyTorch [28] and Transformers [43] libraries, it was trained for 3 epoches;
- FNN trained on fine-tuned RuBERT [20] text embeddings obtained with PyTorch [28]. Text embeddings were calculated by averaging the token vectors of the last hidden state. Age rating was presented as a one-hot numeric array which is the most widely used coding scheme for categorical features [31]. The FNN consisted of three layers including an input layer, a 1024 hidden layer with hyperbolic tangent activation function, and an output layer with softmax activation function. We also used Adam as an optimizer and binary cross-entropy loss. The model was implemented using Keras [13] library;
- CNN trained on Word2Vec embeddings [21]. CNN consisted of four building units including three convolutional units (CU) and a fully connected unit. Each CU contained the following sequence of layers $C - BN - C - BN - P$ where C is a convolutional layer (CL), BN is a batch normalization layer, P is a pooling layer. After every CL the LeakyRelu activation function was applied. At the first CU we used 512 filters 7×7. At the second CL we applied 1024 filters 5×5. As a pooling strategy at each layer we used max polling with a kernel 2×2. The fully connected layer consisted of the following sequence of layers $FL_1 - BN - FN_2$ where FL_1 is a hidden layer with 32 neurons, FL_2 is an output layer. We applied ReLU as an activation function and used stochastic gradient descent with Nesterov momentum and learning rate equal to 5×10^{-2} as optimization parameters. The model was implemented with PyTorch [28].

Table 4. Age-based classification results (%).

Model	Accuracy	F1-score	Precision	Recall
Baselines				
RF (TF-IDF)	85.8	86.37	90	83.03
LSVC (TF-IDF)	83.7	84.01	85.6	82.47
Readability features				
RF baseline + readability	86.5	86.91	89.6	84.37
LSVC baseline + readability	84.9	85.63	90	81.67
RF (readability)	60.6	61.6	63.2	60.08
LSVC (readability)	60.8	61.53	52.6	74.11
Sentiment features				
RF baseline + sentiment	85.4	86.04	90	82.42
LSVC baseline + sentiment	83.8	84.09	85.6	82.63
RF (sentiment)	59.4	62.2	66.8	58.19
LSVC (sentiment)	68	67.01	65	69.15
Lexical features				
RF baseline + lexical	83.2	84.23	89.9	79.23
LSVC baseline + lexical	84.2	84.45	85.8	83.14
RF (lexical)	63.1	64.62	67.4	62.06
LSVC (lexical)	61.8	62.91	64.8	61.13
Grammatical features				
RF baseline + grammatical	85.5	86.26	*91*	81.98
LSVC baseline + grammatical	83.7	84.03	85.6	82.47
RF (grammatical)	56.3	57.86	60.1	55.87
LSVC (grammatical)	59.6	62.1	66.2	58.48
General features				
RF baseline + general	89.8	87.57	90.2	85.09
LSVC baseline + general	87.9	88.01	88.8	87.23
RF (general)	60.8	61.72	63.2	60.31
LSVC (general)	68.8	70.51	74.6	66.85
Publishing attributes				
RF baseline + abstracts	87.4	87.77	90.41	85.28
LSVC baseline + abstracts	85.7	85.89	87.01	84.8
RF baseline + age rating	90.1	90.06	91.21	88.93
LSVC baseline + age rating	89.9	88.94	88.42	89.46
RF baseline + publ. attr	90.4	90.94	96.4	86.07
LSVC baseline + publ. attr	93	92.9	91.6	94.24
All features				
RF baseline + all features	94.9	94.83	93.6	96.1
LSVC baseline + all features	**95.8**	**95.77**	**95**	**96.54**
RF (all features)	94.7	94.67	94.2	95.15
LSVC (all features)	94.2	94.09	92.4	95.85
RF baseline+all features−publ.attr	86.1	86.41	88.4	84.51
LSVC baseline+all features−publ.attr	87.3	87.54	89.2	85.93
RuBERT	*90.5*	*90.16*	84.28	*93.55*
FNN (RuBERT embs + age rating)	94.8	94.78	94.4	95.31
CNN	82.1	80.2	89.6	72.59

The results show that additional features in most cases improve the quality of baselines. According to F1-score, this concerns readability features, general features and publishing attributes. We assume that the advantage of these features is that they describe the text at the document level and allow the model to evaluate the whole text, and not just a fragment. It also can be seen that the using of abstracts and the age rating feature significantly improves the quality of the classification. The best results was obtained by the LSVC model using all considered features (95.8% of accuracy, 95.77% of F1-score, 95% of precision, and 96.54% of recall). These values are shown in bold in Table 4. Among the models that did not use publishing attributes, the best results were shown by RuBERT (90.5% of accuracy, 90.16% of F1-score, and 93.55% of recall) and the LSVC baseline with grammatical features (91% of precision).

Table 5. Example errors.

Original Category	Predicted Category	Age Rating	Genre	Fragment
Adults	Child-ren's	16+	Modern detectives	Morning seeped into the cracks between the curtains. Cheerful, perky, not even annoying. A ray of sun caressingly stroked my cheek. Gently, even somehow intelligently, a green branch knocked on the glass. Alice turned languidly in bed and smiled, happily and lazily. She hadn't even opened her eyes yet. But she already had a reason to smile.[a]
Child- ren's	Adults	12+	Child- ren's detectives; love stories for teens	"I would like you to fight, or kill someone, or do something extraordinary," Vika scolded her boyfriend indignantly, when he appeared with a dead flower to the meeting place second to second. Sometimes Vika was deliberately late in the hope that Petrishchev would be offended and leave and at least some kind of adventure would arise in their relationship. But the young man did not take offense, but waited meekly, pinching the withered color.[b]

[a] "The Queen of the Dead" by Anna Veles (translated from Russian).
[b] "The Flight into the Abyss" by Andrey Anisimov (translated from Russian).

Table 5 shows some error examples. It can be supposed that in order to increase the classification quality, the model needs additional document-level features, as well as features that characterize the plot and characters as a whole.

6 Conclusion

The purpose of the current study was to evaluate different types of features for the task of age-based text classification. The results of this investigation show that features used in text difficulty evaluation can improve the quality of age-based classification. In addition, in this study, we considered publishing attributes (such as book abstracts and age ratings) as classification features. The results showed that the use of these attributes in digital libraries and recommendation systems could significantly improve the quality of machine learning approaches. Our further research will focus on studying other types of features, such as named entity analysis or plot and character features.

References

1. Aivazyan, S.A., Bukhshtaber, V.M., Enyukov, I.S., et al.: Applied Statistics: Classification and Dimension Reduction: A Handbook. Fin. i stat, Moscow (1989)
2. Azpiazu, I.M., Pera, M.S.: Multiattentive recurrent neural network architecture for multilingual readability assessment. Trans. Assoc. Comp. Ling. **7**, 421–436. https://doi.org/10.1162/tacl_a_00278
3. Balyan, R., McCarthy. K.S., McNamara, D.S.: Applying natural language processing and hierarchical machine learning approaches to text difficulty classification. Int. J. Art. Intell. Educ., 1–34 (2020). https://doi.org/10.1007/s40593-020-00201-7
4. Bertills, Y.: Beyond Identification: Proper Names in Children's Literature. Abo Akademi University Press, Turku (2003)
5. Corpus and Baselines for Age-Based Text Clas. https://github.com/oldaandozerskaya/age_based_classification. Accessed 24 Sep 2020
6. Crossley, S.A., Skalicky, S., Dascalu, M., et al.: Predicting text comprehension, processing, and familiarity in adult readers: new approaches to readability formulas. Discourse Process. **54**, 5–6 (2017). https://doi.org/10.1080/0163853x.2017.1296264
7. Cuzzocrea, A., Bosco, G.L., Pilato, G., Schicchi, D.: Multi-class text complexity evaluation via deep neural networks. In: Yin, H., Camacho, D., Tino, P., Tallón-Ballesteros, A.J., Menezes, R., Allmendinger, R. (eds.) IDEAL 2019. LNCS, vol. 11872, pp. 313–322. Springer, Cham (2019). https://doi.org/10.1007/978-3-030-33617-2_32
8. Devlin, J., Chang, M.W., Lee, K. et al.: Bert: pre-training of deep bidirectional transformers for language understanding. arXiv preprint arXiv:1810.04805 (2018)
9. Dogruel, L., Joeckel, S.: Video game rating systems in the US and Europe: comparing their outcomes. Int. Commun. Gaz. **757**, 672–692 (2013)
10. Didegah, F., Thelwall, M.: Which factors help authors produce the highest impact research? Collaboration, journal and document properties. J. Inf. **7**(4), 861–873 (2013). https://doi.org/10.1016/j.joi.2013.08.006
11. Grealy, L., Driscoll, C., Cather, K.: A history of age-based film classification in Japan. Japan Forum (2020). https://doi.org/10.1080/09555803.2020.1778058
12. Glazkova, A.: An approach to text classification based on age groups of addressees. SPIIRAS Proc. **52**(3), 51–69 (2017). https://doi.org/10.15622/sp.52.3
13. Gulli, A., Pal, S.: Deep learning with Keras. Packt Publishing Ltd. (2017)
14. Kim, S.W., et al.: A global comparative study on the game rating system. J. Digital Convergence **1712**, 91–108 (2019)
15. Fed. Law N 436-FZ "On the Protection of Children from Information Harmful to Their Health and Development. http://www.consultant.ru/document/cons_doc_LAW_108808/. Accessed 23 Jul 2020

16. Flekova, L., Stoffel, F., Gurevych, I. et al.: Content-based analysis and visualization of story complexity. In: Vis. sprachlicher Daten, pp. 185–223. Heidelberg: Heid. Univ. Publishing (2018)

17. Juilland, A.G., Brodin, D.R., Davidovitch, C.: Frequency dictionary of French words. Hague, Paris (1971)

18. Hamid, R.S., Shiratuddin, N.: Age classification of the existing digital game content rating system across the world: a comparative analysis. In: Proceedings of KMICe, pp. 218–222 (2018)

19. Korobov, M.: Morphological analyzer and generator for Russian and Ukrainian Languages. In: Khachay, M.Y., Konstantinova, N., Panchenko, A., Ignatov, D.I., Labunets, V.G. (eds.) AIST 2015. CCIS, vol. 542, pp. 320–332. Springer, Cham (2015). https://doi.org/10.1007/978-3-319-26123-2_31

20. Kuratov, Y., Arkhipov, M.: Adaptation of deep bidirectional multilingual transformers for Russian language, arXiv preprint arXiv:1905.07213. 2019

21. Kutuzov, A., Kuzmenko, E.: WebVectors: a toolkit for building web interfaces for vector semantic models. CCIS **661**, 155–161 (2017). https://doi.org/10.1007/978-3-319-52920-2_15

22. Laposhina, A.N., Veselovskaya, T.S., Lebedeva, M.U. et al.: Automated text readability assessment for Russian second language learners. In: Komp. Lingv. i Intel. Tehn., pp. 396–406 (2018)

23. Loper, E., Bird, S.: NLTK: the natural language toolkit, arXiv preprint cs/0205028 (2002)

24. Loukachevitch, N., Levchik, A.: Creating a General Russian Sentiment Lexicon. In: Proc. of LREC-2016, pp. 1171–1176 (2016)

25. Lyashevskaya, O.N., Sharov, S.A.: Frequency Dictionary of the Modern Russian Language (based on the materials of the National Corps of the Russian Language). Azbukovnik, Moscow (2009)

26. Mukherjee, P., Leroy, G., Kauchak, D.: Using Lexical Chains to Identify Text Difficulty: A Corpus Statistics and Classification Study. J. of Biomed. and Health Informatics **23**(5), 2164–2173 (2019). https://doi.org/10.1109/jbhi.2018.2885465

27. Oborneva, I.V.: Automated estimation of complexity of educational texts on the basis of statistical parameters. Pedagogy Cand. Diss, Moscow (2006)

28. Paszke, A. et al.: Pytorch: An imperative style, high-performance deep learning library. In: Adv. in neur. inf. proc. systems, pp. 8026–8037 (2019)

29. Pedregosa, F., Varoquaux, G., Gramfort, A., et al.: Scikit-learn: machine learning in Python. J. Mach. Learn. Res. **12**, 2825–2830 (2011)

30. Piasecki, S., Malekpour, S.: Morality and religion as factors in age rating computer and video games: ESRA, the Iranian games age rating system. Online-Heidelberg J. of Religions on the Int., 11

31. Potdar, K., Pardawala, T.S., Pai, C.D.: A comparative study of categorical variable encoding techniques for neural network classifiers. Int. J. Comp. Appl. **1754**, 7–9 (2017). https://doi.org/10.5120/ijca2017915495

32. Russian National Corpus. https://ruscorpora.ru/new/en/index.html. Accessed 23 Jul 2020

33. Shafaei, M., Samghabadi, N.S., Kar, S., Solorio, T.: Age suitability rating: predicting the MPAA rating based on movie dialogues. In: Proceedings of The 12th Language Resources and Evaluation Conference, pp. 1327–1335 (2020)

34. Sharoff, S.: Meaning as use: exploitation of aligned corpora for the contrastive study of lexical semantics. In: Proceedings of LREC02, pp. 447–452. Las Palmas, Spain (2002)

35. Schicchi, D., Pilato, G., Bosco, G.L.: Deep neural attention-based model for the evaluation of italian sentences complexity. In: 2020 IEEE 14th ICSC, pp. 253–256. https://doi.org/10.1109/icsc.2020.00053

36. Schwarm, S.E., Ostendorf, M.: Reading level assessment using support vector machines and statistical language models. In: Proceedings of ACL 2005, pp. 523–530 (2005). https://doi.org/10.3115/1219840.1219905

37. Solnyshkina, M., Ivanov, V., Solovyev, V.: Readability Formula for Russian Texts: A Modified Version. In: Batyrshin, I., Martínez-Villaseñor, M.L., Ponce Espinosa, H.E. (eds.) MICAI 2018. LNCS (LNAI), vol. 11289, pp. 132–145. Springer, Cham (2018). https://doi.org/10.1007/978-3-030-04497-8_11

38. Solovyev, V., Solnyshkina, M., Ivanov, V., et al.: Prediction of reading difficulty in Russian academic texts. J. Int. Fuzzy Syst. **36**(5), 4553–4563 (2019). https://doi.org/10.3233/jifs-179007

39. Sung, Y.T., Chen, J.L., Cha, J.H., et al.: Constructing and validating readability models: the method of integrating multilevel linguistic features with machine learning. Behav. Res. Methods **47**(2), 340–354 (2015). https://doi.org/10.3758/s13428-014-0459-x

40. Templin, M.C.: Certain Language Skills in Children; Their Development and Interrelationships. Univ. of Minnesota Press, Minneapolis (1957)

41. Text readability rating. http://readability.io/. Accessed 23 Jul 2020

42. Tomina, Y.A.: Objective Assessment of the Language Difficulty of Texts (Description, Narration, Reasoning, Proof). Pedagogy Cand. Diss, Moscow (1985)

43. Wolf, T., Debut, L., Sanh, V., et al.: HuggingFace's Transformers: State-of-the-art Natural Language Processing. ArXiv, arXiv-1910 (2019)

44. Zagoruiko, N.G.: Applied methods of data and knowledge analysis. Izd-vo IM SO RAN, Novosibirsk (1999)

Emotion Classification in Russian: Feature Engineering and Analysis

Marina Kazyulina⬨, Aleksandr Babii(✉)⬨, and Alexey Malafeev⬨

National Research University Higher School of Economics, Nizhny Novgorod, Russia
mskazyulina@edu.hse.ru

Abstract. In this paper, we address the issue of identifying emotions in Russian informal text messages. For this purpose, a new large dataset of text messages from the most popular Russian messaging/social networking services (Telegram, VK) was compiled semi-automatically. Emojis contained in the text messages were used to annotate the data for emotions expressed. This paper proposes an integrated approach to text-based emotion classification combining linguistic methods and machine learning. This approach relies on morphological, lexical, and stylistic features of the text. Furthermore, the level of expressiveness was considered as well. As a result, an emotion classification model demonstrating near-human performance was designed. In this paper, we also report on the importance of different linguistic features of the text messages for the task of automatic emotive analysis. Additionally, we perform error analysis and discover ways to improve the model in the future.

Keywords: Machine learning · Emotion identification · Emotiveness · Sentiment analysis · Natural language processing

1 Introduction

This paper focuses on the development of an integrated approach to detecting emotions expressed in short informal texts, combining linguistic analysis and machine learning. We propose an automatic classifier of text messages based on which emotions are explicitly or implicitly present in the text. Some of real-world applications may include partial automation of linguistic text analysis and psychological testing in such fields as medicine, forensics, and HR management. Moreover, it may provide new capabilities in marketing and especially in targeting. Finally, it may also greatly assist further advancement of suicide prevention mechanisms on social media platforms such as Facebook.

This paper is a much extended and enhanced version of [10]. In comparison with [10], the present paper features additional feature construction and selection with tree-based ensemble models. It also includes feature importance analysis, in particular, analysis of: lexical markers, different types of emotives[1], as well as the use of parts of speech for determining emotion. We also perform error analysis and investigate the reasons

---

[1] By emotives we mean any language units, not only lexis, that are used to express emotions.

© Springer Nature Switzerland AG 2021
W. M. P. van der Aalst et al. (Eds.): AIST 2020, LNCS 12602, pp. 135–148, 2021.
https://doi.org/10.1007/978-3-030-72610-2_10

for most frequent misclassifications. Another contribution of this paper is a substantial review of existing emotion classifiers with feature comparison, as well as links to two manually anonymized datasets[2] and our custom data parser.

The paper is structured as follows: Sect. 2 overviews some recent research in the field of automatic emotion analysis and compares current work to some of the better-known papers; Sect. 3 describes the process of dataset creation; Sect. 4 gives details on the development of our model; Sect. 5 discusses results achieved; finally, Sect. 6 draws conclusions and outlines future work.

2   Related Work

Researchers have achieved good results on image-based emotion recognition [11]. However, classifying textual dialogues based on emotions is a relatively new research area. Another challenge for automated emotion detection is that emotions are complex concepts with fuzzy boundaries and with individual variations in expression and perception.

Emotiveness as an immanent function of language is carried out by the aid of emotive features on different language levels. A number of studies have found that lexical emotive features (dictionaries containing expressive words) are effective in automatic text analysis [14]. Additionally, morphological emotives have been successfully used in the development of a system for predicting the emotiveness of texts [6], which proves that psycholinguistic markers indicate strong emotional reactions of the text author. Among such psycholinguistic markers are the Trager coefficient (ratio of verb count to adjective count, with the normal value close to 1), the coefficient of certainty of action (verb to noun ratio, with the normal value also close to 1), and also the coefficient of aggressiveness (verbs to words in the text overall, with the normal value lower than 0.6). Other indicators at the punctuation and graphic levels, such as exclamation marks, question marks and uppercase words, can also help indicate expressed emotions in the text.

Machine learning algorithms have proven to be highly effective in terms of detection, classification, and quantification of emotions of text in any form [5]. Table 1 shows a number of most relevant studies that take advantage of various machine learning technique to solve the problem of text-based emotion classification for English. While a lot of progress has been made in this field, research into emotion detection in Russian-language texts is still extremely limited. The present work is aimed at addressing this problem for Russian, and some features that we use are language-specific (see Sect. 4).

To develop an automatic classifier of emotions expressed in text messages, it is necessary to adopt a classification of emotions. As of now, there is no generally accepted emotion classification. Ekman's classification [4] appears to be the most widely used. It includes happiness, surprise, sadness, anger, fear, disgust, and contempt. At the current stage of work and given the limitations of our data described in Sect. 3, we removed the categories of disgust and contempt for lack of appropriate data. Moreover, the categories of fear and surprise were merged into one category of uncertainty due to certain similarity of the ways in which these emotions are usually expressed.

---

[2] One with manually tagged messages, another with 50K messages marked up semi-automatically; for more details, please refer to Sect. 3.

Thus, the classifier presented in this paper recognizes five categories of text, according to the emotion most vividly expressed in it: happiness, sadness, aggression, uncertainty, and neutrality (no emotion).

3 Data

Our work is focused on the informal Internet-based discourse of personal messaging. For the purposes of our research, we compiled a new corpus of text messages. The compilation process included 4 stages: 1) Sourcing and parsing of data; 2) Data preprocessing; 3) Data processing; 4) Corpus annotation.

3.1 Parsing and Processing

Working with user data involves addressing the issue of the ethical use of personal information. According to the General Data Protection Regulation (GDPR), all companies operating in the European Union are obliged to provide their users with personal data in the possession of these companies upon request. Personal data includes not only the messages written by user but also the ones they received. This allowed us to request access to and use personal messages and public chats from VKontakte[3] (a social network often used as a messenger) and Telegram[4] (a messenger with some social network features) to create our message corpus. We chose these services because they are highly popular among Russian-speaking people.

A custom parsing algorithm[5] was developed to accommodate the Telegram and VK reporting formats. Regarding the parsed data, statistical population characteristics include people that are 12–35 y.o. with education level from general to postgraduate. The statistical population is represented by the sample of 4584 people. Statistically, most of the parsed messages are 1–9 words long and the average length is 4–5 words. In order to prepare the messages for further work, it is necessary to ensure that they are all anonymized and of the same length. In order to do so, all the messages consisting of more than 9 words were discarded. We also removed tags, hashtags, and messages with no Cyrillic symbols. Furthermore, e-mail addresses and phone numbers were replaced with special tokens <email> and <phone>, respectively. Then the corpus was normalized with regular expressions and lemmatized with the *rnnmorph*[6] lemmatizer. The final dataset of lemmatized messages consists of 1,800,000 messages.

3.2 Corpus Annotation

In order to annotate the messages with emotion labels automatically, it was decided to use emoji ideograms as an objective emotion indicator. Obviously, not all the messages in corpus contain emojis. Thus, it was decided to build 5 classifiers (one for each emotion)

---

[3] https://vk.com.
[4] https://telegram.org.
[5] https://github.com/asbabiy/AIST/blob/master/Parser.py.
[6] https://github.com/IlyaGusev/rnnmorph.

Table 1. Comparison of existing classifiers

Authors	Emotion classification used	Num. of classes	Rationale behind training data	Data	Precompiled emotion lexicons	Method	F1
Klinger 2018 (Baseline) [9]	Sadness, Fear, Joy, Love, Surprise, Anger	6	Hashtags	CrowdFlower, blogs, emotion corpora	−	Logistic regression	0.258
Seyeditabari et al. 2019 [15]	Sadness, Fear, Joy, Love, Surprise, Anger	6	Pretrained word embeddings	CrowdFlower	−	biGRU	0.632
Hasan et al. 2018 [8]	Happy-Active; Happy-Inactive; Unhappy-Active; Unhappy-Inactive	4	Keywords, multiple binary classifiers for labeling	Twitter	+	SVM	0.9
Canales et al. 2016 [2]	Anger, Disgust, Fear, Joy. Sadness, Surprise, Neutral	7	Pretrained word embeddings; WordNet and Oxford synsets	Alm and Aman corpora	+	SVM	0.745
Mohammad, Kiritchenko 2015 [12]	Anger, Disgust, Fear, Joy. Sadness, Surprise	6	Hashtags	Facebook posts	+	SVM	0.595
Chatterjee et al. 2019 [3]	Happy, Sad, Angry, Others	4	Top unigrams; small sets of emojis for each emotion	SemEval-2019	−	?	0.796

(continued)

Table 1. (*continued*)

Authors	Emotion classification used	Num. of classes	Rationale behind training data	Precompiled emotion lexicons	Data	Method	F1
Gupta et al. 2017 [7]	Happy, Sad, Angry, Others	4	Top n-grams; emoticons; sentence embeddings	–	Twitter	SS-LSTM	0.713
Zahiri, Choi 2017 [16]	Sad, Mad, Scared, Powerful, Peaceful, Joyful, Neutral	7	Word embeddings		TV show Friends transcript	SCNN	0.393
Abdul-Mageed, Ungar 2017 [1]	Anger, Disgust, Fear, Joy. Sadness. Surprise	6	Hashtags; synsets	+	Twitter	GRNNs	0.801
This study	Happiness, Sadness, Anger, Uncertainty, Neutral	5	Manually selected sets of emoji; bootstrap for labeling	–	VK, Telegram	Logistic regression	0.718

trained on those objectively labeled messages. Those classifiers then labeled the rest of the data. The details of the process are given below.

First, the emojis that were used less than 40 times overall were removed from the observed data. The rest of the emojis were grouped into 4 sets according to the expressed emotion, which was identified by manually analyzing contexts:

- Joy: 😃, ♥, ☺, 😄, ♡, 😊, ♥, ♥, 🗡, ♥, 😁, 😌, ♥, ♥, ♥, ♥, ♡, ♥, ♥
- Sadness: 😣, 😔, 😕, 😖, 😟, 😦, 😧, ☹, 😞, 😨, 😥, 😖, 😓, 😰
- Anger: 😠, 😡, 👿, 😤, 🗿, 😐, 😫
- Uncertainty: 👀, 😕, 😒, 🙏, 👆
- Neutral messages were annotated manually because they do not explicitly express emotions and include emoji quite rarely

In the next step, all messages containing non-standard (irony, sarcasm) use of emoji were excluded. The number of messages with emoji used in the direct meaning was 4500 out of 11287 messages[7] containing emoji. Whether the use of emoji was direct or non-standard was estimated manually as well. The messages from the final dataset were divided into two sets, training, and validation, containing 2300 and 2200 messages, respectively. Then, in order to obtain five binary classifiers, one for each emotion, either Logistic Regression or Multinomial Naïve Bayes was used. For this purpose, we labeled the expected class as 1 and other emotions as 0 (1-vs-all strategy). The classifiers had F1-score of 90–93%. Finally, after the classifiers predicted emotions for the rest of the text messages, only those messages were selected which were chosen by one classifier only. The resulting dataset[8] of labeled messages contains 19.000–24,000 examples of each emotion and 110.000 items in total.

4 Model Development

The proposed model was obtained with the following steps: feature extraction, model selection, new feature engineering and selection, hyperparameter tuning.

In order to extract features from the labeled data, the *sklearn TfidfVectorizer*[9] was used, yielding 29714 features. After training some traditional machine learning models, the following results were obtained (Table 2).

It should be noted that we used a validation set and the metrics used were normalized for class imbalance. For further work, we chose the logistic regression as it proved to be the best fit for the data and features that we used.

For the purpose of enhancing the model's performance, we experimented with such features as the Trager coefficient, action certainty coefficient, aggressiveness coefficient (see Sect. 2), message length in words, use of uppercase words, number of exclamation

---

[7] https://github.com/asbabiy/AIST/blob/master/msg_with_emoji.csv.

[8] https://github.com/asbabiy/AIST/blob/master/train_data.csv.

[9] https://scikit-learn.org/stable/modules/generated/sklearn.feature_extraction.text.TfidfVectori zer.html.

Table 2. Traditional machine learning models comparison

	Precision	Recall	F1-score
Logistic regression	0.72	0.69	0.70
Naïve bayes	0.73	0.68	0.69
Random forest	0.67	0.65	0.65

marks, number of question marks and number of digits. However, we were unable to improve the F1-score, so the model's performance remained at 0.70 at this stage.

To combat overfitting, the logistic regression function was regularized. The optimal parameters were calibrated using *sklearn GridSearchCV*[10]. The final parameters of the model were: $C = 20$; *solver* = 'lbfgs'; *max_iter* = 100000. As a result, classifier performance (F1-score) reached 0.718 (Table 3).

Table 3. Logistic regression performance

	Precision	Recall	F1-score
Joy	0.90	0.80	0.85
Sadness	0.72	0.59	0.65
Anger	0.50	0.72	0.59
Interest	0.59	0.71	0.64
Neutrality	0.48	0.59	0.53
Weighted average	0.74	0.71	0.718

5 Discussion of Results

In order to interpret the results of the present study, the following steps were taken: 1) comparison with human performance, 2) feature importance analysis, 3) error analysis, 4) search for more features.

5.1 Human Performance

To assess the model against human performance, an experiment was conducted. Four people aged 18–19 years were asked to label a dataset containing 250 random messages from the validation set. As a result, it was found that human annotators perform only slightly better than our model: 0.74 F1-score, on average. In addition, agreement between different annotators is 70.1%. A similar result was obtained by another group of researchers [13]. Thus, we can assume that the proposed classifier has near-human performance.

---

[10] https://scikit-learn.org/stable/modules/generated/sklearn.model_selection.GridSearchCV.html.

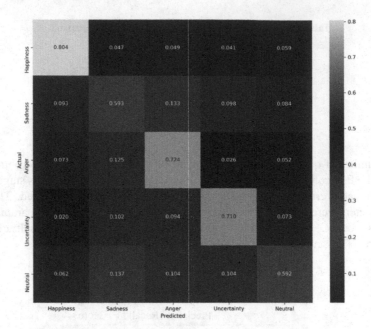

Fig. 1. Confusion matrix for emotion classes

5.2 Feature Importance Analysis

As expected, the classifier performance was different for different emotions (Fig. 1). Indeed, percentage of correctly labeled emotions ranged from 0.592 to 0.804; the Happiness class presented the least challenge, while Sadness and Neutrality were both quite difficult for the classifier. In order to get insights into the reasons for such a difference in performance, feature importance analysis was conducted.

29714 word features were extracted from the training data by TfidfVectorizer. The most important features were chosen using chi-square applied to the TF-IDF feature matrix (with respect to each class). Below are some of the lexical markers estimated to be the most important by the model:

- Happiness: молодец (well done), ура (yay), солнышко (sweetheart), благодарить (to thank), обожать (to adore), любить (to love), добрый (kind), думать (to think), удача (fortune), рад (glad).
- Sadness: скучать (to long for, to feel bored), испортить (to ruin), никогда (never), болеть (to be sick, to hurt), грустный (sad), ничего (nothing), обидеться (to get offended), бедный (poor), жаль (too bad), прощать (to forgive), плакать (to cry), завидовать (to envy), никто (nobody).
- Anger: дебилка (d*mbass), презирать (to despise), вылететь (to get kicked out, to fly out), ненавидеть (to hate), жесть (brutal, rough, creepy), жирно (greasy, brutal), стремный (weird), тупой (stupid), фу (ew), ух (ooh, ugh).
- Uncertainty: *мб* (short of *may be*), *хм* (*erm*), *почему* (*why*), *казаться* (*to seem*), *думать* (*to think*), *сколько* (*how many, how much*), *интересно* (*interesting*),

разве (interrogative particle), *ли* (interrogative particle), *вроде* (particle expressing uncertainty, *close to I think*, *probably*).

- Neutrality: странный (strange, weird), искренний (sincere), уйти (to go away), чисто (purely, simply), география (geography), который (which), знакомый (familiar), дз (short of homework), подняться (to go up), отличаться (to be different).

A few observations can be made. Firstly, there is a drastic difference between Uncertainty and Neutrality on the one hand and the rest of the categories on the other hand in terms of register used. Messages marked as uncertain or neutral rarely contain swear words or explicit language. Anger, in contrast, is characterized by a lot of taboo and offensive word markers. While Uncertainty and Neutrality tend to contain more sophisticated vocabulary (lower-frequency and/or more academic words), this is not true for all other classes. Furthermore, it can be noticed that negative pronouns may be associated with Sadness, while indefinite and interrogative pronouns are present in great numbers in the Uncertainty category. Finally, a lot of verbs seem to serve as Anger markers only in the imperative form. Thus, the grammatical form itself may be an important feature.

However, it is evident that these clusters of selected features are not exactly equal concerning the extent to which they describe the corresponding category. For example, the list of lexical markers for Happiness appears to be more descriptive of the target class than that of Neutrality. This naturally results in Neutrality being less recognizable by our model than Happiness. Thus, in order to provide comprehensive evaluation of each of the groups of features, it was decided to manually analyze the morphological composition and estimate the proportions of the following features in the top 50 correlating word features for each group:

- Denotative emotives - linguistic markers which contain emotional meaning in the very semantic core of it and much more often than not actualize such meaning in a speech (e.g. *ура*).
- Facultative emotives - linguistic markers which contain emotional meaning on the periphery of its semantic structure (usually as connotation) and may or may not actualize such meaning in a speech (e.g. *странный*).
- Potential emotives - linguistic markers which do not contain emotional meaning in its semantic structure and occasionally actualize such meaning in a speech, for example, as a result of secondary nomination. To a certain extent, any word may be qualified as a potential emotive as long as there might exist a communicative situation in which this word would obtain an emotional connotation.

Concerning potential emotives, it needs to be noted that features which fall into this category are hardly of any use in terms of helping to automate emotion identification. That is because they are not universal emotion markers but occasional. Thus, the fact that they appear in the list of the most correlated features may only mean two things: firstly, for people from our current sample this word does carry some emotional meaning, secondly, there is not enough stronger features (denotative or, at least, facultative ones) for this exact category in this exact dataset. Consequently, there is little point in considering these features in further development as we are aiming at building a classifier that would work for messages from a bigger variety of authors than just from those who are represented in the current dataset.

Moreover, a scale of expressiveness was also added, indicating to what extent the words in a group accentuate the corresponding emotion.

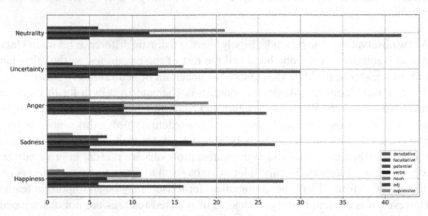

Fig. 2. Feature analysis

It is quite obvious from the graph (Fig. 2) that the denotative emotive rate and the expressiveness rate appear to be the most fluctuating values while the facultative emotive rate seems to be the most stable one among all the categories.

Since there are certain differences in recognition rates for each of the categories (as shown in the confusion matrix above, Fig. 1), it seems logical to conclude that such features as denotative emotives, adjectives and expressiveness are crucial for emotion recognition, as they are the most highly-correlated: the Pearson coefficient values are 0.727, 0.816 and 0.321, respectively. Thus, the more denotative emotives are found for a category, the greater is the probability of its recognition by the model.

5.3   Error Analysis

While Happiness is found to be the most recognizable category, Sadness and Neutrality pose the most challenge. Anger and Uncertainty are placed in between. Figure 3 demonstrates how often the two categories were confused for each other.

Statistics indicate that top three error types are Sadness being confused with other non-positive emotions. Regarding Sadness and Anger confusion, a number of messages in which sad feelings are conveyed contain a lot of expressive words such as swear words, perceived by the model as a feature of the Anger class. Thus, 13.3% Sadness messages were classified as Anger, e.g. *я послан нах*й* (*I'm told to f*ck off*), *с*кааа плачу... первый трек без тома* (*b*tch I'm crying... first track without Tom...*). On the other hand, not all aggressive messages contain explicit language. Hence, they are quite likely to be misclassified as Sadness. For example, *не бейте карину по ноге* (*don't hit Karina on the leg*), *света не общайся со мной больше* (*Sveta don't talk to*

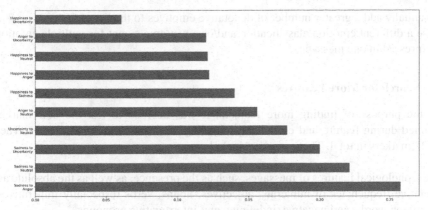

Fig. 3. Pairwise confusion of categories

me anymore), *котьку не обижай* (*don't hurt the kitten*). This error occured in 12.5% of cases with Anger messages.

As for the lack of distinction between Sadness and Neutrality, it is, perhaps, due to the total absence of denotative emotives in the Neutrality class. Consequently, whenever a neutral message contains a potential emotive of Sadness, it is misclassified as such. For instance, *хотя она ж не обижалась* (*though she didn't resent*), *я бы не ставил какие то философские вопросы в проблему* (*I wouldn't put some philosophical issues to the problem*), *ну а что не так* (*so what is wrong*). Such cases amount to 13.7% of all Neutrality items. However, 8.4% of Sadness messages were labeled as Neutrality, too. Since it is noticeable that Neutral markers are plainly random and do not reflect the specifics of a neutral emotional state, it is believed that Sadness may be misclassified as Neutrality only under the following circumstances: firstly, the sad message must have little to no emotive markers of Sadness found by the model, and secondly, there must be several markers of Neutrality. To illustrate, there are a few actual examples: *Не смогли бы записаться и уйти* (*Wouldn't be able to sign up and go*), *Он был не искренним* (*He wasn't sincere*).

Speaking of the reason for which Sadness and Uncertainty tend to be confused, one interpretation is that such emotions are often present in a message simultaneously since it is natural for a person to be upset and uncertain at the same time. Sadness was misclassified as Uncertainty in 9.8% cases (for example, *мне 14 что я вообще здесь делаю* (*I'm 14 what am I even doing here*), *как все сложно* (*everything is so complicated*), while Uncertainty was identified as Sadness in 10.2% cases (e.g. *но мучают сомнения поедем ли* (*but I'm confused if we'll go*), *я хз как это работает* (*I dunno how it works*)).

Taking into consideration all of the above, it is expected that the following steps will help eliminate errors and enhance the model's performance:

- correct the model's behavior concerning explicit language,
- exclude random word features from Neutrality and manually find and add more distinctive markers (not only words, but probably other linguistic levels as well),

- manually add a greater number of denotative emotives to the list of features,
- use a different emotion classification and/or a way to account for multiple emotions expressed in one message.

5.4 Search for More Features

For the purpose of finding more features, it was decided to test the new markers obtained during feature and error analysis, using tree-based ensemble models. There are 27 markers in total. These markers include:

- Morphological features of messages, such as the presence, as well as the absolute and relative frequencies of numerals, adjectives, nouns, verbs (especially imperatives), function words, and negative, indefinite, and interrogative pronouns.
- Formal and lexical markers such as the maximal and average word length in a message, the presence of special words of etiquette like *спасибо*, *пожалуйста* or *простите*, and so forth.
- Graphic/punctuation indicators of expressiveness such as uppercase words, emojis, exclamation and question marks.
- The presence and proportion of explicit language.

Tree-based ensemble models used for detecting significant features include Extra-Trees Classifier, Light Gradient Boosting Machine, and AdaBoost Classifier. By means of *feature_importances_* attribute, the following results were obtained (Fig. 4).

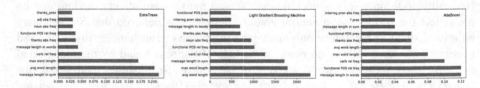

Fig. 4. Feature importance according to tree-based ensemble models

As it is evident from the presented graphs showing top ten important features, the following markers were selected as more significant by all three models: noun count, *спасибо* count, ratio of function words, message length in words, verb ratio, maximal length of a word in the message.

The presence of function words, the number of interrogative pronouns, and message length in symbols seem less significant as they managed to reach top ten of just two tree-based ensemble models out of three. Finally, the presence of *спасибо* and the number of adjectives were also found important by the ExtraTrees model. Thus, this makes a list of features worth implementing in future work.

Apart from this, it is also necessary to work out a way of distinguishing the register of words in a message as it might improve the model's performance by enhancing the recognizability of Neutrality and Uncertainty.

6 Conclusion

In this paper, we addressed the problem of developing an integrated approach to emotion classification combining linguistic analysis and machine learning. We propose an automatic classifier of emotions present in informal text messages in Russian. We also present a new dataset that was based on processing 1800000 messages. A labeled training set was generated semi-automatically: five binary classifiers (one for each emotion) were trained on a smaller manually annotated sample of 4500 messages. The performance of the proposed final classifier of emotions is 0.718 (F1-score). It is estimated to be close to human performance on this task.

Feature importance analysis has shown that denotative emotives, level of expressiveness, morphological composition and stylistic markers play a significant role. However, it is difficult to distinguish between non-positive emotions, as shown in our error analysis. Future work includes reducing the model's mistakes by changing its behavior concerning explicit language, manually excluding insignificant features, and adding important ones such as function words ratio, maximal word length in a message, verb ratio, etc. In addition, it is believed that further study will benefit from feature importance information obtained by using the eli5[11] package. Moreover, we plan to expand the emotion classification by adding the emotions of contempt and disgust and differentiating between fear and surprise. Finally, we plan to implement multi-label classification, as there are cases when the same short text or even a single sentence contains more than one emotion.

Acknowledgments. The work is supported by RSF (Russian Science Foundation) grant 20-71-10010.

References

1. Abdul-Mageed, M., Ungar, L.: Emonet: Fine-grained emotion detection with gated recurrent neural networks. In: Proceedings of the 55th Annual Meeting of the Association for Computational Linguistics, vol. 1: Long papers, pp. 718–728 (2017)
2. Canales, L., Strapparava, C., Boldrini, E., Martnez-Barco, P.: Exploiting a bootstrapping approach for automatic annotation of emotions in texts. In: 2016 IEEE International Conference on Data Science and Advanced Analytics (DSAA), pp. 726–734. IEEE (2016)
3. Chatterjee, A., Narahari, K.N., Joshi, M., Agrawal, P.: Semeval-2019 task 3: emocontext contextual emotion detection in text. In: Proceedings of the 13th International Workshop on Semantic Evaluation, pp. 39–48 (2019)
4. Ekman, P.: Expression and the nature of emotion. Approaches Emot. **3**(19), 344 (1984)
5. Gaind, B., Syal, V., Padgalwar, S.: Emotion detection and analysis on social media (2019). arXiv preprint arXiv:1901.08458.
6. Gudovskikh, D.V., Moloshnikov, I.A., Rybka, R.B.: Analiz emotivnosti tekstov na osnove psikholingvisticheskikh markerov s opredeleniem morfologicheskikh svoystv. Vestnik Voronezhskogo gosudarstvennogo universiteta. Seriya: Lingvistika i mezhkul'turnaya kommunikatsiya (3) (2015)

---

[11] https://eli5.readthedocs.io/en/latest/.

7. Gupta, U., Chatterjee, A., Srikanth, R., Agrawal, P.: A sentiment-and-semantics-based approach for emotion detection in textual conversations (2017). arXiv preprint arXiv:1707.06996.
8. Hasan, M., Rundensteiner, E., Agu, E.: Automatic emotion detection in text streams by analyzing twitter data. Int. J. Data Sci. Anal. 7(1), 35–51 (2019)
9. Klinger, R.: An analysis of annotated corpora for emotion classification in text. In: Proceedings of the 27th International Conference on Computational Linguistics, pp. 2104–2119 (2018)
10. Babii, A., Kazyulina, M., Malafeev. A.: Automatic emotion identification in russian text messages. In: Computational Linguistics and Intellectual Technologies. Papers from the Annual International Conference "Dialogue", no. 19, Supplementary volume, pp. 1002–1010 (2020)
11. Mehendale, N.: Facial emotion recognition using convolutional neural networks (FERC). SN Appl. Sci. 2(3), 1–8 (2020)
12. Mohammad, S.M., Kiritchenko, S.: Using hashtags to capture fine emotion categories from tweets. Comput. Intell. 31(2), 301–326 (2015)
13. Mohammad, S.M., Zhu, X., Kiritchenko, S., Martin, J.: Sentiment, emotion, purpose, and style in electoral tweets. Inf. Process. Manag. 51(4), 480–499 (2015)
14. Pazel'skaya, A.G., Solov'ev, A.N.: Metod opredeleniya emotsiy v tekstakh na russkom yazyke. In Komp'yuternaya lingvistika i intellektual'nye tekhnologii: Po materialam yezhegodnoy Mezhdunarodnoy konferentsii «Dialog» (Bekasovo, 25–29 maya 2011 g.). M.: Izd-vo RGGU, no. 10, p. 17 (2011)
15. Seyeditabari, A., Tabari, N., Gholizadeh, S., Zadrozny, W.: Emotion detection in text: focusing on latent representation (2019). arXiv preprint arXiv:1907.09369.
16. Zahiri, S.M., Choi, J.D.: Emotion detection on tv show transcripts with sequence-based convolutional neural networks. In: Workshops at the Thirty-Second AAAI Conference on Artificial Intelligence (2018)

Generating Sport Summaries: A Case Study for Russian

Valentin Malykh[1]([✉]), Denis Porplenko[2], and Elena Tutubalina[1,3]

[1] Kazan Federal University, Kazan, Russian Federation
valentin.malykh@phystech.edu
[2] Ukrainian Catholic University, Lviv, Ukraine
[3] National Research University Higher School of Economics,
Moscow, Russian Federation

Abstract. We present a novel dataset of sports broadcasts with 8,781 games. The dataset contains 700 thousand comments and 93 thousand related news documents in Russian. We run an extensive series of experiments of modern extractive and abstractive approaches. The results demonstrate that BERT-based models show modest performance, reaching up to 0.26 ROUGE-1F-measure. In addition, human evaluation shows that neural approaches could generate feasible although inaccurate news basing on broadcast text.

Keywords: Sport broadcast · Summarization · Russian language · Neural networks

1 Introduction

Nowadays, sports content is viral. Some events are trendy, watched by billions of people. Every day, thousands of events take place in the world that interest millions of people. The audience of online sports resources is quite broad. Even if a person watched a match, he is interested in reading the news, as there is more information in the news. Therefore, there is a great need for human resources to write this news or several for each sporting event. Media companies have become interested in cutting costs and increasing the quality of news [5]. Some well-known publications such as the Associated Press, Forbes, The New York Times, Los Angeles Times make automatic (or "man-machine marriage" form) generating news in simple topics (routine news stories) and will also introduce research to improve the quality of such news [5].

In this paper, we present the first attempt to apply state-of-the-art summarization models for automatic generation of sports news in Russian using textual comments. Our dataset is provided by a popular website `sports.ru`. The dataset consists of text comments which describe a game at a particular point in time. We explore several state-of-the-art models for summarization [10,13]. We note that recent research often evaluates models on general domain CNN/Daily Mail dataset [8], while the performance on domain-specific datasets, i.e. sport-related,

W. M. P. van der Aalst et al. (Eds.): AIST 2020, LNCS 12602, pp. 149–161, 2021.
https://doi.org/10.1007/978-3-030-72610-2_11

is not well studied. Our dataset differs greatly from the one used for training state-of-the-art approaches. Existing English CNN/DM/XSumm datasets [9,17,21] consist of public news and articles as input documents. The output summaries for them can contain either small summaries in one sentence (first sentence or news headline), summaries written by the same person after reading the input news or summaries which are obtained by the heuristic algorithmic way [16]. Our dataset contains broadcasts at the input and news as output sequences, written by different people in a different context.

In this work, we focus on the generation of news instead of the short summary with a game's results. Sports news describes the score of the match game, extend it with the details of the game including injuries, interview of coaches after the event, the overall picture or situation (e.g., "Dynamo with 14 points takes sixth place in the ranks."). Our contribution is two-fold: (i) we present a new dataset of sports broadcasts and also (ii) we provide the results of current summarization approaches to the news generation task, concluded with a human evaluation of the produced news documents.

2   Related Work

We would like to mention some works that use reinforcement training to solve abstract summarization problems. Paulus et al. proposed to solve the problem of summary generation in two stages: in the first, the model is trained with a teacher, and in the second, its quality is improved through reinforcement learning [20]. Celikyilmaz and co-authors presented a model of co-education without a teacher for two peer agents [3].

For the Russian language, there are recent works on abstract summarization, which mainly appeared in the last year. First of all, this is the work of Gavrilov et al. [4], which presented a corpus of news documents, suitable for the task of generating headings in Russian. Also, in this work, was presented the Universal Transformer model as applied to the task of generating headers; this model showed the best result for Russian and English. Some other works [7,24,25] was based on presented a corpus of news, which use various modifications of models based on the encoder-decoder principle.

Next, we will consider works that are directly related to news generation as a summary of the text. Here we want to highlight a study by the news agency Associated Press [5]. Andreas Graefe, in this study, talks in detail about the problems, prospects, limitations, and the current state in the direction of automatic generating news. In the direction of generating the results of sports events, there is little research. The first is a relatively old study based on the content selection approach performed on a task-independent ontology [1,2].

3   Dataset

For the experiments, we used data provided by http://sports.ru. The data provided in the form of two text entities, these are the comments from a commentator who describes an event and the news. In the provided set, there are 8781 sporting events, and each event contained several comments and news; the news was published both before and after the sporting event. A description of each entity, examples, statistical characteristics, and preprocessing steps are described below.[1]

3.1 Broadcast

The provided data consists of a set of comments for each sporting event. Figure 1 shows examples of the comments. The comments contain various types of information:

- Greetings.
  E.g. "Добрый день" ("Hello"), "хорошего дня любителям футбола" ("good day to the football fans");
- General information about the competition/tournament/series of games. E.g. "подниматься в середину таблицы" ("rise to the middle of the table"), "пятый раз в истории сыграет в групповом турнире" ("the fifth time in history [it] will play in the group tournament");
- Information about what is happening in the game/competition. E.g. "пробил в ближний угол" ("[he] strucks into the near corner", "удар головой выше ворот" ("a head hit above the goal");
- Results/historical facts/plans/wishes for the players. Ex: "0:3 после сорока минут" ("[score in the game is] 0:3 after forty minutes"), "не забивали голов в этом сезоне" ("[they] didn't score a goal this season").

Also, each comment of a game contains additional meta-information: (i) the match identifier, (ii) the names of the competing teams (e.g. Real Madrid, Dynamo, Montenegro), (iii) the name of the league (Stanley Cup, Wimbledon. Men. Wimbledon, England, June), (iv) the start time of the game, (v) an event

Добрый день! Наш сайт поздравляет всех, кто прошедшей зимой с нетерпением считал дни до старта российской премьер-лиги. Наша первая текстовая трансляция чемпионата 2009 поможет Вам проследить за событиями, которые произойдут на стадионе "Локомотив", где одноименная команда принимает гостей из "Химок".
Good day! Our site congratulates everyone who counted the days before the start of the Russian Premier League. Our first text broadcast will help you follow the events that will take place at the Lokomotiv Stadium, where the team of the same name hosts guests from "Khimki".

Fig. 1. Examples of comments for a same sport game.

---

[1] The owner of the dataset approved its publication, so it will be released shortly after the paper is published.

type (e.g. yellow card, goal), (vi) the minute in the sports game when the event occurred, and (vii) comment time.

We sorted by time and merged all the comments for one game into one large text and called it a broadcast. Before merging we cleaned the text from non-text info (like HTML tags). There are 722,067 comments in the dataset, of which we constructed 8,781 broadcasts. In the current study, we used only text information from the commentary. We would like to emphasize that some comments and news contain advertising or non-relevant information. In our experiments, we use some filtering described in Sect. 6.

3.2   News

News is a text message that briefly describes the events and results of a sports game. Unlike a brief summary, news can be published before and after the match. The news that took part in the experiments contains the following information:

- Comments and interviews of a player or a coach. E.g. "ребята отнеслись к матчу очень серьезно, я доволен" ("the guys took the game very seriously, I am satisfied"), "мы проиграли потому, что..." ("we lost because...");
- Events occurring during the game. E.g. "боковой арбитр удалил полузащитника" ("side referee removes midfielder"), "полузащитник «Арсенала» Санти Касорла забил три мяча" ("«Arsenal» midfielder Santi Cazorla scores three goals");
- General information about the competition/tournament/series of games. E.g. "сборные словакии и парагвая вышли в 1/8 финала" ("national teams of Slovakia and Paraguay reached the 1/8 finals"), "выполнить задачу на турнир выйти в четверть финал" ("[they] complete the mission for the tournament to reach the quarter-finals.");
- Game results. E.g. "таким образом, счет стал 1:1" ("thus the score was 1: 1"), "счет в серии: 0-1" ("the score in the series: 0–1");

Мадридский Реал выиграл в 22-м туре чемпионата Испании у Реал Сосьедада (4:1) и довел свою победную серию в домашних играх в этом сезоне до 11 матчей. Подопечные Жозе Моуринью в нынешнем чемпионате еще не потеряли ни одного очка на Сантьяго Бернабеу. Всего победная серия Реала в родных стенах в Примере насчитывает уже 14 встреч. Соотношение мячей 45:9 в пользу королевского клуба.

Real Madrid won in the 22nd round of the championship of Spain against Real Sociedad (4:1) and lead his winning streak in home games this season to 11 matches. José Mourinho's pupils in the current championship have not lost a single point in Santiago Bernabeu. The winning series of Real in the home walls in Example has already 14 meetings. Goal ratio 45: 9 in favor of the royal club.

Fig. 2. A sample news document for a sport game.

Each news document contains additional meta information: (i) title, (ii) time, (iii) sport game identifier. The data set contains 92,997 news documents. Figure 2 shows a sample news document.

4 Metrics

In our research, we used the ROUGE metric, which compares the quality of the human and automatic summary [12]. We have chosen this metric because it shows good statistical results compared to human judgment on the DUC datasets [19]. In ROUGE, a reference is a summary written by people, while the hypothesis (candidate) is an automatic summary. When calculating ROUGE, we can use several reference summaries; this feature makes more versatile comparisons (than comparing with only one reference summary). This metric bases on algorithms that use n-gram overlapping: the ratio of the number of overlap n-gram (between reference and candidate) to the total number of the n-gram in the reference.

$$ROUGE\text{-}N = \frac{\sum_{S \in \{ReferenceSummaries\}} \sum_{gram_n \in S} Count_{match}(gram_n)}{\sum_{S \in \{ReferenceSummaries\}} \sum_{gram_n \in S} Count(gram_n)}, \quad (1)$$

where n stands for the length of the n-gram, $gram_n$, and $Count_{match}(gram_n)$ is the maximum number of n-grams co-occurring in a candidate summary and a set of reference summaries. We used a particular case of ROUGE-N: ROUGE-1, and ROUGE-2. We also use ROUGE-L, a specific case of the longest common sub-sequence in a reference and a hypothesis. The ROUGE metric could be considered a Recall, so we could extend it to Precision and F-measure in common manner. We denote them ROUGE-N-R, $-P$, and $-F$ respectively.

5 Models

5.1 Oracle

This model generates an extractive summary that has the most value ROUGE between broadcast and news. We used the greedy search algorithm: we found the value of custom rouge (ROUGE-1-F + ROUGE-2-F) between each sentence from the broadcast and all the sentences in the news and selected the top 40 sentences. This algorithm stopped working in one of two cases: (1) the number of sentences is greater than the requested upper threshold (40 sentences) or (2) adding the next sentences does not increase ROUGE.

In this series of experiments, we decided to reduce the incoming sequence (broadcast) by applying the extractive approach techniques. We decided to apply the *Oracle* model, which selects sentences (in our case, 40 top sentences) from the broadcast, which have the maximum news relevance (gold reference). We used "bert-base-multilingual-uncased" model as encoder with $max\_pos = 512$. In this experiment, we trained two models that get a short output (the result of the Oracle model) as an input: (i) *OracleA* - a model trained with parameters as in section with parameters as model BertSumAbs and (ii) *OracleEA* - model trained with parameters as model BertSumExtAbs.

5.2   Neural Models

An approach that we utilize in our research proposed [13] is called PreSumm. Yang Liu and Mirella Lapata in [13] proposed to encode the whole document, keeping its sense, to generate a compact conclusion. The abstract summarization is reduced to the neural machine translation problem: an encoder contains a trained model (BERT), and a decoder contains a randomly initialized BERT. If training two models - one pre-trained and other randomly initialized - at the same time with the same parameters, then one model can be overfitting, and the second can be underfitting, or vice versa. The authors use different training parameters (learning rates and warmups).

BertSumAbs is a model for abstractive summarization. This model uses the NMT approach, pre-trained BERT as an encoder, and the randomly initialized transformer in the decoder. We used the abstract BertSumAbs model with bert-base-multilingual-uncased[2] as an encoder and randomly initialized BERT in the decoder. We also trained model *RuBertSumAbs* based on RuBERT model for encoder.

BertSumExtAbs is also using the NMT approach, but unlike BertSumAbs it uses the pre-trained BertSumExt on the extractive summarization task as an encoder. In this experiment, we used the double fine-tune stages for encoder: firstly, we are fine-tuning the model to the extractive summarization task, then we fine-tuning that model on the abstractive task [13]. For the first fine-tune stage, we used BERT "bert-base-multilingual-uncased", learning rate is $2 \cdot 10^{-3}$, dropout is 0.1, max_pos is 512, and 10000 warmup steps. Next, the trained model was used as an encoder for the abstractive summarization task with the same parameters as for BertSumAbs model. Inspecting the preliminary experiments, we realized that max_pos - truncates our incoming sequences; the model trains only 512 of the first broadcast tokens. According to the distribution of token lengths, this is quite small sequences to getting all vital information from the broadcast.

So we introduce *BertSumExtAbs1024* model. For it we used training parameters from previous experiments, with max_pos increased to 1024 and "bert-base-multilingual-uncased" as an encoder model. This model showed better results, compared with previous models with $max\_pos = 512$. The model trained 30,000 steps showed the best results. We hypothesize that we need to select sequences of higher dimensions or reduce the size of the input sequences while preserving the essential meanings and ideas of the entire broadcast.

We found out that generated news incorporated text that does not apply to the sports events; this text in common cases located at the end of the news. In this experiment, we eliminate sentences with such text. We call this model *BertSumAbsClean*. We deleted sentences that contained one of the specific words ("таблица"/"здесь."/"онлайн-трансляц") in broadcasts (source sequence) as well as in the news (target sequence). In broadcasts, a sentence with these words advertises online broadcasts on this site. In the news, sentences that contained

---

[2] https://github.com/google-research/bert/blob/master/multilingual.md

these words referred either to another page or to a visualization (images/table); this information did not help to generate news and increase input sequences. Often such sentences have advertised mobile applications.

There are several works [16,26] showing good results on relatively large amounts of data, and weak results on the small ones. Our dataset contained nearly 8000 data samples, so we decided to increase our data corpus using augmentation. For this experiment, we decided to increase the amount of our data ten times using synthetically generated data based on existing data samples.

Our models for augmentation are based on the work [27]. The idea is to replace words in a broadcast on the words from another model. There are two models: thesaurus and static embedding. Both models receive a word at the input and return a list of words size of 10. Each word in the set is similar to the incoming word: with higher word similarity the higher position in the returned list is correlated. Next, we use a geometric distribution to select two parameters for the model (for each generated sample): the number of words to be replaced in the broadcast and the index of word in the returned list for each word, that should be replaced.

Next we describe two models sing augmentation techniques. For the first model, called *AugAbsTh*, we used a similarity graph model of words from a Russian language thesaurus project called Russian Distributional Thesaurus[3] [18]. The word similarity graph is a distribution thesaurus for the most frequent words of the Russian language, obtained on the embedding of words, which was built on the body of texts of Russian books (12.9 billion words). For the second model, called *AugAbsW2V*, we used word2vec [15] for vectorizing words and the cosine of the angle between the vectors, as a metric for word similarity. As a pre-trained model, we used a model trained on the Russian National Corpus [11]. Since there were several news items related to one broadcast in our dataset, we did not augment the news. We have set random news that was written after a sports game; for broadcasts with less than ten news, we repeated the news. Thus, we got two datasets with sizes of about 80,000 broadcasts each.

5.3   Extractive Models

In our experiments, we apply two models to the implementation of the TextRank algorithm, the first based on the PageRank[4] algorithm, the other on the Gensim TextRank.[5] These approaches differ in the similarity function of two sentences. The PageRank-based model uses cosine distance between vectors of sentences (below we describe the algorithm of getting vector of the sentence); the Gensim model based on the BM25 algorithm. For the PageRank model, we preprocessed the broadcast text. We split the text into sentences using NLTK [14], then split it to words, and lemmatize each word using pymystem3 [22]. To vectorize the words, we used two different pre-trained models: the word2vec model, trained on

---

[3] https://nlpub.mipt.ru/Russian_Distributional_Thesaurus.

[4] http://bit.ly/diploma_pagerank.

[5] https://radimrehurek.com/gensim/index.html.

the Russian National Corpus [11] (*PageRank W2V*), and the FastText model, trained on a news corpus [23] (*PageRank FT*). To vectorize a sentence, we average all the word vectors. Then, we calculated the cosine distance between all sentences, build a similarity matrix, converted it to a graph, and applied the PageRank algorithm. The PageRank parameters remained by default from the library. After that, we selected sentences with a maximum page ranking and formed a summary from them. For the *TextRank* model, we used raw broadcast text and parameter *ratio* = 0.2, which adjust the percentage size of the summary compared to the source text (20% of the sentences from the source text will be in summary).

We also used another extractive approach called the *LexRank* algorithm. We use our implementation of LexRank algorithm[6] since the existing implementation has memory issues. We chose the top 10 sentences, the rest of the parameters used were set by default.[7]

6   Experiments

For our experiments, we selected the news document with minimum length and the ones written after the match. The results of the experiments are presented in Table 1.

All algorithms from the TextRank experiments showed very similar results: ROUGE-1-F is the same in all its variants. ROUGE-2 showed the worst results among all ROUGE metrics. *TextRank* works better than *PageRank W2V* and *PageRank FT*: ROUGE-1-F less than 0.1. We could conjecture that is due to different people describe sports commentary and news in different formats, styles and situations: the commentator describes the emotionally sporting game online, with details; the author of the news, calmly and dryly reports the results or takes an interview from the game player or coach. Therefore, these texts, when comparing, use different words, word forms, expressions. This property leads to an insignificant ROUGE metric based on the overlapping of common words.

The experiment with the *LexRank* approach showed the same result as the TextRank for ROUGE-1-F, higher by 0.01 in ROUGE-L-F and lower by 0.02 in ROUGE-2-F. This algorithm is very similar to TextRank; therefore their results are pretty close to each other.

We cut off long broadcasts and news for neural models: the maximum length of the broadcast was 2500, and the length of the news 200. To train the model, we used the NVIDIA Tesla P100 video card and split our dataset into shards, with a size of 50 examples. Next, we will describe different experiments with different approaches, parameters.

The best ROUGE results show the model that has been trained in 50,000 steps. We noticed that the model tended to overfit after 50,000 iterations. The ROUGE value in this experiment was the highest compared to all extractive

---

[6] https://github.com/DenisOgr/lexrank/pull/1/files.
[7] https://pypi.org/project/lexrank/.

experiments: ROUGE-1-F is greater than 0.13, ROUGE-2-F is 0.07, ROUGE-L-F is 0.13; we made this comparison using the value of the models that showed the highest result, except for the Oracle model.

RuBertSumAbs model showed lower ROUGE results compared to the Bert-SumAbs model: ROUGE-1 F is lower by 0.04, ROUGE-2-F, and ROUGE-L F are less by 0.01 and 0.05, respectively. We assume that the reason for this is the encoder model: "bert-base-multilingual-uncased" generates contextual vectors better than "RuBERT".

The model *BertSumExtAbs1024* did not show significant improvements, compared to the best models, where we used $max\_pos = 512$. The values of ROUGE-1-P and ROUGE-L-P are higher by 0.01. We want to note that this model was trained for 30,000 steps, and this is 20,000 steps lower than BertSumAbs. We have seen that increasing the input sequence from 512 to 1024 did not produce significant improvements, according to the ROUGE metric. We assume that this property of ROUGE metric: the overlap words between summary and "gold" news do not increase while increasing the input sequence in PreSumm approaches.

The *Oracle* models from this experiment showed approximately the same results (among themselves): ROUGE-1-F, ROUGE-2-F are the same, and ROUGE-L-F is 0.01 more for OracleEA than for OracleA. Therefore, we will make a comparison of other models with the best model for ROUGE in this experiment. Also, these models were trained on different numbers of steps: OracleA at 30,000 and OracleEA at 40,000, respectively.

As for *BertSumAbsClean* we have noticed that metric ROUGE decreased compared to previous experiments, and we got the best model by ROUGE, the trained model only 20000 steps (this is the lowest number of training steps in our experiments). Comparing to the oracle models ROUGE-1-F and ROUGE-L-F metrics are less than 0.02, and ROUGE-2-F is less than 0.004. We hypothesize that deleted sentences were increasing our ROUGE: "gold" and generated news had advertisements and "referred" sentences, and they increase the ROUGE.

Both the augmented models indicated significant improve performance of our task and was training on 100000 steps. However, the *AugAbsTh* model showed a higher ROUGE score than the *AugAbsW2V*: the scores of ROUGE-1-F, ROUGE-2-F, and ROUGE-L-F are higher by 0.04. This indicates that using synonyms to generate words in our task is more robust and significantly better than using word2vec embeddings. AugAbsTh model has outperformed the best previous model BertSumExtAbs1024 as well as the oracle models. The score of ROUGE-1-F and ROUGE-2-F are higher by 0.05 and ROUGE-L-F scores higher by 0.07 compared to BertSumExtAbs1024. Comparing with the OracleA model, AugAbsTh has ROUGE-1-F score higher on 0.01, ROUGE-2-F on 0.03, and ROUGE-L on 0.04 accordingly. This suggests that increasing the data corpus using real or "similar to real" data will increase the performance of the models.

Table 1. ROUGE scores from all models.

Method/Metric	ROUGE-1			ROUGE-2			ROUGE-L		
	P	R	F	P	R	F	P	R	F
Oracle	0.2	0.22	0.21	0.02	0.02	0.02	0.18	0.20	0.19
OracleA	0.23	**0.30**	0.25	0.09	0.13	0.10	0.22	**0.28**	0.22
OracleEA	0.23	0.29	0.25	0.09	0.12	0.10	0.21	0.27	0.21
PageRank W2V	0.06	0.15	0.08	0.00	0.00	0.00	0.06	0.15	0.06
PageRank FT	0.06	0.13	0.08	0.00	0.00	0.00	0.06	0.13	0.06
TextRank	0.05	0.17	0.08	0.00	0.01	0.00	0.05	0.16	0.06
LexRank	0.07	0.10	0.08	0.00	0.00	0.00	0.07	0.09	0.06
BertSumAbs	0.19	0.25	0.21	0.07	0.10	0.07	0.17	0.23	0.18
RuBertSumAbs	0.14	0.26	0.17	0.04	0.10	0.06	0.13	0.24	0.13
BertSumExtAbs	0.18	0.25	0.2	0.06	0.10	0.07	0.17	0.23	0.17
BertSumExtAbs1024	0.20	0.25	0.21	0.07	0.10	0.08	0.18	0.23	0.18
BertSumAbsClean	0.18	0.24	0.19	0.07	0.07	0.06	0.18	0.18	0.16
AugAbsTh	**0.26**	**0.30**	**0.26**	**0.12**	**0.14**	**0.13**	**0.23**	**0.28**	**0.25**
AugAbsW2V	0.23	0.26	0.22	0.08	0.10	0.09	0.19	0.25	0.21

7   Human Evaluation

The effectiveness of ROUGE was previously evaluated [6,12] through statistical correlations with human judgment on the DUC datasets [19]. To judge the news, we asked five annotators to rate the news by four dimensions: relevance (selection of valuable content from the source), consistency (factual alignment between the summary and the source), fluency (quality of individual sentences), and coherence (collective quality of all sentences). We chose five random news from different models (with the different number of training steps). We chose only abstractive models since these models have shown better performance compared to the extractive ones. The summary score for each dimension is obtained by averaging the individual scores. The comparison results are displayed in Table 2.

Table 2. Human evaluation results.

Model/Metric	Relevance	Consistency	Fluency	Coherence
OracleA	0.46	0.60	0.78	0.78
BertSumExtAbs1024, 10k steps	0.36	0.58	0.56	0.46
BertSumExtAbs1024, 30k steps	0.26	0.34	0.70	0.54
BertSumAbs	0.28	0.50	0.56	0.58
BertSumAbsClean	0.28	0.56	0.72	0.70

Analyzing the data from Table 2, we want to emphasize that the values of Fluency and Coherence are generally higher than the values of Relevance and

Consistency. This suggests that the models from our experiments generate pretty high-quality and linked sentences, but worse select events from the broadcast. The highest scores of Fluency and Coherence have OracleA and BertSumAbsClean models. News generated by BertSumAbs and OracleA models have the highest scores of Relevance. Concluding this experiment, we did not observe any visual relationships between human judgment and the ROUGE metric. We also want to notice that we received some comments from annotators regarding the quality of the news. Most of the comments were aimed at the fact that the quality of the sentences is pretty good, but the news does not review important events or reviews non-existent events from the broadcast.

8   Conclusion

In this paper, we have investigated the task of generating news based on sports commentary with state-of-the-art approaches for summarization. The main challenges of this novel dataset are: 1) the average size of one document differs greatly from texts in existing general domain corpora [9,17,21], 2) domain of texts is sport-related that includes diverse information about a match game, 3) news are written in a language other than English. Unlike expectations, the state-of-the-art neural models show modest performance for our task. We obtained the maximum value 0.26 by ROUGE-1-F score using BERT as an encoder. We found out that increasing data corpus using text argumentation based on thesaurus gives a substantial improvement: we increase data per ten times, and the ROUGE-1-F score has gone up on 0.05 in the absolute difference in comparison with best no augmentation score.

The quantitative analysis opens up several future research directions. First, we plan to increase the number of documents in our dataset by transforming the comments of the sporting event from audio sources, which are more popular than textual. Second, the effective application of transformers as an encoder suggests continuing experiments with other types of transformers, like GPT-2 or different BERT-based architectures. Finally, we could explore a custom evaluation metric based on the main characteristics of a game: overall score or main events.

We hope that this work will foster the research in text generation in Russian and for narrow domain texts in general.

Acknowledgements. The work of the first author was funded by RFBR, project number 19-37-60027. The final work on the manuscript carried out by Elena Tutubalina was funded by the framework of the HSE University Basic Research Program and Russian Academic Excellence Project "5–100".

References

1. Bouayad-Agha, N., Casamayor, G., Mille, S., Wanner, L.: Perspective-oriented generation of football match summaries: old tasks, new challenges. ACM Trans. Speech Lang. Process. **9**(2), 3:1–3:31 (2012). https://doi.org/10.1145/2287710.2287711

2. Bouayad-Agha, N., Casamayor, G., Wanner, L.: Content selection from an ontology-based knowledge base for the generation of football summaries. In: Proceedings of the 13th European Workshop on Natural Language Generation, pp. 72–81. Association for Computational Linguistics, Nancy, France, September 2011. https://www.aclweb.org/anthology/W11-2810
3. Celikyilmaz, A., Bosselut, A., He, X., Choi, Y.: Deep communicating agents for abstractive summarization (2018)
4. Gavrilov, D., Kalaidin, P., Malykh, V.: Self-attentive model for headline generation. In: Azzopardi, L., Stein, B., Fuhr, N., Mayr, P., Hauff, C., Hiemstra, D. (eds.) ECIR 2019. LNCS, vol. 11438, pp. 87–93. Springer, Cham (2019). https://doi.org/10.1007/978-3-030-15719-7_11
5. Graefe, A.: Graduate school of Journalism. Tow Center for Digital Journalism, C.U.G.S., GitBook: Guide to Automated Journalism (2016). https://books.google.com.ua/books?id=0iPbjwEACAAJ
6. Graham, Y.: Re-evaluating automatic summarization with BLEU and 192 shades of ROUGE. In: Proceedings of the 2015 Conference on Empirical Methods in Natural Language Processing, pp. 128–137. Association for Computational Linguistics, Lisbon, Portugal, September 2015. https://doi.org/10.18653/v1/D15-1013, https://www.aclweb.org/anthology/D15-1013
7. Gusev, I.: Importance of copying mechanism for news headline generation (2019)
8. Hermann, K.M., et al.: Teaching machines to read and comprehend. CoRR abs/1506.03340 (2015). http://arxiv.org/abs/1506.03340
9. Hermann, K.M., et al.: Teaching machines to read and comprehend. In: Cortes, C., Lawrence, N.D., Lee, D.D., Sugiyama, M., Garnett, R. (eds.) Advances in Neural Information Processing Systems 28, pp. 1693–1701. Curran Associates, Inc. (2015). http://papers.nips.cc/paper/5945-teaching-machines-to-read-and-comprehend.pdf
10. Klein, G., Kim, Y., Deng, Y., Senellart, J., Rush, A.M.: OpenNMT: open-source toolkit for neural machine translation. CoRR abs/1701.02810 (2017). http://arxiv.org/abs/1701.02810
11. Kuratov, Y., Arkhipov, M.: Adaptation of deep bidirectional multilingual transformers for Russian language (2019)
12. Lin, C.Y.: ROUGE: a package for automatic evaluation of summaries. In: Text Summarization Branches Out, pp. 74–81. Association for Computational Linguistics, Barcelona, Spain, July 2004. https://www.aclweb.org/anthology/W04-1013
13. Liu, Y., Lapata, M.: Text summarization with pretrained encoders (2019)
14. Loper, E., Bird, S.: NLTK: the natural language toolkit. In: Proceedings of the ACL-02 Workshop on Effective Tools and Methodologies for Teaching Natural Language Processing and Computational Linguistics. ETMTNLP 2002, vol. 1, p. 63–70. Association for Computational Linguistics, USA (2002). https://doi.org/10.3115/1118108.1118117
15. Mikolov, T., Chen, K., Corrado, G., Dean, J.: Efficient estimation of word representations in vector space (2013)
16. Nallapati, R., Zhou, B., dos santos, C.N., Gulcehre, C., Xiang, B.: Abstractive text summarization using sequence-to-sequence RNNs and beyond (2016)
17. Narayan, S., Cohen, S.B., Lapata, M.: Don't give me the details, just the summary! topic-aware convolutional neural networks for extreme summarization. In: Proceedings of the 2018 Conference on Empirical Methods in Natural Language Processing, pp. 1797–1807. Association for Computational Linguistics, Brussels, Belgium, October–November 2018. https://doi.org/10.18653/v1/D18-1206, https://www.aclweb.org/anthology/D18-1206

18. Panchenko, A., Ustalov, D., Arefyev, N., Paperno, D., Konstantinova, N., Loukachevitch, N., Biemann, C.: Human and machine judgements for Russian semantic relatedness. In: Analysis of Images, Social Networks and Texts: 5th International Conference, AIST 2016, Yekaterinburg, Russia, 7–9 April 2016, Revised Selected Papers, pp. 221–235. Springer International Publishing, Yekaterinburg, Russia (2017). https://doi.org/10.1007/978-3-319-52920-2_21
19. Over, P.: An introduction to DUC-2001: intrinsic evaluation of generic news text summarization system (2001)
20. Paulus, R., Xiong, C., Socher, R.: A deep reinforced model for abstractive summarization (2017)
21. Sandhaus, E.: The New York times annotated corpus LDC2008t19 (2008)
22. Segalovich, I.: A fast morphological algorithm with unknown word guessing induced by a dictionary for a web search engine, pp. 273–280 (2003)
23. Shavrina T., Shapovalova, O.: To the methodology of corpus construction for machine learning: « taiga» syntax tree corpus and parser. In: Proceedings of CORPORA2017, International Conference, Saint-Petersbourg (2017)
24. Sokolov, A.: Phrase-based attentional transformer for headline generation. In: Computational Linguistics and Intellectual Technologies (2019)
25. Stepanov, M.: News headline generation using stems, lemmas and grammemes. In: Computational Linguistics and Intellectual Technologies (2019)
26. Tan, J., Wan, X., Xiao, J.: From neural sentence summarization to headline generation: a coarse-to-fine approach. In: Proceedings of the 26th International Joint Conference on Artificial Intelligence, pp. 4109–4115. IJCAI 2017. AAAI Press (2017). http://dl.acm.org/citation.cfm?id=3171837.3171860
27. Zhang, X., Zhao, J., LeCun, Y.: Character-level convolutional networks for text classification (2015)

Automatic Generation of Annotated Collection for Recognition of Sentiment Frames

Yuliana Solomatina$^{(\boxtimes)}$ and Natalia Loukachevitch

Lomonosov Moscow State University, Moscow, Russia

Abstract. While addressing the challenge of sentiment analysis, it is crucial to consider not only the polarity of certain words but also the polarity between them, particularly between the arguments of a predicate. For this purpose, the RuSentiFrames lexicon was created. But the training of the ML model requires an annotated collection of data, and since the manual annotation is laborious and expensive, the automation of the process is preferable. In this paper, we describe a rule-based approach to automatic annotation of semantic roles for the predicates of the RuSentiFrames lexicon. The implementation of the algorithm includes the search of the entities with certain morpho-syntactic features in the order that depends on the case of the entity and is based on calculation of the posterior probabilities of the co-occurrence of a certain case and a certain type of predicate arguments. The results of the algorithm evaluation, based on several different characteristics, were relatively high. The solutions of problematic cases have been suggested and are expected to be implemented in further research.

Keywords: Sentiment analysis · Semantic role labeling · Annotated collection generation

1 Introduction

In order to conduct the semantic analysis of text material, the extraction of semantic relations between the words of the sentence is essential. For this purpose, sets of semantic roles have been elaborated [6, 23] and specific linguistic resources have been created [6, 15], including the frame-based ones (such as FrameNet, introduced in [5]. The earliest approaches to semantic role labeling were rule-based [20, 24]). However, the Machine Learning algorithms are considered preferable nowadays [2, 4, 22].

But the ML models have a significant disadvantage, which is the necessity to annotate large amounts of data manually. Since this process is expensive and laborious, the task of the development of specialized systems for automatic generation of resources with descriptions of semantic structures of predicates (semantic frames), along with the automatic annotation of training samples for extracting such structures, has become especially relevant. Some studies consider simultaneous generation of both semantic structures for predicates and the annotation examples [21], arguing that the building of semantic frames is a complex process which requires thorough expertise, and, consequently, frame structure projects only exist in a small number of languages.

© Springer Nature Switzerland AG 2021
W. M. P. van der Aalst et al. (Eds.): AIST 2020, LNCS 12602, pp. 162–174, 2021.
https://doi.org/10.1007/978-3-030-72610-2_12

In this paper we consider another task setting to generating an annotated collection if compared to unsupervised frame extraction [21]. We suppose that a frame lexicon comprising frames and roles is already created but annotated collections labeled with frame elements should be created automatically. As a frame lexicon, we consider RuSentiFrames [12].

RuSentiFrames is a structured sentiment lexicon for Russian, developed for using in fine-grained sentiment analysis, which additionally includes extraction of sentiment attitudes between participants, positive and negative effects, dependence of sentiment on the reader's position. Currently more than 300 frames are described, the frames comprise more than 7 thousand lexical units, but the annotated dataset allowing for effective extraction of sentiment frames is currently absent. The proposed approach uses morpho-syntactic and statistical analysis of text data.

2 Related Work

In the majority of the studies on semantic role labeling two corpora were used to train the ML-model: FrameNet [5] and PropBank [16]. In the first one consists of the sets of semantic roles, defined specifically for each predicate (such as BUYER, GOODS, SELLER, MONEY for the verbs BUY or SELL). As for the PropBank, it presents the extension of the Penn Treebank corpus, intended for the syntactic parsers. The arguments are numbered from 0 to 5 (where A1 complies with a Prototypical Agent, while Arg1 is a Prototypical Patient) and are labeled with a brief description of their semantic function. Also, the polysemic predicates are given individual frames. And it is important that in the PropBank raising and control predicates can be easily distinguished due to special annotation features. It is important because these predicates carry infinite clauses as there semantic arguments, and the distinction between object and subject raising and control helps define whether the sentential argument is Agent or Patient.

Both corpora have been widely used for the semantic parsing, but the PropBank is considered to be more appropriate due to having a more suitable data structure. With regard to the Russian language, a FrameNet-oriented resource was developed for semantic role labeling – FrameBank [15]. The authors pointed out the importance of taking language-specific features into account, claiming that frames cannot be considered universal (particularly in Russian, syntax is less valuable for SRL as compared with English). The project is aimed at linking a dictionary of valencies and an annotated corpus, processing lexical constructions and the way they are realized in texts [14].

The correlation between syntactic and semantic roles lies in the basis of many studies on automatic semantic role labeling. The method of semantic role recognition, based on syntactic features, was introduced in [7]. For each sentence from the sample a syntactic tree was built and necessary lexical and syntactic features were extracted to train the classifier. Apart from that, the probabilities of the occurrence of a certain semantic role in the position of a certain syntactic argument were calculated. The results of the research proved the usefulness of the syntactic-semantic correlation for the task of semantic role labeling. The concepts proposed in [7] have been developed in further research. For instance, in [2] the authors put forward the idea that the syntax-based approach is quite effective for semantic role labeling but is not robust since it depends entirely on the quality of the syntactic parser. They expanded the list of useful features, adding a few non-syntactic ones (such as the length of the predicate's argument, the distance between the predicate and the argument in both linear and hierarchical structures, etc.), and compared several classifying algorithms. The most appropriate classifier (according to the F-measure) turned out to be the Support Vector Machine ([9, 10, 13]). In other studies (such as [17]) this method also proved to take precedence over other classifying algorithms with regard to the semantic parsing, reaching more than 90% of the Precision, Recall, and F-measure metrics.

In the recent studies the neural networks have been widely used for semantic role labeling. In [4] the architecture of a Convolutional Neural Network, based on the vector representation of words, was introduced with a view to solving the most relevant NLP tasks, including the semantic parsing. Another robust neural network architecture was described in [8]. The authors suggested the model of the associated memory network, which utilizes the inter-sentence attention of associated sentences as a kind of memory to carry out the dependency-based SRL. It is important to mention that the syntactic features were not used during the training of the network, and yet the F-measure was relatively high (85%).

For Russian, neural network approach for SRL was proposed in [19]. The authors used atomic features with word embeddings (instead of feature engineering) to train two network models on the FrameBank corpus. They also processed out-of-domain predicates and concluded that for this task word embeddings are essentials, but not enough to achieve high performance. The idea was further developed in [11], where two separate models were introduced for "known" and "unknown" predicates. The authors showed that this approach could help alleviate the problem with annotation scarcity.

3   RuSentiFrames Lexicon

The RuSentiFrames lexicon is an auxiliary instrument for sentiment analysis in the Russian language, which contains the descriptions of more than 7000 predicates (the majority of them are single verbs and verb collocations) in terms of the polarity of the relations of their arguments [18]. The semantic role representation is similar to the one in the PropBank corpus: the arguments are numbered and are given brief descriptions of their semantics. Also, the frames provide more detailed information about the polarity between the predicate's arguments (such as the effect on a1, caused by a0; the state in which a1 remains after the impact of a0; etc.). The example of the frame for the verb *hope* is presented in Example 1 below.

Example 1. "Надеяться" (to hope)

```
"0_0": {
    "title": [ "надеяться", "возлагать надежду"],
    "variants": ["не терять надежду", "ждать", "опереться", "опираться", "чаять", "те-
шить себя надеждой", "питать надежду", "уповать", "лелеять надежду", "тсшиться
надеждой",  "возлагать надежду", "возложить надежду", "шевелится надежда", "пона-
деяться", "надежда","надеяться", "рассчитывать", "расчет",  "ставка", "ставка на"],
    "comment": "",
    "roles": { "a0": "тот, кто надеется" (the one who hopes), "a1": "то, на что надеются"
(the object of hope) },
    "frames": {"polarity": [ ["a0", "a1", "pos", 1.0] ],
        "value": [ [ "a0","a1", 1.0] ],
        "effect": [ ["a1", "+", 0.7 ] ],
        "state": [ ["a0", "pos", 1.0 ] ]
}
```

The graph "variants" contains sets of synonymic constructions for the basic predicate (which is labeled as "title"), while the graph "roles" refers to the predicate's semantic arguments and gives a short description for each role. The estimation of the polarity between arguments, which is crucial for the task of sentiment analysis, is presented in graphs "polarity" (the attitude polarity of participants to each other), "value" (values of participants), "effect" (positive or negative effect, caused by one participant on the other), and "state" (positive or negative mental state of participants).

4   Method

The process of semantic role labeling consisted of 3 stages: the data preprocessing, the clause and predicate classification and the processing of each class with specific rules.

4.1   Data Pre-processing

We apply our approach of role labeling to news titles, since they are relatively short and have syntactically simple structure. The preliminary processing includes tokenization and normalization. Then among the tokens of each title the retrieval of the predicates from the frames is carried out (including those consisting of more than one token, utilizing the n-gram model). After that, the morpho-syntactic parsing is implemented by means of the DeepPavlov Ru Syntagrus Joint Parsing model [1] with the output in the CoNLL-U format.

4.2   Clause and Predicate Classification

For the semantic role labeling it is important to classify sentences and predicates in a manner that allows us to define specific features for each class, which helps increase the accuracy of the semantic parsing. The predicates are classified depending on the voice (active or passive) and the number of arguments (1 to 3). The classification of the sentences is based on the number of clauses (1 or 2 +), the presence of an infinitive verb form and the dependency type of clauses (the subordinate clauses of time, location and manner were considered equal to independent clauses since they do not impact on the argument structure of the root verb).

4.3 Semantic Role Labeling

As was mentioned above, the processing of each class of verbs and clauses has some specific features. Let us consider the rules which are utilized for SRL on the example of simple clauses without infinitive verb forms.

a0 Extraction. In the majority of the titles a0 is the syntactic subject in the nominative case, marked by "nsubj". But there are cases of the dative subject. The algorithm is as follows: first we search for the canonical subject, if it is not found – search for the dative one, otherwise we state that a0 is not present in the sentence.

a1 Extraction. It is considered that the canonical a1 is the direct object in the accusative case, marked by "obj". If we do not find anything with these features, we search for the genitive object that has the verb as its syntactic head. Then we look for the locative case, after that – the instrumental case. In order not to catch the adjuncts of time and location, the DeepPavlov BERT NER model is used to extract named entities.

a2 Extraction. The canonical a2 is considered to be the recipient in the dative case, marked by "iobj", or the instrument in the instrument, marked by "obl". If neither is found, the search continues for the genitive and locative cases.

a0-a3 Extraction for Four-Argument Predicates. Since there are only four predicates with four participants in the frames, it is more convenient to utilize individual rules for each predicate. Here, to distinguish between the argument, it has been especially useful to take the category of animacy into consideration. For instance, for the verb "to win" the most efficient way to differentiate between the roles "prize" and "defeated one" is to point out that the first one is inanimate as opposed to the second one (and if we only rely on the case category, it might lead to errors because both these roles are usually marked with accusative). To implement these modifications, the animacy annotation has been added to the frames. The argument descriptions were complemented by one of the three tags: "smb" (animate), "smth" (inanimate) or "any" (both animate and inanimate). This also differentiates between several senses of the same verb by putting them into a separate frame, according to animacy distribution within the arguments. Example 2 illustrates the feature distribution for role extraction for the verb "to win".

Example 2. Four argument predicates

"0_4": "title": ["to win"]
"roles": { "a0": "winner" (+nominative case),
        "a1": "prize" ("smth", +accusative case),
        "a2": "defeated one" ("smb", +accusative/genitive case),
        "a3": "sphere of victory" ("smth", +locative case) }

The order of the search was defined based on the calculation of the posterior probabilities of the co-occurrence of a certain case and a certain type of predicate argument, using the information from the frames (Table 1).

Table 1. The probability distribution $p\ (case|arg)$.

	a1	a2
Accusative	0.54	0.18
Dative	0.06	0.31
Instrumental	0.07	0.26
Genitive	0.17	0.14
Locative	0.16	0.1

For the clauses in the passive voice, the algorithm is analogous, but the changes in diathesis are taken into account. During the processing of the other classes, we consider that the subordinate clauses are usually a1, as well as the non-finite clauses.

The example of the output is presented in Fig. 1.

поблагодарить («to thank») - three arguments

{'a0': 'the one who thanks', 'a1': 'recipient', 'a2': 'reason'}

id	lemma	POS	features	head	deprel	
1	Медведев	PROPN	Animacy=Anim|Case=Nom|Gender=Masc|Number=Sing	2	nsubj	Arg0
2	поблагодарить	VERB	Aspect=Perf|Gender=Masc|Mood=Ind|Number=Sing |Tense=Past|VerbForm=Fin|Voice=Act	0	root	
3	экс-президент	NOUN	Animacy=Anim|Case=Acc|Gender=Masc|Number=Sing	2	obj	Arg1
4	Бразилия	PROPN	Animacy=Inan|Case=Gen|Gender=Fem|Number=Sing	3	nmod	
5	за	APD	-	6	case	
6	укрепление	NOUN	Animacy=Inan|Case=Acc|Gender=Neut|Number=Sing	3	nmod	Arg2
7	дружба	NOUN	Animacy=Inan|Case=Gen|Gender=Fem|Number=Sing	6	nmod	
8	и	CCONJ	-	9	cc	
9	сотрудничество	NOUN	Animacy=Inan|Case=Gen|Gender=Neut|Number=Sing	7	conj	

Fig. 1. An example of processing the title "Медведев поблагодарил экс-президента Бразилии за укрепление дружбы и сотрудничество" ("Medvedev thanked the President of Brasil for the strengthening of friendship and cooperation")

In the SRL of all types of sentences it is crucial to consider two fundamental principles, formulated in earlier works.

1. The correlation between the semantic and syntactic structures (the distribution of the syntactic roles provides detailed information about the semantic roles).

2. The theta-criterion: each argument bears one and only one θ-role, and each θ-role is assigned to one and only one argument [3].

5   Evaluation and Analysis of Errors

At first, it should be noted that in the process of the algorithm development the linguistic features of the text material were taken into consideration. Since among the news titles there rarely occur complex syntactic constructions with non-trivial verb order, the algorithm processes such titles fairly well. However, there still are some problematic cases which will be analyzed in the current section.

5.1   Corpus Statistics

In order to see how useful the corpus can potentially be for our task, a large sample of data was processed to see how much needed data we could collect. The analyzed news corpus consists of 157 556 titles, 87 972 of them contain predicates, described in frames, i.e. can be leveraged for annotated collection generation, and 73 017 titles constitute simple contexts (which means they consist of one clause with no infinitive verb forms). The samples fitted for 6313 predicates from the RuSentiFrames lexicon (which contains 7430 predicates), which shows that the corpus is suitable for the task since the for the vast majority of predicates we can collect corresponding samples. In Table 2 the statistics of 10 most frequent verbs in the analyzed news corpus is presented.

Table 2.   10 most frequent predicates.

Verb	Frequency
начать (to begin)	2439
погибнуть (to die)	2046
пострадать (to get hurt)	1967
победить (to win)	1888
подписать (to sign)	1809
создать (to create)	1416
убить (to kill)	1259
помочь (to help)	1102
провести (to undertake)	1023
найти (to find)	866

5.2   Evaluation

In order to estimate the performance of the algorithm, it was tested on 1400 titles from the corpus (1036 of simple contexts and 364 of complex contexts). The error rate was calculated for titles of each type. Among the errors there were both incorrect and missed labels. The errors in simple (1) and complex (2) contexts were analyzed separately.

(1) Греция построит стену на границе с Турцией (Greece will build a wall on the border with Turkey).
(2) Азаров хочет увеличить продуктивность труда в пять раз (Azarov wants to boost productivity fivefold).

The complexity of (2) as compared with (1) which leads to necessity of parsing of several verbs and shared arguments.

5.3 Error Analysis for Simple Contexts

As it has been mentioned above, simple contexts are one-clause titles, in which predicates' arguments are mostly noun phrases.

Out of 1036 titles, 466 contained errors of at least one type. The results of the evaluation for such contexts are presented in Table 3.

Table 3. Error rate for simple contexts.

Polysemy errors	0.351
Errors in a0 extraction	0.113
Errors in a1 extractions for predicates with 2 args	0.274
Confusing a1 and a2 for predicates with 3 args	0.386
Incorrect extraction of a1 or a2 for predicates with 3 args (the one is correct, the other is not)	0.322
Confusing a1, a2 and a3 for predicates with 4 args	0.091

As it is evident from Table 3, the least frequent ones were the errors in four argument predicate titles. Since the rules for such titles were very specific for each verb, considering more features than for other classes of predicates (including animacy), the quality is quite satisfactory. However, due to the fact that there are only four predicates in these class, they occur in the corpus less frequently, which decreases the error rate. In further investigation, the samples must be balanced in size to get more realistic picture.

Another example of relatively low error rate is the a0 extraction because this argument type is the easiest one to identify. The errors here were caused by inaccuracies of the syntactic parser. The most frequent mistakes are related to the differentiation between a1 and a2. For instance, in the title (3) the a1 is mistakenly recognized as a2.

(3) Vladimir Klichko is going to fight with Haye on July 2nd.
    fight - three arguments
    {'a0': 'the one who fights', 'a1': 'the opponent', 'a2': 'what they fight for'}
    a0: Vladimir Klichko

a1:
a2: Haye

Such errors were caused by both syntactic parser errors and individual features of predicates.

Another problem relates to verb polysemy. To label semantic roles accurately, the verb disambiguation must be carried out. It is necessary to not only put aside the sense which is not described in the frames but also to differentiate between the senses within the frames. And sometimes the distinctions between frames are subtle. For example, there are two occurrences of the verb "to win": once in the three-argument category and once in the four-argument category, with the additional role "prize". If the prize is mentioned in the title, it is better to put it in the four-argument class, and if it is not mentioned, the three-argument class is preferential. But in fact, in both cases the sentence semantically fits both classes quite well. However, having two identical frame samples is undesirable, so the titles must be somehow distributed between these two classes. The possible solution might be to balance these samples so that both classes contain examples of the explicitly expressed "prize" and those where this participant is not mentioned.

It is quite evident that the verb polysemy problem is urgent, and the quality of disambiguation has a great impact on the overall quality of sample collection generation. Yet there are many aspects in which polysemy takes place, and to take all of them into account is a separate arduous task. Table 4 represents a subsample of the list of most frequent polysemic predicates from the frames.

Table 4. Most frequent polysemic verbs

	Frame sense	Non-frame sense
Ждать (wait)	To hope	To be inevitable
Разбить (break)	To win	To damage
Выгореть (burn out)	To succeed	To go up in flames
Пожелать (wish)	To want, desire	To say you hope someone has good luck, etc
Почувствовать (feel)	To realize	To experience an emotion
Видеть (see)	To consider, understand	To use one's eyesight
Покорить (subdue)	To defeat, conquer	To induce feeling of love, inspiration, etc

However, the first step to deal with the question of verb polysemy is to utilize the animacy annotation which helps split definite verbs into several frames according to the animacy distribution between the arguments. By doing so we could distinguish between different senses of the same predicate and, consequently, lower the error rate for polysemic predicates. This is the task of the highest priority in the further research.

5.4   Error Analysis for Complex Contexts

For the sentiment attitude extraction task, it is useful to extract sentiments not only from simple finite clauses but also from more complex clauses, including non-finite ones. Since the collection is suitable for both semantic role labeling and attitude extraction systems, making use of complex contexts enables to get as much data as possible out of every title.

The complex contexts were considered those containing subordinate clauses, finite or infinite. Out of 364 title, 189 contained errors of at least one type. The error rate for such contexts is in Table 5 below.

Table 5. Error rate for complex contexts.

Polysemy errors	0.239
Errors in a0 extraction	0.108
Errors in subordinate finite clauses	0.221
Errors in non-finite clauses	0.498

During the evaluation there occurred an issue that subject and object control predicates cannot be differentiated within the algorithm (4). For instance, in the sentence *X promised Y to do Z* (subject control) the verb's "to do" a0 is X, whereas in the sentence *X asked Y to do Z* (object control) this semantic position is occupied by Y. To sort out this complexity, such predicates must be given a special tag, which should be added to the frame structure. The example below (4) illustrates that the a0 for the infinite clause's predicate is wrongfully recognized as the main clause predicate's a0 because the fact that *encourage* is an object control predicate is not considered.

(4)   Poland encourages the USA to help Belarusian dissidents.
      help – three arguments
      {'a0': 'the one who helps', 'a1': 'what a0 helps with', 'a2': 'the one who a0 helps'}
      a0: Poland
      a1:
      a2: Belarusian dissidents

To sort out this problem, the manual annotation of all control predicates has been carried out to classify them based on the control type (subject or object). In the frame collection there occur 180 verbs which can function as control predicates in some contexts, 101 of them are of subject control and the rest 79 – of object control. Table 6 represents the subsample of the current classification.

Depending on whether the predicate belongs to one class or another, the a0 of the non-finite clause may coincide with the main clause's subject or object. In the case of the object control, it can be either direct or indirect object. The entire non-finite clause can also be both a direct or an indirect object for the main clause's predicate. And this

Table 6. Control predicates classification

Subject control	Object control
бояться (to fear)	вдохновить (to inspire)
хотеть (to want)	запретить (to forbid)
запланировать (to plan)	дать приказ (to order)
надеяться (to hope)	молить (to beg)
нравиться (to like)	научить (to teach)
переставать (to stop)	помочь (to help)
отвыкнуть (to wean)	порекомендовать (to recommend)
принести клятву (to swear)	предложить (to suggest)
обещать (to promise)	позволять (to allow)
решить (to decide)	советовать (to advise)

distribution is among predicate's individual characteristics, which must be taken into account while modifying the verb classification.

As it is evident from Table 7, which provides the current quality estimation for titles with non-finite clauses, the predicate classification based on the control type helped lower the overall error rate by 45%. And the extension of the classification with due regard for more specific features is expected to maximize the algorithm's performance.

Table 7. Error rate for complex contexts with infinitives.

Errors in main clauses	0.051
Errors in non-finite clauses	0.040
Confusing a1 and a2 in main clauses	0.185

With regard to titles with 2 + clauses, the errors in subordinate finite clauses were mostly caused by confusion between a1 and a2 since some for predicates sentential arguments are a2 (and not a1, as it is presumed in the algorithm). These predicates must be extracted into an exception list.

In spite of the errors of automatic annotation, the samples for predicates can be collected quite fast after the manual check. Thus, there occurs the necessity of utilizing the automatized procedure.

6   Conclusion

In this paper a rule-based approach to semantic role labeling for automatic generation of the annotated collection of data, based on the RuSentiFrames lexicon, is proposed. In the basis of the algorithm there lie two fundamental principles: the syntactic-semantic correlation and the compliance with structural limits. The implementation of the algorithm included the search of the entities with certain morpho-syntactic features in the order that depends on the morphological case of the entity and is based on calculation of the probabilities of the co-occurrence of each type of case and argument. The results of the algorithm evaluation, based on several different characteristics (the type of the clause, the recognition of each type of arguments, etc.), were relatively high. The annotated titles are planned to be used for automatization of the annotation of more complex sentences from news articles.

However, there occurred some problematic cases. Firstly, the performance of the algorithm strictly depends on the quality of the syntactic parser. Secondly, the verbal polysemy resolution must be carried out in order to minimize the inaccuracies during semantic parsing. Thirdly, the algorithm does not differentiate between direct and indirect objects while parsing complex contexts with non-finite clauses. The solutions to these problems have been suggested and are expected to be implemented in the further research. In addition to that, we plan to refine the verb and clause classification and extract some more informational features which could be useful in differentiating between roles (for example, the correlation between the lexical similarity and the semantic role distribution; the animacy of the arguments, etc.). Since the animacy annotation of the participants for each predicate in the frame collection has already been carried out, it can be utilized to modify the algorithm. Due to the fact that the rule set for four-argument verbs was configured using the animacy contraposition and showed relatively high performance, it is expected that for other classes of predicates this scheme will also be especially useful. Basically, it is quite evident that many semantic roles are associated with one particular animacy class (for instance, Recipient is more likely to be animate while Instrument is inanimate in the vast majority of cases). We expect that this modification will help differentiate between a1 and a2 more clearly.

Thus, by means of the expanded linguistic analysis of the text data with a view to defining informative morpho-syntactic and lexical features of the predicates and their arguments among with the examination and comparison of different classification strategies, it will be possible to elaborate the maximum accurate semantic role labeling system for the Russian language.

Acknowledgements. The reported study was funded by RFBR according to the research project N 20-07-01059.

References

1. Burtsev, M., et al.: DeepPavlov: open-source library for dialogue system. In: Proceedings of the 56th Annual Meeting of the Association for Computational Linguistics-System Demonstrations, pp. 1–6 (2018)
2. Carreras, X., Marquez, L.: Introduction to the CoNLL-2005 shared task: semantic role labeling. In: Proceedings of the Ninth Conference on Computational Natural Language Learning (CoNLL-2005), pp. 152–164. Ann Arbor, MI (2005)
3. Chomsky, N.: Lectures on Government and Binding. Foris Publications, Dordrecht (1981)
4. Collobert, R., Weston, J., Bottou, L., Karlen, M., Kavukcuoglu, K., Kuksa, P.: Natural language processing (almost) from scratch. J. Mach. Learn. Res. **12**, 2493–2537 (2011)

5. Fellbaum, C. (ed.): WordNet: An Electronic Lexical Database. MIT Press, Cambridge (1998)
6. Fillmore, C.: Toward a modern theory of case. In: The Ohio State University Project on Linguistic Analysis, report 13, pp. 1–24. Ohio State University, Columbus (1966)
7. Gildea, D., Jurafsky, D.: Automatic labeling of semantic roles. Comput. Linguist. **28**(3), 245–288 (2002)
8. Guan, C., Cheng, Y., Zhao, H.: Semantic role labeling with associated memory network. In: Proceedings of the 2019 Conference of the North American Chapter of the Association for Computational Linguistics: Human Language Technologies, vol. 1 (Long and Short Papers), pp. 3361–3371. Association for Computational Linguistics, Minneapolis (2019)
9. Joachims, T.: Text categorization with support vector machines: learning with many relevant features. In: Proceedings of the European Conference on Machine Learning, pp. 137–142 (1998)
10. Kudo, T., Matsumoto, Y.: Use of support vector learning for chunk identification. In: Proceedings of the 4th Conference on CoNLL-2000 and LLL-2000, pp. 142–144 (2000)
11. Larionov, D., Shelmanov, A., Chistova, E., Smirnov, I.: Semantic role labeling with pretrained language models for known and unknown predicates. In: Proceedings of the International Conference on Recent Advances in Natural Language Processing, pp. 619–628 (2019)
12. Loukachevitch, N.V., Rusnachenko, N.L.: Sentiment frames for attitude extraction in Russian. In: Proceedings of International Conference Dialog, pp. 541–552 (2020)
13. Lodhi, H., Saunders, C., Shawe-Taylor, J., Cristianini, N., Watkins, C.: Text classification using string kernels. J. Mach. Learn. Res. **2**(Feb), 419–444 (2002)
14. Lyashevskaya O.: Dictionary of valencies meets corpus annotation: a case of Russian frame-bank. In: Proceedings of the 15th EURALEX International Congress, Oslo, Norway, 7–11 August 2012, p. 15 (2012)
15. Lyashevskaya O., Kashkin E.: FrameBank: a database of Russian lexical constructions. In: Proceedings of International Conference on Analysis of Images, Social Networks and Texts, pp. 350–360 (2015)
16. Palmer, M., Gildea, D., Kingsbury, P.: The proposition bank: an annotated corpus of semantic roles. Comput. Linguist. **31**(1), 71–106 (2005)
17. Pradhan, S., Hacioglu, K., Krugler, V., Ward, W., Martin, J.H., Jurafsky, D.: Support vector learning for semantic argument classification. Mach. Learn. **60**, 11–39 (2005)
18. Rusachenko, N., Loukachevitch, N., Tutubalina, E.: Distant Supervision for Sentiment Attitude Extraction. In: Proceedings of the International Conference on Recent Advances in Natural Language Processing (RANLP 2019), pp. 1022–1030 (2019)
19. Shelmanov, A.O., Devyatkin, D.A.: Semantic role labeling with neural networks for texts in Russian. In: Proceedings of International Conference Dialogue, pp. 245–256 (2017)
20. Shelmanov, A.O., Smirnov, I.V.: Methods for semantic role labeling of Russian texts. In: Proceedings of International Conference Dialog, pp. 607–620 (2014)
21. Ustalov, D., Panchenko, A., Kutuzov, A., Biemann, C., Ponzetto, S.P.: Unsupervised semantic frame induction using triclustering. arXiv preprint arXiv:1805.04715 (2018)
22. Zhou, J., Xu, W. End-to-end learning of semantic role labeling using recurrent neural networks. In: Proceedings of the 53rd Annual Meeting of the Association for Computational Linguistics and the 7th International Joint Conference on Natural Language Processing, pp. 1127–1137 (2015)
23. Apresyan, Y.D. Leksicheskaya semantika. Nauka, Moscow (1974)
24. Apresyan, Y.D., Boguslavskij, I.M., Iomdin, L.L., et al.: Lingvisticheskoe obespechenie sistemy ETAP-2. Nauka, Moscow (1989)

ELMo and BERT in Semantic Change Detection for Russian

Julia Rodina[1], Yuliya Trofimova[1], Andrey Kutuzov[2],
and Ekaterina Artemova[1(✉)]

[1] National Research University Higher School of Economics, Moscow, Russia
echernyak@hse.ru
[2] University of Oslo, Oslo, Norway
andreku@ifi.uio.no

Abstract. We study the effectiveness of contextualized embeddings for the task of diachronic semantic change detection for Russian language data. Evaluation test sets consist of Russian nouns and adjectives annotated based on their occurrences in texts created in pre-Soviet, Soviet and post-Soviet time periods. ELMo and BERT architectures are compared on the task of ranking Russian words according to the degree of their semantic change over time. We use several methods for aggregation of contextualized embeddings from these architectures and evaluate their performance. Finally, we compare unsupervised and supervised techniques in this task.

Keywords: Contextualized embeddings · Semantic shift · Semantic change detection

1 Introduction

In this research, we apply ELMo (Embeddings from Language Models) [32] and BERT (Bidirectional Encoder Representations from Transformers) [4] models fine-tuned on historical corpora of Russian language to estimate diachronic semantic changes. We do that by extracting contextualized word representations for each time bin and quantifying their differences. This approach is data-driven and does not require any manual intervention. For evaluation, we use human-annotated datasets and show that contextualized embedding models can be used to rank words by the degree of their semantic change, yielding a significant correlation with human judgments.

It is important to mention that we treat the word meaning as a function of the word's contexts in natural language texts. This concept corresponds to the distributional hypothesis [6].

The rest of the paper is organized as follows: in Sect. 2, we describe previous research on the automatic lexical semantic change detection. In Sect. 3, we

---

J. Rodina and Y. Trofimova—These authors contributed equally to this work.

© Springer Nature Switzerland AG 2021
W. M. P. van der Aalst et al. (Eds.): AIST 2020, LNCS 12602, pp. 175–186, 2021.
https://doi.org/10.1007/978-3-030-72610-2_13

present natural language data used in our research, and the dataset structure. The training of contextualized embeddings and models' architecture is described in Sect. 4. In this section, we also describe and apply various algorithms for semantic change detection. In Sect. 5, we calculate correlation of their results with human judgments to evaluate their performance. In Sect. 6, we summarize our contributions and discuss possibilities for future research.

2   Related Work

The nature and reasons of semantic change processes have been studied in linguistics for a long time, at least since the theoretical work of [2] where the cognitive laws of semantic change were formulated. Later there were other works on categorizing different types of semantic changes [1,40].

With the increasing amount of available language data, researchers started to focus on empirical approaches to semantic change in addition to theoretical work. Earlier studies consisted primarily of corpus-based analysis ([15,29] among many others), and used raw word frequencies to detect semantic shifts. However, there already were applications of distributional methods, for example, Temporal Random Indexing [16], co-occurrences matrices weighted by Local Mutual Information [11], graph-based methods [31] and others.

Among the works focusing on Russian, one can mention a recent book [3] in which 20 words were manually analyzed by exploring the contexts in which they appear and counting the number of their uses for each period. This allowed to picture the history of changes of meanings which words undergo.

2.1   Static Word Embeddings

After the widespread usage of word embeddings [30] had started, the focus has shifted to detecting semantic changes using dense distributional word representations; see [12,17,18,36] and many others. Word embeddings represent meaning of words as dense vectors learned from their co-occurrences counts in the training corpora. This approach has gained popularity as it can leverage un-annotated natural language data and produces both efficient and easy to work with continuous representations of meaning. Comprehensive surveys on research about semantic change detection using 'static' word embeddings are given in [25] and [42].

[3] inspired the first (to our knowledge) publication which applied static word embedding models to Russian data in order to detect diachronic lexical changes [23]. The authors compared sets of word's nearest neighbors and concluded that Kendall's τ and Jaccard distance worked best in scoring changes. More recent work [7] extended this research to more granular time bins (periods of 1 year): they analyzed semantic shifts between static embeddings trained on yearly corpora of Russian news texts from the year 2000 up to 2014. They evaluated 5 algorithms for semantic shift detection and provided solid baselines for future research in Russian language. Note, however, that [7] solved the classification task (whether a word has changed its meaning or not), while in the present

paper we solve the ranking task (which words have changed more or less than the others). Also, the test set we use (*RuSemShift*) is much more extensive and consistent (see Subsect. 3.2).

2.2 Contextualized Word Embeddings

Static word embeddings assign the same vector representation to each word (lemma) occurrence: they produce context-independent vectors at the inference time. However, recent advances in natural language processing made it possible to implement models that allow to obtain higher quality *contextualized* representations. The main difference of contextualized models is that at the inference stage tokens are assigned different embeddings depending on their context in the input data. This results in richer word features and more realistic word representations. There are several recently published contextualized architectures based on language modeling task (predicting the next most probable word given a sequence of tokens) [4,13,32,34].

This provides new opportunities for diachronic analysis: for example, it is possible to group similar token representations and measure a diversity of such representations, while predefined number of senses is not strictly necessary. Thus, currently there is an increased interest in the topic of language change detection using contextualized word embeddings [9,10,14,21,27,28].

[14] used a list of polysemous words with predefined set of senses, then a pre-trained BERT model was applied to diachronic corpora to extract token embeddings that are the closest to the predefined sense embedding. Evolution of each word was measured by comparing distributions of senses in different time slices. In [9] and [10], a pre-trained BERT model was used to obtain representations of word usages in an unsupervised fashion, without predefined list or number of senses. Representations with similar usages then were clustered using k-Means algorithm and distributions of word's usages in these clusters were used in two metrics for quantifying the degree of semantic change: entropy difference and Jensen-Shannon divergence. They also used average pairwise distance, that does not need clusterization and only requires usage matrices from two time periods.

[27] used averaged time-specific BERT representations and calculated cosine distance between averaged vectors of two time periods as a measure of semantic change. [28] tested Affinity Propagation algorithm for usage clusterization and showed that it is consistently better than k-Means. Finally, [21] applied approaches similar to [10], but also analyzing ELMo models and adding cosine similarity of average vectors as a measure. Their algorithms were evaluated on Subtask 2 (ranking) of SemEval-2020 Task 1 [37] for four different languages and strongly outperformed the baselines. We will test each of these clustering methods and cosine similarity of averaged vectors for our task in the present paper.

We decided to compare BERT against ELMo. Most of the above mentioned works used pre-trained BERT language models, but ELMo allows faster training and inference which makes it easier to train diachronic models completely on one's own data.

3 Data

3.1 Corpora

For both BERT and ELMo architectures, we used pre-trained models from prior work. The `RuBERT` model is a Multilingual BERT [4] fine-tuned on the Russian Wikipedia and news articles (about 850 million tokens) [20]. The `tayga_lemmas_elmo_2048_2019` is an ELMo model trained on the Taiga corpus of Russian (almost 5 billion tokens) [39]; it is available from the RusVectores web service [22].

Both models were additionally fine-tuned on the Russian National Corpus (RNC) which contains texts in Russian language produced from the middle of the XVIII to the beginning of the XXI century (about 320 million word tokens in total, including punctuation). Before the fine-tuning, the corpus was segmented into sentences, then tokenized and lemmatized with the Universal Dependencies model trained on the SynTagRus Russian UD treebank [5] using UDPipe 1.2 [41]. This was motivated by recent research [24] which showed that for Russian data, lemmatization improves the results of contextualized architectures on the word sense disambiguation task. Reducing the noise effect of word forms is useful for tracing semantic shifts among all occurrences of the word in the aggregate.

For the purposes of diachronic semantic change detection, the RNC corpus was divided into three parts, with the boundaries between their consequent pairs corresponding to the rise and fall of the Soviet Union, which undoubtedly brought along major social, cultural and political shifts:

1. texts produced before 1917 (pre-Soviet period): 94 million tokens;
2. texts produced in 1918–1990 (Soviet period): 123 million tokens;
3. texts produced in 1991–2017 (post-Soviet period): 107 million tokens.

For extracting time-specific word's embeddings at the inference stage, we used each time-specific sub-corpus separately.

3.2 Evaluation Dataset

We used two human-annotated datasets of Russian nouns and adjectives for evaluation of diachronic semantic change detection methods. They are part of a larger *RuSemShift* dataset[1] [35] and cover both consequent pairs of the RNC sub-corpora: from the pre-Soviet through the Soviet times ($RuSemShift_1$) and from the Soviet through the post-Soviet time ($RuSemShift_2$).

The datasets contain words annotated similar to the DURel framework [38] according to the intensity of the semantic changes they have undergone. Each word was presented to 5 independent annotators along with two context sentences. These pairs of sentences were sampled from the RNC and divided into three equal groups: EARLIER (both sentences are from the earlier time period), LATER (both sentences from the later time period) and COMPARE (one sentence from the earlier period and another from the later period). The annotators

---

[1] https://github.com/juliarodina/RuSemShift.

were asked to rate how different are the word's meanings in two sentences on the scale from 1 (unrelated meanings) to 4 (identical meanings).

There are two measures for quantifying diachronic semantic change based on this manual annotation:

1. ΔLATER which is the subtraction of the EARLIER group's mean difference value from the LATER group's mean difference value:
   $$\Delta\text{LATER}(w) = Mean_{later}(w) - Mean_{earlier}(w);$$
2. COMPARE which is the word's mean difference value in the COMPARE group:
   $$\text{COMPARE}(w) = Mean_{compare}(w).$$

Note that we use two test sets, and thus two pairs of time periods. In each pair, the EARLIER and LATER groups stand for different time periods. Foe example, in $RuSemShift_2$, the EARLIER group corresponds to texts from the Soviet times, and the LATER group corresponds to texts from the post-Soviet times.

The experiments were performed on the filtered versions of the datasets where the words with the annotators' agreement lower than 0.2 were excluded. Their sizes are 48 words ($RuSemShift_1$) and 51 words ($RuSemShift_2$).

4   Experimental Setup

4.1   Contextualized Word Embedding Models

ELMo is a deep character based bidirectional language model (biLM): a combination of two-layer long short-term memory (LSTM) networks on top of a convolutional layer with max-pooling [32]. Having a pre-trained biLM model, we can obtain word representations [32] that are learned functions of the internal state of this model and contain information about word's syntax and semantics as well as its polysemy. The model we used (`tayga_lemmas_elmo_2048_2019`) was trained on the Taiga corpus for 3 epochs, with batch size 192. All the ELMo hyperparameters were left for their default values, except for the number of negative samples per batch which was reduced to 4 096 from 8 192 used in [32]. After fine-tuning on the whole RNC, for each word from the test sets we extracted its contextualized token representations for every occurrence. We used only the top LSTM layer, since higher layer representations are more oriented to learn semantics and capture longer range dependencies [33].

BERT (Bidirectional Encoder Representations from Transformers) makes use of Transformer [43], an attention mechanism that learns contextual relations between words in a sentence. Masked language modeling technique allows bidirectional training of Transformer that allegedly allows for a deeper sense of language context. Training BERT from scratch is a computationally expensive process and demands huge resources. Fine-tuning is a well-known technique for adapting pre-trained BERT models to a specific corpus or task. We fine-tuned

the RuBERT model on the RNC, before producing contextualized word representations. The model was being fine-tuned for 3 epochs, with the batch size 16.
The original RuBERT has 12 heads and its hidden size is 768. Then, for each
word from our test sets, we extracted contextualized embedding for each of its
occurrences during the respective time periods. We used the last BERT layer
representations as token embeddings.

For both ELMo and BERT, this resulted in three matrices of token embeddings, corresponding to three time periods.

4.2 Methods

To evaluate ELMo and BERT in semantic change detection task, we extracted
word embeddings for each word's usage from the RNC sub-corpora corresponding
to the time periods under analysis. Recall that token embeddings are context-
dependent. We applied four different aggregation methods to estimate semantic
change degrees based on the token embeddings. Then we calculated correlation between their outputs and two gold human-annotated measures from the
$RuSemShift_1$ and $RuSemShift_2$: ΔLATER and COMPARE. We discuss the
aggregation methods below.

Inverted Cosine Similarity over Word Prototypes (PRT). The PRT
method outperformed others in [21] and is computationally inexpensive. First,
we average all n token embeddings of a given word in a specific sub-corpus, this
producing a 'prototypical' representation of the word. Then we calculate cosine
distance between the average embeddings u and v from two different sub-corpora
as a measure of semantic change:

$$PRT = 1 - cos(u, v) = 1 - \frac{\sum_{i=1}^{n} u_i \cdot v_i}{\sqrt{\sum_{i=1}^{n} u_i^2} \cdot \sqrt{\sum_{i=1}^{n} v_i^2}} \qquad (1)$$

Clustering. As mentioned above, ELMo and BERT return context-dependent
token embeddings, and we need to compare them in order to estimate semantic
shift. Another way to do it is clustering token embeddings. We can move away
from comparing individual vectors or their averages, and instead look at word
occurrences as belonging to several clusters. It is assumed that these clusters
correspond to word senses. Following the previous work, we tested two clustering methods: Affinity Propagation [8] and k-Means [26]. One needs to manually
set the hyperparameter for the number of clusters for k-Means (we empirically
found 5 to be the best value for our data). At the same time, Affinity Propagation clustering doesn't require a predetermined number of clusters, it is inferred
automatically from data.

We need to compare the resulting clusters of word's usages in two time periods. We used the method proposed in [9]. Namely, we clustered all embeddings of
each word w from the two time periods $(u_1^t, u_2^t, ..., u_n^t)$ and $(u_1^{t+1}, u_2^{t+1}, ..., u_m^{t+1})$.

Then, for each of the time bins t and $t+1$ we calculated the following distribution 2 which indicates the likelihood of encountering the word in a specific sense:

$$p_w^t[k] = \frac{|y_w^t[i] \in y_w^t, \; if \; y_w^t[i] = k|}{N_w^t}, \quad y_w^t \in [1, K_w]^{N_w^t} \tag{2}$$

where y_w^t are the labels after clustering, K is the number of clusters, and N_w^t is the number of w occurrences in the time period t. These distributions are comparable, and we test two different methods for this:

1. **Jensen-Shannon divergence (JSD)**. Here, we compute the JSD between two distributions:

$$JSD = \sqrt{0.5 \cdot (KL(p||m) + KL(q||m))} \tag{3}$$

   where KL is Kullback-Leibler divergence [19], p and q are sense distributions and m is the point-wise mean of p and q. Higher JSD score indicates more intense change in the proportions of clustered word usage types across time periods.
2. **Maximum square**. Here, we assume that slight changes in sense distribution may occur due to noise and do not manifest a real semantic change. At the same time, strong changes in context distribution may indicate serious semantic shifts. Therefore we apply the hand-picked function 4:

$$MS = max(square(p - q)) \tag{4}$$

5   Results

As mentioned above, we extracted all word usages from the RNC diachronic sub-corpora for each word from the filtered *RuSemShift* datasets. Then we produced their contextualized representations and applied the aggregation methods described in Sect. 4. Due to the increasing computational complexity, we have limited the number of usages to 10 000 (by random sampling) for the clustering methods.

Tables 1 and 2 show the Spearman correlation coefficients between the models' and annotators' rankings for $RuSemShift_1$ and $RuSemShift_2$ test sets correspondingly. 'MS' stands for the Maximum Square method. Asterisk indicates statistically significant correlations (p-value > 0.05), bold highlighting indicates best scores for each measure. COMPARE scores in the tables are actually equal to $1 - COMPARE$, because of the nature of this measure: lower COMPARE score means stronger change.

For both datasets, the best aggregation method was the PRT, but clustering was not much inferior. As for the functions for comparing distributions of tokens in clusters, the Jensen-Shannon divergence showed stronger correlation with human judgments than the hand-picked function of maximum square.

If we compare BERT and ELMo results, we can see that they are somewhat similar and it cannot be concluded that one model is better than another (despite ELMo having much less parameters). Also, one can notice from the tables that the models' rankings in general are more correlated with the ranking by the COMPARE measure than by the ΔLATER measure. Thus, the COMPARE measure is easier to approximate. Statistically significant correlations for $RuSemShift_2$ are generally higher that for $RuSemShift_1$. It can be due to lexical specificity of the datasets.

An interesting example of semantic change is the word ' ' ('*failed*'). According to the human annotation, this word has a very strong degree of change when comparing the Soviet period to the post-Soviet period. BERT and ELMo with PRT also place it at the top positions among the words ranked by their semantic change degrees. Indeed, in the Soviet times, ' ' was mostly used in the literal meaning of ' ' – '*a place where the surface collapsed inward*' or figurative meaning '*loss of consciousness*' especially in the collocation ' ' ('*deep dream*'). In the Soviet period, its primary sense shifted to the more common nowadays meaning ('*failed*').

In some cases, the models estimates do not correlate with human judgments. For example, most of our approaches yield high semantic change degree for the word ' ' ('*rain*') when comparing the pre-Soviet to the Soviet period. However, human annotation positions it quite low in the semantic change rankings for this period pair. Whether this is an error of the model or an insufficiency of the dataset, remains yet to be solved.

Table 1. Correlations of models' predictions with $RuSemShift_1$ annotations (change between the pre-Soviet and Soviet time periods).

Table 2. Correlations of models' predictions with $RuSemShift_2$ annotations (change between the Soviet and post-Soviet time periods).

Algorithms		Spearman ρ	
	Measure	ELMo	BERT
PRT	ΔLATER	0.200	0.346*
	COMPARE	0.409*	**0.490***
Affinity/JSD	ΔLATER	**0.406***	0.160
	COMPARE	0.276	0.295*
kMeans/JSD	ΔLATER	0.250	0.270
	COMPARE	0.340*	0.440*
kMeans/MS	ΔLATER	0.060	0.240
	COMPARE	0.380*	0.358*

Algorithms		Spearman ρ	
	Measure	ELMo	BERT
PRT	ΔLATER	**0.300***	0.230
	COMPARE	**0.557***	0.500*
Affinity/JSD	ΔLATER	0.200	0.136
	COMPARE	0.363*	0.408*
kMeans/JSD	ΔLATER	0.200	0.130
	COMPARE	0.535*	0.480*
kMeans/MS	ΔLATER	0.074	0.120
	COMPARE	0.436*	0.420*

6   Conclusion

In this work, we evaluated how semantic change detection methods based on contextualized word representations from BERT and ELMo perform for Russian language data (diachronic sub-corpora of the Russian National Corpus).

In particular, we tested them in the task of automatically ranking Russian words according to the manually annotated degree of their diachronic semantic change.

Pre-trained ELMo and BERT models were fine-tuned on the full RNC corpus to make comparison as fair as possible. Then we applied several algorithms for semantic shift detection: cosine similarity on a word prototypes (PRT) and clustering algorithms together with measures for comparing word's usages distributions.

For the second method, we applied two clustering algorithms: Affinity Propagation and k-Means. K-Means turned out to be generally better than Affinity Propagation. However, the PRT method, using simple cosine similarity between averaged token embeddings (word prototypes) outperformed clustering algorithms in most cases and therefore suits better for this task.

To sum up, we showed that contextualized word representation models have significant correlation with human judgments in diachronic semantic change detection for Russian. Also we found out that there is not much difference between BERT and ELMo contextualized embeddings in this respect and we can't say that one architecture is significantly better than another.

In the future work, it would be interesting to make a more fair comparison between ELMo and BERT on the task of semantic change detection by pre-training both models on identical corpora from scratch. However, this will require significant computational resources in the case of BERT.

Acknowledgments. This research was supported by the Russian Science Foundation grant 20-18-00206.

References

1. Bloomfield, L.: Language. Holt, New York (1933)
2. Bréal, M.: Les lois intellectuelles du langage: fragment de smantique. Annuaire de l'Assocaition pour l'encouragement des tudes grecques en France **17**, 132–142 (1883)
3. Daniel, M., Dobrushina, N.: Dva veka v dvadtsati slovax [Two centuries in twenty words]. Izdatelskij dom NIU VSHE (2016)
4. Devlin, J., Chang, M.W., Lee, K., Toutanova, K.: BERT: pre-training of deep bidirectional transformers for language understanding. In: Proceedings of the 2019 Conference of the North American Chapter of the Association for Computational Linguistics: Human Language Technologies, Volume 1 (Long and Short Papers), Minneapolis, Minnesota, pp. 4171–4186. Association for Computational Linguistics, June 2019
5. Droganova, K., Lyashevskaya, O., Zeman, D.: Data conversion and consistency of monolingual corpora: Russian UD treebanks. In: Proceedings of the 17th International Workshop on Treebanks and Linguistic Theories, Oslo University, Norway, pp. 52–65 (2018)
6. Firth, J.R.: A synopsis of linguistic theory, 1930–1955. Blackwell (1957)
7. Fomin, V., Bakshandaeva, D., Rodina, J., Kutuzov, A.: Tracing cultural diachronic semantic shifts in Russian using word embeddings: test sets and baselines. Komp'yuternaya Lingvistika i Intellektual'nye Tekhnologii: Dialog conference, pp. 203–218 (2019)

8. Frey, B.J., Dueck, D.: Clustering by passing messages between data points. Science **315**(5814), 972–976 (2007)
9. Giulianelli, M.: Lexical Semantic Change Analysis with Contextualised Word Representations. M.S. thesis, University of Amsterdam (7 2019)
10. Giulianelli, M., Del Tredici, M., Fernández, R.: Analysing lexical semantic change with contextualised word representations. In: Proceedings of the 58th Annual Meeting of the Association for Computational Linguistics. Association for Computational Linguistics (2020)
11. Gulordava, K., Baroni, M.: A distributional similarity approach to the detection of semantic change in the Google books Ngram corpus. In: Proceedings of the GEMS 2011 Workshop on GEometrical Models of Natural Language Semantics, Edinburgh, UK, pp. 67–71. Association for Computational Linguistics, July 2011
12. Hamilton, W.L., Leskovec, J., Jurafsky, D.: Diachronic word embeddings reveal statistical laws of semantic change. In: Proceedings of the 54th Annual Meeting of the Association for Computational Linguistics (Volume 1: Long Papers), Berlin, Germany, pp. 1489–1501. Association for Computational Linguistics, August 2016
13. Howard, J., Ruder, S.: Universal language model fine-tuning for text classification. In: Proceedings of the 56th Annual Meeting of the Association for Computational Linguistics (Volume 1: Long Papers), Melbourne, Australia, pp. 328–339. Association for Computational Linguistics, July 2018
14. Hu, R., Li, S., Liang, S.: Diachronic sense modeling with deep contextualized word embeddings: an ecological view. In: Proceedings of the 57th Annual Meeting of the Association for Computational Linguistics, Florence, Italy, pp. 3899–3908. Association for Computational Linguistics, July 2019
15. Jatowt, A., Duh, K.: A framework for analyzing semantic change of words across time. In: Proceedings of the 14th ACM/IEEE-CS Joint Conference on Digital Libraries, pp. 229–238. JCDL 2014. IEEE Press (2014)
16. Jurgens, D., Stevens, K.: Event detection in blogs using temporal random indexing. In: Proceedings of the Workshop on Events in Emerging Text Types, Borovets, Bulgaria, pp. 9–16. Association for Computational Linguistics, September 2009
17. Kim, Y., Chiu, Y.I., Hanaki, K., Hegde, D., Petrov, S.: Temporal analysis of language through neural language models. In: Proceedings of the ACL 2014 Workshop on Language Technologies and Computational Social Science, Baltimore, MD, USA, pp. 61–65. Association for Computational Linguistics, June 2014
18. Kulkarni, V., Al-Rfou, R., Perozzi, B., Skiena, S.: Statistically significant detection of linguistic change. In: Proceedings of the 24th International Conference on World Wide Web, pp. 625–635. WWW 2015, International World Wide Web Conferences Steering Committee (2015)
19. Kullback, S., Leibler, R.A.: On information and sufficiency. Ann. Math. Statist. **22**(1), 79–86 (1951)
20. Kuratov, Y., Arkhipov, M.: Adaptation of deep bidirectional multilingual transformers for Russian language. Komp'yuternaya Lingvistika i Intellektual'nye Tekhnologii: Dialog conference (2019). http://www.dialog-21.ru/media/4606/kuratovyplusarkhipovm-025.pdf
21. Kutuzov, A., Giulianelli, M.: UiO-UvA at SemEval-2020 task 1: Contextualised embeddings for lexical semantic change detection. In: arXiv preprint arXiv:2005.00050 (to appear in Proceedings of the 14th International Workshop on Semantic Evaluation). Association for Computational Linguistics, Barcelona, Spain (2020)

22. Kutuzov, A., Kuzmenko, E.: WebVectors: a toolkit for building web interfaces for vector semantic models. In: Ignatov, D.I., et al. (eds.) Analysis of Images, Social Networks and Texts, pp. 155–161. Springer International Publishing, Cham (2017)

23. Kutuzov, A., Kuzmenko, E.: Two centuries in two thousand words: Neural embedding models in detecting diachronic lexical changes. Quant. Approaches Russ. Lang. 95–112 (2018)

24. Kutuzov, A., Kuzmenko, E.: To lemmatize or not to lemmatize: How word normalisation affects ELMo performance in word sense disambiguation. In: Proceedings of the First NLPL Workshop on Deep Learning for Natural Language Processing, Turku, Finland, pp. 22–28, September 2019

25. Kutuzov, A., Øvrelid, L., Szymanski, T., Velldal, E.: Diachronic word embeddings and semantic shifts: a survey. In: Proceedings of the 27th International Conference on Computational Linguistics, Santa Fe, New Mexico, USA, pp. 1384–1397. Association for Computational Linguistics, August 2018

26. Lloyd, S.P.: Least squares quantization in PCM. IEEE Trans. Inf. Theory **28**(2), 129–136 (1982). http://dblp.uni-trier.de/db/journals/tit/tit28.html#Lloyd82

27. Martinc, M., Kralj Novak, P., Pollak, S.: Leveraging contextual embeddings for detecting diachronic semantic shift. In: Proceedings of The 12th Language Resources and Evaluation Conference, Marseille, France, pp. 4811–4819. European Language Resources Association, May 2020. https://www.aclweb.org/anthology/2020.lrec-1.592

28. Martinc, M., Montariol, S., Zosa, E., Pivovarova, L.: Capturing evolution in word usage: just add more clusters? In: Companion Proceedings of the Web Conference 2020, pp. 343–349. WWW 2020. Association for Computing Machinery (2020)

29. Michel, J.B., et al.: Quantitative analysis of culture using millions of digitized books. Science **331**, 176–82 (2011)

30. Mikolov, T., Sutskever, I., Chen, K., Corrado, G., Dean, J.: Distributed representations of words and phrases and their compositionality. In: Proceedings of the 26th International Conference on Neural Information Processing Systems, vol. 2, pp. 3111–3119. NIPS 2013, Curran Associates Inc. (2013)

31. Mitra, S., Mitra, R., Riedl, M., Biemann, C., Mukherjee, A., Goyal, P.: That's sick dude!: automatic identification of word sense change across different timescales. In: Proceedings of the 52nd Annual Meeting of the Association for Computational Linguistics (Volume 1: Long Papers), Baltimore, Maryland, pp. 1020–1029. Association for Computational Linguistics, June 2014

32. Peters, M., Neumann, M., Iyyer, M., Gardner, M., Clark, C., Lee, K., Zettlemoyer, L.: Deep contextualized word representations. In: Proceedings of the 2018 Conference of the North American Chapter of the Association for Computational Linguistics: Human Language Technologies, Volume 1 (Long Papers), New Orleans, Louisiana, pp. 2227–2237. Association for Computational Linguistics (2018)

33. Peters, M., Neumann, M., Zettlemoyer, L., Yih, W.t.: Dissecting contextual word embeddings: architecture and representation. In: Proceedings of the 2018 Conference on Empirical Methods in Natural Language Processing, Brussels, Belgium, pp. 1499–1509. Association for Computational Linguistics, October 2018

34. Radford, A., Narasimhan, K., Salimans, T., Sutskever, I.: Improving language understanding by generative pre-training (2018), openAI

35. Rodina, J., Kutuzov, A.: RuSemShift: a dataset of historical lexical semantic change in Russian. In: arXiv:2010.06436 (to appear in Proceedings of the 28th Conference on Computational Linguistics (COLING-2020)) (2020)

36. Rosenfeld, A., Erk, K.: Deep neural models of semantic shift. In: Proceedings of the 2018 Conference of the North American Chapter of the Association for Computational Linguistics: Human Language Technologies, Volume 1 (Long Papers), pp. 474–484. Association for Computational Linguistics (Jun 2018)

37. Schlechtweg, D., McGillivray, B., Hengchen, S., Dubossarsky, H., Tahmasebi, N.: Semeval 2020 task 1: unsupervised lexical semantic change detection. In: To appear in Proceedings of the 14th International Workshop on Semantic Evaluation. Association for Computational Linguistics, Barcelona, Spain (2020)

38. Schlechtweg, D., Schulte im Walde, S., Eckmann, S.: Diachronic usage relatedness (DURel): a framework for the annotation of lexical semantic change. In: Proceedings of the 2018 Conference of the North American Chapter of the Association for Computational Linguistics: Human Language Technologies, Volume 2 (Short Papers), New Orleans, Louisiana, pp. 169–174. Association for Computational Linguistics, June 2018

39. Shavrina, T., Shapovalova, O.: To the methodology of corpus construction for machine learning: Taiga syntax tree corpus and parser. In: Proceedings of the CORPORA2017 International Conference. Saint-Petersbourg (2017)

40. Stern, G.: Meaning and Change of Meaning, with Special Reference to the English Language. Wettergren & Kerber (1931)

41. Straka, M., Straková, J.: Tokenizing, POS tagging, lemmatizing and parsing UD 2.0 with UDPipe. In: Proceedings of the CoNLL 2017 Shared Task: Multilingual Parsing from Raw Text to Universal Dependencies, Vancouver, Canada, pp. 88–99. Association for Computational Linguistics, August 2017

42. Tang, X.: A state-of-the-art of semantic change computation. Natl. Lang. Eng. **24**(5), 649–676 (2018)

43. Vaswani, A., et al.: Attention is all you need. CoRR abs/1706.03762 (2017). http://arxiv.org/abs/1706.03762

BERT for Sequence-to-Sequence Multi-label Text Classification

Ramil Yarullin[1,2](✉) and Pavel Serdyukov[3]

[1] National Research University Higher School of Economics,
Pokrovskii bulvar 11, Moscow, Russia
`ramly@ya.ru`
[2] Yandex.Go, 1st Krasnogvardeysky Proezd 19, Moscow, Russia
[3] Yandex, Ulitsa Lva Tolstogo 16, Moscow, Russia

Abstract. We study the BERT language representation model and the sequence generation model with BERT encoder for the multi-label text classification task. We show that the Sequence Generating BERT model achieves decent results in significantly fewer training epochs compared to the standard BERT. We also introduce and experimentally examine a mixed model, an ensemble of BERT and Sequence Generating BERT models. Our experiments demonstrate that the proposed model outperforms current baselines in several metrics on three well-studied multi-label classification datasets with English texts and two private Yandex Taxi datasets with Russian texts.

Keywords: Multi-label text classification · BERT ·
Sequence-to-sequence learning · Sequence generation · Hierarchical text classification

1 Introduction

Multi-label text classification (MLTC) is an important NLP task with many applications, such as document categorization, intent detection in dialogue systems, protein function prediction [23], and tickets tagging in client support systems [13].In this task, text samples are assigned to multiple labels from a finite label set.

In recent years, it became clear that deep learning approaches can go a long way toward solving text classification tasks. However, many of the widely used approaches in MLTC tend to predict the labels as a whole sequence and do not directly use information about dependencies between labels. One of the promising yet fairly less studied methods to tackle this problem is using sequence-to-sequence modeling. In this approach, a model treats an input text as a sequence of tokens and predicts labels in a sequential way taking into account previously predicted labels. Nam et al. [14] used Seq2Seq architecture with GRU encoder and attention-based GRU decoder, achieving an improvement over a standard GRU model [3] on several datasets and metrics. Yang et al. [29] continued this idea by

© Springer Nature Switzerland AG 2021
W. M. P. van der Aalst et al. (Eds.): AIST 2020, LNCS 12602, pp. 187–198, 2021.
https://doi.org/10.1007/978-3-030-72610-2_14

introducing Sequence Generation Model (SGM) consisting of BiLSTM-based encoder and LSTM decoder [6] coupled with additive attention mechanism [2].

In this paper, we argue that the encoder part in a sequence generating MLTC model can be successfully replaced with a heavy language representation model such as BERT. We propose Sequence Generating BERT model (BERT+SGM) and a mixed model which is an ensemble of standard BERT and BERT+SGM models. We show that BERT+SGM model achieves decent results after less than a half of an epoch of training (meaning less than a half of the training data), while the standard BERT model needs to be trained for 5–6 epochs just to achieve the same accuracy and several dozens epochs more to converge.

The key contributions of this paper are:

1. We present the results of Sequence Generating BERT model for MLTC datasets with and without a given hierarchical tree structure over classes and demonstrate its particularly fast learning abilities;
2. We introduce and examine experimentally a novel mixed model for MLTC.
3. We show that the proposed model outperforms baselines on three well-studied MLTC datasets with English texts and two private Yandex Taxi datasets with Russian texts.

2   Related Work and Preliminaries

Let us consider a set $\mathcal{D} = \{(x_n, y_n)\}_{n=1}^{N} \subseteq \mathcal{X} \times \mathcal{Y}$ consisting of N samples that are assumed to be identically and independently distributed following an unknown distribution $P(\mathbf{X}, \mathbf{Y})$. *Multi-class* classification task aims to learn a function that maps inputs to the elements of a label set $\mathcal{L} = \{1, 2, \ldots, L\}$, i.e. $\mathcal{Y} = \mathcal{L}$. In *multi-label* classification, the aim is to learn a function that maps inputs to the subsets of $\mathcal{L}$, i.e. $\mathcal{Y} = 2^{\mathcal{L}}$. In text classification tasks, $\mathcal{X}$ is a space of natural language texts.

In deep learning, a standard pipeline for text classification is to use a base model that converts a raw text to its fixed-size vector representation and then pass it to a classification module that further produces a probability vector over the classes. Typical architectures for base models include different types of recurrent neural networks [3,6], convolutional neural networks [7], hierarchical attention networks [31], and other more sophisticated approaches. These models consider each instance $\boldsymbol{x}$ as a sequence of tokens $\boldsymbol{x} = [w_1, w_2, \ldots, w_T]$. Each token w_i is then mapped to a vector representation $\boldsymbol{u}_i \in \mathbb{R}^H$ thus forming an embedding matrix $U^{T \times H}$ which can be initialized with pre-trained word embeddings [12,16]. Research works of recent years also show that it is possible to pre-train entire language representation models in a self-supervised way. Newly introduced models providing context-dependent text embeddings include BERT [5] and its different modifications, such as XLNet [30] and RoBERTa [10], that succeeded in overcoming several limitations of original BERT. Still, sequence generation abilities of such Transformer-based models for multi-label classification are not yet explored.

A less traditional way to approach MLTC problem and take into account the dependencies between labels is using Seq2Seq modeling. In this framework that first appeared in the neural machine translation [21], there are generally a source input $\mathbf{X}$ and a target output $\mathbf{Y}$ in the form of sequences. We also assume there is a hidden dependence between $\mathbf{X}$ and $\mathbf{Y}$, which can be captured by probabilistic model $P(\mathbf{Y}|\mathbf{X}, \theta)$. Therefore, the problem consists of three parts: modeling the distribution $P(\mathbf{Y}|\mathbf{X}, \theta)$, learning the parameters θ, and performing the inference stage where we need to find $\hat{\mathbf{Y}} = \arg_{\mathbf{Y}} \max P(\mathbf{Y}|\mathbf{X}, \theta)$.

Nam et al. [14] have shown that after introducing a total order relation on the set of classes $\mathcal{L}$, the MLTC problem can be effectively solved as sequence-to-sequence task with $\mathbf{Y}$ being the ordered set of relevant labels $\{l_1, l_2, \ldots, l_M\} \subseteq \mathcal{L}$ of an instance $\mathbf{X} = [w_1, w_2, \ldots, w_T]$. The primary approach to model sequences is decomposing the joint probability $P(\mathbf{Y}|\mathbf{X}, \theta)$ into M separate conditional probabilities. Traditionally, the left-to-right (L2R) order decomposition is:

$$P_\theta(l_1, l_2, \ldots, l_M|\mathbf{X}) = \prod_{i=1}^{M} P_\theta(l_i|l_{1:i-1}, \mathbf{X}) \tag{1}$$

Nam et al. [14] used GRU encoder and attention-based GRU decoder, achieving an improvement over a standard GRU model. Yang et al. [29] continued this idea by introducing Sequence Generation Model (SGM) consisting of BiLSTM-based encoder [19] and LSTM decoder [6] coupled with additive attention mechanism [2]. Wand et al. [22] demonstrated that the label ordering in (1) effects on the model accuracy, and the order with descending label frequencies results in a decent performance on image datasets. Alternatively, if an additional prior knowledge about the relationship between classes is provided in the form of a tree hierarchy, the labels can also be sorted in topological order.

A given hierarchical structure over labels forms a particular task known as hierarchical text classification (HTC). An underlying class hierarchy can help to discover similar classes and transfer knowledge between them improving the overall accuracy of the model. Many of the researchers' efforts to study HTC were dedicated to computer vision applications [17,20,22,27] but these studies potentially can be or have already been adapted to the field of natural language texts. Among the recent works, Peng et al. [15] proposed a Graph-based CNN architecture with a hierarchical regularizer, and Wehrmann et al. [23] argued that mixing an output from a *global* classifier and the outputs from all layers of a *local* classifier can be beneficial to learn hierarchical dependencies. It was also shown that reinforcement learning models with special award functions can be applied to learn non-trivial losses [11,28].

The main difference of the approach used in this work is twofold: (*i*) employing BERT as a pre-trained transformer-based encoder; (*ii*) using a mixed model to bring the best from both standard BERT and sequence-to-sequence BERT models.

3   BERT-Based Models for Multi-label Text Classification

3.1   Multi-label BERT

BERT is a model pre-trained on unlabelled texts for masked word prediction and next sentence prediction tasks, providing deep bidirectional representations

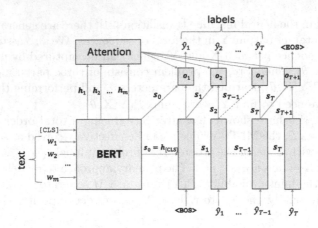

Fig. 1. BERT + SGM. An overview of the model.

for texts. For classification tasks, a special token [CLS] is put to the beginning of the text and the output vector of the token [CLS] is designed to correspond to the final text embedding. The pre-trained BERT model has proven to be very useful for transfer learning in multi-class and pairwise text classification. Fine-tuning the model followed by one additional feedforward layer and softmax activation function was shown to be enough for providing state-of-the-art results on a downstream task [5].

Multi-class classification is a standard downstream task for BERT and was extensively studied in the original work [5]. For examining BERT on the multi-label setting, we change activation function after the last layer to sigmoid. The loss to be optimized is adjusted accordingly to binary cross-entropy loss. The case of multi-label classification was experimentally examined in the work [1].

3.2   BERT Encoder for Sequence Generation

In sequence generation model, we propose to use the outputs of the last transformer block in BERT model as vector representations of words and the embedding of the token [CLS] produced by BERT as the initial hidden state of the LSTM decoder. Generally, this representation is not a good summary of the semantic content of the text [26] and it is often a better practice to pool the sequence of hidden states for all the input tokens. However, in our experiments, the model with this change performs almost the same and takes more epochs to converge.

We obtain a total order relation $(\mathcal{L}, \leq)$ after sorting the set of labels by frequency [22]. If a prior information about the relationship between classes is provided in the form of a tree hierarchy, the labels can also be sorted in topological order. We perform decoding following the work [29] except that in each stage we mask out not only previously predicted labels but also the ones that are 'not greater than' the last predicted label.

The process we follow to predict the final label set $\mathcal{L}_{pred}$ for a text input $\boldsymbol{x}$ consisting of m tokens is described in Algorithm 1 and illustrated in Fig. 1.

Algorithm 1. BERT + SGM

$\mathcal{L}_{pred} \leftarrow \{\}$
$\mathcal{L} \leftarrow \{1, 2, \ldots, L, \texttt{<EOS>}\}$
$[\boldsymbol{h}_1, \boldsymbol{h}_2, \ldots, \boldsymbol{h}_m], \boldsymbol{h}_{\texttt{[CLS]}} \leftarrow \text{BERT}(\boldsymbol{x})$
$s_0 \leftarrow \mathbf{W}_{init} \boldsymbol{h}_{\texttt{[CLS]}}$
$\hat{y}_0 \leftarrow \texttt{<BOS>}$
$t \leftarrow 0$
while $\hat{y}_t \neq \texttt{<EOS>}$ **do**
$\quad t \leftarrow t + 1$
$\quad \boldsymbol{\alpha}_t = \text{softmax}([\boldsymbol{h}_1, \boldsymbol{h}_2, \ldots, \boldsymbol{h}_m]^T \boldsymbol{s}_{t-1})$
$\quad \boldsymbol{c}_t \leftarrow \sum_{i=1}^{m} \alpha_{ti} \boldsymbol{h}_i$
$\quad \boldsymbol{s}_t \leftarrow \text{LSTM}(\boldsymbol{s}_{t-1}, [\hat{y}_{t-1}; \boldsymbol{c}_{t-1}])$
$\quad \boldsymbol{o}_t \leftarrow \mathbf{W}_o \tanh(\mathbf{W}_d \boldsymbol{s}_t + \mathbf{V}_d \boldsymbol{c}_t)$
$\quad \boldsymbol{y}_t \leftarrow \text{masked\_softmax}(\boldsymbol{o}_t, \text{children}(\hat{y}_t))$
$\quad \hat{y}_t \leftarrow \arg\max \boldsymbol{y}_t$
$\quad \mathcal{L}_{pred} \leftarrow \mathcal{L}_{pred} \cup \{\hat{y}_t\}$
return $\mathcal{L}_{pred}$

We train the final model to minimize the cross-entropy objective loss for a given $\boldsymbol{x}$ and ground-truth labels $\{l_1^*, l_2^*, \ldots, l_k^*\} \in \mathcal{L}$:

$$\mathcal{L}_{\text{CE}}(\theta) = -\sum_{i=1}^{k} \log P_\theta(l_i^* | \boldsymbol{x}, l_{1:i-1}^*) \tag{2}$$

In the inference stage, we can compute the objective 2 replacing ground-truth labels with predicted labels. To produce the final sequence of labels, we perform a beam search following the work [25] to find candidate sequences that have the minimal objective scores among the paths ending with the <EOS> token.

3.3   Mixed Model

In experiments, our error analysis has shown that often BERT can predict excess labels while BERT+SGM tends to be more restrained, which suggests that the two approaches can potentially complement each other well. Also, BERT+SGM exploits the information about the underlying structure of labels. Wehrmann et al. [23] in their work propose HMCN model in which they suggest to jointly optimize both local (hierarchical) and global classifiers and combine their final probability predictions as a weighted average.

Following this idea, we propose to use a mixed model which is an ensemble of multi-label BERT and sequence generating BERT models. A main challenge in creating a mixed model is that the outputs of the two models are principally different. In Seq2Seq setup, we suggest to produce a probability distribution over

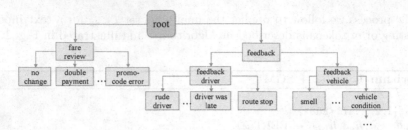

Fig. 2. An example of a subtree of the tree hierarchy over classes in *Y.Taxi Riders* dataset.

the labels by performing element-wise max-pooling operation on softmax outputs of the decoder at each stage $\{y_t\}_{t=1}^{T}$ as per the paper [18]. In our experiments, we found that the probability distributions obtained this way are meaningful and after thresholding can yield predictions with accuracy comparable to the accuracy of BERT+SGM model.

The final predictions $p_{\text{mixed}} \in \mathbb{R}^L$ are computed as a weighted average of probability distributions of both models:

$$p_{\text{mixed}} = \alpha p_{\text{BERT+SGM}} + (1 - \alpha) p_{\text{BERT}} \tag{3}$$

The value of $\alpha \in [0, 1]$ is a trade-off parameter that is optimized on a validation set.

4   Experiments

4.1   Datasets and Preprocessing

Table 1. Summary of the datasets. N is the number of documents, L is the number of labels, W denotes the average number of words per sample $\pm$SD (standard deviation), and C denotes the average number of labels per sample $\pm$SD.

Dataset	N	L	W	C	Structure	Language
RCV1-v2	804 410	103	223.2 ± 206.6	3.2 ± 1.4	Tree	Eng
Reuters-21578	10 787	90	142.2 ± 142.5	1.2 ± 0.7	–	Eng
AAPD	55 840	54	155.9 ± 67.6	2.4 ± 0.7	–	Eng
Y.Taxi Drivers	163 633	374	18.9 ± 22.6	2.1 ± 1.0	Tree	Rus
Y.Taxi Riders	174 590	426	16.2 ± 18.6	3.4 ± 0.8	Tree	Rus

We train and evaluate all the models on three public datasets with English texts and two private datasets with Russian texts. The summary of the datasets' statistics is provided in the Table 1. Preprocessing of the datasets included lower casing the texts and removing punctuation. For the baseline TextCNN and SGM models, we used the same preprocessing techniques as in [29].

Table 2. Results on the five considered datasets. Metrics are marked in bold if they contain the highest metrics for the dataset in their ±SD interval.

	RCV1-v2				Reuters-21578				AAPD			
	HA	miF_1	maF_1	ACC	HA	miF_1	maF_1	ACC	HA	miF_1	maF_1	ACC
TextCNN	0.990	0.829	0.456	0.600	0.991	0.851	0.437	0.827	0.974	0.674	0.445	0.364
HMCN	0.989	0.808	0.546	0.595	–	–	–	–	–	–	–	–
HiLAP	0.990	0.833	0.611	0.607	–	–	–	–	–	–	–	–
EncDec orig.	–	–	–	–	0.996	0.858	0.457	0.828	–	–	–	–
SGM repr.	0.990	0.815	0.428	0.605	0.996	0.788	0.452	0.812	0.974	0.698	0.468	0.372
BERT	**0.992**	0.864	0.556	0.624	**0.997**	**0.899**	**0.534**	**0.857**	0.976	0.713	**0.559**	0.381
BERT+SGM	0.990	0.846	**0.629**	0.602	0.996	0.854	0.467	0.817	0.976	**0.718**	0.496	0.377
Mixed	**0.992**	**0.868**	0.611	**0.631**	0.996	**0.900**	**0.533**	**0.858**	**0.977**	0.719	0.553	**0.397**

	Y.Taxi Drivers				Y.Taxi Riders			
	HA	miF_1	maF_1	ACC	HA	miF_1	maF_1	ACC
TextCNN	0.996	0.610	0.173	0.571	0.994	0.521	0.130	0.381
SGM repr.	0.996	0.629	0.148	0.584	0.993	0.545	0.112	0.399
BERT	0.997	**0.692**	0.226	0.578	0.995	**0.658**	0.153	0.452
BERT+SGM	0.997	0.644	0.196	0.596	**0.997**	0.644	**0.176**	0.465
Mixed	**0.998**	0.681	**0.235**	**0.599**	**0.997**	0.657	0.174	**0.469**

Reuters Corpus Volume I (RCV1-v2) [9] is a collection of manually categorized news stories, with each sample being assigned to labels from one or multiple paths in the class hierarchy. Since there is practically no difference between topological sorting order and order by frequency [14] in multi-path case, we chose to sort the labels from the most common ones to the rarest ones. The original training/testing split [9] is still being used in modern research, but in several other works authors split the data differently [14,29]. To avoid confusion, we decided to be consistent with the original split.

Reuters-21578 is an MLTC dataset with articles from Reuters newswire. We use the standard ApteMod split of the dataset [4].

Arxiv Academic Paper Dataset (AAPD) is a recently collected dataset [29] consisting of abstracts of research papers from arXiv.org categorized into academic subjects.

Y.Taxi Riders is a private dataset from Yandex Taxi client support system consisting of 174k tickets from riders and 426 classes. The dataset was labeled by Yandex Taxi reviewers with one path from tree hierarchy per each sample with an estimated accuracy 76.5 ± 2%. Since in this task there is only one path in the tree to be predicted, we will explore a natural topological label ordering for this dataset. An example of a subtree of the tree hierarchy is provided in Fig. 2.

Y.Taxi Drivers is also a private Yandex Taxi dataset collected in drivers support system. The dataset consists of 164k tickets tagged with 374 labels and has properties similar to the *Y.Taxi Riders* dataset.

4.2 Experiment Settings and Baselines

We use the base-uncased versions of BERT for English texts and the base-multilingual-cased version for Russian texts. The batch size is set to 16. We optimize using Adam [8] with $\beta_1 = 0.9, \beta_2 = 0.99$ and learning rate $2 \cdot 10^{-5}$. For the multi-label BERT, we also use the same scheduling of the learning rate as in the original work by Devlin et al. [5]. We implemented all the experiments in PyTorch 1.2 and ran the computations on a single GeForce GTX 1080Ti GPU. Our implementation is relied on transformers library [26].

To evaluate the performance of the models, we compute hamming accuracy, set accuracy, micro-averaged f_1, and macro-averaged f_1 [11,14,29].

We use a classic two-layer convolutional neural network **TextCNN** [7] as a baseline for our experiments. For comparison, we provide the results of **HMCN** [23] and **HiLAP** [11] models for hierarchical text classification on *RCV1-v2* dataset adopted from the work [11]. The next baseline model is Sequence Generation Model **SGM** [29], for which we reuse the implementation of the authors.[1] For *Reuters-21578* dataset, we also include the results of the **EncDec** model [14] from the original paper on sequence-to-sequence approach to MLTC.

4.3   Results and Discussion

We present the results of the suggested models and baselines on the considered datasets in Table 2.

First, we can see that both BERT and BERT+SGM show favorable results on multi-label classification datasets mostly outperforming other baselines by a significant margin.

In some cases, BERT performs better than the sequence-to-sequence version, which is especially evident on the *Reuters-21578* dataset, and a possible reason might be its small size. However, on *RCV1-v2* dataset the macro-F_1 metrics of BERT + SGM is much larger while other metrics are still comparable with the BERT's results. Also, for both Yandex Taxi datasets in Russian language, we can see that the hamming accuracy and the set accuracy of the BERT+SGM model is higher compared to other models.

In most cases, better performance can be achieved after mixing BERT and BERT+SGM. On public datasets, we see 0.4%, 1.6%, and 0.8% average improvement in miF_1, maF_1, and accuracy respectively in comparison with BERT. On datasets with a tree hierarchy over classes, we observe 2.8% and 1.5% average improvement in maF_1 and accuracy.

We also found that BERT for multi-label text classification tasks takes far more epochs to converge compared to 3–4 epochs needed for multi-class datasets [5]. For *AAPD*, we performed 20 epochs of training; for *RCV1-v2* and *Reuters-21578* – around 30 epochs; for Russian datasets – 45–50 epochs. BERT + SGM achieves decent accuracy much faster than multi-label BERT and converges after 8–12 epochs. Performance of both models on the validation set of *Reuters-21578* during the training process is shown in Fig. 3.

We obtain optimal results with the beam size in the range from 5 to 9. However, the greedy approach with the beam size 1 gives similar results with less than 0.3% difference in the metrics. A possible explanation for this might

---

[1] https://github.com/lancopku/SGM.

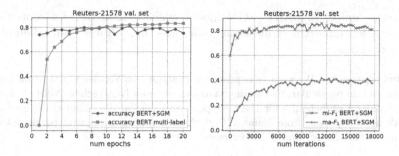

Fig. 3. Performance of BERT and BERT+SGM on *Reuters-21578* validation set during training.

BERT+SGM

we introduce a new language representation model called bert , which stands for bidirectional encoder representations from transformers . unlike recent language representation models , bert is designed to pre - train deep bidirectional representations by jointly conditioning on both left and right context in all layers .

['cs.LG', 'cs.CL']

BERT multi-label

we introduce a new language representation model called bert , which stands for bidirectional encoder representations from transformers . unlike recent language representation models , bert is designed to pre - train deep bidirectional representations by jointly conditioning on both left and right context in all layers .

['cs.LG', 'cs.CL', 'cs.NE']

Fig. 4. Visualization of feature importance for multi-label BERT and BERT+SGM models trained on *AAPD* and applied to BERT paper [5] abstract (cs.LG – machine learning; cs.CL – computation & linguistics; cs.NE – neural and evolutionary computing).

be that, while in natural language generation tasks the token ordering in the output sequence matters a lot and there might be many potential options, label set generation task is much simpler due to a smaller 'vocabulary' size $|\mathcal{L}|$ and indifference to ordering.

A natural question arises as to whether the success of the mixed model is the result of two models having different views on text features. To have a rough idea of how the networks make their prediction, we visualized the word importance scores for each model using the leave-one-out method [24] in Fig. 4. It can be seen from this example that BERT+SGM seems to be slightly more selective in terms of features to which it pays attention. Also, in this particular case, the predictions of sequence generating BERT are more accurate.

5   Conclusion

In this work, we examine BERT and sequence generating BERT on the multi-label setting. We experiment with both models and explore their particular properties for this task. We also introduce and examine experimentally a mixed model which is an ensemble of standard BERT and sequence-to-sequence BERT models.

Our experimental studies showed that BERT-based models and the mixed model, in particular, outperform the considered baselines by several metrics. We establish that multi-label BERT typically needs several dozens of epochs to converge, unlike to BERT+SGM model which demonstrates decent results just after a few hundreds of iterations (one pass through less than a half of the training data.).

Acknowledgments. We want to express our gratitude to Yandex.Go Machine Learning team, and, in particular, to Tatiana Savelieva, Roman Khalkechev, Nikita Seleznev, and Alexander Parubchenko for their support and helpful discussions. We also want to thank all the people from Yandex.Taxi client support team who participated in collecting *Y.Taxi Riders* and *Y.Taxi Drivers* datasets.

References

1. Adhikari, A., Ram, A., Tang, R., Lin, J.: DocBERT: BERT for document classification. CoRR abs/1904.08398 (2019). http://arxiv.org/abs/1904.08398
2. Bahdanau, D., Cho, K., Bengio, Y.: Neural machine translation by jointly learning to align and translate. CoRR abs/1409.0473 (2014)
3. Cho, K., van Merrienboer, B., Bahdanau, D., Bengio, Y.: On the properties of neural machine translation: encoder-decoder approaches. In: Proceedings of SSST-8, Eighth Workshop on Syntax, Semantics and Structure in Statistical Translation, pp. 103–111. Association for Computational Linguistics, Doha, Qatar, October 2014. https://doi.org/10.3115/v1/W14-4012
4. Cohen, W.W., Singer, Y.: Context-sensitive learning methods for text categorization. In: Proceedings of the 19th Annual International ACM SIGIR Conference on Research and Development in Information Retrieval. SIGIR 1996, pp. 307–315. ACM, New York, NY, USA (1996). https://doi.org/10.1145/243199.243278
5. Devlin, J., Chang, M.W., Lee, K., Toutanova, K.: BERT: pre-training of deep bidirectional transformers for language understanding (2018)
6. Hochreiter, S., Schmidhuber, J.: Long short-term memory. Neural Comput. **9**(8), 1735–1780 (1997)
7. Kim, Y.: Convolutional neural networks for sentence classification. In: Proceedings of the 2014 Conference on Empirical Methods in Natural Language Processing, EMNLP 2014, 25–29 October 2014, Doha, Qatar, A meeting of SIGDAT, a Special Interest Group of the ACL, pp. 1746–1751 (2014)
8. Kingma, D.P., Ba, J.: Adam: a method for stochastic optimization (2015). In: Conference Paper at the 3rd International Conference for Learning Representations, San Diego (2015)
9. Lewis, D.D., Yang, Y., Rose, T.G., Li, F.: RCV1: a new benchmark collection for text categorization research. J. Mach. Learn. Res. **5**, 361–397 (2004)
10. Liu, Y., et al.: RoBERTa: a robustly optimized BERT pretraining approach. CoRR abs/1907.11692 (2019)
11. Mao, Y., Tian, J., Han, J., Ren, X.: Hierarchical text classification with reinforced label assignment. In: Proceedings of the 2019 Conference on Empirical Methods in Natural Language Processing, EMNLP (2019)
12. Mikolov, T., Sutskever, I., Chen, K., Corrado, G.S., Dean, J.: Distributed representations of words and phrases and their compositionality. In: Burges, C.J.C., Bottou, L., Welling, M., Ghahramani, Z., Weinberger, K.Q. (eds.) Advances in Neural Information Processing Systems 26, pp. 3111–3119. Curran Associates, Inc. (2013)
13. Molino, P., Zheng, H., Wang, Y.: COTA: improving the speed and accuracy of customer support through ranking and deep networks. KDD (2018)

14. Nam, J., Menc'ia, E.L., Kim, H.J., Fürnkranz, J.: Maximizing subset accuracy with recurrent neural networks in multi-label classification. In: Advances in Neural Information Processing Systems 30: Annual Conference on Neural Information Processing Systems 2017, 4–9 December 2017, Long Beach, CA, USA, pp. 5419–5429 (2017)

15. Peng, H., et al.: Large-scale hierarchical text classification with recursively regularized deep graph-CNN. In: Proceedings of the 2018 World Wide Web Conference on World Wide Web, WWW 2018, Lyon, France, 23–27 April 2018, pp. 1063–1072 (2018). https://doi.org/10.1145/3178876.3186005

16. Pennington, J., Socher, R., Manning, C.D.: GloVe: global vectors for word representation. In: Empirical Methods in Natural Language Processing (EMNLP), pp. 1532–1543 (2014)

17. Salakhutdinov, R., Torralba, A., Tenenbaum, J.B.: Learning to share visual appearance for multiclass object detection. In: CVPR, pp. 1481–1488. IEEE Computer Society (2011)

18. Salvador, A., Drozdzal, M., Giró i Nieto, X., Romero, A.: Inverse cooking: Recipe generation from food images. CoRR abs/1812.06164 (2018)

19. Schuster, M., Paliwal, K.: Bidirectional recurrent neural networks. Trans. Sig. Proc. **45**(11), 2673–2681 (1997). https://doi.org/10.1109/78.650093

20. Srivastava, N., Salakhutdinov, R.: Discriminative transfer learning with tree-based priors. In: Advances in Neural Information Processing Systems 26: 27th Annual Conference on Neural Information Processing Systems 2013. Proceedings of a Meeting Held, 5–8 December 2013, Lake Tahoe, Nevada, United States, pp. 2094–2102 (2013)

21. Sutskever, I., Vinyals, O., Le, Q.V.: Sequence to sequence learning with neural networks. In: Advances in Neural Information Processing Systems 27: Annual Conference on Neural Information Processing Systems 2014, 8–13 December 2014, Montreal, Quebec, Canada, pp. 3104–3112 (2014)

22. Wang, J., Yang, Y., Mao, J., Huang, Z., Huang, C., Xu, W.: CNN-RNN: a unified framework for multi-label image classification. In: 2016 IEEE Conference on Computer Vision and Pattern Recognition, CVPR 2016, Las Vegas, NV, USA, 27–30 June 2016, pp. 2285–2294 (2016). https://doi.org/10.1109/CVPR.2016.251

23. Wehrmann, J., Cerri, R., Barros, R.C.: Hierarchical multi-label classification networks. In: Proceedings of the 35th International Conference on Machine Learning, ICML 2018, Stockholmsmässan, Stockholm, Sweden, 10–15 July 2018, pp. 5225–5234 (2018)

24. Wiegreffe, S., Pinter, Y.: Attention is not not explanation. In: Proceedings of the 2019 Conference on Empirical Methods in Natural Language Processing and the 9th International Joint Conference on Natural Language Processing (EMNLP-IJCNLP), pp. 11–20. Association for Computational Linguistics, Hong Kong, China, November 2019. https://doi.org/10.18653/v1/D19-1002, https://www.aclweb.org/anthology/D19-1002

25. Wiseman, S., Rush, A.M.: Sequence-to-sequence learning as beam-search optimization. CoRR abs/1606.02960 (2016)

26. Wolf, T., et al.: Huggingface's transformers: state-of-the-art natural language processing. arXiv abs/1910.03771 (2019)

27. Yan, Z., et al.: HD-CNN: hierarchical deep convolutional neural networks for large scale visual recognition. In: 2015 IEEE International Conference on Computer Vision, ICCV 2015, Santiago, Chile, 7–13 December 2015, pp. 2740–2748 (2015). https://doi.org/10.1109/ICCV.2015.314

28. Yang, P., Ma, S., Zhang, Y., Lin, J., Su, Q., Sun, X.: A deep reinforced sequence-to-set model for multi-label text classification. CoRR abs/1809.03118 (2018)
29. Yang, P., Sun, X., Li, W., Ma, S., Wu, W., Wang, H.: SGM: sequence generation model for multi-label classification. In: COLING, pp. 3915–3926. Association for Computational Linguistics (2018)
30. Yang, Z., Dai, Z., Yang, Y., Carbonell, J.G., Salakhutdinov, R., Le, Q.V.: XLNet: generalized autoregressive pretraining for language understanding. CoRR abs/1906.08237 (2019)
31. Yang, Z., Yang, D., Dyer, C., He, X., Smola, A., Hovy, E.: Hierarchical attention networks for document classification. In: Proceedings of the 2016 Conference of the North American Chapter of the Association for Computational Linguistics: Human Language Technologies, pp. 1480–1489. Association for Computational Linguistics, San Diego, California, June 2016. https://doi.org/10.18653/v1/N16-1174

Computer Vision

Deep Learning on Point Clouds for False Positive Reduction at Nodule Detection in Chest CT Scans

Ivan Drokin[ID] and Elena Ericheva[✉][ID]

Intellogic Limited Liability Company (Intellogic LLC),
Territory of Skolkovo Innovation Center, Office 1/334/63, Building 1,
42 Bolshoi blvd., 121205 Moscow, Russia
{ivan.drokin,elena.ericheva}@botkin.ai

Abstract. This paper focuses on a novel approach for false-positive reduction (FPR) of nodule candidates in Computer-Aided Detection (CADe) systems following the suspicious lesions detection stage. Contrary to typical decisions in medical image analysis, the proposed approach considers input data not as a 2D or 3D image, but rather as a point cloud, and uses deep learning models for point clouds. We discovered that point cloud models require less memory and are faster both in training and inference compared to traditional CNN 3D, they achieve better performance and do not impose restrictions on the size of the input image, i.e. no restrictions on the size of the nodule candidate. We propose an algorithm for transforming 3D CT scan data to point cloud. In some cases, the volume of the nodule candidate can be much smaller than the surrounding context, for example, in the case of subpleural localization of the nodule. Therefore, we developed an algorithm for sampling points from a point cloud constructed from a 3D image of the candidate region. The algorithm is able to guarantee the capture of both context and candidate information as part of the point cloud of the nodule candidate. We designed and set up an experiment in creating a dataset from an open LIDC-IDRI database for a feature of the FPR task, and is herein described in detail. Data augmentation was applied both to avoid overfitting and as an upsampling method. Experiments were conducted with PointNet, PointNet++, and DGCNN. We show that the proposed approach outperforms baseline CNN 3D models and resulted in 85.98 FROC versus 77.26 FROC for baseline models. We compare our algorithm with published SOTA and demonstrate that even without significant modifications it works at the appropriate performance level on LUNA2016 and shows SOTA on LIDC-IDRI.

Keywords: Image analysis · Computer vision · Computer detection · Computer-assisted · Lung cancer · Pulmonary nodule · False-positive reduction · Point cloud · Chest CT

Supported by BOTKIN.AI, Skolkovo.

W. M. P. van der Aalst et al. (Eds.): AIST 2020, LNCS 12602, pp. 201–215, 2021.
https://doi.org/10.1007/978-3-030-72610-2_15

1   Introduction

Survival in lung cancer (over a period of 5 years) is approximately 18.1%[1]. Early-stage lung cancer (stage I) has a five-year survival rate of 60–75%. A recent National Lung Screening Trial (NLST) study has shown that lung cancer mortality can be reduced by at least 20%, using a high-risk annual screening program with low-dose computed tomography (CT) of the chest [1]. Computerized tools, especially image analysis and machine learning, are key factors for improving diagnostics, facilitating the identification of results that require treatment, and supporting the workflow of an expert [2].

Although CAD systems have shown improvements in readability by radiologists [3–5], a significant number of nodules remained undetected at a low rate of false-positive results, prohibiting the use of CAD in clinical practice. Classification tasks in the medical domain are often a normal vs pathology discrimination task. In this case, it is worth noting the normal class is extremely over-represented in a dataset.

Until recently, CAD systems were built using manually created functions and decision rules. With the advent of a new era of deep learning, the situation has changed. Now, to solve the detection task, CNN models are widely studied. Due to their specificity, CNNs can work efficiently with images and focus on candidate recognition [2]. However, due to architectural limitations, convolution neural networks can only function effectively with data of a strictly specified size. The selection of images of incorrect sizes can significantly reduce the effectiveness of the model for objects that are either too large or too small.

In this work, we have proposed a novel approach for the false-positive reduction stage for CAD systems, based on point cloud networks. These networks require less memory for training and inference, show more stable results when working with multiple data sources and, most importantly, do not impose restrictions on the size of the input image - unlike CNN counterparts.

We have also proposed an algorithm for transforming 3D CT scan data to point cloud. We have developed an algorithm for sampling points from a point cloud, constructed from a 3D image of the candidate region. The algorithm must be able to capture both context and candidate information as part of the point cloud of the nodule candidate.

The goal of the false-positive reduction task is to recognize a true pulmonary nodule from multiple plural candidates, which are received from the detection stage. Both point cloud sampling methods and FPR-model inputs strongly depend on segmentation masks suitable for nodule candidates. Performance evaluation of the FPR is both deeply connected with and dependent on the performance of a detector. In order to ascertain the accuracy of the FPR model performance (investigated separately from the detector) we designed and have described the full process of receiving a special artificial dataset for FPR training and evaluation.

---

[1] 2018 state of lung cancer report: https://www.naaccr.org/2018-state-lung-cancer-report/.

This article consists of the following structure. In the *Sect.* 2 we have provided an overview of major works on lung cancer detection, lungs nodules evaluation and models on point clouds. Under *Sect.* 3 we describe data processing for training both point-cloud-based and baseline models. In the *Sect.* 4 we explain in detail a point cloud sampling policy and, in *Sect.* 5, an augmentation policy. All the experiments we conducted on LIDC-IDRI and LUNA2016 datasets are described under the *Sect.* 6 with a comparison of current SOTA.

2 Previous Work

Currently the conventional pipeline in the screening task for CAD consists of several stages - principally detection and cancer classification. A two-stage machine learning algorithm is a popular approach that can assess the risk of cancer associated with a CT scan [6–10]. The first stage uses a nodule detector which identifies nodules contained in the scan. The second step is used to assess whether nodules are malignant or benign.

Methods to solve false positive reduction tasks separately from full CAD pipelines have been highly favored in recent times. A multicontext 3D residual convolutional neural network (3D Res-CNN) was proposed in [11] to reduce false-positive nodules. In [12] a spatial pooling and cropping (SPC) layer to extract multi-level contextual information of CT data was designed and added to 3D Res-CNN. The method presented in [13] is based on structural relationship analysis between nodule candidates and vessels, and the modified surface normal overlap descriptor to separate low-contrast nonsolid nodules from the candidates. The algorithm proposed in [14] segments lungs and nodules through a combination of 2D and 3D region growing, thresholding and morphological operations. To reduce the number of false positives, a rule-based classifier is used to eliminate obvious non-nodules, followed by a multi-view Convolutional Neural Network. The CNN from [15] is fed with nodule candidates obtained by combining three candidate detectors specifically designed for solid, subsolid, and large nodules. For each candidate, a set of 2D patches from differently oriented planes is extracted. An evaluation of FPR models are performed on independent datasets from the LUNA16[2] and ANODE09[3] challenges and DLCST[16].

Recently, several successful works using point clouds in medical image analysis have appeared. In [25], a segmentation of teeth was presented. In [27], segmentation refinement with false positive reduction by point clouds was proposed. In [26] the authors use point clouds for vertebra shape analysis.

A point cloud is represented as a set of 3D points $\{P_i | i = 1, ..., n\}$, where each point P_i is a vector of its (x, y, z) coordinate in addition to extra feature channels. A neural network, named PointNet, was shown in [22]. It directly works with point clouds, and fully respects the permutation invariance of points in the input. Therefore, the model is able to capture local structures from nearby points, and the combinatorial interactions among local structures. In [23], the authors

---

[2] LUNA16 challenge homepage: https://luna16.grand-challenge.org/.
[3] ANODE09 challenge homepage: https://anode09.grand-challenge.org/.

expand further on the idea to capture local structures induced by the metric space points live in, as well as increasing model ability to recognize fine-grained patterns and generalizability to complex scenes. Finally, in [24], the authors propose a module dubbed EdgeConv suitable for CNN-based high-level tasks on point clouds, including classification and segmentation. Instead of working on individual points like PointNet, however, EdgeConv exploits local geometric structures by constructing a local neighborhood graph and applying convolution-like operations on the edges, connecting neighboring pairs of points (as in graph neural networks).

3   Data and Processing

It is difficult to directly and objectively compare different CAD systems. In [17], an evaluation system was proposed for the automatic detection of nodules on CT images. A large dataset, containing 888 CT scans with annotations from the open LIDC-IDRI database[4], is available from the NCI Cancer Imaging Archive.[5] A detailed description of the data set is given in [18]. This includes a description of the range of possible patients and disease manifestations, as well as the data labels reception process, and an analysis of their ability to support the performance of this method. Any nodule noted by at least one radiologist was considered as positive ground truth. This increased the number of positive ROI in our dataset. In general this led to a decrease in recall at low levels of false positives. Additionally, we took into account all nodules ≤ 3 mm as a false positive. Our reasoning for choosing such a markup is the goal of constructing a pipeline for detecting nodes with maximum recall, including controversial cases. In our opinion, such borderline situations should be detected by an algorithm, and a doctor should have to make a final decision regarding this region. It is in controversial and non-obvious cases that the new CAD systems should be of help to the doctor. Since different series in the dataset can have different slice thickness parameters, and thus different distances between slices, we resampled series space to 1 mm per voxel side.

We described dataset collection for the false positive reduction task as a *Subsect.* 3.1. We have also explained preprocessing algorithms for each model in relevant subsections: for detector in the *Subsect.* 3.2, for baseline FPR in the *Subsect.* 3.3, for point cloud-based in the *Subsect.* 3.4.

3.1   Collecting a Dataset for FPR Task

Since both our point cloud sampling method and FPR-model input strongly depend on segmentation masks suitable for nodule candidates, we have used the LIDC-IDRI dataset as a source for setting up training and testing data for the FPR task. A simple scheme is presented in Fig. 1.

---

[4] LIDC-IDRI Dataset: https://wiki.cancerimagingarchive.net/display/Public/LIDC-IDRI.

[5] TCIA Collections: https://www.cancerimagingarchive.net.

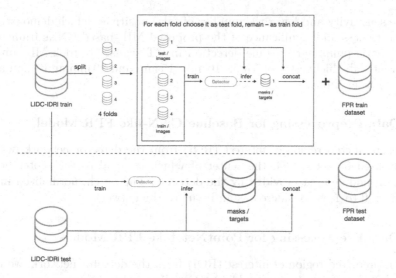

Fig. 1. Full scheme of the artificial dataset for FPR task

We have used DenseNet121-TIRAMISU [20] as a detector for nodule candidates, as well as for train and test datasets for FPR task collection. Firstly, we split LIDC-IDRI datasets into train and test datasets with a 75%–25% ratio. For the creation of the FPR-train dataset we applied 4-fold cross-validation. We split the LIDC-IDRI train dataset into 4 subsets, trained a detector on 3 of these subsets, then inferred the detector on the remaining subset. The data received from the inference phase, including predicted segmentation masks and probabilities, is then appended to the training dataset for the FPR task. We repeated this detector train-infer loop 4 times, consistently considering each of the 4 folds from LIDC-IDRI as a subset to infer. To create the FPR-test dataset we trained the detector on the full LIDC-IDRI train dataset, then inferred the detector on the LIDC-IDRI test dataset. The data received from the inference is then used as a test dataset for the FPR task.

3.2 Data Preprocessing for Detector

The images are presented on the Hounsfield scale.[6] Voxel intensities are limited to an interval from -1000 to 400 HU and normalized in the range from 0 to 1. We used as a detector input a full 2D slice concatenated with 4 Maximum Intensity Projection (MIP) images [19] per sample. MIP images are used by radiologists alongside a complete CT exam in order to improve the detection of pulmonary nodules, especially small nodules. MIP images consist of the superposition of maximum grey values at each coordinate from a stack of consecutive slices. Such a combined image shows morphological structures of isolated nodules and continuous vessels. Experimental results in [19] showed that utilizing MIP images can

---

[6] https://en.wikipedia.org/wiki/Hounsfield_scale.

increase sensitivity and lower the number of false positives, which demonstrates the effectiveness and significance of the proposed MIP-based CNNs framework for automatic pulmonary nodule detection in CT scans. We took MIP images of different slab thicknesses of 5 mm, 10 mm, 15 mm and 20 mm and 1 mm axial section slices as input.

3.3 Data Preprocessing for Baseline CNN-like FPR Model

For each predicted region of interest (ROI) from the detector network, we cut a patch sized $64 \times 64 \times 64$, the center of which is equal to the center of the segmented nodule area provided by the detector. The patch, normalized in the range from 0 to 1, is considered as an input to the network.

3.4 Data Preprocessing for PointNet-Like FPR Models

For each predicted region of interest (ROI) from the detector network, we form a bounding box with 16 mm padding. Bounding boxes here can be any size, matching the binarized segmenting mask obtained from a detector. Thus we can provide context for the candidate and the surrounding neighborhood, with different lung tissues like vessels, pleura, bronchi, etc. To reduce the number of points and exclude the most non-informative points from consideration we selected points only with a HU value from -400 to 400 from this bounding box. This allows us to catch the vast majority of all nodule types and, at the same time, reduces the number of points from 250,000 to 15,000 on average. The input point consists of several features: coordinates of a point, its HU density value, and the probability that it will be predicted by the detector. As a normalization step, we centered coordinates over the center of the segmented nodule area provided by the detector, and in addition translated HU values to a $(-1,1)$ interval.

4   Point Cloud Sampling

A naive resampling approach can invoke the loss of important information and is highly application-dependent. Since 2016, several studies have investigated point cloud analysis [22,23] but still applied a uniform resampling for fixing the number of points. Such an approach does not preserve the finer details of data, which is important for the segmentation tasks. Furthermore, it also ignores the existing strong dependency between the label of each point and its location in the point cloud. For the basis of our proposed point cloud sampling method, we used the method described in [25]. The authors propose a unique non-uniform resampling mechanism that facilitates the network on a fixed-size resampled point cloud, which contains different levels of the spatial resolution, involving both local, fine details and the global shape structure.

We introduced two major differences from this method. First, we change the parameter σ that controls the bandwidth (compactness) of the kernel and depends on the candidate radius r. We choose $\sigma = r/2$. Second, we introduced additional sampling from the nodule area. It should be pointed out that there is a problem of sampling small subpleural candidates with a diameter < 5 mm and an overwhelming surrounding context. The probability of the candidate's point inclusion to the sampled point cloud decreases with the reduction of the candidate's size. To deal with this, we sample additionally and uniformly only from the area corresponding to the segmenting mask provided by the detector. Thus we can guarantee the appearance of candidate points in the sampled point cloud regardless of the candidate's location or size.

The basic intuition that underlies the proposed sampling method is as follows: the farther the point is from the center of the candidate, the less likely it will be selected in the sampling process. To control this, we introduce sigma as half the candidate radius. Obviously, the farther the point is from the candidate, the less likely it is to be classified as a nodule/non-nodule. For example, even if the detector incorrectly selected a part of the vessel as a nodule, the points of interest that are closest to the candidate's points are of greatest interest. This will allow the algorithm to decide that it is a vessel (or bronchus), and not a nodule. Whether or not a point belongs to the candidate is determined by the order of the detector in binarizing its mask. Not only does all of this guarantee a context capture regardless of the size of the original candidate, but uniform sampling also provides us with all the necessary information about the region of interest.

5   Augmentation Techniques

We are dealing with an imbalance between false positives and nodules. In order to avoid overfitting, augmentations are performed at an image level as well as at a point cloud level. At an image level, we apply Gaussian noise blur and Hounsfield units shift. Gaussian noise is added not to a whole slice but rather in accordance with the random generated mask. We have used 0.2 for blur appearance probability and the interval [0.2, 0.8] for the Gaussian filter coefficient. The Gaussian filter coefficient alpha is distributed uniformly.

At the point cloud level we added a rotation (at a randomly chosen degree in the transverse planes) and a random constant coordination shift. This shift presents compression and extension. Balanced random selection from the data set was applied. The augmentation technique is applied only to the training dataset; the test data is utilised as is. The augmentation technique is applied both for CNN-like baselines as well as for PointNet-like FPR models.

6   Experiments

6.1   Experiments on LIDC-IDRI

The goal of the false-positive reduction task is to recognize true pulmonary nodules from a variety of candidates, which are identified at the first step of pulmonary nodule candidate detection. Here we investigated FPR performance separately from the detector results. We considered the simplest and most common architecture of a detector model and used it only as a sampler for creating the FPR train dataset. Performance evaluation of the FPR is deeply connected with and dependent on the performance of a detector. That is why we examined FPR performance scoring only on an artificially received dataset, as described in *Sect.* 3.1.

Table 1. Results of the experiments on LIDC-IDRI. Sensitivity per FP level per exam.

Experiment	0.125 FP	0.25 FP	0.5 FP	1 FP	2 FP	4 FP	8 FP	Mean sens
Li [29]	0.600	0.674	0.751	0.824	0.850	0.853	0.859	0.773
Liao [28]	0.662	0.746	0.815	0.864	0.902	0.918	0.932	0.834
Baseline	0.514	0.600	0.710	0.812	0.893	0.931	0.944	0.772
Pointnet	0.433	0.560	0.693	0.835	0.946	0.977	0.982	0.775
Pointnet w/aug	0.438	0.607	0.698	0.844	0.949	0.990	0.990	0.788
Pointnet++	0.360	0.502	0.640	0.776	0.922	0.972	0.995	0.738
Pointnet++ w/aug	0.356	0.502	0.657	0.789	0.922	0.990	0.995	0.744
DGCNN	0.497	0.628	0.758	0.904	0.969	0.994	0.994	0.821
DGCNN w/aug	**0.545**	**0.679**	**0.842**	**0.971**	**0.990**	**0.995**	**0.995**	**0.859**

We compared several architectures that work with point cloud data and baseline CNN. As a baseline we considered a ResNet3D model [21], trained on the same dataset.

We examined 3 PointNet based models: PointNet [22], PointNet++ [23] and DGCNN [24]. We used FROC [35] as the model's quality criterion, as a natural metric for nodule detection systems. During experiments, we performed several runs for selected architectures and evaluated model performance with and without augmentations during the training procedure. We used an ADAM [36] optimizer with a starting learning rate of 0.001, and trained models during 70 epochs, decreasing the learning rate twice every 10 epochs. We achieved the best FROC of 85.98 using the DGCNN model with augmentation. This result outperforms the 77.26 FROC of the baseline model. Results for all the experiments are shown in Table 1. In this table, Sens./0.125 FP means that models show sensitivity equal to a number from the corresponding table cell at a mean 0.125 False Positives per scan. In the table we have also provided recent published results([28, 29]). Our CNN baseline and our PointNet-like approaches demonstrated a performance matching the level of [29]. At the same time, the best DGCNN with augmentation outperformed [28]. Examples of point cloud samples are provided in Appendix A.

We added the FROC pictures depending on the nodule size. The results show that in a case with large nodules there is no performance degradation. Furthermore, PointNet-like models performed better in such cases compared to a CNN approach (Fig. 2).

To show the effect of the selected sampling method on the performance of the model, we conducted additional experiments with a uniform method of sampling.

Table 2. Results of the experiments on LIDC-IDRI with DGCNN model with augmentation. Sensitivity per FP level per exam. Comparison of the performance of FPR models with various point sampling methods.

Sampling	0.125 FP	0.25 FP	0.5 FP	1 FP	2 FP	4 FP	8 FP	Mean sens
Uniform	0.36	0.52	0.655	0.78	0.902	0.969	0.99	0.7394
(Section 4)	**0.545**	**0.679**	**0.842**	**0.971**	**0.990**	**0.995**	**0.995**	**0.859**

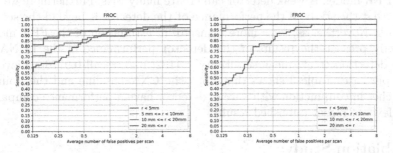

Fig. 2. FROC. left: for baseline CNN 3D model and right: for DGCNN model

Table 2 shows a significant increase in the FROC values for each FP level in cases with proposed sampling, compared to uniform sampling.

6.2 Experiments on LUNA2016 Nodule Detector Track. Benchmark

Table 3. Results of the experiments on LUNA nodule detection track. Sensitivity per FP level per exam

Experiment	0.125 FP	0.25 FP	0.5 FP	1 FP	2 FP	4 FP	8 FP	Mean sens
Liao [28]	0.662	0.746	0.815	0.864	0.902	0.918	0.932	0.834
Zhu [30]	0.692	0.769	0.824	0.865	0.893	0.917	0.933	0.842
Li [29]	0.739	0.803	0.858	0.888	0.907	0.916	0.920	0.862
Wang [31]	0.676	0.776	0.879	0.949	0.958	0.958	0.958	0.878
Khosravan [32]	0.709	0.836	0.921	0.953	0.953	0.953	0.953	0.897
Ozdemir [33]	0.832	0.879	0.920	0.942	0.951	0.959	0.964	0.921
Cao [34]	0.848	0.899	0.925	0.936	0.949	0.957	0.960	0.925
Baseline	0.686	0.811	0.840	0.929	0.972	0.972	0.972	0.883
DGCNN w/aug	**0.725**	**0.832**	**0.901**	**0.933**	**0.945**	**0.945**	**0.945**	**0.8894**

The generally accepted benchmark for the FPR task in nodule detection is the LUNA2016 FPR track competition. We believe that a comparison of our results with the existing benchmark is important for a more transparent reflection on the results obtained and the performance of the model. However, it is also necessary to note that we are unable to use the LUNA2016 FPR track data as a benchmark for the point-cloud-based model. This is because the data on this track does not have the necessary markup. Our FPR model inputs strongly depend on segmentation masks suitable for nodule candidates(see *Sect.* 3.1).

Nevertheless, we are able to evaluate the entire pipeline on the LUNA2016 nodule detection track. We are aware of the heavy dependence of the whole pipeline evaluation on the performance of the detector. FNs cannot be recovered by the FPR model. A weak detector can create many FNs. Furthermore we want to note that the study of the performance of the detector is beyond the scope of this work. Nevertheless, for a more transparent estimation of the FPR models, we briefly present the results of the detector performance of the LUNA2016 nodule detection track: mean sensitivity 0.4132, max recall 0.9776 achieved on 43 FP per case. Table 3 shows that both the CNN baseline and the PointNet approaches in conjunction with a weak detector have a performance comparable to the level of recent published results.

7   Ablation Study

As part of the formulation of the general objective of this work—testing the hypothesis that lung nodules can be effectively represented as a point cloud—we decided that it would be interesting to include the results of an ablation study. First, to clarify the problem: patch classification refers to the classification of objects presented on these patches - the nodule is either represented there or not. Common false-positive detections include pieces of bronchi, blood vessels, fibrosis, etc. In the task of separating these objects from the nodules, information about their shape is critical. As a result, coordinates are needed as data for representation. The coordinates contain information about the shape of the object represented on the patch. Coordinates "structure" points in a point cloud. Speaking more formally, we propose to present objects not as a 3D image, but as a list of points with their own characteristics. The most obvious characteristic of a point is their coordinates (x, y, z). We can expand this list by adding information about the radiological density (HU) and probability of the detector (p). Thus, each example is represented as a list $(x_i, y_i, z_i, hu_i, p_i)i = 1...n$, where n is the number of points in the example. This data is already used by the model as an input. The results presented in Table 4 confirm our assumptions on the importance of HU and p in the context of solving the FPR problem based on PointNet-like models, and presenting samples in the point cloud. Results of such an experiment show that coordinate information for each point is essential for successful nodule/non-nodule classification. This result is consistent with our hypotheses and assumptions.

7.1   Conclusion and Discussion

We have proposed a novel approach for solving False-Positive Reduction tasks for lung nodule detection CAD systems, based on the representation of a nodule candidate as a set of points with known coordinates, radiodensity, and class

Table 4. Results of the experiments on LIDC-IDRI. Sensitivity per FP level per exam. Comparison of the performance of FPR models with various point representation in the Point Cloud.

Experiment	0.125 FP	0.25 FP	0.5 FP	1 FP	2 FP	4 FP	8 FP	Mean sens
Li [29]	0.600	0.674	0.751	0.824	0.850	0.853	0.859	0.773
Liao [28]	0.662	0.746	0.815	0.864	0.902	0.918	0.932	0.834
Baseline	0.514	0.600	0.710	0.812	0.893	0.931	0.944	0.772
DGCNN (xyz + HU + p)	**0.545**	**0.679**	**0.842**	**0.971**	**0.990**	**0.995**	**0.995**	**0.859**
DGCNN (xyz + p)	0.382	0.577	0.756	0.902	0.959	0.979	0.992	0.792
DGCNN (xyz + HU)	0.362	0.467	0.630	0.764	0.862	0.979	0.995	0.723
DGCNN (xyz)	0.362	0.451	0.528	0.646	0.829	0.947	0.995	0.680
DGCNN (HU + P)	0.121	0.211	0.308	0.410	0.597	0.845	0.983	0.496

probability, predicted by a detector. Such representation allows us to use a wide set of models designed to work with point cloud and graph data. Representation of lung tissue as a set of points is quite efficient: a major part of a lung's volume is air, and observation of it can be skipped without missing any important information. This leads us to much more lightweight models compared to traditional CNN 3D, including the usage of the entire patch extracted from a CT scan.

We have provided an extensive comparison with SOTA results from open sources. For a FPR task we compared PointNet-like approaches with widely popular CNN-like models. We have shown that the DGCNN model can outperform CNN 3D at the False Positive reduction task. We have also provided an ablation study as essential for the paper and understanding of the method of the use of point clouds. The results show the importance of such parameters as radiology density (HU) and detector probability (p) in the context of presenting samples in the point cloud.

We have also presented augmentation techniques that lead to better model performance. According to these results, we assume that such a representation and approach can be successfully transferred to nodule detection tasks, and we plan to extend this work and construct the pipeline for nodule detection on point cloud representation on chest CT scans in its entirety. A major point that remains for further research will be the performance of tests on other datasets from different sources.

A   Point Cloud Samples Visualisation

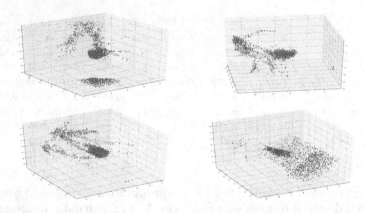

Fig. 3. Examples of point cloud samples for positive candidates (Color figure online)

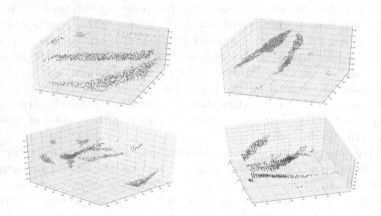

Fig. 4. Examples of point cloud samples for negative candidates (Color figure online)

At Fig. 3 and Fig. 4 red points signify those points that detector marked as nodule candidates. Blue points represent the background. These plots show that point cloud contains all necessary information for successful separation of true positive candidates from false-positive candidates.

References

1. Aberle, D.R., et al.: The national lung screening trial research team: reduced lung-cancer mortality with low-dose computed tomographic screening. N. Engl. J. Med. **365**, 395–409 (2011). https://doi.org/10.1056/NEJMoa1102873

2. Greenspan, H., van Ginneken, B., Summers, R.M.: Guest editorial deep learning in medical imaging: overview and future promise of an exciting new technique. IEEE Trans. Med. Imaging **35**(5), 1153–1159 (2016). https://doi.org/10.1109/TMI.2016. 2553401

3. Sahiner, B., et al.: Effect of CAD on radiologists' detection of lung nodules on thoracic CT scans: analysis of an observer performance study by nodule size. Acad. Radiol. (2010). https://doi.org/10.1016/j.acra.2009.08.006

4. Wu, N., et al.: Deep neural networks improve radiologists' performance in breast cancer screening. IEEE Trans. Med. Imaging (2019). https://doi.org/10.1109/TMI. 2019.2945514

5. Wang D., Khosla A., Gargeya R., Irshad H., Beck A.: Deep learning for identifying metastatic breast cancer. arXiv:1606.05718 (2016)

6. Trajanovski, S., et al.: Towards radiologist-level cancer risk assessment in CT lung screening using deep learning. arXiv:1804.01901 (2018)

7. He, K., Zhang, X., Ren, S., Sun, J.: Deep residual learning for image recognition. In: IEEE Conference on Computer Vision and Pattern Recognition (CVPR), pp. 770–778 (2016). https://doi.org/10.1109/CVPR.2016.90

8. Hunar, A., Sozan, M.: A deep learning technique for lung nodule classification based on false positive reduction. J. Zankoy Sulaimani - Part A **21**, 107–116 (2019). https://doi.org/10.17656/jzs.10749

9. Tang, H., Zhang, C., Xie, X.: NoduleNet: decoupled false positive reduction for pulmonary nodule detection and segmentation. In: Medical Image Computing and Computer Assisted Intervention - MICCAI. Lecture Notes in Computer Science, vol. 11769 (2019). https://doi.org/10.1007/978-3-030-32226-7_30

10. Tang, H., Liu, X., Xie, X.: An end-to-end framework for integrated pulmonary nodule detection and false positive reduction. arXiv:1903.09880, March 2019

11. Zhang, Z., Li, X., You, Q., Luo, X.: Multicontext 3D residual CNN for false positive reduction of pulmonary nodule detection. Int. J. Imaging Syst. Technol. **29** (2018). https://doi.org/10.1002/ima.22293

12. Jin, H., Li, Z., Tong, R., Lin, L.: A deep 3D residual CNN for false positive reduction in pulmonary nodule detection. Med. Phys. **45** (2018). https://doi.org/10. 1002/mp.12846

13. Cao, G., Liu, Y., Suzuki, K.: A new method for false-positive reduction in detection of lung nodules in CT images. In: International Conference on Digital Signal Processing (DSP), pp. 474–479, August 2014. https://doi.org/10.1109/ICDSP.2014. 6900710

14. El-Regaily, S., Salem, M., Aziz, M., Roushdy, M.: Multi-view convolutional neural network for lung nodule false positive reduction. Expert Syst. Appl. (2019). https://doi.org/10.1016/j.eswa.2019.113017

15. Setio, A., et al.: Pulmonary nodule detection in CT images: false positive reduction using multi-view convolutional networks. IEEE Trans. Med. Imaging (2016). https://doi.org/10.1109/TMI.2016.2536809

16. Jesper, P., et al.: The danish randomized lung cancer CT screening trial-overall design and results of the prevalence round. J. Thorac. Oncol. **4**, 608–614 (2009). https://doi.org/10.1097/JTO.0b013e3181a0d98f

17. Armato, S.G., McLennan, G., Bidaut, L., McNitt-Gray, M.F., Meyer, et al.: The lung image database consortium (LIDC) and image database resource initiative (IDRI): a completed reference database of lung nodules on CT scans. Med. Phys. **38**, 915–931 (2011). https://doi.org/10.1118/1.3528204

18. Jacobs, C., van Rikxoort, E.M., Murphy, K., Prokop, M., Schaefer-Prokop, C.M., van Ginneken, B.: Computer-aided detection of pulmonary nodules: a comparative study using the public LIDC/IDRI database. Eur. Radiol. **26**(7), 2139–2147 (2015). https://doi.org/10.1007/s00330-015-4030-7

19. Sunyi, Z., et al.: Automatic pulmonary nodule detection in CT scans using convolutional neural networks based on maximum intensity projection. IEEE Trans. Med. Imaging (2019). https://doi.org/10.1109/TMI.2019.2935553

20. Jégou, S., Drozdzal, M., Vazquez, D., Romero, A., Bengio, Y.: The one hundred layers tiramisu: fully convolutional DenseNets for semantic segmentation. arXiv:1611.09326 (2016)

21. He, K., Zhang, X., Ren, S., Sun, J.: Deep residual learning for image recognition. In: IEEE Conference on Computer Vision and Pattern Recognition (CVPR), pp. 770–778 (2016). https://doi.org/10.1109/CVPR.2016.90

22. Qi, C., Su, H., Mo, K., Guibas, L.: PointNet: deep learning on point sets for 3D classification and segmentation. In: IEEE Conference on Computer Vision and Pattern Recognition (CVPR), pp. 77–85 (2016). https://doi.org/10.1109/CVPR.2017.16

23. Qi, C., Yi, L., Su, H., Guibas, L.: PointNet++: deep hierarchical feature learning on point sets in a metric space. In: NIPS (2017)

24. Wang, Y., et al.: Dynamic graph CNN for learning on point clouds. ACM Trans. Graph. **38** (2018). https://doi.org/10.1145/3326362

25. Zanjani, F., et al.: Deep learning approach to semantic segmentation in 3D point cloud intra-oral scans of teeth. In: Proceedings of Machine Learning Research. Proceedings of The 2nd International Conference on Medical Imaging with Deep Learning, vol. 102, pp. 557–571 (2019)

26. Sekuboyina, A., et al.: Probabilistic point cloud reconstructions for vertebral shape analysis. In: Medical Image Computing and Computer Assisted Intervention - MICCAI. Lecture Notes in Computer Science, vol. 11769 (2019). https://doi.org/10.1007/978-3-030-32226-7_42

27. Balsiger, F., Soom, Y., Scheidegger, O., Reyes, M.: Learning shape representation on sparse point clouds for volumetric image segmentation. In: Shen, D., et al. (eds.) MICCAI 2019. LNCS, vol. 11765, pp. 273–281. Springer, Cham (2019). https://doi.org/10.1007/978-3-030-32245-8_31

28. Liao, F., et al.: Evaluate the malignancy of pulmonary nodules using the 3D deep leaky noisy-or network. IEEE Trans. Neural Netw. Learn. Syst. **30**, 3484–3495 (2017). https://doi.org/10.1109/TNNLS.2019.2892409

29. Li, Y., Fan, Y.: DeepSEED: 3D squeeze-and-excitation encoder-decoder convolutional neural networks for pulmonary nodule detection. arXiv:1904.03501 (2019)

30. Zhu, W., et al.: Deeplung: deep 3D dual path nets for automated pulmonary nodule detection and classification. In: IEEE Winter Conference on Applications of Computer Vision (WACV), pp. 673–681 (2018). https://doi.org/10.1109/WACV.2018.00079

31. Wang, B., Qi, G., Tang, S., Zhang, L., Deng, L., Zhang, Y.: Automated pulmonary nodule detection: High sensitivity with few candidates. In: Medical Image Computing and Computer-Assisted Intervention - MICCAI, Lecture Notes in Computer Science, vol. 11071 (2018). https://doi.org/10.1007/978-3-030-00934-2_84

32. Khosravan, N., Bagci, U.: S4ND: single-shot single-scale lung nodule detection. In: Medical Image Computing and Computer Assisted Intervention - MICCAI (2018). https://doi.org/10.1007/978-3-030-00934-2_88

33. Ozdemir, O., Russell, R., Berlin, A.: A 3D probabilistic deep learning system for detection and diagnosis of lung cancer using low-dose CT scans. IEEE Trans. Med. Imaging (2019). https://doi.org/10.1109/TMI.2019.2947595
34. Cao, H., et al.: Two-stage convolutional neural network architecture for lung nodule detection. IEEE J. Biomed. Health Inform. **24**(7), 2006–2015 (2020). https://doi.org/10.1109/JBHI.2019.2963720
35. Bandos, A., Rockette, H., Song, T., Gur, D.: Area under the free-response ROC curve (FROC) and a related summary index. Biometrics **65**(1), 247–256 (2009). https://doi.org/10.1111/j.1541-0420.2008.01049.x
36. Kingma, D., Ba, J.: Adam: a method for stochastic optimization. In: International Conference on Learning Representations (ICLR). arXiv:1412.6980, December 2014

Identifying User Interests and Habits Using Object Detection and Semantic Segmentation Models

Valeria Volokha and Peter Gladilin(⊠)

ITMO University, Kronverksky 49, Saint-Petersburg, Russia

Abstract. The article describes a software pipeline for identifying and classifying the interests of users in social networks using modern models and deep learning methods. The developed program is able to detect the presence of bad habits (smoking, alcohol), a sporting lifestyle, as well as determine the user's addiction to travel by an available set of photos. The software includes modules that implement deep learning algorithms for the object detection and semantic segmentation of images using the Cascade-R-CNN and DeepLabv3+ models, and the module for converting annotations of the images from COCO, ImageNet, OpenImagesV6 datasets and manually labeled images to the unified format. The models were trained on the created original datasets which include 90200 photos in total. The accuracy of the developed models is from 83.7% up to 86.6% mAP for object detection depending on a specific category of objects and 78.4% pixel accuracy for segmentation.

Keywords: User behaviour · Object detection · Image segmentation · Image processing · Convolutional neural networks

1 Introduction

Among the main tasks facing researchers of social networks, it is possible to single out the development of methods for identifying the psycho-emotional characteristics of users, methods for identifying the interests and habits of a user, as well as patterns of human behavior from the history of his movements, social connections of the user and assessing his informational influence.

The stable popularity of social networks has led to the accumulation in cyberspace of a large amount of data about users, their characteristics, behavior and interactions between them. Thus, data from open sources provide an opportunity to study the patterns of perception and decision-making processes both at the level of individuals and at the level of user groups (communities), including the study of the psycho-emotional characteristics of people as members of society, as well as collective emotions. Vividly expressed personality characteristics influence a person's choice of a profession, hobbies and preferences, for example, in music or cinema. There are studies of the relationship between personality characteristics and the user's choice of multimedia content on their social network profiles [1].

© Springer Nature Switzerland AG 2021
W. M. P. van der Aalst et al. (Eds.): AIST 2020, LNCS 12602, pp. 216–229, 2021.
https://doi.org/10.1007/978-3-030-72610-2_16

The Schwartz psycho-typing method, also known as the "Schwartz value question-naire", or the "Schwartz value test" is used to study the dynamics of changes in personality values in connection with their life problems [2]. This method is based on ten core human values (Schwartz values) and on the theory that all values are divided into social and individual. The method is not widely used in studies of the last decade, being supplanted by the more popular Big-V at the moment [3].

In the framework of the foregoing, it is becoming increasingly relevant to formulate individual-oriented approaches to the formation of personality-oriented recommendations, as well as socially significant projects in the field of education and health care due to the presentation of information about a healthy lifestyle, cultural and social values tailored specifically for the individual.

There are very few works that solve the task of highlighting thematic interests from beginning to end. We can distinguish work [4], which proposes an approach based on the construction of a tree-like hierarchy of user interests. To train the model, the authors of the work did a survey on several hundred users, which consisted of questions aimed at clarifying their interests.

In order to identify topics and interests of users, topic models are often used for semantic analysis of posts [5] and analysis of subscriptions and friendship networks [6] of users to identify interests based on the topics of the user's communities and similar friends. To build a 360-degree user profile, the information obtained by analyzing his posts, subscriptions and social connections must be supplemented by analyzing media content. To highlight topics from the photo, methods of annotating images can be used. There are many methods that are divided into different classes according to the approaches used in them. It can be based on visual spaces, where the pictures and signature are fed independently to the decoder, or the multimodal approach, where the signatures and pictures using encoders are combined in one space, and then using one decoder generate a signature. Models can also be divided into two classes: encoder-decoder and composite architectures [7]. To train models, datasets with annotations and topics highlighted in them can be used (for example, Flickr30k or MS-COCO). In [8], the authors propose a new attention mechanism, which integrates image themes into the attention model. It helps to choose the most important image features. In addition, the features and themes of the images are retrieved by separate networks. In [7] an overview of the above methods is given.

Object detection in images to identify user interests has been used in a large number of works in the last years [9–12]. A large study of methods for analysis of the preferences of Pinterest users is done in [9]. The main approach in this area is based on the formation and clustering of a set of user interests from image annotations based on the objects identified in the image. In particular, this can be done by estimating user interests based on images of his purchases by constructing a weighted average of image features [11]. Approaches to automatic prediction of the user's preferences for hobbies and lifestyle based on an offline analysis of a gallery of photos and videos from mobile device was discussed in [10].

To successfully solve the problem of identifying topics by photo, one will face a number of technical problems, in particular, how to detect parts of images that are important for determining the topic. As a solution, methods for detecting objects, such as RetinaNet in [13], can be used. In addition, it is necessary to highlight approaches for the identification of topics by photos, which do not have annotations, tags, and geolocation. To identify the user's interests in such datasets, one should use methods of objects detection to create tags that indicate markers of interests in certain fields of activity. In this paper we have developed a model that implements the detection of the specific objects in images that can help determine the user's bad habits (cigarettes, alcohol, hookahs, smoke, etc.), their interest in sports (sports equipment, sportswear, etc.), as well as objects related to the background (sky, mountains, sea, etc.) through semantic segmentation model that can identify travelers.

This paper is organized as follows. Section 2 is devoted to a brief review of existing object detection approaches and semantic segmentation models that were used in this research. Section 3 contains information about the classes of the images in the collected dataset for identifying user interests and habits. In Sect. 4 we present the developed method of converting annotations from several datasets to the unified format and the architecture of the implemented classification module. Section 5 describes experimental results and Sect. 6 presents the conclusion on the work.

2   Models

In order to determine the interests and habits of users based on their photos in social networks, one should be able to detect objects in photos. To solve this task we used two computer vision methods: object detection and semantic segmentation. Object detection was used as the main method to find specific objects in the image. Semantic segmentation allowed us to distinguish background classes such as forest, sea, and objects that indicate being indoors. This information can be used to determine individual classes of user interests or to understand the context of the image. There is no need to use panoptic segmentation or instance segmentation because there is no need to process each object of the class separately. The context of an object position in the image is more important for our case.

DeepLabV3+ and Cascade R-CNN models with ResNeSt and Xception backbones were selected for this research. They showed highest results for solving object detection and segmentation problems on various datasets [14]. To choose the best one we have tested different state-of-the-art models and backbones. The main metrics that we have monitored was AP and APs on COCO minival set for object detection models because we need to detect some small objects like cigarettes. Results of comparison presented in the Tables 1, 2 and 3 below. We have compared different models with the same backbone and then compared the best model with different backbones. For semantic segmentation task we monitored mIOU metric on PASCAL VOC 2012 test set (see Table 3).

2.1   ResNeSt

In this work, we use ResNeSt for feature extraction for the Cascade R-CNN object detection model presented in [8]. This model is quite new, but the original work, as well as a number of other studies, show that this model is quite effective and can significantly improve accuracy without losing performance. This model is an improved version of the

Table 1. The comparison of different object detection models on COCO minival set.

Model	Backbone	AP, %	AP_{50}, %	AP_s, %
YOLOv3	*Darknet-53*	*33.0*	*57.9*	*18.3*
RetinaNet	*ResNet-101*	*39.1*	*59.1*	*21.8*
Faster R-CNN	*ResNet-101*	*41.37*	*59.2*	*22.2*
Cascade R-CNN	*ResNet-101*	*42.7*	*61.6*	*23.8*

Table 2. The comparison of Cascade object detection models with different backbones on COCO minival set.

Model	Backbone	AP, %	AP_{50}, %	AP_s, %
Cascade R-CNN	HRNetV2p-W18	*41.3*	*59.2*	*23.7*
Cascade R-CNN	ResNet-101	*42.7*	*61.6*	*23.8*
Cascade HTC	–	*43.2*	*59.4*	*20.3*
Cascade R-CNN	ResNeSt-101	*47.5*	*65.4*	*28.5*
Cascade R-CNN	**ResNeSt-200**	*49.03*	*68.2*	*30.9*

Table 3. The comparison of semantic segmentation models on PASCAL VOC 2012 test set.

Model	Backbone	mIOU, %
Deeplabv3+	ResNet-101	*79.19*
SANet	–	*83.2*
PSPNet	–	*85.4*
DCNAS	–	*86.9*
ExFuse	ResNext-131	*87.9*
Deeplabv3+	**Xception-JFT**	*89.0*

ResNet architecture. The main idea is to partition object maps into separate attention blocks. Each of these blocks divides the map into several groups depending on the channel size, and further into smaller subgroups, where the feature representation of each group is determined by a weighted combination of representations of its sections.

2.2 Cascade R-CNN

When training models using Intersection-over-Unit (IoU) metric, there are a number of problems that cause increased losses and reduced model accuracy. To solve these problems, a multi-stage object detection architecture Cascade R-CNN is proposed in works [8] and [15]. It consists of a sequence of detectors trained with increasing IoU thresholds to be able to solve consistently the problem of false positives. The detectors are trained stage by stage, leveraging the observation that the output of a detector has a good distribution for training the next higher quality detector. The resampling of progressively improved hypotheses guarantees that all detectors have a positive set of examples of equivalent size, which reduces the overfitting problem. A similar cascading procedure is used for output, which allows a more accurate comparison of hypotheses and detector quality at each stage [15, 16]. Unlike the Faster R-CNN architecture, which represents a single stage, the Cascade R-CNN consists of three such stages.

2.3 Xception

In our research, we use a modified Xception to extract features for the segmentation model described in [17]. The original Xception shows that the model demonstrates high accuracy of image classification from ImageNet, as well as a good ratio of accuracy and performance [18]. Also, several works describing various modifications of the original model confirm the effectiveness of this approach [14]. In contrast to the classic model, a deeper version of Xcepton is used as the base. The pulling layers are replaced by a depth-separable convolution with different steps, which allows us to use separable convolution to extract object maps with arbitrary resolution. In addition, other levels of normalization and ReLU activation are added after each 3x3 deep convolution layer.

2.4 DeepLabv3+

Unlike previous versions of DeepLab, DeepLabv3+ is divided into an encoder and a decoder [18]. The first one is responsible for gradually reducing the size of object maps and extracting objects from deeper layers. The second one is responsible for restoring object details and spatial dimensions. Experimental studies have shown that such an approach can effectively solve computer vision problems, including object detection and semantic segmentation, surpassing previous results on some problems [17–19].

3 Datasets

3.1 Classes

To work with the "bad habits" category, we used OpenImagesDatasetV6, which includes classes of various alcohol ("beer", "wine", "alcohol bottle", "bottle opener", "cocktail", "cocktail shaker") and other harmful habits such as smoking. Open datasets contain an insufficient number of examples for every class, so the photos from the Flickr site related to smoking were marked up manually using *annotate.online* service. As a result, we had such classes as "cigarettes", "packs of cigarettes", "hookah", "vape" and "smoke".

In total, the "bad habits" group of classes contains 1760 photos for training and 440 for testing with 12 classes and 160 and 40 photos per class for training and testing respectively. Table 4 shows classes and numbers of images selected for training and testing.

For the "sport" category, we used the COCO dataset classes related to different sports and the global "sport" class. Additionally, we used photos containing people in sportswear, gyms (simulators, various sports equipment), specific sports equipment, and applications for sports. These images were marked up manually using *annotate.online* service. In total, 4000 photos were marked up for the sport category. This group contains 4000 for training and 1000 for testing, 5 classes and 1000 photos per class (see Table 4).

To interpret better the found objects and their meaning and importance we need to know the context where an object exists. The semantic segmentation model has been trained on "Stuff" part of COCO-Stuff dataset, which includes the material and background classes. The following global classes were selected: "Water" ("Water-other", "Water drops", "Sea", "River", "Fog"), "Ground" ("Ground other", "Playing field", "Platform", "Railroad", "Snow", "Sand", "Pavement", "Road", "Gravel", "Mud", "Dirt"), "Sky", "Plant", "Structural", "Building" ("Building-other", "Roof", "Tent", "Bridge", "Skyscraper", "House"), "Textile", "Furniture", "Window", "Floor", "Ceiling", "Wall", "Solid" ("Solid-other", "Hill", "Mountain", "Stone", "Rock", "Wood"). There were 800 photos for training and 200 for testing for each class and in total 83,000 photos were used to train Deeplabv3+.

Table 4. Groups of classes in collected dataset

Group	Train	Test	Total
Bad habits: beer, wine, alcohol, bottle, bottle opener, cocktail, cocktail shaker, *cigarettes, pack of cigarettes, hookah, vape, smoke*	*1760*	*440*	*2200*
Sport:*sportswear, sport equipment, sport shoes,* fitness equipment, sport apps	*4000*	*1000*	*5000*
Background: water, ground, solid, sky, plant, structural, building, textile, furniture, window, floor, ceiling, wall	*66400*	*16600*	*83000*

The bold text in the table indicates the class groups (categories). Regular text shows object classes that are included in this group. The italic text represents the classes that were annotated manually.

3.2 Annotations

Manually annotated images and images from datasets that were used in the work had different annotations. COCO-Stuff had encoded annotations in.json files. ImageNet annotations were stored in.xml files the same as the official PASCAL VOC dataset format. OpenImages annotations were stored in.csv files. We manually annotated images using *annotate.online* service and obtained images with annotations in.json and.png formats. In order to work with different datasets, it was necessary to write dataloaders for each type of annotations or to convert annotations to a single format. In this paper, the second approach is used. We converted annotations to.png format in which every pixel is assigned to a color from a grayscale spectrum and every color represents a class.

To convert all annotations to a single format, we used COCO-Stuff API, some third-party solutions such as voc2coco and openimages2coco libraries and developed script to convert annotations from "annotate.online" to the unified format. The entire annotation conversion process is shown on Fig. 1. The COCO-Stuff API has the function to convert annotations from the.json format to the.png. To convert annotations from OpenImages and ImageNet datasets, we convert annotations to the intermediate format. We used openimages2coco and voc2coc libraries to convert annotations from the intermediate format to COCO style and store them in.json format. Next step was the same as converting from COCO-Stuff dataset. The implemented script converted annotations from.png of *annotate.online* to the necessary format (convert colored pixels to grayscale pixels).

4   Method

We use Xception model and ResNeSt model as backbones for extracting features in order to train classification part of the developed model on the collected datasets. In the case of the Xception model, the fully connected layer and the logistic regression layer are cut off and all the previous ones are frozen. In the case of ResNeSt, Cardinal blocks are frozen, and all further layers are cut off. Two separated Cascade R-CNN models for object detection with ResNeSt model as a backbone have been trained on groups of classes "bad habits" and "sports". The Deeplabv3+ model for semantic segmentation with Xception model as a backbone has been trained on the group of classes "background".

Figure 1 below shows a full pipeline of model pre-processing and evaluation. The pipeline is the same for Cascade R-CNN and Deeplabv3+ models.

The first step in the pipeline is to process the annotations as described in Sect. 3.2 above. As a result, we have all annotations in.png format where the color of every pixel means the class of pixel in the original image. After that, all images and annotations are uploaded to dataloader where they are preprocessed and augmented. Images are rescaled to 400×400 pixels size. We have used various types of augmentation methods like cropping and extracting 380×380 pixels parts from images, flipping along the horizontal and vertical axis, rotating and blurring. Augmentation is used only for the training set, and all artificially produced images are in it. The dataloader returns two loader instances for training and validation sets with 80% to 20% ratio. The training loader shuffles data after each epoch in order to decrease the risk to create batches that were not representative for the overall dataset. Both loaders are passed to the model. For training, max mean IoU metric was used for monitoring the training process and to fine-tune the weights of the neural networks. The cross-entropy was used as the loss function. Models were trained until achieving the stop criteria or the maximum number of epochs (100) with stochastic gradient descent with momentum as optimization algorithm. We are monitoring validation loss curve and if the loss stops decreasing for several epochs in the row then the training stops. It had patience set to 15 epochs.

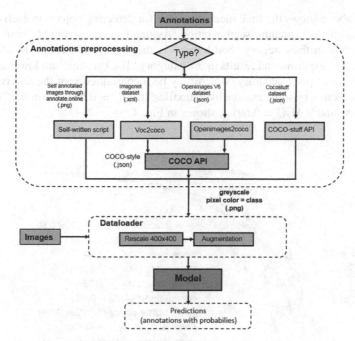

Fig. 1. Method pipeline and data preprocessing

5 Experimental Results

We trained the models of object detection and semantic segmentation in accordance with the method of preprocessing and augmentation of the initial data described in Sect. 4. Figures 2, 3 and 4 show ROC-AUC curves and AUC scores for each class in the category mentioned above. As one can see, object detection results for the selected classes are pretty good and lie in range from AUC $= 0.75$ (for "cigarettes" class) to AUC $= 0.93$ (for "hookah", "wine opener" and "sport apps" classes).

Table 5. Experimental results on the test sample.

Model	Group	mAP/pixel accuracy, %
Cascade R-CNN-1 (ResNeSt)	Bad habits	86.6% (mAP)
Cascade R-CNN-2 (ResNeSt)	Sport	83.7% (mAP)
DeepLabv3 + (Xception)	Background	78.4% (pixel accuracy)

Table 5 above shows the final mean accuracy for detecting objects in each category as well as results of semantic segmentation. As one can see, the mAP metric for the object detection in the category "bad habits" is the highest and equals 86.6% on the test sample. The experimental results in the category "Background" are lower and equal 78.4% in terms of pixel accuracy, which may be a consequence of the relatively poor quality of detection of some classes like "textiles" (AUC = 0.66), "furniture" (AUC = 0.66) and "ground" (AUC = 0.69) as shown in Fig. 4.

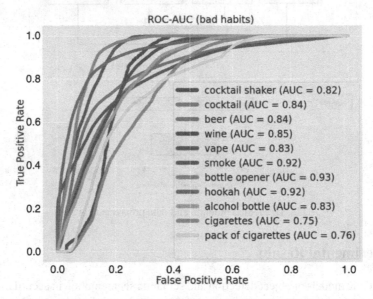

Fig. 2. ROC-AUC curves for the Cascade-R-CNN *(bad habits)* model

Figure 5 shows examples of object detection from "bad habits" category ("cigarettes", "hookah", "bottles" and "wineglasses") and "sport" category ("sport equipment") obtained with developed models based on Cascade R-CNN. Figure 6 shows examples for semantic segmentation by DeepLabV3 + -based model of the input image with areas corresponding to "sky", "clouds", "rocks" and "sea".

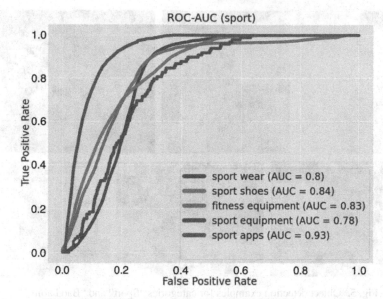

Fig. 3. ROC-AUC curves for the Cascade-R-CNN *(sport)* model

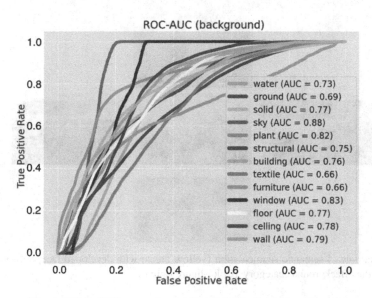

Fig. 4. ROC-AUC curves for the Deeplabv3+ *(background)* model

Fig. 5. Object detection examples for categories "Sport" and "Bad habits"

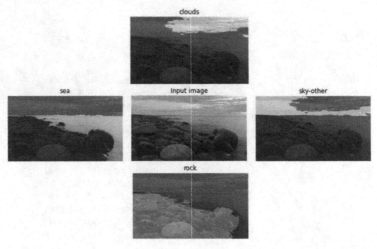

Fig. 6. Example of semantic segmentation (yellow areas) with developed DeepLabv3+ -based model for the "Background" category. (Color figure online)

Table 6 below shows memory usage, training time and inference time of trained model. All numbers were obtained on the singe NVIDIA V100 GPU. As research on user habits will continue, further work on the efficiency of the algorithm will increase the speed of the model.

Table 6. Models memory usage, training and inference time

Model	Memory, GB	Training time, sec/iteration	Inference time sec/image
Cascade R-CNN (1,2)	4	2.6	0.45
DeeplabV3 +	10	1.52	0.52

Table 7 below shows AP metrics of the trained Cascade R-CNN models on MS-COCO dataset and our habits-dataset for some specific objects. For classes "Tennis racket", "Sports ball", "Snowboard" and for class "Wine glass" there were used two separated models trained on "Sports" and "Bad habits" parts of our dataset correspondingly. As one can see, as our dataset was specially developed in order to determine "bad" and "good" habits of the user, the accuracy of the specific object detection for the model trained on it is sufficiently higher than in case of MS-COCO.

Table 7. Accuracies for the detection of some classes for developed Cascade-R-CNN models on MS-COCO and collected dataset.

Dataset	Tennis racket, AP, %	Sports ball, AP, %	Snowboard, AP, %	Wine glass, AP, %
MS-COCO	56.17	56.24	48.03	39.31
Our	59.41	60.33	51.05	58.17

6 Conclusion

We have developed a model that can be used to evaluate markers of interests of social media users, based on the image analysis, using methods for detecting objects and image segmentation. The created pipeline uses parts of large datasets such as OpenImagesDatasetV6, ImageNet, and COCO in order to be able to train models on a sufficient number of marked-up images. In order to be able to identify bad habits of the user, his sports and travel interests, we collected our own dataset, adding and labeling new sets of classes that were not present in large datasets.

For the formation of specific markers of interests, an ontology base should be used, which would make it possible to transfer the set of objects detected for the user to the category of interest, as, for example, was done in the work [12], where BabelNet was used for this purpose. The main result of our work is the creation of a software pipeline that allows the researcher, on the one hand, to use large datasets of images to train their models, and on the other hand, to be able to add new classes of objects that are absent in these datasets (or their number is insufficient), but are important for determining user characteristics and habits. The developed method for converting annotations allows to do it.

In continuation of this work, we plan to expand the set of defined characteristics of a person, his interests, as well as to introduce a model for the classification of emotions, which will require the use of multimodal data from social networks.

Acknowledgments. This research is financially supported by The Russian Science Foundation, Agreement №17-71-30029 with co-financing of Bank Saint Petersburg.

References

1. Burrell, J.: How the machine 'thinks': understanding opacity in machine learning algorithms. Big Data Soc. (2016). https://doi.org/10.1177/2053951715622512
2. Schwartz, S.H.: An overview of the schwartz theory of basic values. Online Readings Psychol. Cult. (2012). https://doi.org/10.9707/2307-0919.1116
3. Bachrach, Y., Kosinski, M., Graepel, T., Kohli, P., Stillwell, D.: Personality and patterns of Facebook usage. In: Proceedings of the 4th Annual ACM Web Science Conference, WebSci 2012 (2012). https://doi.org/10.1145/2380718.2380722.
4. Hua, W., Huynh, D.T., Hosseini, S., Lu, J., Zhou, X.: Information extraction from microblogs: a survey Information extraction from mi- croblogs: a survey. Int J Softw. Inf. **66**(44), 495–522 (2012)
5. Gemp, I., Nallapati, R., Ding, R., Nan, F., Xiang, B.: Weakly semi-supervised neural topic models. In: ICLR (2019)
6. Fang, G., Su, L., Jiang, D., Wu, L.: Group recommendation systems based on external social-trust networks. Wirel. Commun. Mob. Comput. (2018). https://doi.org/10.1155/2018/6709607
7. Hossain, M.D., Sohel, F., Shiratuddin, M.F., Laga, H.: A comprehensive survey of deep learning for image captioning. ACM Comput. Surv. **51**(6), 118 (2019)
8. Zhang, H., et al.: ResNeSt: Split-Attention Networks (2020)
9. Zhai, A., et al.: Visual discovery at Pinterest. In: 26th International World Wide Web Conference 2017, WWW 2017 Companion (2019). https://doi.org/10.1145/3041021.3054201
10. Grechikhin, I., Savchenko, A.V.: User modeling on mobile device based on facial clustering and object detection in photos and videos. In: Morales, A., Fierrez, J., Sánchez, J.S., Ribeiro, B. (eds.) IbPRIA 2019. LNCS, vol. 11868, pp. 429–440. Springer, Cham (2019). https://doi.org/10.1007/978-3-030-31321-0_37
11. Demochkin, K.V., Savchenko, A.V.: User preference prediction in a set of photos based on neural aggregation network. In: Bychkov, I., Kalyagin, V.A., Pardalos, P.M., Prokopyev, O. (eds.) Network Algorithms, Data Mining, and Applications: NET, Moscow, Russia, May 2018, pp. 121–127. Springer International Publishing, Cham (2020). https://doi.org/10.1007/978-3-030-37157-9_8
12. Wieczorek, S., Filipiak, D., Filipowska, A.: Semantic image-based profiling of users' interests with neural networks. In: 4th Working Semantics Deep Learning International Semantics Web Conference 2018 (2018). https://doi.org/10.3233/978-1-61499-894-5-179
13. Lin, T.Y., Goyal, P., Girshick, R., He, K., Dollar, P.: Focal loss for dense object detection. IEEE Trans. Pattern Anal. Mach. Intell. (2020). https://doi.org/10.1109/TPAMI.2018.2858826
14. The latest in machine learning | Papers With Code. https://paperswithcode.com/, Accessed 14 Jul 2020
15. Cai, Z., Vasconcelos, N.: Cascade R-CNN: high quality object detection and instance segmentation. IEEE Trans. Pattern Anal. Mach. Intell. (2019). https://doi.org/10.1109/tpami.2019.2956516

16. Cai, Z., Vasconcelos, N.: Cascade R-CNN: delving into high quality object detection. In: Proceedings of the IEEE Computer Society Conference on Computer Vision and Pattern Recognition (2018). https://doi.org/10.1109/CVPR.2018.00644

17. Chen, L.-C., Zhu, Y., Papandreou, G., Schroff, F., Adam, H.: Encoder-decoder with atrous separable convolution for semantic image segmentation. In: Ferrari, V., Hebert, M., Sminchisescu, C., Weiss, Y. (eds.) ECCV 2018. LNCS, vol. 11211, pp. 833–851. Springer, Cham (2018). https://doi.org/10.1007/978-3-030-01234-2_49

18. Chollet,F.: Xception: deep learning with depthwise separable convolutions. In: Proceedings - 30th IEEE Conference on Computer Vision and Pattern Recognition, CVPR 2017 (2017). https://doi.org/10.1109/CVPR.2017.195

19. Chen, L.C., Papandreou, G., Kokkinos, I., Murphy, K., Yuille, A.L.: Rethinking atrous convolution for semantic image segmentation liang-chieh. IEEE Trans. Pattern Anal. Mach. Intell. (2018). https://doi.org/10.1109/TPAMI.2017.2699184

Semi-automatic Manga Colorization Using Conditional Adversarial Networks

Maksim Golyadkin and Ilya Makarov[✉]

HSE University, Moscow, Russia
myugolyadkin@edu.hse.ru, iamakarov@hse.ru

Abstract. Manga colorization is time-consuming and hard to automate. In this paper, we propose a conditional adversarial deep learning approach for semi-automatic manga images colorization. The system directly maps a tuple of grayscale manga page image and sparse color hint constructed by the user to an output colorization. High-quality colorization can be obtained in a fully automated way, and color hints allow users to revise the colorization of every panel independently. We collect a dataset of manually colorized and grayscale manga images for training and evaluation. To perform supervised learning, we construct synthesized monochrome images from colorized. Furthermore, we suggest a few steps to reduce the domain gap between synthetic and real data. Their influence is evaluated both quantitatively and qualitatively. Our method can achieve even better results by fine-tuning with a small number of grayscale manga images of a new style. The code is available at github.com.

Keywords: Generative adversarial networks · Manga colorization · Interactive colorization

1 Introduction

In the last decade, due to the growth of computing power and the amount of data available, the area of machine learning - deep learning - has experienced a great increase. There have been considerable successes in the ability of computers to understand data of all formats: images, speech, texts.

The success of deep learning models in various fields of science motivates apply them to image colorization task. Over the last few years, many methods using Convolutional Neural Networks (CNNs) [15] and Generative Adversarial Networks (GANs) [8] for the photograph and drawing colorization have been proposed. They exhibited promising results proving the ability of neural networks to colorize images.

In this paper, we consider the task of colorizing the black and white manga. Usually, the manga is created with a pen and ink on white paper. To make an existing manga more attractive, there is a great demand for its coloring. It is done manually using applications such as Photoshop or Autodesk, which is a

© Springer Nature Switzerland AG 2021
W. M. P. van der Aalst et al. (Eds.): AIST 2020, LNCS 12602, pp. 230–242, 2021.
https://doi.org/10.1007/978-3-030-72610-2_17

time-consuming task, so finding an easy way to create realistic coloring for black and white drawings is an urgent task.

The key difference between manga and black and white photographs is that in many cases manga is almost a binary image (some areas may be flooded with a homogeneous grey) and the photographs are grayscale images. Grayscale images contain more information that limits the number of possible colorings (light areas can't be colored in a dark color) and simplifies the implicit solution of the segmentation problem during colorization. Besides, the manga page consists of several panels containing different scenes and it is a natural requirement that their coloring should be homogeneous.

We use deep conditional GAN to perform high-quality end-to-end manga page colorization. Employing conditional input allows the user to manually manipulate coloring for achieving a desirable result.

Since there are no publicly available datasets consisting of image pairs required for supervised learning, we utilize synthetic data. This creates a domain gap between the objects of the training set and manga images. To significantly reduce it, we implement some techniques that do not require auxiliary models.

2   Related Work

Most early methods using CNNs [4,10,14,21] are based on training with a large number of images to map a grayscale image to a color one. Neural networks can learn how to use the information contained in low-level and high-level features to perform colorization. These approaches are fully-automatic and don't require any human involvement. However, since in the presented works the training relies on minimizing the distance between the values of color image pixels and neural network output (MSE or MAE), the colors of the produced image are faded and unnatural. To solve this problem, GANs are used, namely models based on Pix2Pix [11] architecture. The main idea is to add the adversarial loss to the main loss function and use the main neural network as a generator while some additional neural network is used as a discriminator. This approach leads to more plausible and colorful images.

All mentioned approaches have one common drawback: colorization is an ill-posed task. There are several suitable colorings for the majority of black and white images [3] and fully-automatic approaches can obtain only one. Hence, if some object has several suitable colors, the neural network will produce a value close to the average of these colors that leads to degradation of the results. The solution is to use some human-generated hints. These hints can be, for example, sparse images with some pixels labeled with the desired color (color hint) [22] or set of images that provides the requested color distribution (reference images) [9].

In addition, this paper relies on approaches designed for drawing colorization using color hints [5], reference images [17], both of these approaches [7], and text descriptions [13].

3 Method

3.1 Data Preprocessing

Edge Detection. There is no dataset containing pairs of black and white and color manga images, so we had manually colorized images only. Building our dataset by matching pages of black and white and color editions was problematic since corresponding images can only be obtained from different sources, therefore have different sizes and arbitrarily different margins, thus hindering the pixel-wise correspondence essential for training Pix2Pix-like architecture. For this reason, we decided to use synthetic data obtained with an edge detector

Classic edge detectors such as Canny and Sobel operator were found to be unable to obtain an image similar to the black and white manga while retaining the distinguishable text, so we decided to use XDoG [18] algorithm that produces convincing images while retains fine details and text (see Fig. 1). A large number of parameters and the dependence of line thickness on the image size can be considered among the disadvantages of this method.

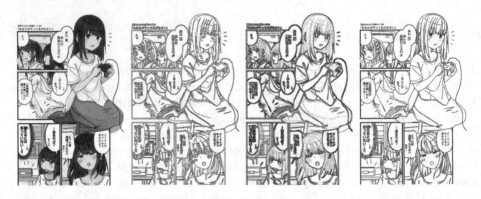

Fig. 1. Comparison of edge detection algorithms: original image, Canny, Sobel, XDoG.

Denoising. On the one hand, since XDoG partially ignores noise, mapping an image without noise to a noisy one can destabilize the learning process. On the other hand, usage of a noisy input image can result in poor colorization since XDoG ignores noise just partially. Therefore, it seems reasonable to use some method that will remove noise while reducing image quality insignificantly.

The neural network model FFDNet [20] was chosen as such a method. This is a convolutional neural network trained to remove additive Gaussian noise via minimization of the mean square error between the pixel values of some image and the output of the network when the input was artificially corrupted with noise.

We used a network that was trained with photographs, but network's high generalization ability enables it to be used with drawings as well.

3.2   Model Overview

We adopt the system proposed in [5] for line art anime drawings colorization with color hints. This work shows the ability of conditional GANs to perform high-quality colorization of drawings being trained on synthetic data generated with XDoG.

Our system takes three objects at the input (see Fig. 2). The first object is a binary, sparse image $x \in \mathbb{R}^{H \times W \times 1}$ generated by XDoG. The second one is the distance field map $d \in \mathbb{R}^{H \times W \times 1}$ as proposed at [17]. It simplifies the learning process since sparse input and color output have different nature and some intermediate representation helps to establish matching.

Fig. 2. Different image representations: original, XDoG, DFM, local features map.

The third input is color hint represented with $h \in \mathbb{R}^{H \times W \times 4}$. During training, it's generated by randomly sampling pixels in the following way:

1. Sample matrix $r \in \mathbb{R}^{H \times W \times 1}$, where $r_{i,j} \sim U(0,1)\ \forall i, j$.
2. Generate binary matrix $b = r > |\xi|$, where $\xi \sim N(1, 0.0005)$.
3. The color hint is $h = \{y * b, b\}$, where $\{\}$ denotes concatenation, $*$ - element-wise product and y is color image.

We add the mask itself to the color hint to distinguish pixels of some color from empty pixels. For example, if the pixel values belong to $[0, 1]$, then 0 corresponds to the black color.

Our architecture consists of 3 fully convolutional networks: local features extractor $\mathcal{E}$, generator $\mathcal{G}$, discriminator $\mathcal{D}$. Extractor use pretrained weights that are never updated during training and generator and discriminator trained simultaneously.

Local Features Extractor. Manga images are quite complex. They contain a variety of objects and are drawn in different styles. Therefore, for high-quality colorization, it is crucial to correctly detect the boundaries and extract semantic features. For this purpose, some pretrained network is used that receives a binary image at the input, and activations of some layer of network are considered as an image description. Moreover, the weights of extractor are frozen since there is a domain gap between synthetic and real data, and such an approach prevents overfitting on synthetic data that leads to reducing the domain gap. In this work, we use SE-ResNeXt-50, activations of *conv3_4* layer are used as local features maps.

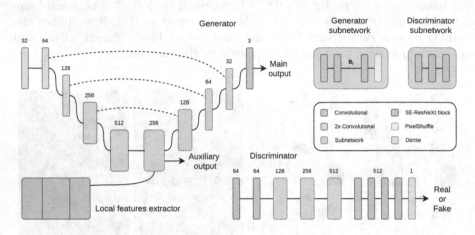

Fig. 3. Model architecture.

Generator. The generator has a UNet-like architecture. Its encoder is built from a sequence of convolutional layers that rapidly reduces the size of the input by 16 times to the size of the local features maps. This simplicity of the encoder is caused by the fact that the local features extractor is used to extract the features, and the main function of the encoder is to propagate color from a color hint.

The decoder is a sequence of 4 subnetworks and a convolutional layer at the end with tanh activation that converts the intermediate result into a final coloring. Each i-th subnetwork consists of B_i SE-ResNeXt blocks with a convolutional layer at the beginning and a sub-pixel convolutional layer at the end. Each subnetwork increases the height and width of the input by 2 times. Moreover, dilated convolutions are added to all subnetworks. It enables us to increase the receptive field without increasing the calculation cost. All activation functions are LeakyReLU except the last one. In this work B_i values are equal to 20, 10, 10, 7 respectively.

The output of the first subnetwork is also used for computing the auxiliary image using an additional subnetwork consisting of a sequence of transposed convolutions. The loss function is calculated using this image to improve the gradient propagation.

This generator is much deeper than most existing gans. The work [5] shows that the large receptive field and increased capacity of the model have a positive impact on its quality and helps produce more natural images.

The main and auxiliary outputs are tensors of size $H \times W \times 3$ that are an estimation of coloring in RGB format.

Discriminator. It is used for adversarial learning. Receives real color images and generator outputs and estimates the probability that the input image is real. It comprises a sequence of convolutional layers and SE-ResNext blocks with a sigmoid activation function at the end as shown in Fig. 3. The spectral normalization [16] is applied to the convolutional layers.

3.3 Loss

To train the model on producing analogous coloring for similar patterns and to diversify the range of colors used, we employ a combination of several loss functions.

L1 Loss. In order to enforce the generator to produce similar coloring to the ground truth, we use a pixel level L1 metric that calculates the distance between output image $\mathcal{G}(x, d, h, \mathcal{E}(x))$ and ground truth image y:

$$\mathcal{L}_{L1}^{\mathcal{G}} = ||\mathcal{G}(x, d, h, \mathcal{E}(x)) - y||_1 \tag{1}$$

Perception Loss. To ensure similarity not only at the pixel level but also at the structural level, we use perceptual loss presented in [12]. It's an L2-loss based on the distance between some CNN feature maps for generated and ground truth images. We compute it as follows:

$$\mathcal{L}_{per}^{\mathcal{G}} = \frac{1}{chw}||\mathcal{V}(\mathcal{G}(x, d, h, \mathcal{E}(x))) - \mathcal{V}(y)||_2^2, \tag{2}$$

where c, h, w denotes the number of channels, height, width of features maps and $|| \cdot ||_2$ denotes Frobenius norm. We use VGG-16 pretrained on ImageNet [6] as a CNN, $\mathcal{V}$ denotes activation after 4th convolutional layer.

Adversarial Loss. To diverse colorings, we use adversarial loss that computed in the following way:

$$\mathcal{L}_{G_{adv}} = \mathbb{E}_{x,h}[\log\left(1 - \mathcal{D}(\mathcal{G}_{main}(x, d(x), h, \mathcal{E}(x))))\right], \tag{3}$$

$$\mathcal{L}_{D_{adv}} = -\mathbb{E}_y[\log \mathcal{D}(y)] - \mathbb{E}_{x,h}[\log\left(1 - \mathcal{D}(\mathcal{G}_{main}(x, d(x), h, \mathcal{E}(x))))\right], \tag{4}$$

where $\mathcal{G}_{main}(\cdot)$ and $\mathcal{G}_{aux}(\cdot)$ are main and auxiliary output of the generator correspondingly.

White Color Penalty. Our model tends to leave objects uncolored if it fails to recognize them. There can be a lot of such objects because of the domain gap, so we penalize the network if the pixel value is close to the maximum corresponding to the white color. $\mathcal{L}_{white}$ is calculated as the mean square of the channel-wise sums (we assume that values of $\mathcal{G}$ output belong to $[0, 1]$).

Overall Loss Function. Combining all above terms, we set loss functions for generator and discriminator:

$$\mathcal{L}_G = \lambda_{L1}(\mathcal{L}_{L1}^{G_{main}} + \lambda_{aux}\mathcal{L}_{L1}^{G_{aux}}) + \mathcal{L}_{per}^{G_{main}} + \mathcal{L}_{G_{adv}} + \mathcal{L}_{white} \quad (5)$$

$$\mathcal{L}_D = \mathcal{L}_{D_{adv}} \quad (6)$$

We set $\lambda_{L1} = 10$, $\lambda_{aux} = 0.9$ to force the generator to reconstruct the original image rather than deceiving discriminator.

4   Experiments

In this section, we describe the collected dataset. Then we discuss the learning and inference processes and evaluate model performance.

4.1   Dataset

Web-scraping was used to extract images of manually colored manga: One Piece, Akira, Bleach, JoJo's Bizarre Adventure, Dragon Ball, Naruto. We also collected artistic anime drawings from the dataset Danbooru2019 [1]. We made use of images that had the tag *colorized*. It means that these images were produced from line art by manual coloring. Such images usually have distinct black edges helping to synthesize a higher quality XDoG image. Overall, 59164 color images were collected. In addition, we collected images of such black and white manga as Fullmetal Alchemist, Lone Wolf and Cub, Wolf and Spice, Pandora Hearts, Phoenix to evaluate the performance of the model and perform fine-tuning. These titles are not included in the training set and have different drawing styles, thus allowing an objective assessment to be made.

The images are often uploaded to the Internet in JPEG format with a high compression factor to reduce the memory consumption, but this causes the images to have artifacts that distort their content, thus affecting the learning process. To restore images, FFDNet was applied.

We used the following method to synthesize a monochrome image:

1. Resize the image to be with the shortest side equals $512k$, where k randomly selected from $\{2, 3, 4\}$.
2. Apply XDoG algorithm with parameters $\tau = 0.95$, $\varphi = 90$, $\epsilon = 0$, $k = 4$, $\sigma = 0.5$ and binarization using the Otsu method to the intermediate image.
3. Resize the image to be with the shortest side equals 512.

We need to resize the image to the uniform size in order to perform training. Increasing the image size helps to preserve fine details and text more efficiently when applying the XDoG algorithm and produce monochrome images with better quality.

4.2   Training Setup

The following models with pretrained weights are used for training:

- VGG-16 pretrained on ImageNet for evaluating perceptual loss.
- SE-ResNeXt-50 pretrained on monochrome drawings for tag prediction.
- FFDNet pretrained on photographs for denoising.

The weights of the generator and the discriminator are initialized with Xavier initialization. The training takes 20 epochs using the ADAM optimizer with hyperparameters $\beta_1 = 0.5$, $\beta_2 = 0.9$ and batch size is 4 since limited computational resources. Longer training leads to overfitting, increasing the domain gap. During the first 15 epochs, the learning rate of the generator and the discriminator is equal to 10^{-4} and $4 * 10^{-4}$ respectively, whereas during the last 5 epochs it is 10^{-5} and $4 * 10^{-5}$. We use an imbalanced learning rate to keep the ratio of the discriminator and generator updates equal to 1:1 since the utilization of spectral normalization slows down discriminator learning as shown in [19]. One-sided label smoothing is applied for discriminator training.

During training random cropping to 512×512 size and random horizontal flipping are applied. Moreover, we use random color shift before XDoG algorithm application.

We generate the mask for color hint with probability 0.5 the way described above, with probability 0.49 the mask contains all zeros and with probability 0.01 the mask contains all ones. The use of an empty mask provides quality network performance without hints, whereas a full mask allows the network to correctly deal with a large number of colored pixels.

The neural networks are implemented using the PyTorch library for Python. Image augmentations are performed with the Albumentations [2]. DFM images are generated using the library Snowy.

The training takes about 60 h on a device with a single NVIDIA GTX 1080TI graphics card and a Xeon® E5-2696 CPU.

4.3   Fine-Tuning

Regrettably, the XDoG image and the black and white manga image differ sufficiently, so that a network trained with one type of image cannot work properly with the other. To even greater regret, XDoG of black and white image and XDoG of color one also slightly different. Despite that visually they are almost indistinguishable, the neural network is capable to find features at XDoG of the color image that contain the information about the pixel intensity of the original image, that may not be presented at black and white images. As a result,

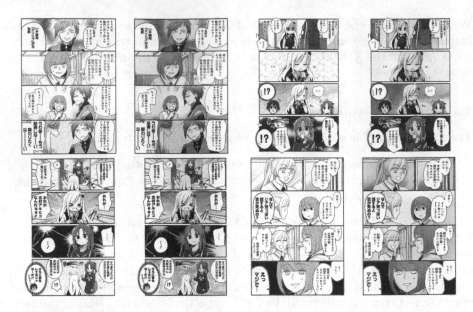

Fig. 4. Example of coloring without user interaction.

a domain gap is made between the training set and the images handled during inference.

To address this problem, we utilize fine-tuning with XDoG images of black and white manga. The same preprocessing procedure is applied to these images as to the color ones. During 5000 iterations, the network is trained with XDoG of color images while learning rate equal to $4 * 10^{-5}$ and 10^{-5} for the discriminator and generator. Every fifth iteration generator receives XDoG of black and white image and an empty color hint. Here, we apply the same loss function to the discriminator and use $\mathcal{L}_{G_{adv}}$ with a learning rate 10^{-6} for the generator.

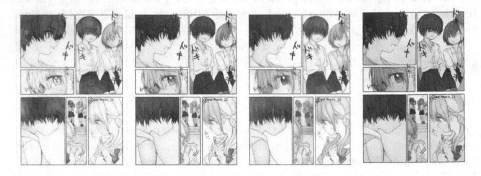

Fig. 5. The influence of white color penalty and fine-tuning on the manga that was not represented in the training and fine-tuning set: AlacGAN [5], without penalty and fine-tuning, without fine-tuning, our model.

This method enhances the quality of coloring both for images of the fine-tuning set and for the others while not requiring any labeling. Therefore, if the model is intended to be applied to the big set of images of one style, it is reasonable to perform fine-tuning using images of that set.

With a longer learning time, the coloring for images of fine-tuning set spoils due to the absence of the $L1$ or perceptual loss. For the rest of the images, the positive effect disappears as the network overfits for particular images.

4.4 Inference

Summing up the above mentioned, the algorithm for colorizing black and white manga is as follows:

1. The user inputs a black and white image of the manga page of size $H \times W$.
2. If the image corrupted with noise or JPEG artifacts, FFDNet with user-defined noise level σ is applied.
3. This image is used to generate XDoG and DFM images, as described earlier. Their size is $512 \times (W * \frac{512}{H})$, if $H \leq W$, and $(H * \frac{512}{W}) \times 512$ otherwise.
4. Zero tensor is assigned to the color hint.
5. The coloring for an empty color hint is generated, and it is used as a color image to create a nonempty color hint using the method described above (optional).
6. The acquired objects are used to produce the coloring with generator.
7. If the result is unsatisfactory, the color hint is modified by the user and go to step 6.

4.5 Results

As shown in Fig. 4, the model is capable to perform quality colorization even without color hint. The model does not paint spaces between panels if they are not filled in the source. Also, it's able to distinguish different objects on images of different styles (all exhibited images, except one, belong to manga that not represented in the training set). The network is especially effective at recognizing heads, sky, water and grass. Characters within a page have homogeneous coloring if their appearance doesn't differ drastically. Moreover, the model can qualitatively colorize the manga double-page spreads, which have a different aspect ratio than the standard images.

However, objects that the network cannot recognize with confidence are colored either in skin color or green-yellow color or even are not colored at all. Pix2Pix models tend to prefer one color, so this behavior was not unexpected. A network may also colorize text clouds, if they have a shape or content that differs from those that are contained in the training set.

The color hint is used to fix the improper coloring. Using color hint enables to color arbitrary objects in arbitrary colors, it is possible to color even small objects (see Fig. 6). However, if Step 5 of inference isn't applied, colors propagated in color hint can be indirectly distributed to neighboring objects (for example, changing the shade in the direction of the propagated color). That does not allow to change the color of one small object without adding pixels to the color hint for neighboring objects and impedes obtaining a perfect coloring.

Fig. 6. Example of coloring with user interaction.

We use this flaw as an advantage by applying Step 5 of the inference. If the color propagated through the color hint spreads to neighboring objects, we can fill the improper white using initial coloring. In this case, such color spreading does not occur since the majority of objects on the image are already covered with pixels in the color hint. Moreover, the coloring artifacts can be removed. Nevertheless, this approach increases the computational time for obtaining the initial coloring and requires additional actions to modify coloring since we need to remove pixels of the color hint first on the object we want to recolorize.

The FID is used for the quantitative evaluation of the model performance. We calculate the distance between the set of color images of manga and the set of colorings generated by the model without user interaction. Colored images are represented by a holdout set of our manually colored images dataset, and colorings are generated with the same chapters of black and white editions and some samples are taken from the fine-tuning set. The size of both sets is 4000. For comparison, we have trained AlacGAN with our dataset using the training procedure described at [5]. In addition, we perform an ablation study to demonstrate the effectiveness of the proposed methods in bridging the domain gap. The results are presented in Table 1. As can be seen, the numerical metric mostly corresponds to the visual impression. The obtained results confirm the validity of the proposed approaches, as well as evidence the negative impact of prolonged fine-tuning. It can be noticed that AlacGAN outperforms our model without white color penalty and fine-tuning according to FID metric, but these approaches are also applicable to it. However, as can be seen in Fig. 5, AlacGAN tends to leave more uncolored objects, so we stick with our approach.

Table 1. Comparison of different models with FID metric. Notation: 1 - adversarial loss, 2 - white color penalty, 3 - fine-tuning, 4 - Step 5 of the inference, 5 - extended fine-tuning

Model	FID
Without 1–5	34,18
Without 2–5	33,79
AlacGAN	33,16
Without 3–5	31,25
Without 4–5	25,43
Without 5 (out model)	**25,05**
Without 4, with 5	29,06
With 1–5	25,74

5 Conclusion

In this paper, we proposed an approach for coloring manga page images with color hints using deep neural networks. To do this, a dataset consisting of the manually colored manga was collected, cleaned of noise and converted to monochrome images to create pairs for supervised learning. The proposed model was successfully trained on that dataset. We also suggested some approaches to reduce the domain gap.

The obtained results show that neural networks are able to produce high-quality and natural coloring of an entire manga page while maintaining homogeneity between the panels, and the use of color hint provides the ability to change the coloring and fix errors. However, due to the use of synthetic data and the removal of noise using a neural network, the model proved to be sensitive to the input and dependent on a large number of data preprocessing parameters. The results can be significantly improved by using a more advanced way of transforming a color image into black and white or by creating a proper dataset to avoid the use of synthetic data.

References

1. Branwen, G.: Danbooru 2019: A large-scale crowdsourced and tagged anime illustration dataset (2020)
2. Buslaev, A., Iglovikov, V.I., Khvedchenya, E., Parinov, A., Druzhinin, M., Kalinin, A.A.: Albumentations: fast and flexible image augmentations. Information **11**(2), 125 (2020)
3. Charpiat, G., Hofmann, M., Schölkopf, B.: Automatic image colorization via multimodal predictions. In: Computer Vision - ECCV 2008 (2008)
4. Cheng, Z., Yang, Q., Sheng, B.: Deep colorization. In: 2015 IEEE International Conference on Computer Vision (ICCV) (2015)

5. Ci, Y., Ma, X., Wang, E.A.: User-guided deep anime line art colorization with conditional adversarial networks. In: Proceedings of the 26th ACM International Conference on Multimedia (2018)
6. Deng, J., Dong, W., Socher, R., Li, L., Kai, L., Fei-Fei, L.: Imagenet: a large-scale hierarchical image database. In: 2009 IEEE Conference on Computer Vision and Pattern Recognition (2009)
7. Furusawa, C., Hiroshiba, K., Ogaki, K., Odagiri, Y.: Comicolorization: semi-automatic manga colorization. In: SIGGRAPH Asia 2017 Technical Briefs (2017)
8. Goodfellow, I., et al.: Generative adversarial nets. In: Advances in Neural Information Processing Systems 27. Curran Associates, Inc. (2014)
9. He, M., Chen, D., Liao, E.A.: Deep exemplar-based colorization. ACM Trans. Graph. 37(4), 1-16 (2018)
10. Iizuka, S., Simo-Serra, E., Ishikawa, H.: Let there be color!: joint end-to-end learning of global and local image priors for automatic image colorization with simultaneous classificatfion. ACM Trans. Graph. 35(4), 1-11 (2016)
11. Isola, P., Zhu, J., Zhou, T., Efros, A.A.: Image-to-image translation with conditional adversarial networks. In: 2017 IEEE Conference on Computer Vision and Pattern Recognition (CVPR) (2017)
12. Johnson, J., Alahi, A., Fei-Fei, L.: Perceptual losses for real-time style transfer and super-resolution. In: European Conference on Computer Vision (2016)
13. Kim, H., Jhoo, H., Park, E., Yoo, S.: Tag2pix: Line art colorization using text tag with secat and changing loss. In: Proceedings of the IEEE International Conference on Computer Vision (2019)
14. Larsson, G., Maire, M., Shakhnarovich, G.: Learning representations for automatic colorization. In: European Conference on Computer Vision (ECCV) (2016)
15. Lecun, Y., Bottou, L., Bengio, Y., Haffner, P.: Gradient-based learning applied to document recognition. In: Proceedings of the IEEE (1998)
16. Miyato, T., Kataoka, T., Koyama, M., Yoshida, Y.: Spectral normalization for generative adversarial networks. In: International Conference on Learning Representations (2018)
17. Shi, M., Zhang, J.Q., Chen, S.Y., Gao, L., Lai, Y.K., Zhang, F.L.: Deep line art video colorization with a few references (2020)
18. Winnemöller, H., Kyprianidis, J.E., Olsen, S.C.: Xdog: An extended difference-of-gaussians compendium including advanced image stylization. Comput. Graph. 36(6), 740-753 (2012)
19. Zhang, H., Goodfellow, I., Metaxas, D., Odena, A.: Self-attention generative adversarial networks. In: Proceedings of the 36th International Conference on Machine Learning (2019)
20. Zhang, K., Zuo, W., Zhang, L.: Ffdnet: Toward a fast and flexible solution for CNN-based image denoising. IEEE Trans. Image Process. 27(9), 4608–4622 (2018)
21. Zhang, R., Isola, P., Efros, A.A.: Colorful image colorization. In: Computer Vision - ECCV 2016 (2016)
22. Zhang, R., Zhu, J.Y., Isola, E.A.: Real-time user-guided image colorization with learned deep priors. ACM Transactions Graph (2017)

Automated Image and Video Quality Assessment for Computational Video Editing

Konstantin Lomotin[1]([✉]) [iD] and Ilya Makarov[1,2] [iD]

[1] HSE University, Moscow, Russia
`iamakarov@hse.ru`
[2] Samsung-PDMI Joint AI Center, St. Petersburg Department of Steklov
Institute of Mathematics, St. Petersburg, Russia

Abstract. We study non-reference image and video quality assessment methods, which are of great importance for computational video editing. The object of our work is image quality assessment (IQA) applicable for fast and robust frame-by-frame multipurpose video quality assessment (VQA) for short videos.

We present a complex framework for assessing the quality of images and videos. The scoring process consists of several parallel steps of metric collection with final score aggregation step. Most of the individual scoring models are based on deep convolutional neural networks (CNN). The framework can be flexibly extended or reduced by adding or removing these steps. Using Deep CNN-Based Blind Image Quality Predictor (DIQA) as a baseline for IQA, we proposed improvements based on two patching strategies, such as uniform patching and object-based patching, and add intelligent pre-training step with distortion classification.

We evaluated our model on three IQA benchmark image datasets (LIVE, TID2008, and TID2013) and manually collected short YouTube videos. We also consider interesting for automated video editing metrics used for video scoring based on the scale of a scene, face presence in frame and compliance of the shot transitions with the shooting rules. The results of this work are applicable to the development of intelligent video and image processing systems.

Keywords: Image quality assessment · Video assessment · Transition assessment · DIQA · Complex scoring · Automated video editing

1 Introduction

With the development of mobile devices and social networks, the number of photos and videos produced every day has grown dramatically. An ordinary

---

Ilya Makarov—This research is partially based on the work supported by Samsung Research, Samsung Electronics.

person shooting a video, as a rule, does not have the skills of a professional cameraman or photographer. Therefore, the result of shooting may have various defects. Moreover, in professionally filmed material defects and distortions are also possible. To identify them, one must manually view all the footage and use the editor to correct or cut the failed frames. Frame success is a largely subjective concept, but some parameters can be estimated automatically. The following metrics are considered in this paper (all metrics are also applicable to individual images):

1. Frame quality as an image. When shooting and saving the frame, various defects may appear: flare, defocus, compression artifacts, blur.
2. The scale of the scene in the shot. Traditionally, in cinematography shots can be divided into close up, medium and long shots. Their sequencing affects the aesthetics and perception of the entire video.
3. The presence of certain objects in the frame. Due to limited computing resources, it was decided to use face recognition. This metric can be useful when filming various events when it is important to capture the participants.

In this study, models were trained and tested on three commonly used in IQA benchmark datasets: LIVE [20], TID2008 [17] TID2013 [16]. The entire described framework accuracy cannot be assessed objectively since there are no datasets containing videos marked up with subjective scores frame-by-frame. Five short (from 3 to 15 min) amateur clips (party, hiking in nature, walk in the city) from the YouTube were downloaded for demonstration and manual evaluation of the framework. These videos contain distortions (blur, bad exposure, compression artifacts) and different types of shot.

The purpose of this work is to create a flexible universal framework for evaluating videos and images, which allows us to identify defective frames. The emphasis is on the flexibility of the system architecture. First, we propose three ways to improve the existing state-of-the-art IQA algorithm. Then we describe the steps of video estimation and discuss the obtained results. The code is available at [anonymized].

2   Related Work

2.1   IQA

Image quality assessment is a computer vision task, whose goal is to predict the subjective quality score for a given image. Digital images are subjected to a vast specter of distortions from the moment they are taken to the moment they are consumed: compression artifacts, Gaussian blur, white noise, etc. IQA methods can be classified into three groups, depending on the availability of a reference image: no-reference or blind IQA (NR-IQA, BIQA), reduced-reference IQA (RR-IQA), full-reference IQA (FR-IQA). The most frequently used kind of target quality value is the mean opinion score (MOS). We took TID2008 and TID2013 datasets with $MOS \in [1,9]$, where 9 corresponds to highest image quality, and LIVE dataset with $MOS \in [0, 100]$, where 100 stands for the most severe image distortion and the worst quality. For this study, MOS values were standardized to the TID range and meaning.

Since reference images are not available in the majority of cases, NR-IQA has the greatest practical importance due to the general applicability. Many recently proposed NR-IQA methods are based on machine learning algorithms, such as support vector machines (SVMs) [24] and artificial neural networks (ANN), to perform image quality score regression. Although raw images can be used for quality assessment by deep learning algorithms [8], one of the most successful approaches is explicit extraction of features called natural scene statistics (NSS) [12–14,19]. This group of methods is based on the assumption that natural images have statistical regularity affected by distortions. In many studies numerical features engineered this way are combined with deep neural networks [2,10] to achieve higher model accuracy. The approach Deep Image Quality Assessor (DIQA) [9] uses reference images to train the full-CNN model to generate an objective error map as an intermediate target. This is also known as layer pre-training. First, hidden layers of the model are tuned to extract features for generating the difference map between reference and distorted images. Then, the pre-trained layers are used to train the regressor, which maps the distorted image to the subjective quality score.

Deep ANN methods provide significant opportunities for enhancement due to the complexity and capacity of a deep neural estimator. The same feature selection layers could be used for subjective quality score regression and for distortion type classification. The aim of this work is to propose novel approaches to NR-IQA, compare model quality metrics to the baseline method, and achieve reasonable video processing performance. Namely, it is expected to reach the higher quality prediction accuracy compared to the other state-of-the-art methods, and achieve higher video processing performance compared to the baseline algorithm. We propose the following enhancements:

1. Uniform patching for performance improvement;
2. Object detection-based patching;
3. Distortion type recognizing;

2.2 Shot Boundary Detection

To assess an individual shot (a sequence of frames presenting a scene), it is necessary to recognize the transitions. These are usually sharp cuts. Sometimes transitions are also considered a smooth change of scenes, such as semi-transparent, fade in, fade out, and wipes. Shot boundary detection (SBD) methods are aimed at solving this problem. These methods are divided into two classes: based on the analysis of color and pixel brightness [1,6,15,23] and a group of deep learning-based methods [7,21]. The TransNet model, introduced in [21], uses small 3D dilated deep CNN cells of $3 \times 3 \times 3$, which allows the extraction of temporal features with a relatively small number of trainable parameters. Due to this, the compactness of the model and high speed is achieved with an average F1 score of 0.94. This model also implements the end-to-end concept, namely, it allows one to obtain shot boundaries without additional preprocessing operations and explicit feature computation. The implementation of this method was used in the current work.

2.3 Shot Type Classification

For large-scale video assessment, the shot type classification algorithm is used. Generally, 3 to 7 types of plans are distinguished in a scene.

The problem of recognizing the type of shot often arises in the automatic analysis of records of sports events and the automation of processes in cinematography. There are several approaches to solving this problem, depending on the subject area. In sporting events, information about colors that can be used is often known - the field is green, and the uniform of the players contrasts with it. The fraction of the field color in the frame may serve as a feature for the shot type classification [5, 22]. This approach lacks robustness and is suitable only for certain subject areas. Another approach is proposed by the authors of [4].

On the one hand, this approach is more flexible than color-based and can be expanded by describing other objects. On the other hand, if the content of the scene is not known in advance, this approach cannot be applied. In [18], the authors of the work use a model pre-trained on the ImageNet dataset, and then fine-tuned on assembled and tagged frames from various films to classify frames into six classes of proximity. The test result was 0.91 accuracy score, which is an impressive result. This approach has several advantages compared to other methods: it is an end-to-end solution that does not require additional processing steps, the model has high robustness and is independent of the subject area, high quality classification. However, it tends to fail on ambiguity model, caused by black-and-white frames, low-contrast frames and some artistic techniques used in the shooting.

Despite this, the method was used in this study. The movie dataset has not been published, nevertheless, the authors published an open-source Python implementation of the algorithm and fine-tuned neural network weights.

2.4 Object Detection

The generic object detection task introduces many challenges. The most general is the trade-off between accuracy and efficiency. For both requirements, there are different approaches. The most famous of them have appeared recently and are based on deep learning. According to [11], the most modern architectures such as RCCN, ResNet and VGGNet are the main detector models. These models and their modifications have high accuracy and relatively high speed. The disadvantage of all these models is their size. For example, the YOLO model, trained on the COCO dataset, has a size of about 250 Mb. The solution is to compress the models. One of the simplest methods is the quantization of weights. Further, fine-tuning of the deep network requires significant computational resources which were unavailable in this study. One possible solution is to use a pre-trained model. However, the majority of models, which are compatible with the proposed framework, lack "human" category. There is no doubt, it has high importance for the task of video assessment. Thus, for the step of object detection and recognition we use a quite basic model, mostly for demonstrative purposes.

Among the fast and lightweight non-neural network detectors, the Viola-Jones detector [58] s effectively implemented in the OpenCV cross-platform library, which provides wide possibilities for its application in real applications. Moreover, a cascade detector can be used to recognize a set of different objects. Due to the modular architecture of the proposed evaluation framework, this detector is selected for the object detection, however, it can be replaced with any other.

3   DIQA Enhancements

As improvements to the baseline DIQA, several new approaches have been tested. All methods were evaluated separately.

1. Uniform patching for performance improvement. The core hypothesis is when analyzing an image using the DIQA method, sufficient features can be retrieved not from the whole image, but from its evenly taken fragments.
2. Content-based patching. The key idea is to develop the previous approach by increasing the amount of information in patches. To do this, a detector model predicts bounding boxes for detected objects. These borders are used to cut patches.
3. Distortion type recognizing. An additional stage of pre-training the deep layers of the network to build up a classifier type distortion. This can increase the accuracy of the assessment by deriving more informative features.

3.1   Experiment Setup

For the training and evaluation of all proposed solutions, three datasets were used: TID2008, TID2013, and LIVE (Table 1). These image sets are commonly used in IQA studies. Each image has information about the type and severity of distortion, as well as a mean opinion score (MOS). This rating was obtained by averaging the ratings obtained from a survey of multiple subjects. In TID datasets, all images have a fixed size of 512×384 pixels, but images from the LIVE dataset vary in size.

Table 1. Description of the datasets.

	TID2008	TID2013	LIVE
Reference images	25	25	29
Distorted images	1700	3000	982
Distortion types	17	24	5

The accuracy of the algorithms was measured by the Spearman Rank Order Correlation Coefficient (SROCC) and Pearson Linear Correlation Coefficient (LCC) metrics between the predicted and real MOS sequences.

Hardware setup for performance measurement: CPU Intel Core i5-8600, 16 Gb RAM, GPU NVidia RTX 2070.

3.2   Uniform Patching

As described in [3], patches can be used to evaluate quality using deep CNN. It is assumed that the accuracy of predictions will fall slightly, but the speed of work will increase.

In this work, uniform patching is parametrized with two values: side size of the square patch S_p and by the number of patches per side N_p. The result is a numeric tensor of shape $(N_p, S_p, S_p, 3)$ if the number of color channels is 3. In this particular example, the resulting amount of data is 46% of the original image. Thus, more than half of the image data is discarded. The resulting score is computed as the weighted average of the patch ratings. The algorithm adjusts distances between patches to fit the edge patches to image corners, preserving same distances along vertical and horizontal axe

The center patch is supposed to hold most of the information. Hence, it should bring more influence on the model and on the result. For this purpose, patch weights are calculated to give the center patch twice more weight compared to the side patches in such a way that the sum of the weights is equal to 1.

3.3   Content-Based Patching

Uniform patching dramatically reduces the amount of data to process. However, it may lose many high-frequency regions and catch much low-frequency data. The idea of content-based patching is motivated by two reasonable assumptions:

1. The scene contains objects, and the quality of these objects matters more, than the quality of surroundings.
2. Image regions with the objects contain much high-frequency information that could be used for MOS prediction.

As a detector model, we selected the MobileNet SSD network pre-trained with images from the COCO dataset. It is compressed to 4 Mb of disk space, and one detection cycle takes about 30 ms.

To increase the robustness of detection, we suggest to take a fixed number of objects with the highest probability (for which the detector has the highest certainty), and to build a common bounding box around their bounding boxes. The number of selected images is empirically set up to 3.

However, the detector is not perfect. For some scenes, it fails to detect the main objects and results with a set of minor bounding boxes. An additional heuristic may help us to handle this. Let A_{obj} be the area of the resulting detected bounding box, and A_{full} is the area of the full-size image. If $\frac{A_{obj}}{A_{full}} < T_d$, patching strategy is switched to the uniform. In other words, if the detector fails to detect reasonable objects, patches are extracted uniformly from the entire image. For the benchmark image datasets, the most appropriate T_d tends to be approximately equal to 0.3. This value is closely related to the shot scale and could be considered as a tunable parameter.

3.4 Distortion Type Recognizing

Distortion type classifier (DTC), trained on the same hidden layers as a quality regressor, introduces two improvements. Firstly, an additional step in pre-training can change the output of feature selection layers. Training of the subjective score predictor may result in more accurate scoring. Second, determining the type of distortion is additional functionality that can be used for the automatic correction of distorted images.

For each dataset with $N_D T$ distortion types two DTC structures were evaluated:

1. A sequence of dense layers with 128, 64 and $N_D T$ outputs (S_1).
2. A sequence of dense layers with 512, 256, 128, 64 and $N_D T$ outputs (S_2).

The output layer had a softmax activation function, the rest of the layers had ReLu. As a loss function binary cross-entropy was employed. Targets (types of distortion) were taken from the datasets.

This enhancement exclusively takes place in the training step. It does not introduce additional inference calculations and does not affect quality score prediction speed.

3.5 Results and Discussion

All the proposed approaches were evaluated on the test subset of the datasets. The results were aggregated to Table 2. All models were pre-trained with 5 epochs of error map generation and fine-tuned with 50 epochs of subjective score prediction. Patching strategies were trained and tested with $N_p = 3$ and $S_p = 64$ and relative center patch coefficient 2.

Table 2. Evaluation results of all proposed IQA enhancements on image dataset.

Enhancement	TID2008		TID2013		LIVE	
	SROCC	LCC	SROCC	LCC	SROCC	LCC
Baseline DIQA	0.494	0.582	0.442	0.557	0.879	0.876
Uniform patching	0.576	0.649	0.498	0.547	0.958	0.948
Content-based patching	0.500	0.579	**0.667**	**0.709**	**0.977**	**0.979**
DTC S_1	**0.611**	**0.661**	0.622	0.661	0.890	0.890
DTC S_2	0.512	0.577	0.519	0.482	0.910	0.911

TID correlation values are much lower than presented in the related studies. The possible reason is the implementation-specific optimization described in [9] and applied in this study. An input shape of the model was fixed to allow faster GPU computations. This could force some images to be unnaturally reshaped by breaking the aspect ratio. However, this work is focused on the relative score increase. For TID2008, an additional pre-trained step added about 20% to the baseline accuracy. However, classification accuracy and F1-score appear to be insignificantly low. Thus, it was decided to discard the classifier model, keeping the pre-training stage as it has an influence on the IQA result. For TID2013 and LIVE datasets the content-based patching assessment method outperforms the baseline method by approximately 40% and 11% accordingly. The interesting feature is that content-based patching almost did not change the baseline correlation scores for the TID2008. This could point to the detector's weakness.

Thus, object detection for smart IQA is the most accurate approach comparing to the baseline DIQA. Moreover, the modular structure of the model fits in the entire video assessment framework architecture.

4   Video Scoring and Summarization Framework

4.1   The Architecture Overview

The proposed algorithm for the complex video assessment consists of several steps: shot boundary detection, shot type classification, image quality assessment, object detection, transition assessment and score aggregation.

The flowchart is shown in Fig. 1. The video is divided into shots using the SBD algorithm. Then each shot is individually assessed by all scoring algorithms. After completing all estimates, the gained information is collected in one score for each frame within each shot.

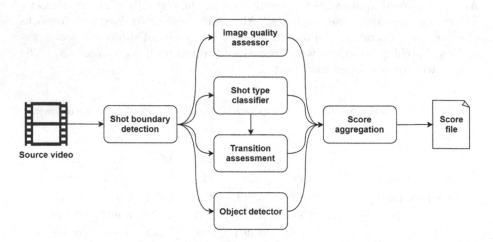

Fig. 1. The general scheme of the assessment framework.

The key feature of the proposed architecture is modularity. Modularity allows you to remove, add, and replace algorithm steps depending on the application. For example, to evaluate user videos shot on a smartphone, the recognition of the type of shots and the estimation of transitions may be redundant. Another example, for nature photography, face recognition can be replaced by the recognition of natural objects. Another advantage of this architecture is the ability to run evaluation components in parallel. Further all the stages of the assessment will be considered in detail.

4.2 Shot Boundary Detection

To determine which frames are separated by transitions, the TransNet model is used. It allows you to define transition types such as sharp cuts and gradual dissolve transitions. At the input, the model receives a sequence of frames in the format of 24 bits per pixel. The dimension of the input network layer is 48×27 pixels. This allows in most cases to load all the videos into RAM for more efficient processing. As a result, a list of shot boundaries is collected from the probabilities, which is input to all evaluation algorithms.

4.3 Shot Type Classifier

To determine the scale of the shot was used the method implemented in [18] and described in the previous study. The approach is based on training a shot type classifier based on the pre-trained ResNet network and is aimed at cinema production. The classification accuracy is about 0.91 F1-score, which can be considered as a result of a very accurate model. The mistakes are made analyzing scenes with ambiguity, insufficient lighting, or special artistic techniques are used.

In this work, frames are classified into 6 types: extreme close up (ECU), close up (CU), medium close-up (MCU), medium shot (MS), long shot (LS, also mentioned as wide shot, WS) and extreme wide shot (EWS).

To handle the inaccuracies of the STC model, the aggregation of all received scores takes into account all the predicted probabilities. In general, the accuracy of this shot type detection method depends on how the feature extraction network was pre-trained. A special STC that is more resistant to distortion can be trained.

4.4 Transition Assessment

Transitions between shots are an important component of the video, which greatly affects its perception by the viewer. Human psychology is designed in such a way that shot scales given in certain order seems more aesthetic to us. These patterns are outlined in cinematography manuals. The guideline [3] provides many rules regarding exposure, lighting, camera angle, and other aspects that affect the final frame quality and artistic value of the scene being filmed. In this paper, heuristics are used that allow the use of available information about transitions. The "good" transitions are $ECU \leftrightarrow MCU$, $CU \leftrightarrow MS$, $MCU \leftrightarrow LS$, $MS \leftrightarrow EWS$, $MCU \leftrightarrow CU$ and $MS \leftrightarrow LS$.

A shot may represent a scene with several plans. A certain fraction F_{ST} of frames lying around the borders of the shot is taken, and the most common among them type is considered, and this value is used in transitions scoring. The issue of choosing the frame percent remains open. For shots with a long duration, it can be small, and for shots of a few seconds it can reach up to 100%. In this work, the value $F_{ST} = 0.3$ is used, which means that is about a third of the frames around transition are taken. The fraction of "good" transitions is treated as a transition quality score for the video

4.5   Score Aggregation

After all stages of analysis, for each frame there is collected data: image quality score, attractive transition percent, frame scale types, and the number of detected faces. There are currently no well-known algorithms for complex video and image rating composition. This is a field for future work.

In this work, the aggregation of data into a single assessment occurs in two steps. In the first step, for each frame, the estimate is collected by the formula

$$S_{Total} = (P_{ST} \cdot W) \otimes S_{IQA} + N_F \tag{1}$$

where S_{Total} – final frame score; P_{ST} – vector with predicted probabilities for each shot type; W – fixed vector of penalty weights; S_{IQA} – predicted image quality score; N_F – amount of detected face.

In the general case, if frames are processed by batches or by shots, P_{ST} may be a matrix, N_F and S_{IQA} – a vector. For this reason, the multiplication of weighted probabilities by IQA score is denoted as element-wise. Weight coefficients are 0.8 for close-up scales (ECU, CU, MCU), 0.9 for medium shots, and 1 for long shots (LS, EWS). The purpose of these values is to introduce stricter requirements for scenes filmed in detail, reducing the impact of each scale probability.

In the second step, scores of F_{ST} fraction of frames taken from around "good" transitions are multiplied by the upscaling coefficient. This value is taken 1.2 in this study. The frames, forming scenes with aesthetic transitions will more likely have a high score.

The open issue is how to deal with object detection data. It could provide wide opportunities for flexible tuning for the framework user. Categories of objects could be marked as "negative" or "positive" to give higher scores to scenes with desired objects and exclude scenes with unwanted categories. However, the default behavior is undefined. The usage of semantics-based scoring should depend on the situation

4.6   Postprocessing

Raw scores, predicted by the assessment system, may slightly differ over frames within one scene, keeping the overall trend although. An attempt to select the best frames with such scoring would result in an extremely fragmented video with multiple shots containing only a few frames. This problem could be solved by applying a smoothing filter. The scores are considered as a signal in time, an the use of the Butterworth filter with order 4 leads to a sufficiently smooth score trend.

After two propagations of filter (to avoid biasing), scores are ready for video summarization (Fig. 2). The task is to select M frames with the highest score values, such that $\frac{M}{FPS} = T_r$, where FPS – original video framerate (the most common options are 25, 30 and 60 frames per seconds) and T_r – the required duration of the video summary in seconds. This step is implemented with the usage of the K-th percentile of the scores. The algorithm consists of three steps:

1. Compute $K = 1 - \frac{T_r}{N_{frames} \cdot FPS}$, where N_{frames} – the total amount of frames in a video.
2. Compute K-th percentile.
3. Discard all frames with score values less than the computed percentile.

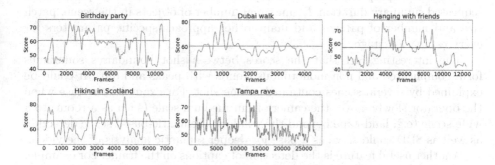

Fig. 2. The 70-th percentile of the smoothed scores for the test YouTube videos.

Examples of developed application work can be seen in Fig. 3. It demonstrates the best and the worst frames from three videos: "Birthday party" (stable camera, mostly domestic scenes; many faces.), "Dubai walk" (artistic techniques, such as filming reflections in water and abstract scenes; text labels and titles; sever exposure and blur distortions) and "Hiking in Scotland" (the video was taken by the action camera; high quality; misty scenes; mostly wide shots with nature scenes; almost no faces).

5   Discussion and Future Work

For testing purposes, five short videos with various scene settings and distortion types were downloaded from YouTube. The download script is available in the project GitHub repository. The frame resolution is unified to 360×640 pixels. From the results of manual testing of the application, it was noticed that the

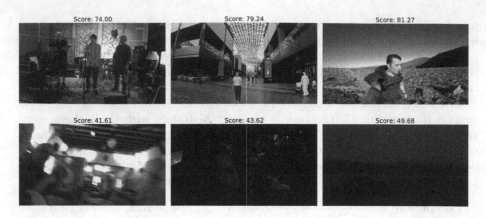

Fig. 3. The upper row shows the best frames examples, the bottom row—the worst frames.

accuracy of the proposed framework depends on a set of factors and parameters: requested summary duration T_r, maximal number of objects in the scene, patch size and number of patches and many other approach-specific parameters for every estimation stage.

The interesting fact that the scores between shot boundaries sometimes form complicated and non-monotonous trends with peaks and dips. It could be explained by several scenes contained in one shot. For example, the case when the operator slowly moves the camera from medium-scale (such as a room) to a wide scene (e.g. landscape beyond the window). Semantic-based scene detection as well as SBD could serve as a base for deeper semantic analysis of the video.

Another useful feature is the detection of captions on the frame. For example, a scene with congratulation is present in the "Birthday party" test video. IQA algorithm has scored it poorly, as frames from this scene contain nothing but the cloudy background and white letters with curved shapes. Passed to the DIQA algorithms, they are treated as low-contrast and distorted images. It could be compensated by giving higher scores to the frames with detected text or, in general, any specific semantic feature.

6   Conclusion

In this work, three approaches to enhancement of the state-of-the-art image quality assessment method are introduced. They are aimed at accuracy and functionality improvements and at general enhancement. As a result of the experiment, it was found that content-based patching outperforms the baseline DIQA method by from 40% to 11% depending on the test dataset. Performance gain relative to the original algorithm is 26%.

Further, the universal and flexible framework for complex video assessment and summarization was introduced. The proposed method has a high potential for specialization. It is possible to enable and disable specific components and to tune parameters for each of them to achieve the most accurate match to different subject fields. Some of assessment steps are independent and can be performed in parallel. Generally, generated summaries are perceived better-quality and more aesthetic, than the original videos.

References

1. Amel, A.M., Abdessalem, B.A., Abdellatif, M.: Video shot boundary detection using motion activity descriptor. arXiv preprint arXiv:1004.4605 (April 2010). http://arxiv.org/abs/1004.4605

2. Bosse, S., Maniry, D., Wiegand, T., Samek, W.: A deep neural network for image quality assessment. In: 2016 IEEE International Conference on Image Processing (ICIP). pp. 3773–3777. IEEE (September 2016). https://doi.org/10.1109/ICIP.2016.7533065, http://ieeexplore.ieee.org/document/7533065/

3. Brown, B.: Cinematography: Theory and Practice: Image Making for Cinematographers and Directors. Taylor & Francis (2016). https://books.google.ru/books?id=GiQlDwAAQBAJ

4. Cherif, I., Solachidis, V., Pitas, I.: Shot type identification of movie content. In: 2007 9th International Symposium on Signal Processing and Its Applications, pp. 1–4. IEEE (February 2007). https://doi.org/10.1109/ISSPA.2007.4555491, http://ieeexplore.ieee.org/document/4555491/

5. Ekin, A., Tekalp, A.: Robust Dominant Color Region Detection with Applications to Sports Video Analysis. In: Proceedings 2003 International Conference on Image Processing (Cat. No. 03CH37429) (2003)

6. Ferman, A., Tekalp, A.: Two-stage hierarchical video summary extraction to match low-level user browsing preferences. IEEE Trans. Multimedia 5(2), 244–256 (2003). https://doi.org/10.1109/TMM.2003.811617, http://ieeexplore.ieee.org/document/1208494/

7. Hassanien, A., Elgharib, M., Selim, A., Bae, S.H., Hefeeda, M., Matusik, W.: Large-scale, Fast and Accurate Shot Boundary Detection through Spatio-temporal Convolutional Neural Networks. arXiv preprint arXiv:1705.03281 (May 2017). http://arxiv.org/abs/1705.03281

8. Kang, L., Ye, P., Li, Y., Doermann, D.: Convolutional Neural Networks for No-Reference Image Quality Assessment. In: The IEEE Conference on Computer Vision and Pattern Recognition (CVPR) (2014)

9. Kim, J., Nguyen, A., Lee, S.: Deep CNN-based blind image quality predictor. IEEE Trans. Neural Netw. Learn. Syst. 30(1), 11–24 (2019). https://doi.org/10.1109/TNNLS.2018.2829819

10. Li, Y., et al.: No-reference image quality assessment with shearlet transform and deep neural networks. Neurocomputing 154, 94–109 (2015). https://doi.org/10.1016/j.neucom.2014.12.015, https://linkinghub.elsevier.com/retrieve/pii/S0925231214016798

11. Liu, L., et al.: Deep learning for generic object detection: a survey. Int. J. Comput. Vis. 128(2), 261–318 (2020). https://doi.org/10.1007/s11263-019-01247-4, http://link.springer.com/10.1007/s11263-019-01247-4

12. Mittal, A., Moorthy, A.K., Bovik, A.C.: No-reference image quality assessment in the spatial domain. IEEE Trans. Image Process. 21(12), 4695–4708 (2012). https://doi.org/10.1109/TIP.2012.2214050

13. Mittal, A., Soundararajan, R., Bovik, A.C.: Making a "Completely Blind" image quality analyzer. IEEE Signal Process. Lett. **20**(3), 209–212 (2013). https://doi.org/10.1109/LSP.2012.2227726

14. Moorthy, A.K., Bovik, A.C.: Blind image quality assessment: from natural scene statistics to perceptual quality. IEEE Trans. Image Process. **20**(12), 3350–3364 (2011)

15. Pass, G., Zabih, R., Miller, J.: Comparing images using color coherence vectors. In: Proceedings of the fourth ACM international conference on Multimedia - MULTI-MEDIA 1996. pp. 65–73. ACM Press, New York, USA (1996). https://doi.org/10.1145/244130.244148, http://portal.acm.org/citation.cfm?doid=244130.244148

16. Ponomarenko, N., et al.: Image database TID2013: Peculiarities, results and perspectives. Signal Process. Image Commun. **30**, 57–77 (2015). https://doi.org/10.1016/j.image.2014.10.009,   https://linkinghub.elsevier.com/retrieve/pii/S0923596514001490

17. Ponomarenko, N., Lukin, V., Zelensky, A., Egiazarian, K., Carli, M., Battisti, F.: TID2008-a database for evaluation of full-reference visual quality assessment metrics. Adv. Mod. Radioelectronics **10**(4), 30–45 (2009)

18. Somani, R.: AI For Filmmaking: Recognising Shot Types with ResNets (2019). https://rsomani95.github.io/ai-film-1.html

19. Saad, M.A., Bovik, A.C., Charrier, C.: Blind image quality assessment: a natural scene statistics approach in the DCT domain. IEEE Trans. Image Process. **21**(8), 3339–3352 (2012). https://doi.org/10.1109/TIP.2012.2191563

20. Sheikh, H.R., Wang, Z., Cormack, L., Bovik, A.C.: LIVE image quality assessment database release 2 (2005). http://live.ece.texas.edu/research/quality (2005)

21. Souček, T., Moravec, J., Lokoč, J.: TransNet: a deep network for fast detection of common shot transitions. arXiv preprint arXiv:1906.03363 (June 2019). http://arxiv.org/abs/1906.03363

22. Tabii, Y., Djibril, M.O., Hadi, Y., Thami, R.O.H.: A new method for video soccer shot classification. In: VISAPP (1), pp. 221–224 (2007)

23. Zhang, H., Kankanhalli, A., Smoliar, S.W.: Automatic partitioning of full-motion video. Multimedia Syst. **1**(1), 10–28 (1993). https://doi.org/10.1007/BF01210504, http://link.springer.com/10.1007/BF01210504

24. Zhang, P., Zhou, W., Wu, L., Li, H.: SOM: semantic obviousness metric for image quality assessment. In: The IEEE Conference on Computer Vision and Pattern Recognition (CVPR) (2015)

Ensemble-Based Commercial Buildings Facades Photographs Classifier

Aleksei Samarin[1,3]([✉]) and Valentin Malykh[2,3]

[1] Saint-Petersburg State University, Saint Petersburg, Russia
[2] Kazan (Volga Region) Federal University, Kazan, Russia
[3] Suprema Labs, Berlin, Germany

Abstract. We present an ensemble-based method for classifying photographs containing patches with text. In particular, the proposed solution is suitable for the task of classification the images of commercial building facades by the type of provided services. Our model is based on heterogeneous ensemble usage and analysis of textual and visual features as well as special visual descriptors for areas with English text. It should be noted that our classifier demonstrates remarkable performance (0.71 in F_1 score against 0.43 baseline result). We also provide our own dataset containing 3000 images of facades with signboards in order to provide complete classification benchmark.

Keywords: Ensemble · Image classifier · Visual and textual features · Signboard image descriptor

1 Introduction

Among other problems of classifying images with text, one can single out the problem of classifying commercial buildings facades photographs by the type of provided services. That problem is of significant importance in the field of applied marketing and pattern recognition [1–4]. Unfortunately that type of an automatic signboards classification is extremely difficult due to the large number of factors such as varying shooting conditions, visual distortions, unique signage fonts and styles [1]. This variety of adverse factors leads to the need to maximize the use of all information suitable for classification. In order to implement that idea we combine several different classifiers that make decisions according to different feature types.

The first type of classifiers that we use in our ensemble is general images classifiers. At the present time, many computer vision methods have achieved outstanding results in the classification of general images [5–8]. However dissimilar objects images classification problem significantly differs from classification of photographs of facades of commercial buildings with advertising signboards by visual properties of presented objects. It should be noted that signboards are often have similar shape, that sharply distinguishes them from dissimilar objects from general datasets [9,10]. Despite this circumstance, general images

© Springer Nature Switzerland AG 2021
W. M. P. van der Aalst et al. (Eds.): AIST 2020, LNCS 12602, pp. 257–265, 2021.
https://doi.org/10.1007/978-3-030-72610-2_19

classifier is useful for extraction context from background that allows to draw conclusions about the type of provided services in some cases. We also use general visual features extractors for independent style analysis of advertising signs that is significant for overall performance of our system (Sect. 5).

Another important feature type for business type recognition is textual data that is placed on an advertising signs and contains the most useful information in some cases. Currently there are a lot of approaches for extraction textual information from input images [11–13]. However different fonts sizes and styles tends to mistakes in the output of text recognition engines. Misspelling can be compensated for by the use of noise-resistant embeddings [14] but pure text-based classifiers also do not provide an appropriate performance in grouping photographs of facades by the type of business (Sect. 5).

We also use CNN-based image encoder for evaluation of visual descriptors of image patches with signboards. As a result we take into account posters style features and contours-based visual characteristics of presented text. Usage of that information significantly improves performance of our ensemble (see in Sect. 5).

Thus we introduce an ensemble-based method for classification of photographs of commercial buildings facades with advertising signs by the type of provided services. Our method demonstrates better performance than the baseline [1]. We also provide a new dataset that contains 3000 images grouped into 4 classes by a business type in order to provide complete benchmark for a performance comparison of classification methods.

2   Problem Statement

The problem of classification of photographs of commercial buildings facades with advertising signboards by the type of business can be formalized as follows. Each input image Q containing advertising sings should be associated with one of groups $C = C_i$, where $i \in [0, N]$ according to the type of provided services. Considering photograph can be significantly influenced by various shooting conditions (angles, lighting and colour balance), noise, sun glare, defocus areas, and other artifacts arising in the process of image capturing. Moreover, the style and design of posters can vary significantly, as a presented text size and fonts type.

3   Proposed Method

We use a similar architecture as described in [1] for commercial building facades classification. As in [1] we use CNN-based [5,6,8,15–17] module for general features extraction from background. And we also use such encoders for an independent estimation of styles and colors of text areas that are presented on signboards. It should be noted that we use prior text detection for better OCR performance and pure visual analysis of presented posters. For textual information analysis we obtain embeddings for succeeding analysis. Finally we use all mentioned feature types by a classifier that groups images into 4 classes according to a type of provided business. All of the modules of the proposed pipeline are presented in Fig. 1.

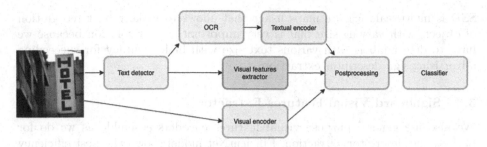

Fig. 1. General pipeline of the proposed solution.

3.1 Background Visual Feature Extractor

As a backgrounds descriptors we investigated intermediate representation of several CNN architectures: VGG [15], Inception [5], ResNet [6], MobileNet [16], EfficientNet [8], NASNet [17], DenseNet [18], InceptionResNet [19]. We selected that networks because of remarkable results demonstrated by them in general images classification problems. We also noticed that combined descriptors (Fig. 2) allow to obtain better performance in our pipeline. We analyzed several configurations of the encoder (Sect. 6) and selected the best one (InceptionResNet + EfficientNet) according to our experiment results. Final scheme of our background feature extractor presented in Fig. 2. The first branch of our combined features extractor - InceptionResNet architecture that contains Inception feature extractors and residual connections. The second branch is presented with EfficientNet B4 that is based on several key ideas (for example, scaling method) and outperforms MobileNet and ResNet architectures on Imagenet dataset.

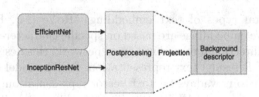

Fig. 2. Background descriptor generation scheme.

3.2 Text Detector

We tested several detectors (SSD [20], YOLO [21], EAST [12], CTPN [11]), TextBoxes [22]) for text areas localization and selected the best one (SSD with ResNet backbone) in order to provide better performance of our combined model (Sect. 6).

The SSD architecture is based on convolutional neural network usage for fixed-size bounding boxes and scores prediction. One of the key concepts of the

SSD is multi-scale feature maps usage that allows to make robust recognition of objects with varying size that is very important for our solution because we have to detect labels with various text size with high accuracy for succeeding advertising sign descriptor extraction.

3.3   Signboard Visual Features Extractor

We also use general purpose visual features encoders ensemble as we do for background descriptor extraction. EfficientNet model shown the best efficiency in our experiments (Sect. 3.4). We used the same scheme as we introduced for background encoder (Sect. 6) with the difference that instead of EfficientNet B4 and InceptionResNet we use EfficientNet B3 and EfficientNet B2 combined feature extractor. This architecture allowed to estimate special visual text features better than investigated counterparts.

3.4   Text Recognition Engine

In order to achieve better performance in OCR we compared several text recognition engines: TesseractOCR [13], modified CRNN [23], EasyOCR (CRAFT [24] based OCR tool). Modified CRNN shown the best performance in the context of our pipeline (Sect. 6). This version of the CRNN architecture uses EfficientNet-like feature extractor and a proposed multi-head attention mechanisms for better character localization and higher recognition accuracy that is of great importance in the context of our task, when the symbols of one word can have different sizes and styles. However attention mechanism allows to estimate different parts of an each symbol in order to get better performance in overall characters recognition.

3.5   Text Embeddings

We analyzed several types of text embeddings: RoVe [14], BERT [25], and ALBERT [26]. RoVe embeddings are based on specific character-level representation followed by bidirectional recurrent neural network, this design is tailored for better noisy (as OCR result) text representation. BERT model could be considered standard de-facto nowadays for text vector representation, while ALBERT being a direct descendant of BERT is significantly different in its internal structure. The important feature of both BERT and ALBERT models is used BPE representation [27]. This representation while not explicitly designed to handle noisy texts is able to represent virtually any input string. Given that and the significant capacity of named models (more than 100 million parameters for BERT) one could expect a model to handle the OCR text output. In the presented model we used BERT which have the best results in our experiments.

4   Datasets

In our work we used two datasets. The first dataset was presented in [1] and contains ~360 images of commercial buildings facades grouped into 4 classes (hotels, restaurants, shops and 'other'). Each category contains approximately the same number of samples (~90 photographs per class). Unfortunately that dataset contains relatively small number of elements, so we compare our approach on it only.

In addition, we collected and make publicly available a new dataset, which we call Commercial Facades Dataset (CFD)[1]. It contains 3886 images (Fig. 3) divided into the same categories as [1]. The class distribution is even for three of four classes, with one class (hotels) being slightly smaller than the others (886 vs 1000) in the collected dataset. All of the images were collected from open sources. This dataset was used for ablation study and comparison.

We used a train-validation-test split ratio of 10:1:1 for [1] dataset and the same one for CFD. In both cases class balance was ensured in all parts of the split.

a)   b)   c)   d)

Fig. 3. Examples of images from CFD dataset: a) photograph of a hotel's facade; b) photograph of a store's facade; c) photograph of a restaurant's facade; d) an element from 'other' category.)

5   Results

In our experiments the following configuration showed the best results: it includes combined background feature extractor with EfficientNet B4 and InceptionResNet. The best configuration of our classifier also contains text detector based on modified SSD [20] detector with a ResNet [6] bottom. The text recognition engine of our solution is presented with the modified RCNN [23] model and the signboard visual feature extractor is presented with a combined model that uses EfficientNet B2 and EfficientNet B3 features.

We also provide a comparison of the best configuration of our model and the combined advertising sign classifier on CFD and dataset presented in [1]. Results of the comparison are presented in Table 1. Thus the best configuration of our ensemble significantly overcomes [1] on CFD and a dataset proposed in [1].

---

[1] We used https://Flickr.com and chose only the images with 'commercial use and modifications allowed' licensing. The dataset is available here: https://github.com/madrugado/commercial-facades-dataset.

Table 1. Metrics (P - precision, R - recall, F - F-measure) for our system and combined classifier [1] on CFD and a dataset presented in [1].

Model	Dataset					
	CFD			Malykh *et al.* [1]		
	P	R	F	P	R	F
Combined classifier [1]	0.72	0.31	0.43	0.36	0.18	0.24
Ours	**0.87**	**0.6**	**0.71**	**0.83**	**0.32**	**0.46**

6   Ablation Study

Below we provide an ablation study for our modular architecture. In order to find the best configuration of our model we investigated all of the possible variations of each module. The results of ablation study are presented in Table 2.

Table 2. F-measure for our system configured with different module variations on CFD (top-5 results for each module at max).

Model variation	F_1 score
Ours	**0.71**
Background feature extractor type	
EfficientNet B4 + EfficientNet B2	0.64
EfficientNet B4 + EfficientNet B3	0.68
EfficientNet B3 + InceptionResNet	0.66
EfficientNet B2 + EfficientNet B3	0.68
Signboard feature extractor type	
EfficientNet B3 + InceptionResNet	0.61
EfficientNet B4 + InceptionResNet	0.65
EfficientNet B4 + EfficientNet B2 t	0.64
EfficientNet B3 + EfficientNet B4	0.67
Text detector type	
YOLO	0.60
CTPN	0.69
EAST	0.68
TextBoxes	0.70
OCR engine	
TesseractOCR	0.58
EasyOCR	0.63
Text embedding type	
RoVe	0.61
ALBERT	0.69

For background visual feature extractor we experimented with several combinations of architectures mentioned in Sect. 6. Top 5 results that we managed to obtain using declared background features extractor and arbitrary combinations of other elements of our pipeline. EfficientNet B4 + InceptionResNet combined feature extractor shown the best performance in terms of our model. The same investigation we provide for visual feature extractor of a signboard area. EfficientNet B3 and EfficientNet B2 demonstrated best result. The text detection module in the ours best configuration was presented with SSD with ResNet backbone. After text detection we used modified CRNN for text recognition and BERT embeddings in order to achieve the best performance for our model (0.71 in averaged $F1$ score).

Conclusion

We proposed the ensemble-based method for classifying images of commercial building facades with advertising signs by the type of provided services. The presented architecture is based on usage of several neural network classifiers. Each classifier is responsible for an estimation of a specific feature type, that allows us to explicitly analyze each factor containing significant information about the type of business. Our model demonstrated better performance than considered baselines (Sect. 5). Moreover we presented our own dataset that contains 3000 images in order to provide a complete performance benchmark in the context of the stated problem. It should be noted that the presented solution architecture is extensible and can be easily modified in order to support new feature types. Thus the further research of the proposed approach can be focused on the study of new feature types and classifiers that can be included in the proposed ensemble for overall performance improvement of the considered system.

References

1. Malykh, V., Samarin, A.: Combined advertising sign classifier. In: van der Aalst, W.M.P., et al. (eds.) AIST 2019. LNCS, vol. 11832, pp. 179–185. Springer, Cham (2019). https://doi.org/10.1007/978-3-030-37334-4_16
2. Intasuwan, T., Kaewthong, J., Vittayakorn, S.: Text and object detection on billboards. In: 2018 10th International Conference on Information Technology and Electrical Engineering (ICITEE), pp. 6–11, July 2018
3. Zhou, J., McGuinness, K., O'Connor, N.E.: A text recognition and retrieval system for e-business image management. In: Schoeffmann, K., et al. (eds.) MMM 2018. LNCS, vol. 10705, pp. 23–35. Springer, Cham (2018). https://doi.org/10.1007/978-3-319-73600-6_3
4. Watve, A., Sural, S.: Soccer video processing for the detection of advertisement billboards. Pattern Recogn. Lett. **29**(7), 994–1006 (2008)
5. Szegedy, C., et al.: Going deeper with convolutions. In: 2015 IEEE Conference on Computer Vision and Pattern Recognition (CVPR), pp. 1–9, June 2015

6. He, K., Zhang, X., Ren, S., Sun, J.: Deep residual learning for image recognition. In: 2016 IEEE Conference on Computer Vision and Pattern Recognition (CVPR), pp. 770–778, June 2016

7. Andrew, G., et al.: Mobilenets: efficient convolutional neural networks for mobile vision applications, April 2017

8. Tan, M., Le, Q.: Efficientnet: rethinking model scaling for convolutional neural networks, May 2019

9. Lin, T.-Y., et al.: Microsoft COCO: common objects in context. In: Fleet, D., Pajdla, T., Schiele, B., Tuytelaars, T. (eds.) ECCV 2014. LNCS, vol. 8693, pp. 740–755. Springer, Cham (2014). https://doi.org/10.1007/978-3-319-10602-1_48

10. Deng, J., Dong, W., Socher, R., Li, L., Li, V., Fei-Fei, L.: Imagenet: a large-scale hierarchical image database. In: 2009 IEEE Conference on Computer Vision and Pattern Recognition, pp. 248–255, June 2009

11. Tian, Z., Huang, W., He, T., He, P., Qiao, Y.: Detecting text in natural image with connectionist text proposal network. In: Leibe, B., Matas, J., Sebe, N., Welling, M. (eds.) ECCV 2016. LNCS, vol. 9912, pp. 56–72. Springer, Cham (2016). https://doi.org/10.1007/978-3-319-46484-8_4

12. Zhou, X., et al.: East: an efficient and accurate scene text detector, April 2017

13. Smith, V.: An overview of the tesseract ocr engine. In: Ninth International Conference on Document Analysis and Recognition (ICDAR 2007), vol. 2, pp. 629–633, September 2007

14. Malykh, V.: Robust word vectors for russian language. In: Proceedings of Artificial Intelligence and Natural Language AINL FRUCT 2016 Conference, Saint-Petersburg, Russia, pp. 10–12 (2016)

15. Simonyan, K., Zisserman, A.: Very deep convolutional networks for large-scale image recognition. CoRR, abs/1409.1556 (2014)

16. Sandler, M., Howard, A., Zhu, M., Zhmoginov, A., Chen, L.: Mobilenetv 2: inverted residuals and linear bottlenecks. In: 2018 IEEE/CVF Conference on Computer Vision and Pattern Recognition, pp. 4510–4520 (2018)

17. Zoph, B., Vasudevan, V., Shlens, J., Le, V.: Learning transferable architectures for scalable image recognition, pp. 8697–8710 (2018)

18. Huang, G., Liu, Z., Van Der Maaten, L., Weinberger, K.Q.: Densely connected convolutional networks. In: 2017 IEEE Conference on Computer Vision and Pattern Recognition (CVPR), pp. 2261–2269 (2017)

19. Szegedy, C., Ioffe, S., Vanhoucke, V., Alemi, A.: Inception-v4, inception-resnet and the impact of residual connections on learning. In: AAAI Conference on Artificial Intelligence (2016)

20. Liu, W., et al.: SSD: single shot multibox detector. In: Leibe, B., Matas, J., Sebe, N., Welling, M. (eds.) ECCV 2016. LNCS, vol. 9905, pp. 21–37. Springer, Cham (2016). https://doi.org/10.1007/978-3-319-46448-0_2

21. Redmon, J., Divvala, S., Girshick, R., Farhadi, A.: You only look once: unified, real-time object detection, June 2015

22. Liao, M., Shi, B., Bai, X., Wang, X., Liu, W.: Textboxes: a fast text detector with a single deep neural network, November 2016

23. Sang, D., Cuong, L.: Improving crnn with efficientnet-like feature extractor and multi-head attention for text recognition, pp. 285–290, December 2019

24. Baek, A., Lee, B., Han, D., Yun, S., Lee, H.: Character region awareness for text detection, pp. 9357–9366, June 2019

25. Devlin, J., Chang, M.-W., Lee, K., Toutanova, K.: Bert: pre-training of deep bidirectional transformers for language understanding, Actober 2018
26. Lan, Z., Chen, M., Goodman, S., Gimpel, K., Sharma, P., Soricut, R.: Albert: a lite bert for self-supervised learning of language representations, September 2019
27. Sennrich, R., Haddow, B., Birch, A.: Neural machine translation of rare words with subword units. In: Proceedings of the 54th Annual Meeting of the Association for Computational Linguistics (Volume 1: Long Papers), pp. 1715–1725 (2016)

Social Network Analysis

Linking Friends in Social Networks Using HashTag Attributes

Olga Gerasimova$^{(\boxtimes)}$ and Viktoriia Syomochkina

HSE University, Moscow, Russia
ogerasimova@hse.ru

Abstract. Social networks are an integral part of modern life. They allow us to communicate online and exchange all kinds of information. In this paper, we consider the social network Instagram and its hashtags as a key tool for finding relevant information and new friends. The aim of our work is an empirical analysis of hashtags for posts in Instagram with certain locations. We obtain database of users of the Instagram network and collect a dataset of posts for three Far Eastern cities. Then, we build a friendship graph, for which we solve the link prediction problem. We show that both, structural and attributive graph information, such as hashtags, is important to achieve best quality.

Keywords: Social network · Link prediction · Hashtag

1 Introduction

In our research, we study hashtags, their popularity and benefit of using them for finding new friends. We have chosen Instagram[1] as one of the most popular hashtag-sharing platforms. This social network is an American service launched in 2010 for uploading photos and videos, which can be organized with hashtags and geotags. To browse content by different tags, users can follow information enthralling them, check current trends or find their needs.

1.1 Hashtags

Concept of hashtags was created on Twitter[2] in 2007. A goal was to retrieve information as efficient as possible. To evolve in 2011, Instagram also began using hashtags to help users find specific posts, popular peoples, advertised services, and so on. The next step of Instagram progress was a way to follow not only personal pages, but also hashtags, which allows to show relevant topics in user's feeds. Nowadays, almost each Instagram post contains multiple hashtags, see Fig. 1.

O. Gerasimova—The article was prepared within the framework of the HSE University Basic Research Program.

[1] https://www.instagram.com/.
[2] https://twitter.com.

© Springer Nature Switzerland AG 2021
W. M. P. van der Aalst et al. (Eds.): AIST 2020, LNCS 12602, pp. 269–281, 2021.
https://doi.org/10.1007/978-3-030-72610-2_20

Fig. 1. Example of an Instagram post with hashtags.

Moreover, Instagram stimulates peoples to create specific hashtags, rather than using generic words, to diversify tendencies and draw attention to posts. As we said, Instagram users can create "trends" based on hashtags. The trends usually highlight a specific topic for posting information on. For example, hashtags can promote an election campaign, raise social issues or reflect food preferences.

1.2 Main Contribution

Analyzing the conditions that cause people to subscribe to each other on social networks is a well-known task for developing recommendation systems. It is clear that people like to communicate with those who have similar interests. Hashtags can be considered as one of the forms of interests description. By hashtags, it is easy to search for posts with desired information, and, hence, for people who made these posts.

So, we are going to use hashtags for a link prediction task on friendship graphs based on data scraping from the Instagram.

The social network is represented by the graph $G(V, E)$, where V is a set of nodes corresponded people, and E is a set of edges - interactions between them. We obtained three friendship graphs for three Russian Far Eastern cities: Blagoveshchensk, Khabarovsk, Vladivostok. We are interested in comparison of users behavior in cities of different size, geographical, cultural and economical meanings in the same region.

We compared similarity-based and several machine learning methods for link prediction. As model features, we considered both models with only hashtags information and combined data (structural features and hashtags). We obtained that adding hashtags to structural features increases accuracy of prediction.

1.3 Organization

The structure of the paper is as follows. In Sect. 2, we consider works in related fields. Next, in Sect. 3, we describe methods of our datasets collection. Section 4 contains empirical analysis on hashtags. In Sect. 5, we concentrate on link prediction task on Instagram graphs using structural features and hashtag attributes. Finally, in Sect. 6 we make the conclusion and discuss future work.

2 Related Work

We made an overview of related papers in several directions. Firstly, we considered hashtag-based works. Secondly, we focused on researches related with Instagram. Finally, we mentioned papers about link prediction and the approaches that use social interests and network structure to recommend social relations.

2.1 Researches on Hashtags

In our research, we concentrate our attention on hashtags, which can be used as a powerful tool for different tasks.

There are many works about certain hashtags, when their usage creates a new tendency, supports a protest or motivates to do something.

In [15,23,24,26], some events became a motivation for investigations. The work [24] is inspired by a hostage incident in Sydney in December 2014 and is devoted to solidarity towards Muslims in Australia, in which author explored the nature of Twitter networks based on the hashtag #illridewithyou. Authors of [15] analyzed an online movement by women being sexually harassed and sharing their stories with the hashtag #metoo. Another research focused on elections to the constituent assembly in Venezuela and classified tweets with the #Maduro hashtag. The authors identified the explicit and implicit topics related to this important political event [23]. In [26], content analysis of tweet tagged with #rescue in the 2017 North Kyushu Heavy Rain disaster was investigated. Using only hashtags on Twitter, authors of [11] predicted the election results in India.

There are researches focusing on emotions' analysis. For example in [30], an image classifier to define Non-Suicidal Self-Injury (NSSI) or non-NSSI images was developed for posts, which contain a hashtag #selfharm. Another case is devoted to explore how users conceptualize trust for understanding ideas that help people to negotiate trust. The researchers sourced the Twitter data for the hashtag #trust to analyze the vocabulary used alongside the given hashtag.

It is also interesting to explore data in the entertainment sectors (movie, music, food, fitness and etc.). Authors of [13] chose to analyze the Twitter hashtag #avengersendgame in framework of sentiment analysis task. Their approach aimed to classify the posts with a given hashtag into positive and negative type. In [32], the aim of the study is to analyze users' emotional states in terms of their musical preferences. Authors collected datasets with #nowplaying tweets and retrieved affective context hashtags included in the same tweets. As for food analysis, in [33] there is a study on gender differences in terms of using hashtags for photos tagged with #Malaysianfood or in [20] the hashtags #fitness, #brag and #humblebrag are examined for exploring self-presentation strategies.

So, analyzing hashtags is a hot topic that promotes to appear many new researches related with them.

Different recommender systems based on hashtags' semantic analysis were constructed [9]. This research focuses on the classification of semantic words using a user's hashtag data and co-occurrence hashtag information. Understanding the meaning of a hashtag is one of the ways to learn latent semantic expressions of words. In [27], it was proposed a semi-supervised sentiment hashtag embedding model, which preserves both semantic and sentiment hashtags distribution.

2.2   Researches on Instagram

There are a lot of various researches around Instagram, because it is a enormous data source for a wide range of tasks.

In [22], Müngen et al. have found most effective Instagram posts, instead of users, focused on fuse motif analysis. Authors of [4] suggested method of matching of Instagram images and topics obtained aid topic modelling of hashtags. In [8], another way based on the HITS algorithm and the principles of collective intelligence for matching of Instagram hashtags and corresponding visual content of the images was presented. This method allowed to obtain noise-free training datasets for content-based image retrieval.

Also hashtags-based method for specific community detection problem was proposed in [5].

There are various important studies on brand mentioning practice of influencers [31], in which authors suggested a neural network-based model for post classification according to (non-)sponsorship parameter.

Among political related researches, it is clear there are also investigation in Instagram. For example in [3], Zahra Aminolroaya and Ali Katanforoush proposed novel ideas of hashtag diffusion for Iranian communities in Instagram during the last legislative election in Iran.

In travel segment, a model-based location recommender system for designing a location's profile helps to recommend locations based on user preferences[21].

To sum up, we can identify the following task that are solved on Instagram data: automatic image annotation [4,8], topic modelling [4], community detection [3,5], user/post/image classification [30,31], information diffusion modeling [3], sentiment analysis [13], recommender system [21], and others.

We focus on the structure of the user network that connect Instagramm users and their content as a core goal of our research.

2.3   Link Prediction

In network analysis, one of the main tasks is a link prediction [18], the objective of which is to predict the pairs of nodes that will be connected by the link or not in the next state of the network. Being important task, the link prediction has become significant in various fields. Hence, many different methodologies to solve it have been suggested such as similarity-based [2], maximum likelihood models [7], probabilistic models [17], or based on deep learning [6].

There are applications of link prediction in most domains from classical task of prediction social relationships [34], web linking [1] to developing various recommender systems [16,28].

Systematic surveys on link prediction methodologies were described in [12, 14,19,25,29].

3 Data Collection

3.1 Methods

To perform the link prediction task, we collected data from the Instagram social network based on the geotags of chosen locations.

After detailed study on extracting data from Instagram, the following solutions were proposed: (1) the Instagram API, (2) searching for posts by hashtags of given cities through a web page, (3) creating a bot.

As for the first method, there is a problem of missing important attribute information in the Instagram API, in particular, the location id attribute. This parameter has been removed since October 2019. The significant drawback of the second way is that it misses many posts without city hashtags, but tagged with the given locations. As a result, the third method was chosen, because it allows to solve our task on obtaining data.

To automatically collect posts, a data retrieval bot was written. The following technologies were chosen to implement the bot: Java, spring, springjpa, MySQL. The MySQL DBMS was used to store the database. Based on the dumped database, we obtained graphs in the form of edges lists and sets of users' hashtags.

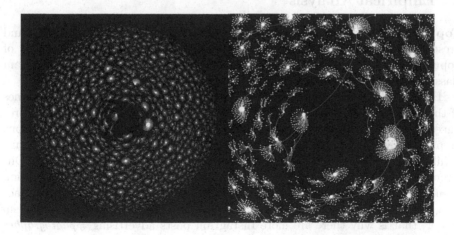

Fig. 2. Visualisation of the Blagoveshchensk graph.

3.2 Datasets

Graphs. We built three friendship graphs based on geotags for the following locations: Blagoveshchensk, Khabarovsk, Vladivostok. The Blagoveshchensk graph is illustrated via the Gephi[3] visualization platform, see the Fig. 2.

We collected posts with the given locations. Next, we identified users who created these posts and built friendship graphs. We meant that two users are friends if they follow each other. We removed users with posts not containing any hashtags. Table 1 gives a summary statistics of all the three datasets.

Hahstags Preprocessing. All hashtags were converted to lowercase. Also, we removed all the symbols except letters and numerals.

Table 1. Summary statistics of datasets.

	Blagoveshchensk	Khabarovsk	Vladivostok
Num. of users	15 262	36 807	51 430
Num. of links	14 119	34 118	47 523
Num. of hashtags	440 872	1 173 889	1 504 696
Num. of posts		2 695 509	

4 Empirical Analysis

Top Hashtags. We analyzed most popular hashtags in our three cities and presented our results in Table 2. We generalized and defined several topics of popular hashtags. Hashtags were translated from Russian except the Instagram class and were ordered by popularity from top to bottom within categories.

Because of the idea for data collection, many popular hashtags contain names of the cities. The current quarantine situation related with COVID-19 also produces many relevant tags. Also, general hashtags about seasons, celebrations or popular services are retrieved. We have noted several interesting patterns. While Vladivostok is a seaport, #sea is the most popular in this city. Different seasons preferences are identified. Among these cities, the celebration #9may is more often mentioned in Blagoveshchensk, possibly due to the fact that Blagoveshchensk is a border town. In addition, Vladivostok is also a tourism center, that is why there are more Instagram posts advertising #photographer services than #manicure, unlike other cities.

---

[3] https://gephi.org.

Table 2. Most popular hashtags in three datasets.

Topic	Blagoveshchensk	Khabarovsk	Vladivostok
Instagram	#instagood	#instakhv	#instagood
	#repost	#instagood	#repost
	#followme	#repost	#followme
General	#family	#love	#sea
	#love	#family	#love
	#beautiful	#beautiful	#beautiful
Locations	#blg	#khv	#vladivostok
	#blagoveshchensk	#khabarovsk	#vl
	#blaga	#khv27	#vdk
Seasons	#summer	#spring	#summer
	#spring	#summer	#spring
	#winter	#autumn	#autumn
Celebrations	#newyear	#newyear	#newyear
	#9may	#birthday	#birthday
	#birthday	#8march	#8march
Services	#manicureblagoveshchensk	#manicurekhabarovsk	#photographervladivostok
	#nailsblagoveshchensk	#manicure	#manicure
	#gelpolishblagoveshchensk	#photographerkhabarovsk	#manicurevladivostok
Quarantine	#stayhome	#stayhome	#quarantine
	#selfisolation	#quarantine	#stayhome
	#stayinghome	#selfisolation	#selfisolation

Distributions. We plot the distribution of the number of hashtags in each post for the united dataset of posts for three locations in Fig. 3 (left). There are around $\sim$ 50% posts included up to 10 hashtags. The upper limit for the number of hashtags per Instagram post is 30. So, we can see an increase of around 30, because there are many sponsored ad posts with the maximum number of tags. However, there are a few posts with exceeding limit of the number of hashtags.

Figure 3 (right) illustrates distributions of the number of users according to the number of shared hashtags for three graphs. It can be seen that most people use quite small set of hashtags. In the next Fig. 4, we plots the distributions of the number of times hashtag was shared in all posts, i.e. we analyze the frequency of using hashtags. There are a lot of specific hashtags, which are popular among narrow circle of users. Also, there are general widespread hashtags, but their number is much smaller. It is clear and logical that both distributions look like the power law.

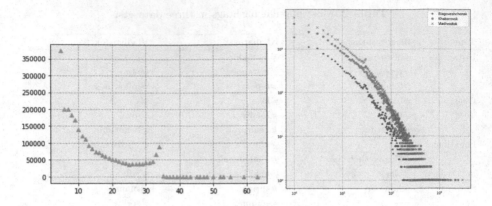

Fig. 3. Left: distribution of the number of hashtags in each post in all three datasets, x-axis corresponds to the number of hashtags and y-axis – number of posts. Right: distributions of the number of times a hashtag is used (loglog scale), x-axis corresponds to the number of users and y-axis – number of hashtags.

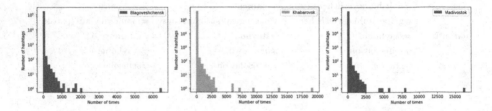

Fig. 4. Distributions of the number of times a hashtag is used (loglog scale).

5   Experiments

We consider the link prediction as a binary classification task: we predict unit for edges, which appear in the next state of the network, and zero for non-existent edges.

Training Settings and Metrics. To receive examples for the negative class, we applied the negative sampling strategy. We randomly choose non-existent edges such that the number of them would be approximately the same, as the number of existent edges for a balanced sample. Moreover, we aggregated results over five negative samplings. We split our data into training and test sets as 70% and 30%, respectively.

We use standard classification performance metrics for evaluating quality such as Accuracy (Acc.), F1-score (micro, macro), Log-Loss, ROC-AUC. The smaller Log-Loss is better, whereas the greater other metric is better.

Hashtag Vectorization. The bag of words method was used to vectorize information about users' hashtags. We had more than 160, 350 and 424 thousand different hashtags for Blagoveshchensk, Khabarovsk, Vladivostok locations, respectively. According to the chosen vectorization method, the vector dimension describing the user's hashtags should be equal to the number of different hashtags. It is reasonable to reduce the dimension choosing hashtags that were shared more than once, in other words, by more than one user. The resulting dimensions are approximately 40, 93 and 112 thousand hashtags, respectively. We can see that there are a lot of hashtags used by only one user in our datasets.

Features. In our problem statement, each user is characterized by a set of hashtags and connections with other users. So, we used two types of features: binary vectors describing used hashtags and binary vectors corresponding to a set of friends (rows from the graph adjacency matrix).

5.1 Similarity-Based Models

As a basic model for link prediction, we used Cosine Similarity (CS) and Jaccard Index (JI) metrics.

$$CS(v_1, v_2) = \frac{v_1 \cdot v_2}{\|v_1\| \|v_2\|} \quad JI(v_1, v_2) = \frac{|v_1 \cap v_2|}{|v_1 \cup v_2|}$$

These metrics were calculated for the test set containing existent edges and non-connected pairs of nodes from negative sampling set. The threshold for prediction was obtained on training step. We used a grid search from 0 to 1 with a step of 0.0001. We predicted existence of edge if metric value was greater than corresponding threshold chosen for each metric.

We made three experiments to calculate the similarity metrics with different feature spaces: vectors with links data only, vectorized hashtags only, and combined users' features as concatenated binary vectors with vectorized hashtags and information about friends. We got a statistically significant increase in accuracy by ~ 2–4 % adding information about hashtags to graph information, as shown in Table 3. We can see that prediction accuracy using only hashtags is greater than using graph information, which support our hypothesis that users of Instagram are linked by their interests. However, this approach did not show sufficiently high accuracy, that is why, we also looked at more advanced techniques using machine learning models.

Table 3. Accuracy on test data for similarity-based models.

	Model	Blagoveshchensk	Khabarovsk	Vladivostok
JI	Friends	0.513	0.5	0.515
	Tags	0.537	**0.534**	0.54
	Friends+tags	**0.538**	0.533	**0.542**
CS	Friends	0.499	0.5	0.5
	Tags	0.537	0.536	0.538
	Friends+tags	**0.538**	**0.539**	**0.54**

5.2 Machine Learning Models

We considered machine learning (ML) models for binary classification task such as Logistic Regression (LogReg) and Extreme Gradient Boosting (XGB).

As model features, we decided to use node2vec network embeddings [10] with random walks parameters $p, q = (1, 1)$, dimension of the embedding $d = 64$, length of walks $l = 30$, and number of walks per node equaled $n = 200$. Also, we made experiments with combining node2vec embeddings and vectorized hashtags.

Edge Functions. To receive edge feature, we applied specific component-wise functions to node features for source u and target v nodes of a given edge. This model was suggested in [10], in which four functions for such edge embeddings were presented:

$$\text{Average:} \quad \frac{u+v}{2} \qquad \text{Weighted} L_1: \quad |u-v|$$
$$\text{Hadamard:} \quad u \cdot v \qquad \text{Weighted} L_2: \quad (u-v)^2$$

In Table 4 and Table 5, we presented values of quality metrics on test data for some experiments on the giant connected component (GCC) of the Blagoveshchensk graph. All results are significantly better than for similarity-based models.

Table 4. Comparison of ML models on GCC of the Blagoveshchensk graph, part 1.

		Average					Hadamard				
		Acc.	F1-micro	F1-macro	Log-loss	ROC-AUC	Acc.	F1-micro	F1-macro	Log-loss	ROC-AUC
node2vec	LogReg	0.745	0.745	0.735	8.82	0.745	**0.964**	**0.964**	**0.964**	1.26	**0.964**
	XGB	0.953	0.953	0.953	1.64	0.953	0.953	0.953	0.953	1.64	0.953
node2vec+#	LogReg	0.803	0.803	0.799	6.8	0.803	**0.964**	**0.964**	**0.964**	1.26	**0.964**
	XGB	**0.956**	**0.956**	**0.956**	**1.51**	**0.956**	0.953	0.953	0.953	1.64	0.953

Table 5. Comparison of ML models on GCC of the Blagoveshchensk graph, part 2.

		WeightedL_1					WeightedL_2				
		Acc.	F1-micro	F1-micro	Log-loss	ROC-AUC	Acc.	F1-micro	F1-micro	Log-loss	ROC-AUC
node2vec	LogReg	0.839	0.839	0.837	5.55	0.839	0.836	0.836	0.834	5.67	0.836
	XGB	0.843	0.843	0.839	5.42	0.843	0.843	0.843	0.839	5.42	0.843
node2vec+#	LogReg	**0.964**	**0.964**	**0.964**	**1.26**	**0.964**	**0.887**	**0.887**	**0.886**	**3.9**	**0.887**
	XGB	0.956	0.956	0.956	1.51	0.956	0.858	0.858	0.855	4.91	0.858

For features with node2vec embeddings only, XGB showed better quality in terms of all metrics for Average, WeigtedL_1 and WeigtedL_2 edge functions than LogReg. However, the situation is opposite for the Hadamard edge function. If we add features with vectorized hashtags, we obtain better results for LogReg with all edge functions except Average. Generally, combined features lead to higher quality, which supports our claim and core idea of the paper.

6 Conclusion and Discussion

In the framework of this paper, we made a survey of different works on hashtags to show the motivation for hashtag-based researches, to present hashtags as a powerful tool for describing the preferences of social media users, influencing public opinion and disseminating information. While we focused on the Instagram, we also decided to describe a wide range of research opportunities based on this social network.

For our study, we created a database based on Instagram posts tagging with chosen locations. Constructing Instagram friendship graphs, we solved link prediction task using different feature types. In addition, we made general empirical analysis of hashtags to be aware of their usage trends at Russian Far Eastern cities.

For the future work, we aim to study another methods of graph embedding, which can contribute to link prediction problem from both, hashtag similarity and structural information feature engineering. We are also going to look at additional features related to social profiles of Instagram users for our problem of friend recommendations.

References

1. Adafre, S.F., de Rijke, M.: Discovering missing links in wikipedia. In: Proceedings of the 3rd International Workshop on Link Discovery, pp. 90–97 (2005)
2. Adamic, L.A., Adar, E.: Friends and neighbors on the web. Soc. Netw. **25**(3), 211–230 (2003)
3. Aminolroaya, Z., Katanforoush, A.: How Iranian Instagram users act for parliament election campaign? a study based on followee network. In: 2017 3th International Conference on Web Research (ICWR), pp. 1–6. IEEE (2017)

4. Argyrou, A., Giannoulakis, S., Tsapatsoulis, N.: Topic modelling on Instagram hashtags: An alternative way to automatic image annotation? In: 2018 13th International Workshop on Semantic and Social Media Adaptation and Personalization (SMAP), pp. 61–67. IEEE (2018)
5. Bejandi, S.A., Katanforoush, A.: How unseen communities of Instagram users are revealed using the real-valued collocations of hashtags. In: 2017 IEEE 4th International Conference on Knowledge-Based Engineering and Innovation (KBEI), pp. 0487–0491. IEEE (2017)
6. Berg, R.V.D., Kipf, T.N., Welling, M.: Graph convolutional matrix completion. arXiv preprint arXiv:1706.02263 (2017)
7. Clauset, A., Moore, C., Newman, M.E.: Hierarchical structure and the prediction of missing links in networks. Nature **453**(7191), 98 (2008)
8. Giannoulakis, S., Tsapatsoulis, N.: Filtering Instagram hashtags through crowd tagging and the hits algorithm. IEEE Trans. Comput. Soc. Syst. **6**(3), 592–603 (2019)
9. Gorrab, A., Kboubi, F., Ghezala, H.B., Le Grand, B.: New hashtags' weighting schemes for hashtag and user recommendation on twitter. In: 2017 IEEE/ACS 14th International Conference on Computer Systems and Applications (AICCSA), pp. 564–570. IEEE (2017)
10. Grover, A., Leskovec, J.: Node2vec: Scalable feature learning for networks. In: Proceedings of the 22Nd ACM SIGKDD International Conference on Knowledge Discovery and Data Mining, KDD 2016, pp. 855–864. ACM, New York, NY, USA (2016)
11. Gupta, S., Singh, A.K., Buduru, A.B., Kumaraguru, P.: Hashtags are (not) judgemental: The untold story of lok sabha elections 2019. arXiv preprint arXiv:1909.07151 (2019)
12. Haghani, S., Keyvanpour, M.R.: A systemic analysis of link prediction in social network. Artif. Intell. Rev. **52**(3), 1961–1995 (2017). https://doi.org/10.1007/s10462-017-9590-2
13. Handayanto, R.T., Setiyadi, D., Retnoningsih, E., et al.: Corpus usage for sentiment analysis of a hashtag twitter. In: 2019 Fourth International Conference on Informatics and Computing (ICIC), pp. 1–5. IEEE (2019)
14. Hasan, M.A., Zaki, M.J.: A survey of link prediction in social networks. In: Aggarwal C. (eds.) Social Network Data Analytics, pp. 243–275. Springer, Boston (2011) https://doi.org/10.1007/978-1-4419-8462-3_9
15. Hassan, N., Mandal, M.K., Bhuiyan, M., Moitra, A., Ahmed, S.I.: Can women break the glass ceiling?: An analysis of #metoo hashtagged posts on twitter. In: 2019 IEEE/ACM International Conference on Advances in Social Networks Analysis and Mining (ASONAM), pp. 653–656. IEEE (2019)
16. He, Q., Pei, J., Kifer, D., Mitra, P., Giles, L.: Context-aware citation recommendation. In: Proceedings of the 19th International Conference on World Wide Web, pp. 421–430. WWW 2010. ACM, New York, NY, USA (2010)
17. Heckerman, D., Meek, C., Koller, D.: Probabilistic entity-relationship models, PRMs, and plate models. Introduction to statistical relational learning, pp. 201–238 (2007)
18. Liben-Nowell, D., Kleinberg, J.: The link-prediction problem for social networks. J. Assoc. Inf. Sci. Technol. **58**(7), 1019–1031 (2007)
19. Lü, L., Zhou, T.: Link prediction in complex networks: a survey. Phys. Stat. Mech. Appl. **390**(6), 1150–1170 (2011)

20. Matley, D.: This is not a #humblebrag, this is just a #brag: the pragmatics of self-praise, hashtags and politeness in instagram posts. Discourse Context Media **22**, 30–38 (2018)
21. Memarzadeh, M., Kamandi, A.: Model-based location recommender system using geotagged photos on Instagram. In: 2020 6th International Conference on Web Research (ICWR), pp. 203–208. IEEE (2020)
22. Müngen, A.A., Kaya, M.: Quad motif-based influence analyse of posts in Instagram. In: 2017 2nd International Conference on Advanced Information and Communication Technologies (AICT), pp. 51–55. IEEE (2017)
23. Niklander, S.: Content analysis on social networks: Exploring the #maduro hashtag. In: 2017 International Conference on Computing Networking and Informatics (ICCNI), pp. 1–4. IEEE (2017)
24. Rathnayake, C., Suthers, D.D.: Networked solidarity: An exploratory network perspective on twitter activity related to #illridewithyou. In: 2016 49th Hawaii International Conference on System Sciences (HICSS), pp. 2058–2067. IEEE (2016)
25. Raut, P., Khandelwal, H., Vyas, G.: A comparative study of classification algorithms for link prediction. In: 2020 2nd International Conference on Innovative Mechanisms for Industry Applications (ICIMIA), pp. 479–483 (2020)
26. Sato, S.: Analysis of tweets hashtagged "#rescue" in the 2017 north Kyushu heavy rain disaster in japan. In: 2018 5th International Conference on Information and Communication Technologies for Disaster Management (ICT-DM), pp. 1–7. IEEE (2018)
27. Singh, L.G., Anil, A., Singh, S.R.: She: sentiment hashtag embedding through multitask learning. IEEE Trans. Comput. Soc. Syst. **7**(2), 417–424 (2020)
28. Su, Z., Zheng, X., Ai, J., Shang, L., Shen, Y.: Link prediction in recommender systems with confidence measures. Chaos Interdisc. J. Nonlinear Sci. **29**(8), 083133 (2019)
29. Wang, P., Xu, B.W., Wu, Y., Zhou, X.: Science China Information Sciences **58**(1), 1–38 (2014). https://doi.org/10.1007/s11432-014-5237-y
30. Xian, L., Vickers, S.D., Giordano, A.L., Lee, J., Kim, I.K., Ramaswamy, L.: #selfharm on Instagram: quantitative analysis and classification of non-suicidal selfinjury. In: 2019 IEEE First International Conference on Cognitive Machine Intelligence (CogMI), pp. 61–70. IEEE (2019)
31. Yang, X., Kim, S., Sun, Y.: How do influencers mention brands in social media? sponsorship prediction of Instagram posts. In: Proceedings of the 2019 IEEE/ACM International Conference on Advances in Social Networks Analysis and Mining, pp. 101–104 (2019)
32. Zangerle, E., Chen, C.M., Tsai, M.F., Yang, Y.H.: Leveraging affective hashtags for ranking music recommendations. IEEE Trans. Affect. Comput. **12**, 78–91 (2018)
33. Zhang, Y., Baghirov, F., Hashim, H., Murphy, J.: Gender and Instagram hashtags: A study of #malaysianfood. In: Conference on Information and Communication Technologies in Tourism (2016)
34. Zhang, Y.: Language in our time: an empirical analysis of hashtags. In: The World Wide Web Conference, pp. 2378–2389 (2019)

Emotional Analysis of Russian Texts Using Emojis in Social Networks

Anatoliy Surikov[✉][iD] and Evgeniia Egorova[iD]

ITMO University, St. Petersburg, Russia

Abstract. Sentiment analysis is a key task in natural language processing and has a wide range of real-world applications. Traditional methods classify "plain texts" as "positive" and "negative". We propose a method that is significantly different from traditional approaches. In addition to "plain text", we have analyzed ways to express emotions in a message and a variety of emotional indicators, emoticons and emojis. The proposed model with emotional indicators to predict text polarity improves the prediction accuracy of sentiment classes by 6% compared to traditional models. The model uses an original data set marked up according to an extended list of Plutchik emotion classes. Therefore, the model predicts 8 independent sentiment classes: "joy", "sad", "distaste", "fear", "anger", "surprise", "attention" and "trust". The novelty of the research also lies in using the Russian language data set. The model considers national linguistic, semantic and semiotic features of social environment.

Keywords: Emotional analysis · Social networks · Natural language processing

1 Introduction

Digital footprint analysis using machine learning methods in social networks is a compelling and knowledge-based task. To analyze text digital footprints, the required size of labeled data set is collected, and machine learning methods are applied. However, insufficient data make digital footprint analysis impossible. The VKontakte, a Russian-language social network lacks the collected open labeled data sets. For this research, in particular, a large dataset of 6340 short text messages was marked manually, which took a lot of time and effort.

The research aims to build a model to detect emotions in Russian-language texts. The other task of text sentiment analysis is to form and choose prediction classes. A combination of several approaches is used to predict emotions in texts. Therefore, various emotions scales that exist in psychology were considered. As a result, Plutchik's scale of emotions [12] was selected as a standard as a formally described in scientific sources.

The rest of the paper is organized as follows. In Sect. 2, we overview the papers related. In Sect. 3, we describe our data for this study. The proposed

Supported by The Russian Science Foundation.

W. M. P. van der Aalst et al. (Eds.): AIST 2020, LNCS 12602, pp. 282–293, 2021.
https://doi.org/10.1007/978-3-030-72610-2_21

method is described in Sect. 4. Section 5 contains all results of the emotional analysis research of the suggested model and a comparison of the suggested model with the baseline model. The conclusion is in Sect. 6.

2   Related Work

Text sentiment analysis in [3, 4, 7] focuses on emotion classes rather than multi-dimensional emotions. In our research, a sentiment data set was collected and marked up into 8 brightly expressed emotions in each message. Emotional indicators (emoticons) for the advanced sentiment analysis were considered.

The issue of emotion prediction using supervised machine learning methods with SNoW architecture [2] is explored in [1]. The paper aims to classify the emotional closeness of sentences in fairy tales for appropriate expressive reproduction of text-to-speech synthesis. Initial experiments on a data set of 22 fairy tales show encouraging results using the bag-of-words [3] approach to classify emotional and non-emotional content with some dependence on parameter settings.

Habibeh Naderi et al. [4] proposed a regression system to infer emotion intensity of a tweet. They developed a multi-aspect feature learning mechanism to capture the most discriminative semantic features of a tweet, as well as the information on emotions conveyed in every word. Their method offers several types of feature groups: a tweet representation learned by an LSTM network [5], word embeddings [6] word and character n-grams, features derived from various sentiment and emotion lexicons, and hand-crafted features. An SVR regressor [6] is then trained over the full set of features. Performance assessment of ensemble feature sets on data sets reach a Pearson correlation of 72% on tweet emotion intensity prediction.

A method for classifying tweets into emotional categories based on the NRC vocabulary [8] was proposed in [7]. The authors conducted an experiment using two approaches: (1) The word-centroid model [10] creates word vectors from tweet-level attributes (for example, unigrams and Brown clusters) by averaging all tweets with a target word, (2) word embedding represents continuous, low-dimensional word vectors trained on a document basis. The results show that the expanded lexicon achieves major improvements over the original lexicon when classifying tweets into emotional categories.

Robert Plutchik, a psychologist, created an emotion scale [12] with 8 basic emotions: "joy", "trust", "fear", "surprise", "sad", "attention", "anger", and "distaste". Here, joy is opposite to sadness, fear is opposite to anger, distaste is ortagonal to trust, and anticipation opposes surprise. Plutchik's emotion scale is the most referenced in scientific research, for example in NRC lexicon [4].

Various problems and solutions related to sentiment analysis are summarized in [9]. The key problems for automatic systems are described, as well as the algorithms, functions and data sets used in sentiment analysis. Several manual and automatic approaches to create a vocabulary related to valency and emotions are described. Sentiment analysis is described at the sentence level.

3   Experiments

3.1   Data

The RuSentiment corpus [17] was used for our research. The RuSentiment is a data set of posts from a social network Vkontakte, and these posts had already been marked in five classes "positive", "neutral", "negative", "speech", and "skip". For our study we used posts with class marks "positive" and "negative", 10558 posts in total, including:

– with emotional indicators: 6957
– without emotional indicators: 3598

Emojis, emoticons and punctuation marks that express emotions are referenced as emotional indicators in our research. The data set of emotional indicators, their code, values, meanings in English, and sentiment assessments, and emotion classes was used, including:

– Emoji: 277
– Emoticons: 125
– Punctuation marks that express emotions: 20

This RuSentiment corpus was marked up at 8 emotions classes ("joy", "sad", "surprise", "distaste", "anger", "trust", "attention", "fear") described in Table 1. Experts have manually marked up 6340 posts. Data collection on minority classes is ongoing. The article have presented intermediate results. The data set was annotated by 4 experts manually. The annotation was performed following specific rules for experts. Any emotion from Plutchik's scale that could be highlighted by experts in a short text was marked with positive markers. Not highlighted emotions were marked with negative markers. Most often (in 77% of cases), the experts distinguished from one to three prevailing emotions in a post. In a relatively small number of texts (9%), experts did not select any pronounced emotional posts. The binary markup was averaged and used to generate the final data set. In most cases (91%), an expert can easily identify dominant emotions using the Plutchik's emotion scale in a short text from a social network.

Data prepossessing consist of several steps:

1. Remove posts without emotional indicators.
2. Convert posts to lowercase.
3. Remove posts containing less than ten words.
4. Remove contents including the urls, mentions, and emails.
5. Remove punctuation marks.
6. Remove words that are not in Russian.
7. Change emotional indicators to their text meaning.
8. Lemmatizing posts using Yandex Mystem [18].

Table 1. Description of emotions.

Emotion	Description
Joy	enjoyment, happiness, relief, bliss, delight, pride, thrill, ecstasy
Sad	grief, sorrow, gloom, melancholy, despair, loneliness, depression
Trust	friendliness, trust, kindness, affection, love, devotion
Distaste	contempt, disdain, scorn, aversion, distaste, revulsion
Fear	anxiety, apprehension, nervousness, dread, fright panic
Anger	fury, outrage, wrath, irritability, hostility, resentment, violence
Surprise	shock, astonishment, amazement, astound, wonder
Attention	acceptance, care, attentiveness, courtesy

3.2   Data Analysis

The ratio of emotion classes to all posts in the data set was analyzed after data
prepossessing stage (Fig. 1 and Fig. 2). Positive emotions prevailed over negative
emotions in posts, and "fear" appears to be least often. Most often, emotional
indicators were used to express emotions of joy.

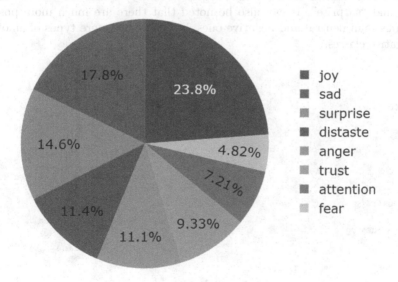

Fig. 1. The ratio of emotions in the data set.

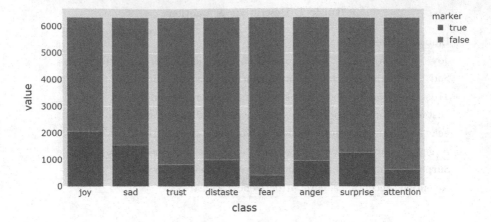

Fig. 2. Emotions Distributions in the data set.

Emotions have a strong relation with sentiment classes (Fig. 3). For example, the positive class (above the diagonal) consists of such emotions as "joy" and "trust". The negative class (below the diagonal) comprises "sad", "distaste" and "anger". The neutral class (in the diagonal vicinity) consist of "fear", "attention", and "surprise". It can also be noted that there are much more positive emotions than neutral and negative ones, but there are more types of emotions in negative classes.

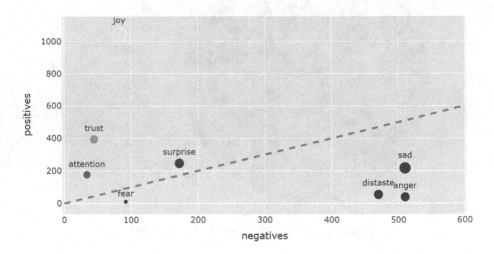

Fig. 3. Location of emotions in the space of sentiment classes.

The distribution of emotional indicators in posts according to emotion classes allocated for these posts are demonstrated in Fig. 4. The presence of an emotional indicator in a particular class means that in more than 50% of posts that have this indicator, this class of emotion is highlighted. It is clearly seen that people are more willing to use emotional indicators, expressing joy and sad, and such an emotional class as 'attention' has not received a single emotional indicator.

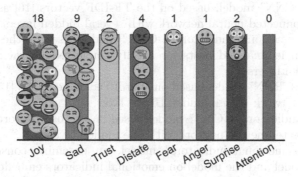

Fig. 4. Distribution of emotional indicators in posts according to the classes of emotion.

A positive correlation between "distaste" and "anger" is demonstrated on the graph (Fig. 5), A negative correlation is also noted among "joy", "sad", "anger", and "distaste". Therefore, this means that negative emotions oppose positive emotions.

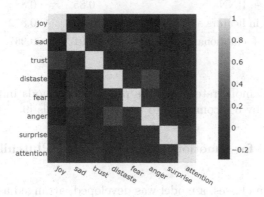

Fig. 5. Correlation matrix of emotions.

4   Methods

4.1   Alternative Method of Sentiment Analysis Using Emojis and Emoticons

The Alternative method of sentiment analysis using emojis and emoticons was applied to analyze not only "plain text", but also to study how important emotion indicators in sentiment analysis. The final result is presented in Table 2. The study was focused on a set of posts from the social network Vkontakte. The models were built to predict two independent classes: "negative" and "positive". Five models were trained and their metrics on a test data set were calculated:

1. "TF-IDF + CNN" model, based on the TF-IDF vectors [16] as embeddings for a fully connected neural network with several hidden layers.
2. "TD-IDF + emotional indicators + CNN" model based on the TF-IDF vectors built on lemmatized posts for a fully connected neural network with several hidden layers.
3. "Word2Vec + RNN" model based on Word2Vec embeddings [15] for a recurrent neural network with a hidden LTSM layer.
4. "emotional indicators + CNN" model based on the bag-of-words vectors for a fully connected neural network with one hidden layer.
5. "Word2Vec+emotional indicators" model, an ensemble consisting of the word2vec model and the model on emotional indicators embeddings.

Table 2. Text polarity prediction accuracy.

Model	Accuracy	Roc curve area
TF-IDF + CNN	0.83	0.863
TD-IDF + emotional indicators + CNN	0.89	0.928
Word2Vec + RNN	0.85	0.884
Emotional indicators + CNN	0.85	0.872
Word2Vec + emotional indicators	0.91	0.937

The research demonstrated that the proposed methods improve prediction of polarity classes by 6% compared to traditional models [9].

4.2 The Model for Emotional Analysis Using Plutchik's Emotions Scale

To predict emotion classes, a model was developed, arranged as follows:

1. First, lemmatized texts with highlighted emotional indicators were vectorized using the classic TF * IDF approach (emotional indicators are included in the dictionary along with ordinary lemmas)
2. Embeddings, constructed this way, act as predictors at the input of a fully connected neural network, trained to classify them into 8 independent classes corresponding to emotions on Plutchik's scale.

The target model pipeline is demonstrated (see Fig. 6). As a result, a vector of 8 values is obtained, which demonstrates the probability of a given text message to belong to emotional classes.

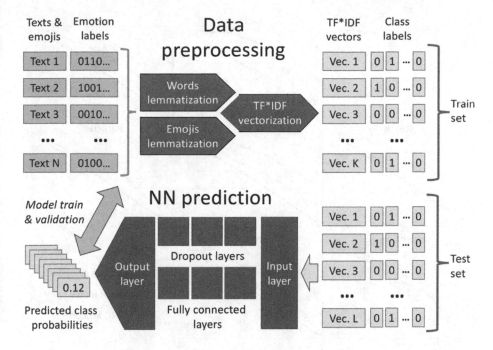

Fig. 6. Target model pipeline.

5 Results of Experiments

The test set included 1268 posts and the training set was 5072 posts. Validation results are shown in Table 3. As expected, due to strong imbalance of emotion classes, accuracy is considered to be inefficient metric to evaluate this model. More reasonable metrics are the area under the ROC curve and the area under the Precision-Recall curve (see Fig. 7), which indicate that the model shows a satisfactory quality of prediction in three emotional classes: "joy", "sad", and "fear".

Table 3. Results of model validation for all emotional classes.

Class of emotion	Roc curve area	Precision–recall area
Joy	0.83	0.68
Sad	0.79	0.56
Trust	0.68	0.25
Distaste	0.68	0.25
Fear	0.8	0.45
Anger	0.67	0.32
Surprise	0.63	0.27
Attention	0.59	0.13

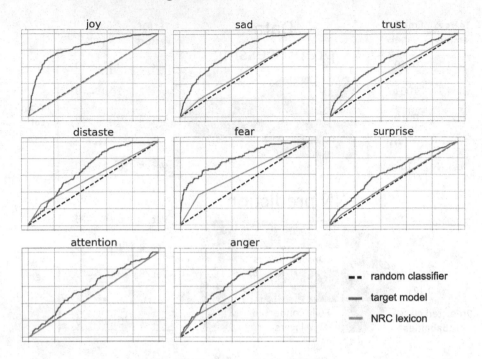

Fig. 7. Roc curves for 8 emotions classes.

The model was trained and validated for 8 independent classes corresponding to emotions on Plutchik's scale. The model is a fully connected neural network with several hidden layers, which also include dropout layers. The TF-IDF vectors were used as embeddings, that are based on texts from the original data set.

As a baseline was model based on the NRC lexicon [8]. The NRC Emotion Lexicon is a list of English words and their associations with eight basic emotions on the Plutchik's emotion scale and two sentiments (negative and positive). The NRC Emotion Lexicon was translated in Russian by machine translation. The vocabulary consist 11641 lemmas from NRC lexicon. The model was built on this vocabulary. As emotional classes in the text, we used the averaging of the vectors of emotional classes included in the lemmas.

It should be noted that the emotional indicators available in the texts (emoticons and emoji) were processed in a special way. Emotional indicators were lemmatized and considered by the TF-IDF model as lemmas. For this purpose, a lemmatizer of emotional indicators was developed and normalized. Thus, various emoticons and emojis with the same semantic content are reduced to one final lemma.

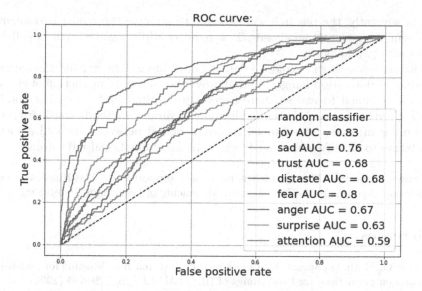

Fig. 8. ROC model for emotions classes.

The model demonstrates that not all classes of emotions are predicted equally well. For example, the "joy", "sad", and "fear" classes are predicted steadily, showing about 80% ROC-AUC (Fig. 8), and the "attention" class is the worst predicted (less than 60% ROC-AUC). This is indirectly related to the data set features: for the "joy" and "sad" classes, it contains 1977 and 1324 markers of the selected class. It gives a balanced sample that does not need special balancing. There are few positive markers for other classes of emotions. Therefore, artificial oversampling to balance classes was performed. Oversampling in data analysis is a technique used, on the one hand, to adjust the class distribution of a data set. On the other hand, it creates a situation of "information hunger" for teaching the neural network to predict these classes.

6 Conclusion and Future Work

The article provides an insight into emotional analysis. The Plutchik's emotion scale to predict emotions in short texts from social networks was applied. The original method of accounting for emotional indicators was described. Moreover, the model was developed that considered the linguistic, semantic and semiotic features of the Russian language texts.

The model developed in this research can be used to compile a variety of model ensembles for social network analysis. The model presented can also be applied to consider emotional dominants of a user's behavior in social networks and make predictions about their character.

Consequently, the research will continue to improve the model performance for 5 remaining emotional classes. This requires additional research in collaboration with psychologists.

Further research will explore a data set of training data, expand the data set to reduce "information hunger" and improve the quality of prediction of classes. It is also planned to attract more experts to markup posts. Another notable area for further research is the application of various methods considering the word order in texts, for example, recurrent neural networks, as well as other approaches to classify lemmatized texts containing emotional indicators.

Acknowledgments. This research is financially supported by The Russian Science Foundation, Agreement 17-71-30029 with cofinancing of Bank Saint Petersburg.

References

1. Alm, O., Roth, D., Sproat, R.: Emotions from text: machine learning for text-based emotion prediction. In: Proceedings of HLT/EMNLP, pp. 579–586 (2005)
2. Roth, D.: The SNoW learning architecture. Technical report UIUCDCS-R-99-2101, UIUC Science Department (May 1999)
3. Chen, M., Weinberger, K.Q., Sha, F.: An alternative text representation to TF-IDF and bag-of-words. In: Proceedings of the Conference on Information and Knowledge Management (CIKM 2012) (2012)
4. Naderi, H., Soleimani, B.H., Mohammad, S., Kiritchenko, S., Matwin, S.: Deep-Miner at semEval-2018 task 1: emotion intensity recognition using deep representation learning. In: Proceedings of the 12th International Workshop on Semantic Evaluation, pp. 305–312. Association for Computational Linguistics (ACL) (2018)
5. Hochreiter, S., Schmidhuber, J.: Long short-term memory. Neural Comput. **9**(8), 1735–1780 (1997)
6. Basak, D., Pal, S., Patranabis, D.C.: Support vector regression. Neural Inf. Process. Lett. Rev. **11**(10), 203–224 (2007)
7. Bravo-Marquez, F., Frank, E., Mohammad, S.M., Pfahringer, B.: Determining word-emotion associations from tweets by multi-label classification. In: Proceedings of the 2016 IEEE/WIC/ACM International Conference on Web Intelligence. Omaha, NE, USA, pp. 536–539 (2016)
8. Mohammad, S.M., Turney, P.D.: C emotion lexicon (NRC Technical Report) (2013)
9. Mohammad, S.M., Bravo-Marquez, F.: Emotion intensities in tweets. In: Proceedings of the Sixth Joint Conference on Lexical and Computational Semantics (*Sem). Vancouver, Canada (2017)
10. Radev, D.R., Jing, H., Stys, H., Tam, D.: Centroid-based summarization of multiple documents. Inf. Process. Manage. **40**(6), 919–938 (2004)
11. VK API method. https://vk.com/dev/methods. Accessed 4 June 2020
12. Plutchik, R.: Emotion: A Psychoevolutionary Synthesis. Harper and Row, New York (1980)
13. Segalovich, I.: A fast morphological algorithm with unknown word guessing induced by a dictionary for a web search engine. In: MLMTA, pp. 273–280 (2003)
14. Pelevina, M., Arefyev, N., Biemann, C., Panchenko, A.: Making sense of word embeddings. In: Proceedings of the 1st Workshop on Representation Learning for NLP, pp. 174–183. Association for Computational Linguistics (2017)

15. Maas, A.L., Daly, A.E., Pham, P.T., Huang, D., Ng, A.Y., Potts, C.: Learning word vectors for sentiment analysis. In: Proceedings of the 49th Annual Meeting of the Association for Computational Linguistics: Human Language Technologies, vol. 1, pp. 142–150. Association for Computational Linguistics (2011)
16. Ramos, J.: Using TF-IDF to determine word relevance in document queries. In: ICML (2003)
17. Rogers, A., Romanov, A., Rumshisky, A., Volkova, S., Gronas, M., Gribov, A.: RuSentiment: an enriched sentiment analysis dataset for social media in Russian. In: Proceedings of the 27th International Conference on Computational Linguistics (COLING 2018), Santa Fe, NM, USA, 20–26 Aug 2018, pp. 755–763 (2018)
18. Yandex MyStem. https://yandex.ru/dev/mystem/. Accessed 4 June 2020

Community Detection Based on the Nodes Role in a Network: The Telegram Platform Case

Kseniia Tikhomirova[(✉)] and Ilya Makarov[(✉)]

HSE University, Moscow, Russia
katikhomirova@edu.hse.ru, iamakarov@hse.ru

Abstract. The paper studies the community detection problem on Telegram channels. The dataset is received from TGStat service and includes the information of 58k forwards between 100 politician Telegram channels. We implement modern clustering approaches to solve the problem of missing social links. Our study is based on a combination of structural features with strategy-based attributes, including indicators designed according to the nodes' role in a network. Authors provide ten novel indicators, which are calculated for each network's member per each message in order to vectorize a Telegram channel with regard to its strategy of information spread and the way of contacting other channels. Authors construct a metric-based graph of channel relations and cluster channels representations using network science techniques. Obtained results are studied using quantitative and qualitative analysis showing promising results in applying joint network-based and KPI-based models for the stated problem.

Keywords: Gatekeeping theory · Clustering media · Community detection · Media embedding

1 Introduction

Telegram channels were introduced in September 2015 as a tool for "broadcasting" inside the Telegram platform. The mission of the platform is to ensure freedom of speech, which is close to almost all technology companies nowadays [10]. Anonymity and plenty of voices make it challenging and essential to analyze social connections between Telegram channels, zones of their influence and similarity.

The communication processes among Telegram channels differ from processes in other social media: Facebook, Twitter, and YouTube. Channels do not include data of their audience: connections between them and information flow. The lack of information leads to the problem of community detection. To solve it, we rely on gatekeeping theory, filter bubbles and introduce the new approach based on them.

© Springer Nature Switzerland AG 2021
W. M. P. van der Aalst et al. (Eds.): AIST 2020, LNCS 12602, pp. 294–302, 2021.
https://doi.org/10.1007/978-3-030-72610-2_22

2   Related Work

The work concerns to a communication network that represents people's relationships in terms of communication processes. It has a horizontal structure, and all related network members are equal participants in communication [2]. They act simultaneously as sources and recipients of messages. In the case of Telegram, we recognize three possible roles of nodes in a network: an author, a distributor, and a reader.

An author creates a message or evidence, which is a start point for a new communication process or a new level of a previous one. A distributor forwards messages making other people see it. Finally, readers are the endpoints of spreading. They receive a message and do nothing with it among the current network. With the current network, we suppose the Telegram platform.

To make it simple, we describe these roles as levels. Level A is for authors, level B is for distributors, and level C is for readers. They are available for any node of a network in any combination. Levels lie in one plane and are not a real hierarchy. For instance, an author can be connected with a reader directly without a distributor. Moreover, the same node can be on different levels depends on the situation.

The problem we face here is that Telegram does not provide researchers with information of the readers' level. This data is useful for structure analysis in the case of social media [1].

2.1   Levels of Gates

Being on some level means to have different access to information and power. The closeness to content production gives more information and actions to do. First of all, it applies to the level of distributors. In our case, they are public channels that forward other ones. In this situation, they become not only distributors but gatekeepers as well.

Communicators are not equal in gatekeeping. Through the reduction of official gatekeepers, accountability, professionalism, and expert information decline [3]. Cultural gatekeepers are necessary to help people filter through what is reliable and vital [5]. These gatekeepers can influence some amount of people. After that, they can influence further ones. This way, levels of gates are created.

2.2   Filter Bubbles as Clusters

Social homophily can be a substantial contributing factor to the emergence of filter bubbles and echo chambers and, consequently, group polarization effects [9]. That is why gatekeepers and information asymmetry create filter bubbles. We suggest that the structure of Telegram channels and their audience can be defined as plenty of filter bubbles: found gates or borders of bubbles, we can describe the network structure.

2.3 Hypotheses

Based on the theory we suppose the following statements. For an author, the possible strategy is to maximize reach and make this reach a qualitative one. For a distributor or a gatekeeper, the goal is to find and spread unique content. Summarizing this part, we have the following hypotheses.

Hypothesis 1. The environment is essential for information spreading. Not only network influencers or hubs distinguish information flow. It depends on the whole structure, which can be described as bubbles. Knowing these communities allows us to know about communication processes and connections. Vice versa is true as well.

Hypothesis 2. A role describes not only a place in a network. It concerns a strategy. We suppose each node in a network to have a role depending on the situation. Moreover, playing a role means to lead some strategy and tactics. It works here as in management. To measure the result, for example, the efficiency of strategy, managers have KPI - key performance indicators. Here we can compute KPIs for authors and distributors, suggesting that it can describe their role.

Hypothesis 3. The strategy of a node can be identified with its KPIs. Estimating performance leads to an understanding of what a node is suitable for. This approach covers conscious and unconscious strategies. Knowing the strategy can help us to discover the upper level - the network structure itself.

3 Methodology

The code for this and following sections is available on GitHub:
https://github.com/tikseniia/telegram-clusterization.

Trying to find groups of Telegram channels, we can see this problem in two ways: as clusterization or as a link prediction. The problem with clusterization is that we want to group observations without having the necessary data of readers' level. Our idea is to use data of previous communications between distributors and authors to define their interests which means unseen connections and, after that, find their structure.

The approach is inspired by agent-based modeling (ABM). In ABM, interactions between individual agents are simulated as a consequence of rules that were set by the researcher. The effects of interest are usually systemic and appear on the macro-level [4]. That means, the dependent variables of interest are often properties of the society and not of individuals.

Knowing members' rules they follow, we can identify the structure of the network. Actions reproduce structures. Whether we are talking about pressures or incentives, the result is ultimately the creation or maintenance of social structures. Relationship structures are further solidified as rules, norms, and ways of doing things.

3.1 Indicators

In this paper, we introduce seven indicators for evaluating the network members' behavior. These indicators or KPIs are separated into two groups "author's" and "distributor's." It allows us to observe the network participants from both sides of interest.

Author indicators are the following: "my strength" and "their weakness." "My strength" shows how many nodes in a network forward a message. "Their weakness" concerns how many other messages forwarders repost on average.

Distributor indicators are: "uniqueness", "fertility", "humility" and "verse humility", "diversity". Uniqueness shows how many nodes in a network repost the same message. Fertility: how many messages one of the distributors have forwarded for time. Humility: how many times one has forwarded the particular author. Humility verse: how many times the particular author has forwarded one distributor. Diversity: how many authors at all one has forwarded.

All the indicators above are combined in three high-level KPIs: author's KPI, distributor's KPI, and total indicator, which combines everything.

3.2 Dataset

For this research, we received the dataset of 58341 forwards between 100 Telegram channels from TGStat. The sample includes channels concerning Russian politics, which authors are famous politicians, media persons, organizations (including media ones), and anonymous Telegram channels.

The forwards are collected from the period from 31/12/2018 till 17/10/2019. Forwards includes messages of three groups: a clear forward (24 225 messages), a forward with a post (16 521), and a channel mention (17 595).

3.3 Bipartite Graph

As a start point of analysis, we build a simple graph of channels where nodes are channels, and edges show the forward existence. The number of nodes equals the number of channels in our sample - 100. The number of edges between them is 3028.

However, with this approach, we are losing the information of connection strength between nodes. That is why the simple graph was supplemented with weights. The problem with this graph is that it does not tell us anything about network structure behind forwards. It does not count the direction of connections and nodes' roles. Moreover, it allows us to cluster nodes only by their closeness in terms of structure. It cannot be the same as the closeness of views or strategies.

The second approach is a bipartite graph. In the case of Telegram channels, we have two groups: authors of messages ("Authors") and their forwarders ("Forwarders"). Between these groups, we observe 4643 edges. It means that some channels mentioned themselves.

3.4   Features Extraction

In the analysis, we rely on historical data of previous interactions between users. This approach was successfully used for information flow prediction [11,12]. To distinguish communities, we use data about messages, not about edges. For this purpose, we define unique messages in the dataset according to their text. This way, we received 29 902 messages.

For every network member, we are counting a vector of estimations. Each estimation belongs to a node reaction for a message of our dataset, so the vector's length is 29 902. This way, we can compare users in a network from a cascade perspective [7].

To define communities in a network, we need to find similar or close nodes. Benson et al. (2019) as a similarity evaluator use the number of common neighbors and the Jaccard similarity of the neighbor sets [1]. We, based on the idea of strategy, use seven indicators. Some of them show the similarity mentioned above. Below we provide calculations for each indicator.

My strength and their weakness represent the effectiveness of an author. If a member is a mentioned channel, we count these indicators. "My strength" is the number of reposts of a message. "Their weakness" is the sum of the forwarders' neighbors from the bipartite graph. This way, it is the number of other mentions at all. If a member is not mentioned, these indicators equal zero.

Humility and humility verse demonstrate the strength of the connection between the message author and a forwarder. Humility shows how many times one node forwards another one. It equals the weight of the edge between a node as a forwarder to a mentioned channel. Humility verse shows the opposite situation when this node becomes a mentioned channel, and another one is a forwarder. If this node did not mention a message, these indicators equal zero.

Uniqueness, fertility, and diversity concern the quality of the distributor's selection. Uniqueness shows how many nodes in a network repost the same message, and equals "my strength" of an author. However, in this case, we suppose that a more significant value tells about imperfect selectivity. Fertility is the total number of forwarded messages, which equals the sum of edges weights. The same as for uniqueness, we suppose that the lower value demonstrates a higher level of a distributor. Diversity equals the number of neighbors for a node as a forward. Diversity is supposed to be better with a higher value.

We use not only seven mentioned indicators but their combinations into author's, distributor's and total KPIs.

$$Author'sKPI = mystrength/theirweakness \qquad (1)$$

This way, we receive a balance between strength and weakness. It means a more objective estimation of the author's performance.

$$Distributor'sKPI = (uniqueness/fertility) * (humility/diversity) \qquad (2)$$

It is also a balanced estimation that shows our expectations of distributor's behavior. This KPI represents the direct dependence with uniqueness and humility, and verse one with fertility and diversity. This way, we try to increase opportunities to find gatekeepers whose preferences can be determined.

$$TotalKPI = author'sKPI - distributor'sKPI \qquad (3)$$

In total KPI, we combine everything. Author's and distributor's KPIs are taken with different signs to separate nodes into two groups. If the value of the total KPI is low, then the behavioral pattern is closer to the distributor's one.

We estimate the closeness between vectors as a cosine distance. For our analysis, it is essential to have channels at different distances. My strength, their weakness, uniqueness, author's, distributor's, and total KPIs suit it perfectly. These cosine distances are used as adjacency matrices for the further clusterization. Interestingly, using indicators which are connected with neighbors' number makes all channels be away from each other.

4 Results

In this paper, we compare two clusterization algorithms: spectral clustering and Louvain [6]. Both of them were used in recent papers for community prediction in networks [7,8].

Table 1. Internal evaluation of models' quality.

model	modularity	silhouette_score	calinski_harabaz_score	davies_bouldin_score
spectral_weigthed	−0.0183529	0.0148307	2.91602	2.44996
louvain_weigthed	0.335296	−0.169756	3.93369	2.70981
spectral_my_strength	1.79732	0.00232392	1.03317	1.80411
louvain_my_strength	0.0779005	0.0863935	22.314	1.29734
spectral_their_weakness	1.76902	−0.00266084	1.00009	3.48916
louvain_their_weakness	0.0751391	0.0552681	17.7253	1.49726
spectral_uniqueness	1.62898	0.0260392	1.21571	1.39766
louvain_uniqueness	0.0650882	0.207633	17.7294	1.21126
spectral_author_kpi	1.8445	−0.00347799	1.0148	4.13686
louvain_author_kpi	0.278538	0.112184	8.19885	1.16404
spectral_dist_kpi	1.58625	0.0242084	1.14377	0.920112
louvain_dist_kpi	0.135027	0.252768	16.0904	1.34273
spectral_total_kpi	1.68748	−0.00207765	1.11468	1.72686
louvain_total_kpi	0.481642	0.168621	24.6475	1.26312

Looking at modularity estimation (Table 1), we see that Louvain clustering gives not stable clusters. It is evident because this algorithm decides that every node of a network creates an independent cluster. For spectral analysis, we defined the necessary number of clusters to be five; that is why the result is better. However, most of the nodes go to one cluster. The rest four have from one to five members. We need to mention that the modularity value higher than

1 here is observed because it is counted on the graph with weights lower then 1. Weights here equal cosine distance metrics.

Moreover, we count the following metrics: the silhouette score, Calinski-Harabasz score, and Davies Bouldin score. All of them evaluate the internal network structure and the strength of clusters connections. The Silhouette score reflects how similar a point is to the cluster it is associated with. The Calinski-Harabasz index compares the variance between-clusters to the variance within each cluster. The higher the score, the better the separation is. The Davies-Bouldin index is the ratio between the within-cluster distances and the between cluster distances and computing the average overall the clusters.

In our case, only the Calinski-Harabasz index shows valuable value, which can be interpreted as a quality clusterization. However, it works only for Louvain clusterization that, as we mentioned, did not group users in any community.

Table 2. External evaluation of models' quality.

model	rand	homogeneity	adjusted_mutual_info	fowlkes_mallows_score
spectral_my_strength	−0.0250306	0.0511688	0.00421431	0.70913
louvain_my_strength	0.0164394	0.0625062	0.00223105	0.28965
spectral_their_weakness	−0.122413	0.050485	−0.0162526	0.606886
louvain_their_weakness	0.00483757	0.0721738	−0.00147661	0.272592
spectral_uniqueness	0.119401	0.148283	0.108473	0.760084
louvain_uniqueness	−0.0211666	0.216441	−0.00541081	0.220809
spectral_author_kpi	−0.0704647	0.103896	0.0361316	0.547032
louvain_author_kpi	0.0198206	0.117768	−0.00387399	0.250986
spectral_dist_kpi	0.00351347	0.0466625	0.005682	0.732886
louvain_dist_kpi	−0.00832401	0.256632	−0.0033223	0.20803
spectral_total_kpi	−0.0207983	0.0831373	0.0368166	0.710339
louvain_total_kpi	−0.0181329	0.0902386	−0.00415316	0.28128

Furthermore, we compare models with models based on the simple graph with weighted edges, which was introduced before (Table 2). This way, we want to estimate how different our feature extraction approach is from an ordinary one. We calculate the rand index, homogeneity score, adjusted mutual info score, and Fowlkes-Mallows score. These metrics estimate the similarity of two clusterings. In our case, they are different. The first three estimators consider clusterization with modernized features to be random compared to the simple graph clusters. Only Fowlkes-Mallows score for spectral clustering shows the middle-level similarity.

5 Discussion

This paper is dedicated to community analysis among Telegram channels. Relying on gatekeeping theory, filter bubbles description, and Castells' communication model, we proposed the method for clustering. It is based on the idea that every node in a network has its strategy, and we can describe it through a performance indicator.

The proposed method differs from the classical approach of using structure-based features. However, quality metrics indicate a weak structure. Louvain modularity shows the low quality of clusters' structure: every node becomes a separated cluster. Spectral clustering with the Laplacian matrix produces some clusters. However, most channels rest as one big cluster.

The reasons of this result can be the following: the selection of indicators describing nodes behavior, the chosen metric of a distance - cosine one, or clusterization algorithms. To check the real reason, we can step back and test the proposed approach on random data, similar to Telegram. In this case, we could control external conditions and find out how network members behave under our conditions with seven indicators. After that it is also possible to test our findings on data with available audience data, for example, on Twitter.

To summarize, we consider this work as the beginning step in studying community detection among Telegram channels. Hypotheses match theory backgrounds not only from classical studies but from new papers as well. The methodology part provides an approach based on existing ones in the community detection field. Further investigations allow us to deepen our understanding of communication processes on the Telegram platform and produce sustainable algorithms of community detection in the case of the connection's absence.

References

1. Benson, A., Kleinberg, J.: Link prediction in networks with core-fringe data. In: Proceedings of the World Wide Web Conference, WWW 2019, pp. 94–104, (2019)
2. Castells, M.: The Power of Communication. Publishing House of the Higher School of Economics, Moscow (2009)
3. Chin-Fook, L., Simmonds, H.: Redefining gatekeeping theory for a digital generation redefining gatekeeping theory for a digital generation. McMaster J. Commun. **8**, 7–34 (2011)
4. Flache, A., et al.: Models of social influence: towards the next frontiers. J. Artif. Soc. Soc. Simul. **20**(4), 1–31 (2017)
5. Keen, A.: The Cult of the Amateur: How Blogs, Myspace, Youtube, and the Rest of Today's User-Generated Media are Destroying Our Economy, Our Culture, and Our Values. Doubleday, New York (2008)
6. Pedregosa, F., et al.: Scikit-learn: machine learning in python. JMLR **12**, 2825–2830 (2011)
7. Prokhorenkova, L., Tikhonov, A., Litvak, N.: Learning clusters through information diffusion. In: The Web Conference 2019 - Proceedings of the World Wide Web Conference, WWW 2019, pp. 3151–3157 (2019)

 8. Prokhorenkova, L., Tikhonov, A.: Community detection through likelihood optimization: in search of a sound model. In: The Web Conference 2019 - Proceedings of the World Wide Web Conference, WWW 2019, pp. 1498–1508 (2019)
 9. Sunstein, C.R.: # Republic: Divided Democracy in the Age of Social Media. Princeton University Press, Princeton (2018)
10. Vos, T.P., Russell, F.: Theorizing journalism's institutional relationships: an elaboration of gatekeeping theory. J. Stud. **20**(16), 2331–2348 (2019)
11. Wu, T., Chen, L., Xian, X., Guo, Y.: Evolution prediction in multi-scale information diffusion dynamics. Knowl.-Based Syst. **113**, 186–198 (2016)
12. Zhu, H., Yin, X., Ma, J., Hu, W.: Identifying the main paths of information diffusion in online social networks. Physica A **452**, 320–328 (2016)

Study of Strategies for Disseminating Information in Social Networks Using Simulation Tools

Alexander Usanin[1](✉), Ilya Zimin[1](✉), and Elena Zamyatina[1,2](✉) (iD)

[1] Perm State University, Perm, Russian Federation
[2] National Research University Higher School of Economics, Perm, Russian Federation

Abstract. The paper presents simulation tools for investigation not only the structural characteristics of social networks in order to study information dissemination strategies, but also the dynamic characteristics of this process. A feature of this software system is not only the ability to work with virtual social networks, but also with data from real networks. To collect data about a real social network, a special software agent has been developed. An ontological approach is used to store data about real network.

Keywords: Social network · Dynamic modeling · Static modeling · Software agent · Information dissemination · Virtual social network · Real social network · Process mining · Social Networks Mining

1 Introduction

Social networks have become an integral part of human life. There are about 70 million active users of social networks, each of whom spends 136 min on average viewing social networks per day in Russia [1] (the world average is 144 min per day). According to the Russian news agency, the largest social network in the world – Facebook - was used by 1.52 billion people daily at the end of 2019, and the number of active monthly users was 2.32 billion people [2]. Currently, the monthly number of active users of Facebook has increased by 5.6% and reached 2.45 billion [3]. Social networks are used not only by simple users, but also by specialists in various fields of science, production, marketing, and education. So, when analyzing social networks, economists receive information about transactions, about the influence of others on human behavior, while political scientists investigate the formation of political preferences. Marketers use social networks to promote advertising and products [4–7].

However, social networks are also actively used by scammers. Fraudsters, as well as terrorists, use networks to spread false, harmful, or even life-threatening information [8–11].

The study was carried out with the financial support of the Russian Foundation for Basic Research in the framework of the research project No. 18-01-00359.

W. M. P. van der Aalst et al. (Eds.): AIST 2020, LNCS 12602, pp. 303–315, 2021.
https://doi.org/10.1007/978-3-030-72610-2_23

Thus, there is a need to study social networks, to study strategies that either contribute to the promotion of information, or, conversely, prevent the spread of unwanted information blocking it [12–14].

One may use here the methods of Social Networks Mining, which is mainly based on the structural characteristics of social networks. Using SNA tools (Social Network Analyses), the following metrics can be obtained: mutual orientation, homogeneity, transitivity of connections, etc.) [4, 5, 14]. These research methods can also be called *static*.

However, structural characteristics do not adequately reflect some aspects of the behavior of users of a social network, they cannot identify causal relationships [16], the ability to determine which events affect the improvement of information dissemination, or, conversely, information blocking. In this case, simulation methods should be used, which can be called *dynamic* (for here we study the behavior of network users in period of time, the change in the structure and metrics of social networks in period of time too). So, to block the spread of malicious information, there are various strategies, including immunization strategies, and they can be investigated using simulation tools [14]. Most often virtual social networks are explored. Virtual social networks may be represented as a graph $G = (V, E)$, where V is the set of vertices. Each vertex represents a network user or group of network users, and the set of E edges are connections between users. Random graphs (Erdéshi-Renyi, Bollobosi-Albert, etc.) [17–19] are used to describe virtual social networks. It is well known that these random graphs are the most adequate models of real social (OSN, Online Social Network) networks. On the other hand, M. Gatti, a member of the research organization "IBM Research", in her paper [16], describes the study of data obtained from real networks.

The authors of this paper offer simulation modeling tools that would allow using both static and dynamic research methods, and the subject of research can be either virtual networks or real ones. Earlier in [14], the description of the simulation system Triad.Net was described, which implements both static and dynamic methods for studying OSN.

The next step in our research is the creation of software tools that allow us to explore not only virtual online networks, but also real ones. To study real social networks, it is necessary to solve the problem of collecting information about users of a real social network and using it in the simulation system (Triad.Net). In the future, the work will be structured as follows: first we plan to consider related works, then we are going to give a description of the simulation system, simulation model and special program tools for collecting data about a real social network. Then we will show how to store data about real social network and how to use saved data for simulation experiments.

2 Simulation of Online Social Networks

Traditionally there are three main groups of tasks on social networks: monitoring and analysis of the network; network prediction and information management [4].

Let us consider several works in which tasks related to social networks are dealt with. As mentioned above, social networks are investigated in order to study strategies that affect either the speed of information dissemination or the prevention of the dissemination of information (malicious information).

In [7], the authors studied several strategies for disseminating knowledge in the collaboration network of the academic center. For this purpose, the authors developed a dynamic model using the Monte Carlo method according to the Markov chain scheme. The network structure in this study is static (the number of agents and the connections between them do not change), but the process of disseminating knowledge is dynamic.

The simulation was carried out according to the following scenario: at each moment of time for each pair of related agents it turned out whether they were in contact during this period of time or not, after which, if the contact occurred, some of the information was transferred to one of the agents by the other. Upon reaching a certain level of awareness, the agent joined in the dissemination of knowledge. The knowledge dissemination strategy in this study refers to which agents will initially disseminate knowledge. Four strategies were considered: (1) the first 5 agents selected by degree of centrality; (2) 5 agents who published the most work; (3) the first 5 agents selected according to the indicator of intermediate centrality (betweenness centrality); (4) 5 central agents in clusters. To assess the effectiveness of strategies, two main indicators were considered: the share of knowledgeable agents at certain intervals of time and the amount of time required to spread knowledge between a specific share of agents. This model was tested on a collaboration network of a science center. The model was built on the basis of bibliometric information: the coauthors were identified; the number of published works was determined. he authors obtained the following results: a scenario in which agents were selected on the basis of centrality in clusters had the greatest impact on the dissemination of knowledge.

In [23], for the first time, a conceptual model was developed for the distribution of competing rumors in social networks, motivated by the desire to understand the spread and survival of false ideas in new channels of social networks. The authors consider the following assumptions/constructions: (1) rumors are competing, (2) rumors are directed, (3) rumors are spread by the agent through interaction, (4) rumors in the network develop in parallel. In this regard, the work is aimed at understanding the spread of rumors, taking into account the characteristics of the agent (such as reputation and effort) and the dynamics of interaction when considering competing rumors. An agent model was developed in the NetLogo language, the structure of the social network in the model-based model is the random graph Erdéshi-Renyi. Each agent is characterized by 3 parameters: reputation (mainly determined by the number of user connections), effort (the amount of energy the agent spends on the interaction at a certain point of time, trying to influence his peer) and the threshold (minimal influence from another agent, required to change the degree of belief in rumors). Interactions occur when two agents meet; as a result, a change in the state of the model can occur with some probability. The concept of the model is that there are 5 classes of people for each rumor $x \in \{A, B\}$: N_x for neutral, S_x for active supporters, R_x for hidden supporters, C_x for active opponents and L_x for hidden opponents. Each of these five subclasses is believed to represent a belief position in this rumor. This division into classes was once proposed in [15], the author of which developed two models for the dissemination of ideas—NSRL and NSCRL. The authors constructed several simulation models trying to test their hypothesis that the presence of a small percentage of supporters of rumors will ensure the survival of these rumors.

In [16], an egocentric network (that is, a network that considers a node (person) as a focal point and its adjacency) of Barack Obama was built and analyzed. The authors set themselves the goal of analyzing the dissemination of information with a unique site as a source, they took open data from a real social network Twitter and modeled an egocentric network focused on Obama, consisting of 24,526 nodes (Obama's followers at a distance of no more than 3) with 5.6 million tweets. This data set allowed them to model the behavior of users on the network when reading and publishing messages related to the two main candidates in the 2012 elections: Barack Obama and Mitt Romney. The authors shared all the tweets on the topic/ mood, where there are two main topics: "Obama" and "Romney" and two main feelings: "Positive" and "Negative". The paper examined several scenarios in which the authors changed the behavior of influential users to assess their impact. A multi-agent approach was applied. The modeling process is iterative and consists of six phases: (1) uploading data from a real social network; (2) a classification of topics and moods of tweets extracted from the sample data; (3) sets of samples for each user are created from previously classified data; (4) each user behavior model is built from previously created sets; (5) models are used as input for the simulator. Authors of the work used the SMSim simulator, developed in the Java programming language, as a simulation environment. Having considered interesting papers mentioned above, we tried to create a special software tool - a social network simulator, with which it would be possible to explore online networks, both virtual and real, and apply both static and dynamic methods.

In our previous paper [14], the simulation system Triad.Net was discussed. It allows to simulate virtual social networks, study the structural characteristics of these networks, evaluate their effect on the dissemination (or, conversely, preventing of the dissemination) of information, change user behavior. In this paper we consider new components of Triad.Net for real OSN simulation. Triad.Net meets the requirements that, in our opinion, a social network simulator must meet. So it is important for simulation system to have: (1) software (and linguistic) tools for constructing Internet graphs [17–19]; (2) software (and linguistic) tools for researching Internet graphs; (3) software (and linguistic) tools for simulation of the behavior of the users (in this case, agent-based modeling is a suitable paradigm); (4) software (and linguistic) tools to work with large volumes of data [20], to work with large graphs introducing social networks. Let's discuss the Triad.Net simulation system.

3   Presentation of the Simulation Model in Triad.Net

The simulation system Triad.Net was developed on the basis of the Triad [21, 22] computer-aided design and simulation system, which was intended for the design and simulation of computer systems. Triad.Net has a three-level representation of the simulation model: M = (STR, ROUT, MES), where STR is the structure layer, ROUT is the *routine* layer, MES is the *message* layer. A layer of *structures* is a collection of objects interacting with each other by sending messages.

Each object has *poles* (input and output), which serve, respectively, for receiving and transmitting messages. The basis of the representation of the layer of structures is a graph. Separate objects (here, users of social network and communities) should be considered as the nodes of a graph. The arcs of the graph define the relationships between objects. The structure layer is a parameterized procedure and allows you to flexibly change the number of nodes in the graph, etc. One may change the input parameter both before the start and during the simulation experiment. In the second case, the override is performed in the simulation condition.

Model objects act according to a specific scenario, which is described using a routine. A *routine* is a sequence of e_i events planning each other (E is a set of *events*; many routine events are partially ordered in model time). The execution of an event is accompanied by a change in the state of the object. The state of the object is determined by the values of the *variables* of the routine. Thus, the simulation system is event-driven. A routine, like an object, has input and output poles. The input poles are used for receiving messages, and the output poles are for transmitting them. The input event e_{in} has a name. All messages that arrive at the input poles of the routine are processed by the input event. To send a message, a special out operator (*out* <message> *through* <pole name>) is used. A set of routines defines a ROUT routine layer. The message layer (MES) is intended to describe messages of complex structure.

As was mentioned earlier, the model includes the description of the structure, routines and messages. The collection of information during the simulation experiment is carried out by *information procedures*. They work like sensors and monitor: changes in the values of variables, the arrival and sending of messages, and the execution of events. The list of information procedures is specified in a special program unit called 'the conditions of simulation'. The *conditions of simulation* determine the initial conditions during the simulation experiment, the moment of completion of the simulation, determine the algorithm for the study of the simulation model.

To create a simulation model of a real social network, it is necessary to collect information about the network and save it. For this purpose, an intelligent *agent* has been developed in the Triad.Net simulation system that extracts information from a real social network. Information extracted from a real network is stored in *ontologies* and *logs*, and then this information is used to build a simulation model.

4   An Agent for Data Collection in Real Social Network

To implement the agent, two groups of requirements were put forward. Firstly, these are the requirements for the data collection: (1) it is necessary to collect data on the topology of the social network—the relationship of users and their characteristics should be stored in the ontology; (2) it is necessary to collect data on user actions; (3) all events must have a time marker so that it can be used during modeling; (4) all collected data should be saved in a format that will be understood both by the modeling environment and by means of visualization and data analysis; (5) The data should be relevant at any given time. Secondly, these are requirements for the agent: (1) the agent must build an ontology along some limited part of the social network; (2) the agent must monitor the actions performed by users and store information in the event log; (3) the number of agent requests to the social network should be minimized; (4) the researcher can expand the studied group of people at any time; (5) the agent must collect data over a long period, and at any time, the researcher can obtain the information collected; (6) upon termination of the agent with an error, the data should not be lost; (7) Agent installation and start-up should be automated. We give more detail to those decisions that are taken to implement the agent. It was decided that the object of the study would be the VKontakte social network. As a result of the agent's work, the simulation system should receive data about users that should be stored in ontologies, and data about user behavior - in event logs. We'll take a closer look at the structure of ontologies and event logs.

One may see a structure of ontology (see Fig. 1). An ontology has the following set of classes: (1) Person - a class representing a user of a social network. It contains the user identifier in a particular network; (2) Community - a class representing the community that the user is subscribed to. It contains the community identifier in a specific social network; (3) Activity - a class representing the interests of the user. It contains the identifier of activity in a specific social network.

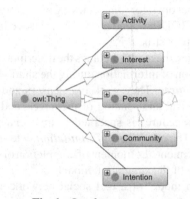

Fig. 1. Ontology structure

Relations used in the ontology: (1) hasFirstName - relation from the user (Person) to the string, indicates the username. That is, Person hasFirstName Ivan, will mean that the user name is Ivan; (2) hasSecondName - relation from the user (Person) to the string, indicates the user's last name. That is, Person hasSecondName Ivanov, will mean that the user has the last name Ivanov; (3) hasGender and hasBitrthDate - indicate the gender and date of birth of the user, respectively, etc.

OWL was chosen to describe ontologies.

The following types of events will be placed in the event log:

- the user has logged on to the network (online);
- the user saw a record of the community or his friend (post_seen);
- the user rated the post (post_liked);
- The user or community has copied the post to their wall (post_copied);
- post or community added a new post to the wall (post_add)
- the user is offline (offline).

One may see a structure of an event log:

```
<trace>
<event>
<string key = "concept:name" value = "online"/>
<string key = "org:resource" value = "44239068"/>
<string key = "user:name" value = "44239068"/>
<date key = "time:timestamp" value = "2020-04-18T13:21:49.698502 + 00:00"/>
</event>
<event>
<string key = "concept:name" value = "post_seen"/>
<string key = "org:resource" value = "44239068"/>
<string key = "post:id" value = "-31480508_617130"/>
<string key = "post:type" value = "post"/>
<string key = "post:is_ads" value = "0"/>
<date key = "post:date" value = "2020-04-18T11:35:34 + 00:00"/>
<string key = "owner:id" value = "53182060"/>
<date key = "time:timestamp" value = "2020-04-18T13:22:15.578177 + 00:00"/>
</event>
<event>
<string key = "concept:name" value = "offline"/>
<string key = "org:resource" value = "44239068"/>
<date key = "time:timestamp" value = "2020-04-18T13:25:04.260628 + 00:00"/>
</event>
</trace>
```

The event or resource is the user or community identifier. All events have a time marker. To study and verify the data obtained as a result of the program, Protégé [24] and ProM [25] were chosen, since they are the most common, as well as free. The agent is written in Python. To test the agent's performance, consider the group of users of the VKontakte network, Group A. A group of 4th year students at the Faculty of Mechanics and Mathematics of State University. All representatives of this group have accounts in the selected social network, which means they are of interest for modeling. As an Internet community, it was decided to study groups related to university and city topics, since most of the selected users are subscribed to them. A total of 17 people and 10 communities. The time period will be taken 3 h. Consider the results of the agent. Figure 2 shows the user data. This male user was born on January 27. You can see that he is friends with users whose identifiers are 5382060, 44239068, 361950485, 132549939 and 135282929. He is subscribed to communities with identifiers 39080597, 159146575, 63708206, 57846937 and 147286578. He likes programming and humor.

The information presented in this way allows it to be processed by both a computer and a person. This data will be exported to a simulation system.

Let us check the agent's ability to update information received from the social network.

After that, user 5382060 adds 361950485 to friends. After some time, the ontology is updated. An updated result is shown at Fig. 4.

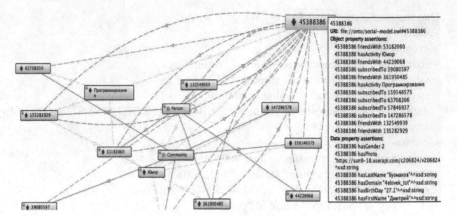

Fig. 2. Information about user

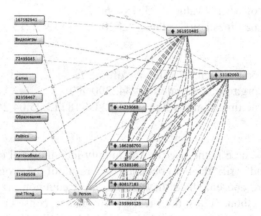

Fig. 3. Ontology before friend's adding

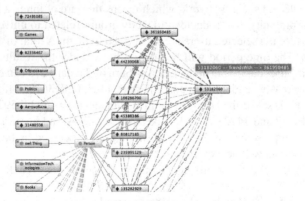

Fig. 4. Ontology after adding a friend

Then, we considered a group of users from Perm who were most active when copying information about COVID-19.

In addition to ontology, a log of events that took place in a selected section of the social network was collected. Figure 5 shows how often these or other events occurred.

All events		
Total number of classes: 6		
Class	Occurrences (absolute)	Occurrences (relative)
post_seen	4123	62,131%
online	800	12,055%
offline	795	11,98%
post_add	721	10,865%
post_copied	176	2,652%
post_liked	21	0,316%

Fig. 5. A log of events

5 Simulation of Real Social Networks in Triad.Net

To simulate real social networks in the Triad.Net simulation system, the TriadRealSo-cialNet module was developed, which allows: (1) to read data from ontologies that store information about real social networks; (2) perform analysis of social networks. The module corresponds to the agent modeling paradigm. Individual users and groups of the social network are represented as an autonomous agent (in our case, the requirements for the agent - autonomy, ability to social impact, reactivity and preventiveness are fulfilled), this will allow you to easily change the parameters in the behavior of a particular entity for further research.

The interaction of the TriadRealSocialNet module with the data acquisition subsystem is carried out through two files: 1) an ontology file, 2) an event log file. It provides the ability to download files from both local and remote (cloud) storage. After reading the input data, a routine describing its behavior is automatically put on each entity of a social network (user of a social network, user community) represented as a graph. You can make a change to the routine. When creating a simulation model of a real network, it is possible to perform the following actions: build a graph, add/remove vertices (users of social networks and communities), add/change events, connect information procedures (collect information about a simulation experiment), run a simulation experiment in a certain period of time, to present a report on the simulation.

The application graphical editor window is shown at Fig. 6.

The tools on the sidebar on the right provide us with all the features for working with real social networks. The "Data loading" panel allows you to read data from the ontology and the log file. The "Users" and "Communities" panels allow viewing/ editing information about users and communities, respectively. The Modeling panel allows you to build a network in a graphical editor, clear the network, set the simulation interval, connect information procedures and simulation conditions, start the simulation process and generate a simulation report.

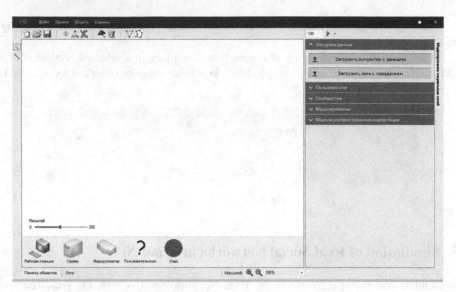

Fig. 6. A window of a graphical editor

The network is structured as follows: all communities are displayed along the left edge of the panel, and users are displayed in a circle to the right of communities (ring). An example of a constructed network is presented in Fig. 7.

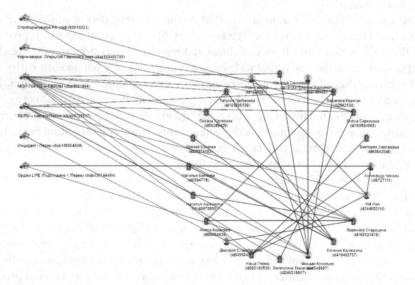

Fig. 7. Building a network in a graphical interface

Simulation of 8,434 events took 3,431 s on a PC with the following characteristics: AMD FX-8370 processor (8 cores, 4 GHz per core) and 16 GB of RAM. The experiment was performed with the operating system Windows 10 Pro x64.

The simulation results are displayed in a double window: the event log and the results of information procedures (see Fig. 8).

Simulation results

Time	Element	Message
3.195939	Михаил Кузнецов (id4548907)	User 'Михаил Кузнецов' (id4548907) Looked at post
3.195939	Михаил Кузнецов (id4548907)	User 'Михаил Кузнецов' (id4548907) is online
3.195941	Оксана Халявина (id50289429)	User 'Оксана Халявина' (id50289429) Looked at post
3.200001	Михаил Кузнецов (id4548907)	User 'Михаил Кузнецов' (id4548907) Looked at post
3.200004	Орджо LIFE (Подслушано г. Пермь) (club136144494)	Community 'Орджо LIFE (Подслушано г. Пермь)' (club136144494) Added post
3.200005	СтопКоронавирус.РФ (club193013221)	Community 'СтопКоронавирус.РФ' (club193013221) Added post
3.200007	Оксана Халявина (id50289429)	User 'Оксана Халявина' (id50289429) Looked at post
3.200021	Михаил Кузнецов (id4548907)	User 'Михаил Кузнецов' (id4548907) Looked at post
3.200041	Михаил Кузнецов (id4548907)	User 'Михаил Кузнецов' (id4548907) Looked at post
3.200139	Виктория Харламова (id63642548)	User 'Виктория Харламова' (id63642548) is offline
3.200325	Наталья Беляева (id6394778)	User 'Наталья Беляева' (id6394778) Looked at post

Results of information procedures

IP	Element	Result	Description
IPCountEvent(OnlineEventCount)	Вероника Старицина (id160121416)	60	Count of network appearances
IPCountEvent(PostSeenEventCount)	Вероника Старицина (id160121416)	959	Posts viewed
IPCountEvent(PostCopiedEventCount)	Вероника Старицина (id160121416)	14	Count of reposts
IPCountEvent(OfflineEventCount)	Вероника Старицина (id160121416)	60	Count of network exits
IPCountEvent(PostLikedEventCount)	Вероника Старицина (id160121416)	8	Count of likes
IPCountEvent(PostAddEventCount)	Вероника Старицина (id160121416)	2	Posts added
IPCountEvent(OnlineEventCount)	Евгения Коляскина (id419492757)	87	Count of network appearances
IPCountEvent(PostCopiedEventCount)	Евгения Коляскина (id419492757)	12	Count of reposts
IPCountEvent(OfflineEventCount)	Евгения Коляскина (id419492757)	87	Count of network exits
IPCountEvent(PostAddEventCount)	Евгения Коляскина (id419492757)	1	Posts added
IPCountEvent(OnlineEventCount)	Михаил Кузнецов (id4548907)	120	Count of network appearances
IPCountEvent(PostSeenEventCount)	Михаил Кузнецов (id4548907)	1172	Posts viewed

Fig. 8. The results of simulation experiment

The structural characteristics of all nodes can be seen at Fig. 9.

	A	B	C	D	E	F	G	H
1	Vertex	Degree	Strongly connected component numbe	Clustering coefficien	Centrality degree	Centrality as mediation	Density of centrality	Eigenvector of centrality
2	СтопКоронавирус.РФ (club193013221)	2	1	0	0,079999998	0,017771319	2,84	0,088179958
3	Коронавирус. Оперштаб Пермского края (club193450705)	2	1	0	0,079999998	0,013077007	2,8	0,089883978
4	МОЙ ГОРОД — ПЕРМЬ! (club3551694)	10	1	0,155555556	0,400000006	1	1,68	0,484405766
5	59.RU – новости Перми (club29725517)	7	1	0,142857143	0,280000001	0,416228861	2,04	0,323946965
6	Инцидент | Пермь (club149304838)	1	1	0	0,039999999	0	3,08	0,036468429
7	Орджо LIFE (Подслушано г. Пермь) (club136144494)	2	1	0	0,079999998	0,022018552	2,64	0,089599963
8	Вероника Старицина (id160121416)	6	1	0,133333333	0,239999995	0,362803161	2,08	0,282630799
9	Евгения Коляскина (id419492757)	5	1	0,4	0,200000003	0,167653948	2,44	0,231688405
10	Михаил Кузнецов (id4548907)	7	1	0,142857143	0,280000001	0,445177197	1,92	0,330133785
11	Валентина Томилова (id249318807)	2	1	1	0,079999998	0	2,88	0,097359896
12	Наша Пермь (id505160539)	5	1	0,1	0,200000003	0,420140803	2,12	0,170196992
13	Дмитрий Старобогатов (id5095045)	1	1	0	0,039999999	0	3,04	0,060542767
14	Анюта Ковалёва (id66854638)	2	1	0	0,079999998	0,008494467	2,72	0,122789554
15	Наталья Ашихмина (id148970850)	3	1	0,666666667	0,119999997	0,026824629	2,2	0,234427407
16	Наталья Беляева (id6394778)	1	1	0	0,039999999	0	3,4	0,049678528
17	Ксения Шилова (id20274153)	2	1	1	0,079999998	0	2,56	0,153840463
18	Оксана Халявина (id50289429)	3	1	1	0,119999997	0	2,52	0,195800288
19	Татьяна Грибанова (id103205139)	3	1	1	0,119999997	0	2,52	0,195800288
20	Иван Грачев (id1228991)	2	1	0	0,079999998	0,008941544	2,64	0,102361132
21	Наталья Санникова (id1913375)	2	1	1	0,079999998	0	2,36	0,174981681
22	Артем Нарванов (id2730422)	1	1	0	0,039999999	0	2,88	0,070819746
23	Василиса Коротун (id2982134)	5	1	0,2	0,200000003	0,23728618	2,04	0,290511701
24	Елена Саркисова (id189594563)	4	1	0,166666667	0,159999996	0,057114102	2,8	0,128979992
25	Виктория Харламова (id63642548)	1	1	0	0,039999999	0	3,08	0,036468429
26	Александр Чиклин (id6727111)	3	1	0,333333333	0,119999997	0,026824631	2,72	0,087260809
27	Wit Pon (id244552115)	4	1	0,166666667	0,159999996	0,149547324	2,32	0,146983497

Fig. 9. Structural characteristics of all nodes

Let us choose one post published by the club136144494 community and see how long it would take to spread it on the web. Information dissemination was implemented using the IC model [15]. After first simulation experiment, we changed simulation model excluding the node with the highest degree of centrality. It is a user with the id45548907 identifier. Second simulation experiment showed that the number of steps for which the post reached users in most cases decreased (but some of numbers remained unchanged) (see Fig. 10).

Node	Before node exclusion	After node exclusion
id160121416	4	5
id419492757	4	∞
id4548907	1	-
id249318807	3	∞
id505160539	2	3
id5095045	5	6
id66854638	3	4
id148970850	2	6
id6394778	4	∞
id20274153	3	∞
id50289429	4	∞
id103205139	3	∞
id1228991	2	3
id1913375	2	∞
id2730422	2	5
id2982134	3	5
id189594563	4	∞
id63642548	3	4
id6727111	1	1
id244552115	2	2

Fig. 10. A number of steps to users before and after excluding club136144494

6   Conclusion

So, this paper demonstrates the viability of the simulation tool Triad.Net for solving problems related to the research of social networks, including the tasks of disseminating various information.

Using the proposed tools, the researchers can study both virtual social networks and real ones, use static methods (structural characteristics of OSN) and dynamic (simulation). To study real networks, special software was developed - an intelligent agent that extracts the necessary information from a social network and places it in an ontology and special log. The ontology stores information about users and their relationships with other users (or communities). The special log stores information about user behavior. To create the log, Process Mining (ProM) methods were used. To build a simulation model in the Triad.Net environment software tools were also developed which helped extract information from ontologies and logs. Researchers can build the model automatically.

In the future, the authors of this paper intend to implement a multi-model approach for social media research.

Acknowledgements. The reported study was funded by RFBR and the Krasnodar Region Administration, project number 19-47-230003.

References

1. Globlee. Dannye ob ispolzovanii socialnyh setei v Rossii. http://www.advertology.ru/articl e147249.htm
2. Socialnaya set Factbook. https://ria.ru/20190204/1550239388.html
3. Socialnye seti - eto osnova SMM. https://sendpulse.com/ru/support/glossary/social-media-marketing (2020)

4. Gubanov, D., Chkhartishvili, A.: A conceptual approach to the analysis of online social networks. Large-Scale Syst. Control, 222–236 (2013)
5. Davydenko, V.A., Romashkina, G.F., Chukanov, S.N.: Modelirovanie sotsial'nkh setei. Vesntik TSU, 68–79 (2005)
6. Zhao, N., Cheng, X., Guo, X.: Impact of information spread and investment behavior on the diffusion of internet investment products. Phys. Stat. Mech. Appl. **512**, 427–436 (2018)
7. Kang, H., Munoz, D.: A dynamic network analysis approach for evaluating knowledge dissemination in a multi-disciplinary collaboration network in obesity research. In: Proceedings of the 2015 Winter Simulation Conference. Huntington Beach, pp. 1319–1330 (2015)
8. Ilieva, D.: Fake news, telecommunications and information security. Int. J. Inf. Theor. Appl. **25**(2), 174–181 (2018)
9. Yang, D., Liao, X., Shen, H., Cheng, X., Chen, G.: Dynamic node immunization for restraint of harmful information diffusion in social networks. Phys. Stat. Mech. Appl. **503**, 640–649 (2018)
10. Bindu, P.V., Thilagam, P.S., Ahuja, D.: Discovering suspicious behavior in multilayer social networks. Comput. Hum. Behav. **73**, 568–582 (2017)
11. Tumbinskaya, M.V.: Protection of information in social networks from social engineering attacks of the attacker. J. Appl. Inform. 88–102 (2017)
12. Filippov, P.B.: Use and implementation of personal data protection in social networks of the Internet. J. Appl. Inform. 71–77 (2012)
13. Dang-Pham, D., Pittayachawan, S., Bruno, V.: Applications of social network analysis in behavioural information security research: concepts and empirical analysis. Comput. Secur. **68**, 1–15 (2017)
14. Dmitriev, I., Zamyatina, E.: How to prevent harmful information spreading in social networks using simulation tools. In: van der Aalst, W.M.P., et al. (eds.) AIST 2019. CCIS, vol. 1086, pp. 201–213. Springer, Cham (2020). https://doi.org/10.1007/978-3-030-39575-9_21
15. Zimin, I.M., Zamyatina, E.B.: Issledovanie algoritma dinamicheskoi immunizacii uzlov dlya ogranicheniya rasprostranenia vredonosnoi informacii v socialnyh setya: Matematika i mezdisciplinarnye issledovania (2019). https://www.psu.ru (2019)
16. Gatti, M., et al.: Large-scale multi-agent-based modeling and simulation of microblogging-based online social network. In: Alam, S.J., Parunak, H.V.D. (eds.) MABS 2013. LNCS (LNAI), vol. 8235, pp. 17–33. Springer, Heidelberg (2014). https://doi.org/10.1007/978-3-642-54783-6_2
17. Raigorodskii, A.M.: Random graph models and their application. In: Proceedings of Moscow Institute of Physics and Technology (State University), pp. 130–140 (2010)
18. Buckley, P., Osthus, D.: Popularity based random graph models leading to a scale-free degree sequence. Discrete Math. **282**(1–3), 53–68 (2004)
19. Watts, D., Strogatz, S.: Collective dynamics of 'small-world' networks. Nature **393**, 440–442 (1998)
20. Mikov, A., Zamyatina, E., Kozlov, A., Ermakov, S.: Some problems of the simulation model efficiency and flexibility. In: Proceedings of 8-th EUROSIM Congress on Modelling and Simulation EUROSIM 2013, Cardiff, Wales, United Kingdom, 10–13 of September, pp. 532–538 (2013)
21. Zamyatina, E.B., Mikov, A.I.: Programmnye sredstva sistemy imitatsii Triad.Net dlya obespecheniya ee adaptiruemosti i otkrytosti. Informatizatsiya i svyaz', pp. 130–133 (2012)
22. Chaitanya, E.Y., Stephen, E.C.: An agent based model of spread of competing rumors through online interactions on social media. In: Proceedings of Winter Simulation Conference, pp. 3985–3996 (2015)
23. Protege. https://protege.stanford.edu/products.php. Accessed 15 July 2020
24. ProM. http://www.promtools.org php. Accessed 15 July 2020

Data Analysis and Machine Learning

Advanced Data Recognition Technique for Real-Time Sand Monitoring Systems

Artem Appalonov[1](✉) ⓘ, Yulia Maslennikova[1,2] ⓘ, and Artem Khasanov[2] ⓘ

[1] Kazan Federal University, 18 Kremlyovskaya Street, 420008 Kazan, Russia
[2] SONOGRAM LLC, 59/1 Magistralnaya Street, 420108 Kazan, Russia

Abstract. Sand production in oil and gas wells is a serious issue for the petroleum industry around the world. The commonly used non-intrusive sand monitoring systems are based on - acoustic emission measurement techniques. This research presents advanced data recognition techniques that can significantly improve the accuracy of sand monitoring. At the first step, factor analysis was used to identify key acoustic features of sand particles. Then, the following machine learning techniques have been applied: support vector machines, logistic regression, random forest method and gradient boosting. For training and testing the recognition system we used the acoustic database obtained in the laboratory of the oilfield service company SONOGRAM LLC (Kazan, Russia). The database consisted of acoustics signals from sand particles impacting on the inside and outside of a pipe wall in various scenarios (dry and wet gas, different flow rates, etc.). It was shown that the use of support vector machines with the Gaussian kernel reduces false positives compared with the algorithm that is based on ultrasound power peaks detection.

1 Introduction

During the last few decades, sand production has become a serious issue for the operators in the oil and gas industry. When sand is being produced from an unconsolidated reservoir it lowers the hydrocarbon production rate and increases maintenance costs. Solids in a produced fluid also represent a serious hazard to the wellhead equipment and its surroundings [1]. These can be natural solids, such as sand grains from unconsolidated rocks, and/or man-made particles of proppant injected in the reservoir during hydraulic fracturing.

In recent years, the most promising particle detection methods have been based on the principles of acoustic emission (AE) measurements. When a solid particle strikes a pipe wall, a part of the induced energy dissipates as elastic waves, which propagate through the pipe material and can be detected by a suitable AE sensor. This approach provides a very high temporal resolution and highly sensitive to particle impacts. A detailed review of the application of AE in sand monitoring systems is presented in many works [1–4].

This work was supported by the Russian Government Program of Competitive Growth of Kazan Federal University.

© Springer Nature Switzerland AG 2021
W. M. P. van der Aalst et al. (Eds.): AIST 2020, LNCS 12602, pp. 319–330, 2021.
https://doi.org/10.1007/978-3-030-72610-2_24

Popular wellhead sand monitoring systems measure the ultrasonic signal that is generated by particles impacting on the inside of the pipe wall, just after the bend where the sensor is located. Sand particles in a fluid flow are commonly represented as power peaks in the ultrasonic frequency range [1]. Before the measurement can be done, it is critical that such monitoring systems are calibrated in various flow rates and fluid compositions in order to find the adequate threshold for the ultrasonic power peaks. Often such systems show many positives in wet gas flow or during rain when water droplets impact the outer pipe wall [4]. For this reason, there have been many attempts over the past ten years to develop methods to recognize the particle-pipe collisions in the acoustic signal data using modern Machine Learning (ML) techniques. Many studies [1,3,5,6,8] demonstrate that an acoustic sensor designed with the assistance of computational intelligence may provide a simple, robust and practical real time solution to identifying solid particles in gas flow. A classical artificial neural network was successfully used to model the non-linear relationships between the characteristics of the acoustic signal and the flow [1]. The study [8] provides principal component analysis, which was combined with a kernel density statistics method to assess the results of a qualitative analysis of the monitored sand using vibration sensors. The paper [6] demonstrates an ML system that was developed for locating sand production intervals by analyzing AE data.

This paper proposes an advanced approach that combines an acoustic sensor with ML recognition system to construct a real-time sand detection methodology by analyzing high-resolution AE signals. The acoustic sensor is used for its simplicity, non-intrusiveness, and low cost. The main part of the paper consists of the following sections: Sect. 2, which describes the modeling methods for the proposed ML recognition system. Section 3 presents the experimental setup and measurement techniques used in the study. Section 4 provides and discusses the empirical results obtained in the laboratory, and finally, Sect. 5 concludes the paper.

2   Methods

This section presents an advanced approach for acoustic data recognition in a real-time sand-monitoring system. Figure 1 demonstrates a flow chart of this approach. It is divided into three sections, namely: the data input section, the recognition section, and the outputs. The data input section prepares the signal data for the ML recognition model. It pre-processes the acoustic signal data and extracts the significant acoustic features to reduce the dimension of the original acoustic signal.

The acoustic signal produced by a sand particle impacting on a metallic pipe contains specific information in a restricted time interval. To extract this information a Short-time Fourier transform (STFT) was applied. STFT is a Fourier-related transform used to determine the sinusoidal frequency and phase content of local sections of a signal as it changes over time [9]. As a first step, the original acoustic signal was divided into shorter segments of equal length and then the Fourier transform was applied separately on each of these shorter segments (without overlapping). Each segment of the signal was normalized. The normalization used in this work is the min-max normalization formulated in Eq. (1). This formula scales a sample x in the interval $[-1, 1]$. In Eq. (1), x_{max} is the maximum value of the sample and x_{min} is the minimum value of the sample.

$$y = 2\frac{x - x_{min}}{x_{max} - x_{min}} - 1. \tag{1}$$

In the discrete-time case, the STFT is defined by [9]:

$$Y(m, f) = \sum_{n=-\infty}^{\infty} y(n)g(n - m)e^{-j2\pi fn}, \tag{2}$$

where $y(n)$ is the original signal, and $g(n)$ is a window function centered around zero. In our case, the width of $g(n)$ was close to the duration of a sand particle impact, and the STFT was performed using the Fast Fourier Transform (FFT), so both variables m and f were discrete and quantized. The magnitude squared of the STFT yields the spectrogram representation of the Power Spectral Density (PSD) $|Y(m, f)|^2$. Vectors of the PSD matrix were used as input features for the ML recognition system.

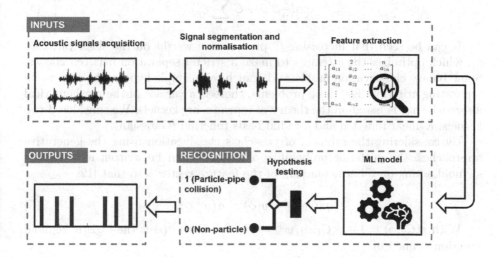

Fig. 1. Flow chart of the acoustic data recognition system.

The recognition system consists of a machine learning model and hypothesis testing. In our research, we focus on the binary classification problem in which output can take either of two values only: 1 (particle-pipe collision) and 0 (all other events). The final section "outputs" provides a binary index for each sample of the original signal. We selected the following ML approaches: Support vector machine, logistic regression, random forest and gradient boosting. The details of these computational approaches are given below.

An SVM classifies the data by finding the best hyperplane that separates all data points of one class from those of the other class. The best hyperplane for an SVM is the one with the largest margin between the two classes. The margin defines the maximal width of the slab parallel to the hyperplane that has no interior data points. This is because the investigated data might not allow for a separating hyperplane. So we used an SVM with a soft margin, which means a hyperplane that separates many, but not all data points. The data for training is a set of points (vectors) x_j along with their categories y_j. For some dimension d, the $x_j \in R^d$, and the $y_j = \pm 1$. The equation of a hyperplane is [10]:

$$f(x) = x'\beta + b = 0, \tag{3}$$

where $\beta \in R^d$ and b is a real number. The following problem defines the best separating hyperplane (i.e., the decision boundary): find β and b that minimize $\|\beta\|$ such that for all data points (x_j, y_j), $y_j f(x_j) \geq 1$.

There are two standard formulations of soft margins: L^1- and L^2-norm problems. Both involve adding slack variables ξ_j and a penalty parameter C (also known as a regularization coefficient). We used the following L^2- norm problem [11]:

$$\min_{\beta,b,\xi}(\frac{1}{2}\beta'\beta + C\sum_j \xi_j^2). \tag{4}$$

It can be seen that increasing C places more weight on the slack variables ξ_j, which optimizes the attempts to make a stricter separation between classes. Some binary classification problems do not have a simple hyperplane as a useful separating criterion. For those problems, there is a variant of the mathematical approach that is based on the theory of reproducing kernels. We considered two kernels: a linear function and a radial basis function (Gaussian).

By considering the problem of two-class classification using the generative approaches, the posterior probability of class C_1 can be written as a logistic sigmoid acting on a linear function of the feature vector ϕ so that [12]

$$p(C_1|\phi) = y(\phi) = \sigma(w^T\phi). \tag{5}$$

With $p(C_2|\phi) = 1 - p(C_1|\phi), w^T = [\beta_0, \beta_1]$. Here $\sigma(\cdot)$ is the logistic sigmoid function defined by

$$\sigma(a) = \frac{1}{1 + \exp(-a)}. \tag{6}$$

In the terminology of statistics, this model (5) is known as logistic regression (LR). LR is a widely used technique because it is very efficient and does not require too many computational resources. LR provides a statistical analysis model that attempts to predict precise probabilistic outcomes based on independent features. For high dimensional datasets, this may lead to the model being over-fit on the training set, which means overstating the accuracy of predictions on the training set and thus the model may not be able to predict accurate results for the test set. This usually happens when the model is trained on a little amount of training data with lots of features. So for high dimensional datasets, regularization techniques should be considered to avoid over-fitting (but this makes the model complex). Additionally, nonlinear problems can't be solved with the LR model since it has a linear decision surface [12].

Random forest (RF) is a supervised learning algorithm. The "forest" it builds, is an ensemble of decision trees, which are usually trained with the "bagging" method. The general idea of the bagging method is that a combination of learning models improves the overall result. The RF algorithm adds additional randomness to the model while growing the trees. Instead of searching for the most important feature while splitting a node, it searches for the best feature amongst a random subset of features. This results in a wide diversity that generally results in a better model. There is the "number of trees" hyperparameter, which is just the number of trees the algorithm builds before taking the maximum voting or taking the average of predictions. The main limitation of RF is that a large number of trees can make the algorithm too slow and ineffective for real-time predictions. In general, these algorithms are fast to train but quite slow to create predictions once they are trained. A more accurate prediction requires more trees, which results in a slower model [13].

Gradient boosting (GB) is a machine learning technique for regressing and classifying problems, which produces a prediction model in the form of an ensemble of weak prediction models, typically decision trees. It builds the model in a stage-wise fashion as other boosting methods do, and it generalizes them by allowing optimization of an arbitrary differentiable loss function. The GB model will continue improving to minimize all errors. This can overemphasize outliers and cause overfitting. It must use cross-validation to neutralize. Being computationally expensive, GBs often require many trees (>1000) which can be time and memory exhaustive [14].

3   Experimental Setup

We used special procedures to collect acoustic signals generated by the collisions of sand particles with a metallic pipe wall. There were three main parts of the experimental setup: the particle-pipe collision system, the sand-gas carrying flow system, and the high-precision acoustic Data Acquisition System (DAQ). The experimental setup is in the technology center of the oilfield service company SONOGRAM LLC (Kazan, Russia). Figure 2 shows a schematic representation of the test facility. The particle-pipe collision system used a 30 mm-pipe with a wall thickness of 1.5 mm. The acoustic sensor was a piezoelectric cylinder crystal 304L with a diameter of 12.8 mm and thickness 2.1 mm.

The acoustic sensor was mounted on the outer surface of the steel pipe at the place of the bend where the interactions of the sand particles with the pipe wall have the highest frequency. The signal recorded from the sand particles

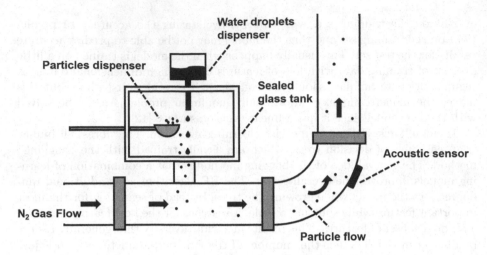

Fig. 2. Schematic representation of the sand particle detection test facility.

that have impacted the pipe wall was very weak and distributed in the high-frequency domain. Consequently, the acoustic sensor was connected to an NI PCI-4462 and a charge amplifier ZET 440 that is specially used for piezoelectric detectors. The resulting signal was digitized with a sampling rate of 204.8 kHz and 24-bit digitization. Sand particles and water droplets were added to the gas flow stream using a dispenser system. The solids were made of quartz proppant with a density of $2.65\,\mathrm{g/cm^3}$ of two uniform granularities: 20/40 and 30/60 mesh (see Fig. 3).

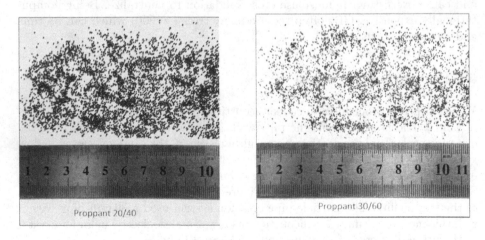

Fig. 3. Proppant samples used in the experiment: 20/40 mesh (left), 30/60 mesh (right).

The nitrogen (N_2) flow was supplied from a compressed gas cylinder because a rotary-screw compressor generates a high intensity acoustic noise. The gas flow velocity and a pressure drop were metered using a gas flow meter and a manometer. A special grounding system was developed to eliminate additional interference to the acoustic sensor (see Fig. 2).

The experiments covered five scenarios mainly (the conditions of each scenario are shown in Table 1). Experiment-1 tested the collision of sand particles with the pipe wall in a dry gas flow of different flow velocities. Experiment-2 assessed the effect of water droplets impacting the pipe wall when carried by the gas stream at various flow velocities. Experiment-3 and Experiment-4 evaluated the detection of AE signals when sand particles and water droplets impact on the other side of a the pipe wall. Experiment-5 aimed at evaluating the level and acoustic properties of the AE signals generated by the gas flow of various velocities. Each test condition was repeated at least 3 times. Figure 4 shows typical acoustic signals from the impingements of sand particles (20/40 mesh) on the pipe wall of the bend at various gas flow velocities: $\Delta P = 1$ atm (Fig. 4a)

Table 1. Experiments design and conditions

Number	Experiment	Sand size, mesh	ΔP, atm
1	Dry gas flow + Sand particles	20/40, 30/60	0.5; 1; 2; 3
2	Gas flow + Water Droplets	–	0.5; 1; 2; 3
3	Water Droplets outside of the pipe ("rain")	–	–
4	Sand particles outside of the pipe	20/40 30/60	–
5	Dry gas flow	–	0.5; 1; 2; 3

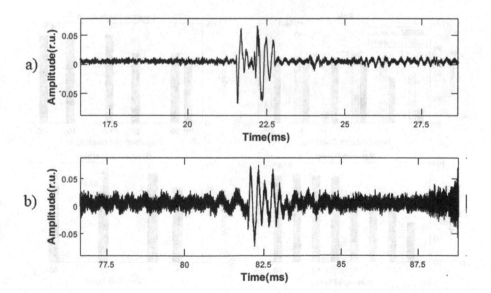

Fig. 4. A typical acoustic signal from particles (20/40 mesh) impingements in gas flow of two velocities: $\Delta P = 1$ atm (a) and $\Delta P = 3$ atm (b).

and $\Delta P = 3$ atm (Fig. 4b). It is clear that higher background noise corresponds to the higher gas flow velocities.

4   Results and Discussion

A series of experiments were conducted using the test facility described in Sect. 3. The observed database was pre-processed to select the time intervals with impacts of sand particles, water droplets, and other events. Then the corresponding acoustic signals were divided by samples of 216 points. STFT was used to extract the PSD feature vectors. Additionally, the PSD feature vectors were post-processed by averaging the powers of neighbouring frequencies using the Daniell method. The Daniell method leads to a more accurate representation of the true PSD and enables clear identification and rejection of deterministic noise peaks [15].

The total of 680 54-dimensional feature vectors were randomly divided. 80% were set aside for training and the remaining for testing. The training samples were used for building the recognition system and the testing samples for testing the performance of the built system. This dataset partitioning is usually acceptable when the dataset is sufficiently large [12].

The results of the training for different ML recognition systems are shown in Fig. 5. It was found that the Gaussian kernel function provides higher accuracy of the recognition in comparison with the linear kernel function. The best accuracy for the SVM model after the training was 98.1% for a regularization coefficient of 5. The logistic regression-based model showed an accuracy of 97.7%. Models

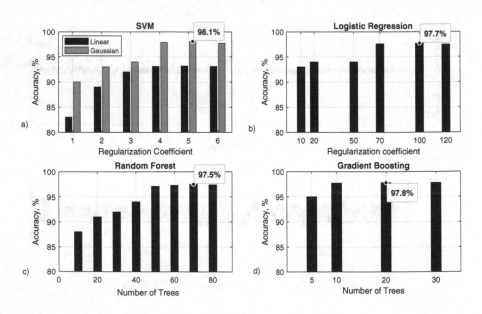

Fig. 5. Results of training machine learning algorithm.

based on Random forest and Gradient boosting showed similar results for the number of trees 70 and 20 respectively.

The results of quantitative error analyses of the suggested ML models are presented in Table 2. The most preferred relation between false positives (7%) and false negatives (9%) was observed for the support vector machine model (SVM in Table 2). This model is much better in comparison with the algorithm based on the analysis of ultrasound energy peaks that shows false positives 35% of the time (see "Energy Peaks" in Table 2). The RF and GB models demonstrated approximate close results but both methods are more computationally demanding in comparison with the SVM model.

Table 2. Quantitative error analysis of ML models accuracy

Test dataset		
ML model	False positives, %	False negatives, %
SVM	**7**	**9**
Logistic regression	9	13
Random Forest	16	10
Gradient Boosting	15	9
Energy Peaks	**35**	**10**

Figure 6 demonstrates the recognition system outputs the case of sand particles (20/40 mesh) impacting the pipe wall in dry gas flow with the following delta pressures: $\Delta P = 1$ atm (time interval from 2 to 5 s), $\Delta P = 2$ atm (from 5 to

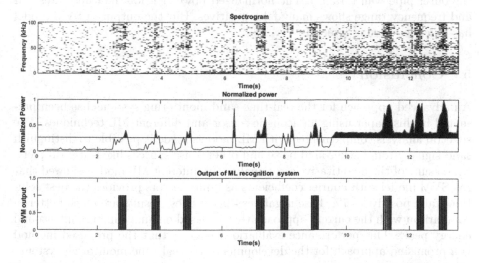

Fig. 6. Spectrogram (top), normalized power log calculated >20 kHz (middle), and the output of SVM model (bottom) in Experiment-1 (proppant 20/40 mesh, gas flow of $\Delta P = 1$ atm, $\Delta P = 2$ atm, and $\Delta P = 3$ atm).

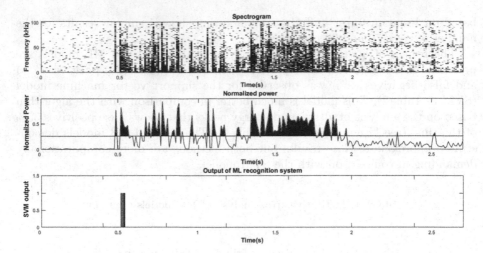

Fig. 7. Spectrogram (top), normalized power log calculated >20 kHz (middle), and the output of SVM model (bottom) in Experiment-3: water droplets impacted on the outside of a pipe.

9.5 s), and $\Delta P = 3$ atm (from 9.5 to 13.5 s). High frequency peaks (>50 kHz) are clearly seen on the spectrogram and the normalized power log that was calculated for ultrasound range >20 kHz. This range is typical for popular wellhead sand monitoring systems. The threshold value significantly depends on the gas flow velocity. The output of the Support Vector Machine (SVM) model shows all the detected time intervals (total number - 9) of the particles impingement. Figure 7 demonstrates the same set of graphic plots when water droplets impacted on the outer pipe wall ("rain"). The normalized power log for the same threshold and frequency range shows many false positives. But the output of SVM model has much less wrong intervals.

5   Conclusion

An advanced approach for the real-time sand monitoring systems has been presented in this paper using an acoustic sensor and different ML techniques. The spectral analysis complemented with ML models made it possible to distinguish sand signals from background noise and other noise sources like water droplets. The results of the quantitative error analysis of different ML models showed that the SVM model with Fourier coefficients as input vectors provides the best solution (false positives - 7%, false negatives - 9%). These results are much better in comparison with the current approach that is based on analyzing the ultrasound energy peaks This performance evaluation suggests that the proposed method is a promising approach for the development of a real-time monitoring system.

Further research and development work are required to enhance the technology. The recognition system should be trained using additional data with a broad particle size distribution to evaluate the threshold of sensitivity (minimal detected particle size, velocity, and concentration). More powerful recognition techniques like deep learning neural networks would help establish a greater degree of accuracy not only for qualitative analysis (particle-laden or particle-free flow) but also for quantitative outputs like estimated sand flow rate, sand concentration and/or particle size. We plan to determine the effectiveness of the proposed technique when applied to oil and gas well data.

Acknowledgements. The authors are thankful to Lilia Spirina and Vladimir Bochkarev (SONOGRAM LLC, Russia) who provided insight and expertise that greatly assisted the research. And we are grateful to Antoine Elkadi (TGT Abu Dhabi, UAE) for the help in the preparation of this paper.

References

1. Haugsdal, T.: The most efficient use of acoustic sand monitors. Lessons learned from many years of operation. Society of Petroleum Engineers (2017). https://doi.org/10.2118/185888-MS
2. Gupta, A., et al.: Getting the best out of online acoustic sand monitoring system: a practical method for quantitative interpretation. In: International Petroleum Technology Conference (2016). https://doi.org/10.2523/IPTC-18688-MS
3. El-Alej, M.E.: Monitoring sand particle concentration in multiphase flow using acoustic emission technology. World Acad. Sci. Eng. Technol. **7**(6), 1–7 (2014)
4. Lee, P.Y., Kasper, S.F., Quinn, C.: The 7 sins of managing acoustic sand monitoring systems. Society of Petroleum Engineers (2017). https://doi.org/10.2118/189213-MS
5. Bougher, B.B.: Machine learning applications to geophysical data analysis. T. University of British Columbia (2016). https://doi.org/10.14288/1.0308786
6. Aslanyan, I., Maslennikova, Y., Minakhmetova, R., Aristov, S., Sungatullin, L., Giniyatullin, A.: Determination of sand production intervals in unconsolidated sandstone reservoirs using spectral acoustic logging. Society of Petroleum Engineers, 25 October 2019. https://doi.org/10.2118/196445-MS
7. Aminu, K.T., McGlinchey, D., Chen, Y.: Optimal design for real-time quantitative monitoring of sand in gas flowline using computational intelligence assisted design framework. J. Petrol. Sci. Eng. **177**, 1059–1071 (2019). https://doi.org/10.1016/j.petrol.2019.03.024
8. Wang, K., Liu, G., Li, Y., Wang, G., Feng, K., Yi, L.: Vibration sensor approaches for experimental studies of sand detection carried in gas and droplets. Powder Technol. **325**, 386–396 (2019). https://doi.org/10.1016/j.powtec.2019.04.067
9. Sejdić, E., Djurović, I., Jiang, J.: Time-frequency feature representation using energy concentration: an overview of recent advances. Digit. Signal Process. **19**(1), 153–183 (2009). https://doi.org/10.1016/j.dsp.2007.12.004

10. Hastie, T., Tibshirani, R., Friedman, J.: The Elements of Statistical Learning. SSS. Springer, New York (2009). https://doi.org/10.1007/978-0-387-84858-7
11. Christianini, N., Shawe-Taylor, J.: An Introduction to Support Vector Machines and Other Kernel-Based Learning Methods. Cambridge University Press, Cambridge (2000)
12. Bishop, C.: Pattern Recognition and Machine Learning. Springer, New York (2006)
13. Breiman, L.: Random forests. Mach. Learn. **45**, 5–32 (2001). https://doi.org/10.1023/A:1010933404324
14. Jerome, H.: Friedman: greedy function approximation: a gradient boosting machine. Ann. Stat. **29**, 1189–1232 (2001)
15. Marple Jr, S.L.: Digital Spectral Analysis: With Applications. Prentice Hall Signal Processing Series. Prentice-Hall, Upper Saddle River, p. 492 (1987)

Human Action Recognition for Boxing Training Simulator

Anton Broilovskiy$^{(\boxtimes)}$ (iD) and Ilya Makarov$^{(\boxtimes)}$ (iD)

HSE University, Moscow, Russia
albroilivskiy@edu.hse.ru, iamakarov@hse.ru

Abstract. Computer vision technologies are widely used in sports to control the quality of training. However, there are only a few approaches to recognizing the punches of a person engaged in boxing training. All existing approaches have used manual feature selection and trained on insufficient datasets. We introduce a new approach for recognizing actions in an untrimmed video based on three stages: removing frames without actions, action localization and action classification. Furthermore, we collected a sufficient dataset that contains five classes in total represented by more than 1000 punches in total. On each stage, we compared existing approaches and found the optimal model that allowed us to recognize actions in untrimmed videos with an accuracy 87%.

Keywords: Action recognition · Action classification · Untrimmed video · Temporal convolution

1 Introduction

Sport undoubtedly takes an important role in everyday human life because it helps to promote health and maintain physical fitness. Like any of the popular areas, sport is always changing and developing. In each sport, new approaches elaborating to improve the training process and equipment. Computer vision systems along with other technologies have made sports more accessible and cheaper for ordinary people. For example, The Fastest Fist [1] is a boxing simulator that tracks the user's actions through virtual reality technology.

Professional sports also present a vast field of opportunities for the application of new technologies. For example, there is one popular type of training in the box named "shadow boxing". This training is widespread among boxers since performing without a partner. It's aimed to work out different types of punches, increase your endurance and speed of blows. During such training, the coach carefully monitors the athlete and corrects his mistakes. It is not a new sphere for technology applications. The first step to replacing the trainer with an automatic system was the work [2]. This work focuses on classifying common boxing punches during training without a partner. The main problem in this field is that all actions come from boxer are very similar. Often non-professional athletes can't distinguish different types of punches.

W. M. P. van der Aalst et al. (Eds.): AIST 2020, LNCS 12602, pp. 331–343, 2021.
https://doi.org/10.1007/978-3-030-72610-2_25

In this article, we decided to continue working in this direction. Our goal is to develop a neural network-based solution for punches recognition in a continuous video. This solution is aimed to help boxers during training.

To achieve this goal, we create a hand-crafted dataset with five different types of punches. We have implemented several approaches for action recognition on video based on SlowFast [4] approach and the ResNet model [5]. The resulted pipeline contains three stages. In the first stage, the model is responsible for clearing the frame stream of frames that do not hold actions. The second stage contains a model that divides a continuous stream of frames into separate actions. Finally, in the third stage model classifies actions.

As a result, we have a model that can recognize punches in untrimmed videos with F_1 score 0.88. Also, the model for classifying actions showed the quality of classification 0.94. The quality of the entire pipeline in untrimmed video is 87%. All code can be found on github page https://github.com/anton-br/box_simulator.

2   Related Work

There are two approaches to solve the problem of recognizing actions in the video. The first approach is based on manual feature generation, while the second uses neural networks.

2.1   Feature Based Approach

The work [3] was the pioneer in the application of computer vision for boxing. This work describes an approach for punches classification based on SVM and Random Forest classifiers. The main aim of this work is to demonstrate that these approaches can recognize punches on image sequences of eight elite boxers. The dataset consisted of four punches with a view of the boxer from above with 192 boxing punches in general. As a result, the authors trained a model that correctly recognized the punch's type with 96% accuracy. The main disadvantage of this approach is the small size of the dataset.

Two years later, the authors improved on previous achievements in the paper [2]. In this work, the authors continued the development of a robust framework for boxing punches recognition. They expand the dataset from eight to fourteen different boxers. Also, the authors add depth information. Thus, the new dataset contains 605 examples. In this work, instead of information about the position of hands, authors use the features based on the trajectory, shape descriptors, and arm skeleton. As a result, they bet a previous result accuracy 97.3% with a hierarchical SVM classifier. The main drawback of feature-based approaches is a slow operation speed. There is another approach for video analysis based on neural networks that work significantly faster than previous methods.

2.2   Neural Network-Based Approach

One of the first approaches to video processing using neural networks is work [6]. The authors provided four approaches for video classification based on temporal information from consecutive frames. The best performance shows the Slow fusion approach and bets all existed feature-based methods on 20%. This work shows that single-stream convolutional neural networks can learn powerful functions from slightly labeled data. Continuing the achievements of the previous article, the authors in [7] decided to use two parallel networks instead of a single one. The first network uses the single frame as an input and the second one uses a multi-frame optical stream that allows taking into account the time context. However, these networks are trained separately, so they can't aggregate information from each other.

Other popular approaches are based on 3D convolutions. One fused two different inputs by 3D convolutions [8] or by using 3D convolutions on time dimension [9]. The work [10] proves that 3D networks show the best results among all other approaches. Based on previous work, article [4] introduced a new state of the art model. Their idea is to use a single stream network with two different frame rates. The first branch is a Slow pathway. It takes only a few frames from one second of the video. The second branch uses almost every frame and is named Fast pathway. The main idea of this approach is to learn how to find both long actions, that last more than a few seconds, and fast actions, that sometimes last less than even a second. Also, these pathways are connected, so they can aggregate each other's information. In our work, we aimed to recognize short actions, so this approach is one of the most suitable for solving the problem of classifying actions. It is widely known that 3D convolutions contain many parameters. To get rid of this problem work [11] suggests simplifying 3D convolution operations by factorization 3D filters into separate spatial and temporal components. Finally, the work [12] allows defining the boundaries of action in an untrimmed video.

This work is aimed to recognize different punches from boxing. A relatively small number of researchers consider this particular area, so this article corrects this situation. The main contributions are summarized as follows:

1. We have collected a dataset that contains a variety of boxing punches taken from the front view. Since this field is too narrow, there are no from-view datasets with boxing punches.
2. Different punches from boxing have no strong differences between them, so this work is also aimed at developing a network that will be resistant to these differences.
3. A three-staged pipeline has been developed and shows strong results on real data.

The rest of the paper is organized as follows. The third section describes the collected dataset and data preprocessing. In the fourth section, one can find a description of all used models and the way how we train them. The fifth section describes the achieved results. The conclusion is presented in the sixth section.

3 Dataset Description

One of the significant contributions of this article is the creation of a representative dataset containing various boxing punches[1]. We selected the most common punches from boxing that can be recognized by a person who is not a professional in this sport. Most hits in boxing are made with the far hand. We have selected the three most popular hits: "Punch", "Cross" and "Uppercut". These are basic and distinguishable punches. Also, we added one short-handed strike named "Jab". This strike is very similar to the "Punch" since is also a direct hit but with the near hand. In addition to strikes, we added the "Defence" class. The resulted dataset contains five different classes. Examples of all actions are shown in Fig. 1.

(Punch)   (Cross)   (Uppercut)   (Jab)   (Defence)

Fig. 1. (Punch). Direct hit with the far hand. **(Cross)**. Side kick with the far hand. **(Uppercut)**. Low blow with the far hand. **(Jab)**. Direct hit with the near hand. **(Defence)**. Locked hands or a long pause without blows.

We used continuous shooting mode while recording the dataset. All video recordings last from two to six minutes on two different backgrounds. In each video, several people take turns entering the frame. The person in the frame uses punches from the given list randomly. While the person is in the frame, he applies the punches without pauses. In total, two men and one woman participated in the creation of the dataset. They also changed their clothes between shots. As a result, we recorded five videos with an overall length of about 16 min. In total, we have 29334 frames and 1009 actions.

We also calculated statistics for the collected dataset, is shown in Table 1. You can see that the average length of all actions is about 30 frames, and the minimum and maximum values do not deviate much from this value. Based on these statistics, we choose the 30-frame window as the main window for classifying actions.

We divide these actions into train, validation, and test sets in the following proportions: 0.7, 0.1, 0.2. Based on these proportions we use 700 actions for training, 103 for validation, and 206 for testing. For validation and testing, we used several untrimmed parts of videos.

---

[1] https://yadi.sk/d/aLSOf8klnUgSRQ.

Table 1. Dataset description.

Punch name	Total number	Average length	Min length	Max length
Punch	201	26.32	14	45
Cross	188	31.04	15	45
Uppercut	253	29.48	14	51
Jab	238	25.64	12	46
Defence	129	28.35	10	102
Total	1009	28.17	10	102

Before we train neural networks, we cropped the persons and reshape the images to 224×224. During training, we perform three types of data augmentation:

- Randomly deleting or zooming in on an image.
- Mirror image.
- Random color altering by applying a random color variation.

4 Models

The main approach consists of three stages. In the first stage, we use Action Splitting Network (ASN) to delete parts of the video that do not contain actions. The second stage is aimed to split a video into separate actions. The third stage is an action classifier on a trimmed video. All these three stages represent the full pipeline of video processing. For training, we use GPU Nvidia 2080 Ti with 11 GB.

4.1 Action Splitting Network

To detect frames in which no action occurs, we set this problem as a frame-by-frame classification. This network classifies frames by batches directly inside the main pipeline, right before action localization and classification. To solve this problem, we use ResNet50 [5], because this network contains a small number of parameters and allows to achieve good quality. This network receives an input set of frames, usually 60 to 90 frames long. Further, based on the prediction of the network frames, that have not contained actions are removed from further processing.

One of the main problems during the training of the ASN model was the class imbalance. Due to the specifics of the prepared dataset, the number of frames without actions is approximately ten times less than frames containing actions. To get rid of this problem we used the sampling method that chooses frames with and without actions in a way that during the training process both classes appear equally often.

We use Binary Cross Entropy (BCE) loss for training. The model trained with batch size 4 and a length of the frames sequence equal to 90. For optimization, we pick Stochastic Gradient Descent (SGD) optimizer with a learning rate 0.001.

4.2   Action Classification

The quality of the entire pipeline depends on the correct classification of action therefore it is important to achieve good quality at this stage. To reach this goal we used one of the best approaches for classifying actions called SlowFast. As the main network, we used ResNet50 with 3D convolutions. Also, we compare 3D convolutions with spatiotemporal convolutions [11]. To make a representative comparison, we added another well-established network named I3D.

All classification models were trained on prepossessed sets of frames using Multi-Class Cross Entropy loss. Each sequence contains only one action. The length of frame sets for training equal to the average length of all actions - 30 frames. If the action is more than 30 frames, we use only the first 30 frames. If the action is less than 30 frames, the frame set is copied the required number of times, so that set contains exactly 30 frames. All models are trained with batch size 12. We use Adam optimizer with a learning rate 10^{-3} and cosine annealing schedule with weight decay 10^{-6}.

4.3   Action Localization

Action Localization in Untrimmed Video. The goal of this network is to evaluate starting and ending probabilities for each frame. The backbone of an action localization network is the paper [12]. Based on this approach we construct the network that received set of prepossessed frames with size 224×224 as an inputs and outputs two probability sequence $P_s = \{p_{t_n}^s\}_{n=1}^l$ and $P_e = \{p_{t_n}^e\}_{n=1}^l$. Where $p_{t_n}^s$ and $p_{t_n}^e$ is a respectively starting and ending probabilities for time t_n. In contrast to the article we do not consider probability that particular frame contains action or actionness probability, because all frames without actions were rejected by ASN. The resulted model is shown in Fig. 2.

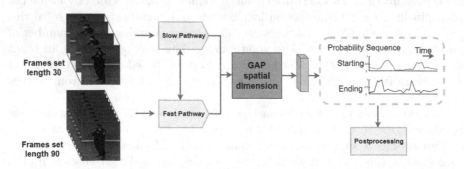

Fig. 2. The architecture of action localization on the untrimmed video network. Backbone is a SlowFast network that predicts the starting and ending probabilities for each frame. After preprocessing, we receive the indices of action boundaries.

In the field of action localization, a popular approach is to use pretrained networks. We compared the two approaches with learning the entire model from scratch and using a classification network from the previous section as a base model.

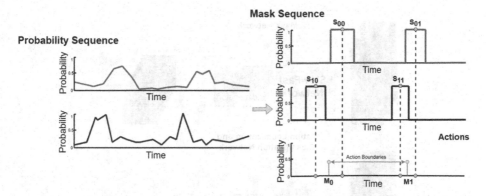

Fig. 3. Postprocessing procedure. Firstly we binarize the probabilities by discarding small probabilities. Also, we combine the high probabilities located closer than ten frames to one cluster. Then we calculate the midpoint for each cluster. Resulted boundaries are middle points between the closest midpoints of different classes.

An important task for this approach is to post-process data because it is often difficult to determine exactly when the action begins and ends. Since there is no pause between actions because of the task statement, first of all, we discard moments with height probability of start or end if there is no height probability in the opposite class. Also, we know that all actions last at least ten frames, so all probabilities greater than the threshold τ are combined into a single cluster if they are less than ten frames away. After that, for each cluster the midpoint S_{ij} is calculating, where i is 0 for start class, and 1 for end class j is a number of cluster. Then, a common midpoint M_j is calculated for each pair of clusters of a different class by the following formula $M_j = |S_{0j} - S_{1j}|$. This point is the end of the previous action and the point $M_j + 1$ is the beginning of the next one. Figure 3 describes the postprocessing procedure.

Since the model returns two probability vectors, we used the sum of BCE loss from both outputs. To achieve a more stable result, the actual true labels for the start and end points were expanded to three frames instead of one.

This model trained with a batch size 4, a length of the frames sequence equal to 90, and optimized using SGD with an initial learning rate of 0.01 and momentum 0.9. Also, we used the cosine annealing schedule with weight decay 10^{-6}.

4.4 Action Localization and Classification

This section describes models that were developed to combine networks for action classification and localization into a single network. All architectures are shown on Fig. 4.

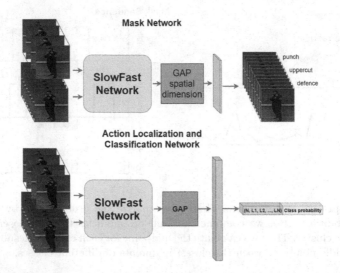

Fig. 4. Mask network. The architecture aimed to predict the class for each frame. **ALCN** predicts the vector that might be splitted into two parts. The first part represents the number of actions in the given frame set (N), the next N numbers show the length of each action. Second part contains $N * 5$ numbers. Where $P_i...P_{i+5}$ shows the probability for each class for i-th part of frame set with length L_i.

Mask Network. This network is also based on the SlowFast [4] approach with Resnet50 as a backbone. In contrast to the network for classification, here the Global Average Pooling layer is applied only for the spatial dimension and does not affect the temporal dimension. The main idea is to predict a mask that contains the classes for each frame.

The following approach was used to train this model. At first, we randomly sample the action. Then if it is shorter than the required length of frames set, the sequence is shifted to random offset. Thus, at least one whole action is located in a random place in the frame set. This sampling helps to fully the actual splitting when all frame sets are split by time.

Multi-class Cross Entropy was used to train this model using SGD with a learning rate 0.01 and batch size equal to 4.

Action Localization and Classification Network. This approach is also a variation of SlowFast with a modified network head. In this case, only the fully connected layer changes. To predict the number, length, and class of actions, the output vector is constructed as follows. $(N, L_1, L_2, ..., L_N, P_{10}, ..., P_{15}, ..., P_{N0}, ..., P_{N5})$. Here N is a number of actions on the input frame set. L_i is the length of each action in frames. P_{ij} probability that i-th action has j-th class. There is one more parameter N. This parameter is responsible for the maximum possible number of actions within a single frame set. As far as we know, the minimum length for one action is ten, then this number is calculating as follows: $N = \frac{FL}{10}$, where FL is the length of a frame.

We use Two types of losses to train this model. Smooth L1 loss [13] is aimed at solving the regression problem, which consists of predicting the number of actions and the length of each of them. Multi-Class Cross Entropy loss is used to solve the multi-class classification problem to predict the class for each action.

$$L(\hat{y}_i, y_i) = L_1 + 2 * L_c \tag{1}$$

Where L_c is a multi-class Cross Entropy loss and L_1 is a Smooth L1 loss. In the total loss formula 1 the weight of L_c is twice as high because in our experiments it was found out that such combinations give better improvements for action classification.

To optimize this model, we used SGD with a learning rate of 0.01 and the cosine annealing schedule with weight decay 10^{-8}. The overall frame set's length was 90 and the batch size is equal to 4.

5   Results

Actions Splitting. ROC-AUC score [14] was selected as the quality metric. This metric is suitable for the reason, that it is insensitive to the inequality of classes. We use ResNet50 networks with different types of convolutions to solve this problem. As the result, ResNet with spatiotemporal convolutions reaches 0.85 RocAUC while ResNet with 3D convolutions only 0.84 RocAUC. Also, replacing convolutions allows training the model an hour faster than using 3D convolutions.

Most errors that have been done during validation occurs at the junctions of classes. When a person completes a hit and starts swinging the next one, the model can recognize these frames as a "no action" frame.

Table 2. Comparison between different methods for action classification.

Model	Accuracy	Training time (h)
ALCN	0.25	12
MaskNet	0.36	16
I3D	0.76	**8.5**
SlowFast 3D	0.9	12.5
SlowFast SP	**0.94**	10

Action Classification. To compare the results of classification models, we used accuracy as a quality metric, since there is no strong class inequality. All results are shown in the Table 2.

Experiments have shown that the SlowFast model with spatiotemporal convolutions achieved the best result, but it took an hour and a half more to train than the fastest model - I3D. The worst result was shown by the Action Localization and Classification Network (ALCN).

For a more detailed error analysis, we calculate the confusion matrix for each model and one can find them in Fig. 5.

Although the SlowFast model with spatiotemporal convolutions shows a good result based on the confusion matrix, the most likely error for this model occurs when classifying *Uppercut* class as a *Defence*. We assume that at the final point of the *Uppercut* the position of the hand is similar to the position of the hands during *Defence*.

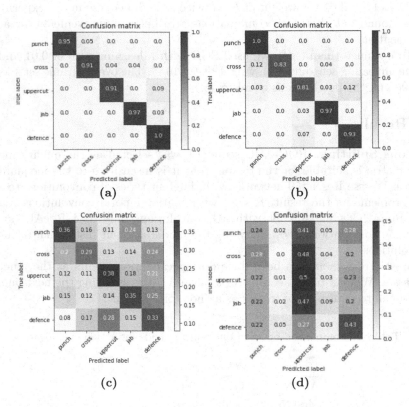

Fig. 5. Confusion matrices for: **(a).** SlowFast model with spatiotemporal convolutions. **(b).** SlowFast with 3D convolutions. **(c).** Mask Network. **(d).** Action Localization and Classification Network.

Action Localization in Untrimmed Video. For action localization, we developed three approaches. To compare them we use the F1 score. Before proceeding to evaluate the quality of algorithms, we should define *TP, FP, FN* for each of the algorithms. For Action localization in untrimmed video network (StEnd), *TP* prediction means that the distance between predicted and true centers is less than the determined distance. This distance is denoted as *D*. In this work, we analyze how quality depends on the given distance, where this distance *D* is changed from 5 to 3 frames. Prediction is considered as *FP* if the distance between this prediction and real position is greater than *D*. *FN* counted when there are no predictions close than *D* for a particular label.

The calculation of these components for ALCN is very similar. *TP* prediction means that the value of the predicted length is closer to the real length than *D*. It is important to notice that if there are several actions in the frame set, the next time prediction value is calculated as the sum of all the previous ones. Prediction is considered as *FP* when the model predicts more actions than in labels. *FN* counted when the model predicts fewer actions than in labels.

All results are shown in Table 3. The table shows the values of the F1 metric for various values of *D* from list $[5, 4, 3]$ where different lengths are indicated by @. One can notice that the best score shows the StEnd approach trained from scratch. Despite that it trains almost five times longer than the StEnd model fine-tuned with the classification model, StEnd from scratch gives significantly better results.

Table 3. Results for different models that aimed to solve the action localization task.

Model	F_1@5	F_1@4	F_1@3	Training time (h)
ALCN	0.16	0.13	0.1	12
Feature based StEnd	0.7	0.63	0.52	**4.5**
StEnd	**0.88**	**0.86**	**0.79**	20

Another important parameter for the StEnd model is τ, which was mentioned in the Model section. The following parameter were selected during validation: $\tau = 0.02$ for the Stand from scratch and $\tau = 0.015$ for the Feature-based StEnd. Based on the results, we can see that if we *D* equal to 3, then the value of the metrics drops significantly. This is because the initial response length is three frames, so in this case, an offset from the correct response by at least one frame will result in an error. The error value allows us to conclude that the generalization capacity of the StEnd network is high enough to be used to localize actions on the video.

Prediction in Untrimmed Video. To predict actions on the entire video, it has been cut into parts that are 90 frames long. For each part, first of all, frames that do not contain actions are dropped by ASN. Then the rest frames are divided into separate actions using the StEnd network. Next, each prediction is supplemented or trimmed to 30 frames and go to the classification network for prediction.

The accuracy of the resulted pipeline reached 0.87. The entire prediction time took 44.9 s. We use a video that lasts 135 s. Based on this, we can conclude that the operation of this pipeline is quite close to the prediction in real-time.

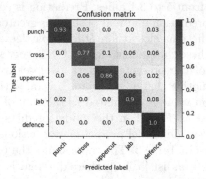

Fig. 6. Confusion matrix for entire prediction.

As a result of the untrimmed video prediction, one can notice that the quality has not changed significantly. Based on the confusion matrix, which is shown in Fig. 6, we can conclude that the action that most sensitive to borders is *Cross*, since the quality of *Cross* classification has dropped.

6   Conclusion

In this work, we present a novel dataset for action recognition in the boxing domain and build the pipeline that contains three independent models for action recognition in untrimmed videos.

The dataset contains more than 1000 actions for five different types of punches. There are three people in the dataset: two men and one girl recorded in two locations. Between videos, clothes also changed. The collected dataset is representative and sufficient to train new models that will be able to recognize actions on the new data.

Trained models have proven that they have sufficient generalizing ability to predict actions in a continuous video. Thus, for a full video, the final accuracy was 87%.

References

1. Beijing Skyline Interaction Technology Co, The Fastest Fist. https://store.steampowered.com/app/544540/The_Fastest_Fist/

2. Kasiri, S., Fookes, C., Sridharan, S., Morgan, S.: Fine-grained action recognition of boxing punches from depth imagery. Comput. Vis. Image Underst. **159**, 143–153 (2017). https://doi.org/10.1016/j.cviu.2017.04.007
3. Kasiri, S., Fookes, C., Sridharan, S., Morgan, S., Martin, T.: Combat sports analytics: boxing punch classification using overhead depth imagery, pp. 4545–4549. IEEE (2015)
4. Feichtenhofer, C., Fan, H., Malik, J., He, K.: SlowFast networks for video recognition (2019)
5. He, K., Zhang, X., Ren, S., Sun, J.: Deep residual learning for image recognition (2015)
6. Karpathy, A., Toderici, G., Shetty, S., Leung, T., Sukthankar, R, Fei-Fei, L.: Large-scale video classification with convolutional neural networks. In: CVPR (2014)
7. Simonyan, S., Zisserman, A.: Two-stream convolutional networks for action recognition in videos. In: NIPS (2014)
8. Feichtenhofer, C., Pinz, A., Zisserman, A.: Convolutional two-stream network fusion for video action recognition (2016)
9. Tran, D., Bourdev, L., Fergus, R., Torresani, L., Paluri, M.: Learning spatiotemporal features with 3D convolutional networks (2014)
10. Carreira, J., Zisserman, A.: Quo vadis, action recognition? A new model and the kinetics dataset (2017)
11. Tran, D., Wang, H., Torresani, L., Ray, J., LeCun, Y., Paluri, M.: Quo vadis, action recognition? A new model and the kinetics dataset (2017)
12. Lin, T., Zhao, X., Su, H., Wang, C., Yang, M.: BSN: boundary sensitive network for temporal action proposal generation. In: Ferrari, V., Hebert, M., Sminchisescu, C., Weiss, Y. (eds.) ECCV 2018. LNCS, vol. 11208, pp. 3–21. Springer, Cham (2018). https://doi.org/10.1007/978-3-030-01225-0_1
13. Huber, P.: Robust estimation of a location parameter. Ann. Stat. **53**, 73–101 (1964). https://doi.org/10.1214/aoms/1177703732
14. McClish, D.K.: Analyzing a portion of the ROC curve. Med. Decis. Making **9**, 190–195 (1989). https://doi.org/10.1177/0272989X8900900307

Bayesian Filtering in a Latent Space to Predict Bank Net Income from Acquiring

Evgeny Burnaev[✉][iD]

Skolkovo Institute of Science and Technology, Skolkovo 121205, Russia
e.burnaev@skoltech.ru
http://www.adase.group

Abstract. A macro stress-testing and scenario analysis is an important part of an official assessment of any bank regarding its safeguarding and stability. There is a lack of efficient tools for scenario analysis to model uncertainty of the bank financial indicators depending on the main macro-economic parameters. In this work we present a new model for prediction of the bank financial indicators. We develop an approach to filtering in a latent space capable of modeling dependence of a huge cross-section of the indicators on the set of macro-economic parameters. We demonstrate a superior ability of our model to predict bank net income from acquiring compared to standard predictive models.

Keywords: Filtering · Latent dynamics · Net income · Acquiring

1 Introduction

Banking performance depends on the macroeconomic situation, characterized by interbank currency exchange rates, etc. For example, in Fig. 1 we demonstrate how ruble interbank rates depend on time. We observe a significant increase around end of 2014. In Fig. 2 we demonstrate how currency interbank rates depend on time. By vertical lines we denote moments of significant deposits churn following significant change in the currency interbank rates.

It is obvious that to assess safeguarding and stability of any bank a reliable macro stress-testing and scenario analysis is needed. We need efficient tools to assess uncertainty of the bank financial indicators depending on the main macro-economic parameters when playing out various macro-economic scenarios [7].

Often a cross-section of indicators of interest has a very high dimension and nonlinearly depends on a set of macro-economic parameters. Moreover, available historic datasets are limited in size compared to the dimension of the cross-section. Fortunately, indicators in the cross-section are often significantly dependant and so instead of estimating parameters of a predictive model via averaging across trajectories we can compensate this to some extent by utilizing averaging across cross-sectional dimension. The task becomes more complex not only due to limited amount of data, but also due to the dependence between the indicators is nonlinear. Thus standard linear models (e.g. VARMAX) and other similar tools from econometrics for modeling of panel data [12] provide poor accuracy.

© Springer Nature Switzerland AG 2021
W. M. P. van der Aalst et al. (Eds.): AIST 2020, LNCS 12602, pp. 344–355, 2021.
https://doi.org/10.1007/978-3-030-72610-2_26

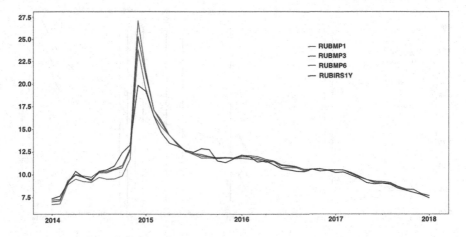

Fig. 1. Ruble interbank rates.

To overcome these challenges we develop an approach to filtering in a latent space capable of modeling dependences between components in a huge cross-section of indicators, and how the main macro-economic parameters influence these dependencies. We turn the problem of modeling high-dimensional non-linear dynamics of the indicators into the problem of identifying a low-dimensional latent state space, in which locally linear dynamics can be defined. The low-dimensional representation helps to compensate a lack of sufficiently long historical realizations of time-series compared to their original dimension. In order to learn such a latent space we use a generative model based on a variational autoencoder (VAE) [10]. The resulting dynamic model in the latent space thanks to the probabilistic nature of VAE allows to define a probability distribution over trajectories in the state space, and is trained in the unsupervised way. The induced probability distribution can be used to estimate uncertainty of predictions.

To illustrate efficiency of our approach we consider a problem of predicting bank net income from acquiring that is important for assessing a bank capital adequacy and liquidity risks. We demonstrate a superior predictive ability of our model compared to standard predictive models.

We concentrate to solve the prediction problem on a "vintage" level. Here by vintage we mean economic units grouped by some categorical characteristics (e.g. segment, territory, characteristics of a banking product) and united by some time interval (e.g. a month when an agreement starts). Usually, a bank is interested in the forecast either on a vintage level, or on a level when several vintages are grouped with respect to some categorical variables.

In this work we predict the bank net income from acquiring $x_{t,j}$ within each of the ~ 50 asset groups j, formed by belonging to a specific segment, territory and customer's bank affiliation, for a 12-month period ahead from the current date t: $(x_{t+1,j}, \ldots, x_{t+12,j})$. Within the group, vintages differ only by a month when

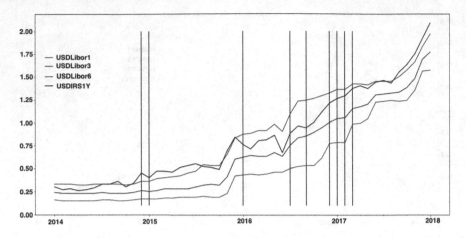

Fig. 2. Currency interbank rates vs. time.

Table 1. Asset groups, formed by belonging to a specific segment, territory and customer's bank affiliation.

Code num.	Segment	Regional bank	Client	Num. of
0	Clients of XXX	ZZZ bank	Non-YY	∼100
1	Clients of XXX	ZZZ bank	YY	∼150
⋮	⋮	⋮	⋮	⋮
∼ 50	Clients of ZYX	XYZ bank	YY	∼200

an agreement is signed. Table 1 gives a summary of all groups. The challenge is to construct the model which makes such predictions depending on the current macro-economic indicators for the considered year.

As an error metric we use

$$L(x, \hat{x}) = \frac{1}{\#\{\text{Code num.}\}} \sum_{j \in \text{Code num.}} \frac{\|x_j - \hat{x}_j\|_1}{\|x_j\|_1}. \tag{1}$$

Here $x_j = (x_{1,j}, \ldots, x_{12,j})$ is a monthly bank net income from acquiring for some asset group j and year, $\hat{x}_j$ is a prediction of x_j, $\| \cdot \|$ is L_1-norm. We average values of error metric (1) with respect to different years.

The main challenge is related to extremely high-dimensionality of the time-series: to predict the bank net income from acquiring within each of the ~ 50 asset groups j we first need to forecast dynamics of the multi-dimensional time series $x_t \in \mathbb{R}^{n_x}$, describing evolution of monthly incomes of $n_x > 7000$ vintages depending on the current macro-economic indicators for the considered year.

As can be seen from Fig. 5, Fig. 6 and Fig. 7, the behaviour of time-series is nonlinear, i.e. a lot of historical data is needed for model identification. However, historical monthly data is available only starting from 2014, which is not enough.

Fortunately, time series are dependent due to territorial proximity and/or similar business focus of organizations, so we can utilize this property to compensate for lack of sufficiently long realizations of time-series and replicate the nonlinear behavior, observed in the historical data.

The article has the following structure. In Sect. 2 we discuss related works. In Sect. 3 we describe an idea of a latent space dynamics. In Sect. 4 we provide a strict formulation of the proposed latent dynamics model. In Sect. 5 we discuss results of computational experiments and in Sect. 6 we make conclusions.

2 Related Works

Standard approaches to make predictions in the considered case is to utilize VARMAX and/or other standard linear state-space models for panel data [3]. However, as we already discussed, significant nonlinear behaviour and dependencies between components of the high-dimensional state space require modern approaches.

We propose to view the sequence of $x = \{x_t\}_{t=0}^{T}$ as a realization of a Markov chain, i.e. $p(\{x_t\}_{t=0}^{T}) = p(x_0) \prod_{t=1}^{T} p(x_t|x_{t-1})$. There are two possible approaches in this case for inducing a nonlinear dynamics and obtaining a computationally efficient model:

- Learn the dynamics in some latent space $z = \{z_t\}_{t=1}^{T}$,
- Learn the whole model as a transition operator over the state space.

This brings a connection with several works. [13] proposed to interpret the stochastic Markov chain $q(z|x) = q(z_0|x) \prod_{t=1}^{T} p(z_t|z_{t-1}, x)$ as a variational approximation, where $\{z_t\}_{t=0}^{T-1}$ is a set of auxiliary random variables. [9] proposed to generate MCMC chain in the latent space and then decode it. [16,17] proposed a method to learn the dynamics in the latent space for sequence prediction. [11] proposed to learn a multi-head GAN (with a number of heads, corresponding to the number of layers in the target neural network), which sample from noise the sequence of convolution filters. [14] proposed to learn a generative adversarial network as a markovian transition operator.

Thus these works consider different approaches to nonlinear dynamics. However, the models use complex approximators based on deep neural networks [8], which are not efficient in our case due high-dimensionality ($n_x > 7000$) and small size of the available historical time-series realization. These challenges turn out to be even more severe when the whole model is represented as a transition operator over the initial state space.

To attack described challenges we have to develop a compact nonlinear model similar to linear state-space models, which are beneficial for modeling panel data in econometrics. The model should be able to take into account dependence of a huge cross-section of indicators on a set of macro-economic parameters.

3   Latent Space Dynamics

Time series are dependent due to territorial proximity and/or similar business focus of organizations. To replicate this property by the model we can assume that there exist a latent space with dimension $n_z \ll n_x$, such that behaviour of time-series components depends on the dynamics in the latent space and

- if the time-series are close in the latent space, then they should have similar predictions,
- predictions of the model should be different for distant latent points.

Schematic example of the latent space dynamics is in Fig. 3.

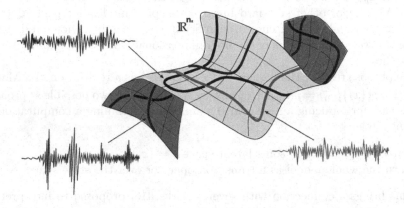

Fig. 3. Latent space dynamics. Picture credit to [15].

We denote the historical dataset as $\mathcal{D} = \{x_t, u_t, x_{t+1}\}_{t=0}^T$, where

- $x_t \in \mathbb{R}^{n_x}$, $n_x \gg 1$—time-series at moment t (income values for vintages),
- u_t—characteristics of the macro-economic situation at time t; so here u_t can be considered as some kind of a control variable.

Using dataset $\mathcal{D}$ we should construct a model that takes as input current values of state x_t and control value u_t, and predicts future value x_{t+1}.

In order to specify the model we impose the following assumptions:

- Dynamics of x_t is complex and nonlinear,
- We can find a representation $z_t \in \mathbb{R}^{n_z}$, $n_z \ll n_x$, such that

$$x_t = f(z_t), \quad z_{t+1} = A(z_t)z_t + B(z_t)u_t + C(z_t). \tag{2}$$

Here $f(\cdot)$ is some nonlinear mapping (decoder), which we model using a shallow fully-connected neural network (FCNN) [8], and z_t follows a nonlinear generalization of the standard linear state-space dynamics [3].

4   Probabilistic Model

To perform inference for model (2) and estimate its prediction uncertainty we use the probabilistic framework [2]. By adopting the approach from [16] we get the following probabilistic model

$$x_t = f(z_t) + \xi = W_x g_\theta^d(z_t) + w_x + \xi_x, \ \xi_x \sim \mathcal{N}(0, \Sigma_x),$$
$$z_{t+1} = A(z_t)z_t + B(z_t)u_t + C(z_t) + \xi_z, \ \xi_z \sim \mathcal{N}(0, \Sigma_z), \ z_0 \sim \mathcal{N}(0, I). \qquad (3)$$

Here $f(\cdot)$ plays a role of a decoder (cf. VAE) and is modeled by a shallow FCNN with parameters (W_x, w_x) and θ. Representations for $A(\cdot)$, $B(\cdot)$ and $C(\cdot)$, defining the dynamics, have the form

$$A(z) = W_A g_\psi^t(z) + w_A, \ B(z) = W_B g_\psi^t(z) + w_B, \ C(z) = W_C g_\psi^t(z) + w_C,$$

where $g_\psi^t(\cdot)$ is also a shallow FCNN with parameters ψ. The control variable u_t is modeled by a nonlinear embedding of the macro-economic indicators into some d_u-dimensional space. The embedding is done by a shallow FCNN with parameters γ, i.e. $u_t = g_\gamma^e(F_t)$, where F_t are features, calculated from the macro-economic indicators, known up to and including moment of time t.

Using historical realization $\{x_t\}_{t=0}^T$ we want to estimate parameters

- (W_x, w_x), (W_A, w_A), (W_B, w_B), and (W_C, w_C) of (f, A, B, C),
- θ, ψ and γ of $g_\theta^d(\cdot)$, $g_\psi^t(\cdot)$ and $g_\gamma^e(\cdot)$ respectively,

and posterior distribution $p(z_0, \ldots, z_T | x_0, \ldots, x_T)$. A straightforward approach based on MLE maximization

$$\mathcal{L}(D) = \sum_{(x_t, u_t, x_{t+1}) \in \mathcal{D}} -\log p(x_t, u_t, x_{t+1}) \to \max_{parameters} \qquad (4)$$

is intractable, since there is no explicit expression for the likelihood. Due to the same reason we can not calculate $p(z_t | x_1, \ldots, x_t)$.

To attack this challenge we use the variational inference approach [2]. We introduce approximate posterior

$$p(z_t | x_1, \ldots, x_t) \approx q_\phi(z_t | x_t) = \mathcal{N}(\mu_t, \Sigma_t),$$

where

$$\mu_t = \mu(x_t) = W_\mu g_\phi^e(x_t) + w_\mu, \ \Sigma_t = \mathrm{diag}(\sigma_t^2), \ \log \sigma_t = W_\sigma g_\phi^e(x_t) + w_\sigma,$$

where $g_\phi^e(\cdot)$ is also represented by FCNN with parameters ϕ. Here $\mu(x)$ plays a role of an encoder (cf. VAE). Thus, from (3) we get an approximate dynamics

$$z_t | x_t \sim q_\phi(z|x) = \mathcal{N}(\mu_t, \Sigma_t), \ \hat{z}_{t+1} | x_t \sim \hat{q}_\psi(\hat{z}|z, u) = \mathcal{N}(A_t \mu_t + B_t u_t + C_t, D_t),$$
$$\hat{x}_{t+1} | \hat{z}_{t+1} \sim p_\theta(x|z) = \mathcal{N}(f(\hat{z}_{t+1}), \Sigma_x),$$

where $A_t = A(z_t)$, $B_t = B(z_t)$, $C_t = C(z_t)$ and $D_t = A_t \Sigma_t A_t^\top + \Sigma_z$.

In can be proved that the likelihood (4) is lower bounded by the Evidence Lower Bound (ELBO) [2], which has the form

$$\mathcal{L}(\mathcal{D}) = - \sum_{(x_t, u_t, x_{t+1}) \in \mathcal{D}} \log p(x_t, u_t, x_{t+1}) \geq \sum_{(x_t, u_t, x_{t+1}) \in \mathcal{D}} \mathcal{L}^{bound}(x_t, u_t, x_{t+1}),$$

where

$$\mathcal{L}^{bound}(x_t, u_t, x_{t+1}) = \mathbb{E}_{\substack{z_t \sim q_\phi \\ \widehat{z}_{t+1} \sim \widehat{q}_\psi}} \left[-\log p_\theta(x_t|z_t) - \log p_\theta(x_{t+1}|\widehat{z}_{t+1}) \right] \tag{5}$$

$$+ KL(q_\phi || \mathcal{N}(0, \mathbf{I})). \tag{6}$$

In practice we optimize the regularized ELBO with some regularization $\lambda > 0$

$$\sum_{(x_t, u_t, x_{t+1}) \in \mathcal{D}} \mathcal{L}^{bound}(x_t, u_t, x_{t+1}) + \lambda KL\left(\widehat{q}_\psi(\widehat{z}|\mu_t, u_t)||q_\phi(z|x_{t+1})\right) \to \max_{parameters}. \tag{7}$$

The target functional in (7) can be interpreted as follows. The first term in (5) evaluates accuracy of the compression/decompression in/from the latent space: to which extent x_t is accurately recovered from z_t. The second term in (6) estimates accuracy of prediction based on the latent space dynamics: to which extent x_{t+1} is accurately predicted by decoding z_{t+1}, predicted from z_t. The term in (6) is a regularization: since we work with distributions, as a regularization we use a Kullback-Leibler divergence, i.e. $KL(q_\phi(z_t)|\mathcal{N}(0, \mathbf{I}))$; in a Gaussian case it is somewhat close to L_2 regularization. The second term in (7) estimates proximity of the latent variable, predicted by (3), with the value of the latent variable, estimated by the variational approximation.

5   Results

5.1   Architecture Description

In Fig. 4 we provide the overall scheme of the model, tailored to predict bank net income from acquiring on the vintage level depending on the current macro-economic indicators for year t. In this case x_t is a 12-dimensional vector, components of which are equal to monthly incomes of the vintage. Let us provide a quantitative description of the components of the proposed model.

As input to $g_\gamma^e(\cdot)$ we use a vector $F_t \in \mathbb{R}^{144}$ of features for year t, generated in a sliding window from the macro-economic indicators, including exchange rates and interbank interest rates. As features we use running mean, standard deviation and maximum, log-returns, and other similar statistics.

Using three fully-connected layers we map these features into an embedding vector of dimension 18 in order to model control u_t. Here is a description of these fully-connected layers:

1. (fc1): Linear(in_features = 144, out_features = 72, bias = True)

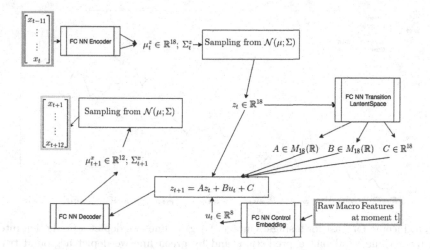

Fig. 4. Scheme of the model.

2. (fc2): Linear(in_features = 72, out_features = 36, bias = True)
3. (fc3): Linear(in_features = 36, out_features = 18, bias = True)

As input x_t to $g_\phi^e(\cdot)$ we use a history of returns for a considered vintage for the last 12 months relative to the current moment. We map this data by four fully-connected layers into the latent space of dimension 18. Here is a description of these fully-connected layers:

1. (fc1): Linear(in_features = 12, out_features = 24, bias = True)
2. (fc2): Linear(in_features = 24, out_features = 72, bias = True)
3. (fc3): Linear(in_features = 72, out_features = 36, bias = True)
4. (fc4): Linear(in_features = 36, out_features = 24, bias = True)
5. (mu_fc): Linear(in_features = 24, out_features = 18, bias = True)
6. (log_sigma_fc): Linear(in_features = 24, out_features = 18, bias = True)

The decoder module $g_\theta^d(\cdot)$ has the same mirror-symmetrical architecture.

Let us describe the module $g_\psi^t(\cdot)$ that is used when performing transition in the latent space. Its aim is to calculate matrices A, B, C for current latent vector z_t. The module is described by the following fully-connected layers:

1. (fc1_z): Linear(in_features = 18, out_features = 36, bias = True)
2. (fc2_z): Linear(in_features = 36, out_features = 72, bias = True)
3. (fc3_z): Linear(in_features = 72, out_features = 96, bias = True)
4. (fc4_z): Linear(in_features = 96, out_features = 156, bias = True)
5. (fc_z_A): Linear(in_features = 156, out_features = 324, bias = True)
6. (fc_z_C): Linear(in_features = 156, out_features = 18, bias = True)
7. (fc_z_B): Linear(in_features = 156, out_features = 324, bias = True)

Thus matrices A and B have dimension 18×18, and $C \in \mathbb{R}^{18}$.

Taking into account such architecture, it follows that the computational complexity of the model is comparable to that of standard neural network-based approaches such as Recurrent Neural Network models [8].

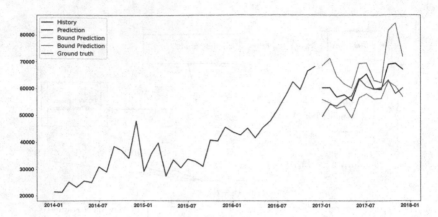

Fig. 5. Income forecast for a single vintage. By gray lines we depict a prediction interval; by red line we depict a prediction, and by green line we depict a ground truth trajectory. (Color figure online)

5.2   Accuracy of Forecasting

To evaluate performance of the proposed approach we used a proprietary dataset with values of bank income from acquiring on a vintage level.

In Fig. 5 and Fig. 6 we provide examples of forecasts on a vintage level. We can observe that the forecasts follow ground truth time-series rather accurately despite their significantly nonlinear behaviour. For example, if we use standard techniques for time-series analysis, we would just predict a linear growth of the time-series in Fig. 5 after the last known historical value. The proposed model is capable to predict the unexpected drawdown around 2017-01 better: thanks to the nonlinear dynamics in the latent space we can take into account the dependence of incomes between different vintages and the influence of macro-economic indicators.

In Fig. 7 we demonstrate how the proposed model forecasts bank net income from acquiring on a group level. We can observe that the forecast follows ground truth time-series very accurately.

We compared our model to several standard approaches including VARMAX model [3], XGBoost model [4] and RNN-GRU model (Recurrent Neural Network model with GRU cell) [5]. In case of the XGBoost model we follow a standard machine learning pipeline:

- as input we use the same macro-economic features as in case of $g_\gamma^e(\cdot)$ (see Subsect. 5.1) plus historical values x_{t-1} of the considered time-series,
- the model is represented as a regression from these input features for the period $t-1$ to the output value x_t equal to the bank net income from acquiring for a 12-month period from the current date.

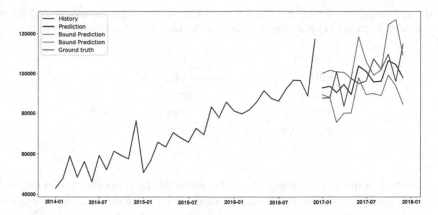

Fig. 6. Income forecast for a single vintage. By gray lines we depict a prediction interval; by red line we depict a prediction, and by green line we depict a ground truth trajectory. (Color figure online)

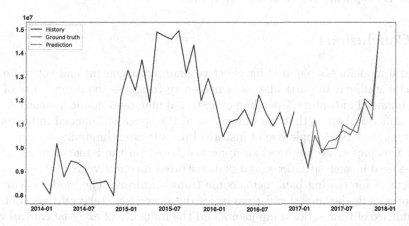

Fig. 7. Net income from acquiring forecast on a group level. By red line we depict predictions, and by green line we depict a ground truth trajectory. (Color figure online)

For each model we conducted a grid search over all tunable hyper-parameters on the held-out set of vintages (with 10% of all vintages). The hidden dimension of the Recurrent layer is chosen from $\{5, 10, 20, 40\}$. To estimate errors we used a standard cross-validation procedure for time-series (so-called rolling-origin-recalibration evaluation procedure, see [1]). The architecture, described in Sect. 5.1, was obtained by this search.

As a result we obtained that the proposed model has error (1) significantly smaller compared to standard XGBoost, RNN-GRU and VARMAX models, see Table 2.

Table 2. Comparison of our model to VARMAX model, XGBoost model and RNN-GRU model. Formula (1) was used to calculate errors of predictions.

Method	Relative error
Filtering in a latent space	14%
RNN-GRU	22%
XGBoost	25%
VARMAX	47%

Moreover, a qualitative study of the behavior of the proposed model depending on the macro-economic indicators showed that the model accurately approximates all major economic crises occurred in the considered historical period: the sharp changes in the values of the macro-economic indicators, observed in the historical data, lead to the predictions of changes in the bank financial indicators that nicely replicate the observed behaviour.

6   Conclusions

Scenario modeling is required for effective bank management and control of its financial stability. In particular, it is necessary to assess the uncertainty of the bank financial indicators depending on external macro-economic indicators. The main difficulty here is that the dimension of the space of financial indicators is very high, but the sample size of historical data is often limited.

In this paper, we proposed an approach based on the latent dynamics that can be used in such situations, and demonstrated its effectiveness by solving the problem of forecasting bank net income from acquiring. The model is capable to represent high-dimensional time-series data, and can take into account the dependence of time-series components and the influence of external control variables on them. Thanks to the Bayesian framework we can construct predictive intervals.

The potential limitation of the method is that the latent space dynamics is markovian (see Eq. (2)), i.e. it is difficult to represent time-series data with longer temporal dependences. To address this limitation an interesting research direction could be to introduce attention-based mechanisms in the model, which are capable to represent long memory phenomena [6].

Acknowledgement. The author would like to thank Sergey Strelkov, Ksenia Gubina, Denis Orlov (Sberbank, Treasury), and Evgeny Egorov (Skoltech) for fruitfull discussions and computational experiments respectively. The work was supported by the RFBR grant 20-01-00203.

References

1. Bergmeir, C., Benítez, J.M.: On the use of cross-validation for time series predictor evaluation. Inf. Sci. **191**, 192–213 (2012). data Mining for Software Trustworthiness
2. Bishop, C.M.: Pattern Recognition and Machine Learning. Information Science and Statistics. Springer, Heidelberg (2006)
3. Casals, J., Garcia-Hiernaux, A., Jerez, M., Sotoca, S., Trindade, A.: State-Space Methods for Time Series Analysis: Theory, Applications and Software. Chapman and Hall/CRC Monographs on Statistics and Applied Probability. CRC Press, Boca Raton (2018)
4. Chen, T., Guestrin, C.: XGBoost: a scalable tree boosting system. In: Proceedings of the 22nd ACM SIGKDD International Conference on Knowledge Discovery and Data Mining. KDD 2016, pp. 785–794. ACM, New York (2016)
5. Cho, K., et al.: Learning phrase representations using RNN encoder-decoder for statistical machine translation. In: Proceedings of the 2014 Conference on Empirical Methods in Natural Language Processing (EMNLP), pp. 1724–1734. Association for Computational Linguistics, Doha (2014)
6. Du, S., Li, T., Yang, Y., Horng, S.J.: Multivariate time series forecasting via attention-based encoder-decoder framework. Neurocomputing **388**, 269–279 (2020)
7. van den End, J.W., Hoeberichts, M., Tabbae, M.: Modelling scenario analysis and macro stress-testing. DNB Working papers 119, Netherlands Central Bank, Research Department, November 2006. https://ideas.repec.org/p/dnb/dnbwpp/119.html
8. Goodfellow, I., Bengio, Y., Courville, A.: Deep Learning. MIT Press, Cambridge (2016). http://www.deeplearningbook.org
9. Habib, R., Barber, D.: Auxiliary variational MCMC. In: International Conference on Learning Representations (2019). https://openreview.net/forum?id=r1NJqsRctX
10. Kingma, D.P., Welling, M.: An introduction to variational autoencoders. Found. Trends Mach. Learn. **12**(4), 307–392 (2019)
11. Ratzlaff, N., Fuxin, L.: HyperGAN: a generative model for diverse, performant neural networks. arXiv preprint arXiv:1901.11058 (2019)
12. Rose, C., et al.: Introduction to Time Series and Forecasting. STSSTS. Springer, New York (2002). https://doi.org/10.1007/b97391
13. Salimans, T., Kingma, D., Welling, M.: Markov chain monte carlo and variational inference: bridging the gap. In: International Conference on Machine Learning, pp. 1218–1226 (2015)
14. Song, J., Zhao, S., Ermon, S.: A-NICE-MC: adversarial training for MCMC. In: Advances in Neural Information Processing Systems, pp. 5140–5150 (2017)
15. Taylor, K.M., Procopio, M.J., Young, C.J., Meyer, F.G.: Estimation of arrival times from seismic waves: a manifold-based approach. Geophys. J. Int. **185**(1), 435–452 (2011). https://doi.org/10.1111/j.1365-246X.2011.04947.x
16. Watter, M., Springenberg, J., Boedecker, J., Riedmiller, M.: Embed to control: a locally linear latent dynamics model for control from raw images. In: Advances in Neural Information Processing Systems, pp. 2746–2754 (2015)
17. Yildiz, C., Heinonen, M., Lahdesmaki, H.: ODE VAE: deep generative second order ODEs with Bayesian neural networks. arXiv preprint arXiv:1905.10994 (2019)

Gradient-Based Adversarial Attacks on Categorical Sequence Models via Traversing an Embedded World

Ivan Fursov[1]([✉]), Alexey Zaytsev[1], Nikita Kluchnikov[1], Andrey Kravchenko[2], and Evgeny Burnaev[1]

[1] Skolkovo Institute of Science and Technology, Moscow, Russia
{ivan.fursov,a.zaytsev,n.kluchnikov,e.burnaev}@skoltech.ru
[2] DeepReason.ai, Oxford, UK
andrey.kravchenko@deepreason.ai

Abstract. Deep learning models suffer from a phenomenon called adversarial attacks: we can apply minor changes to the model input to fool a classifier for a particular example. The literature mostly considers adversarial attacks on models with images and other structured inputs. However, the adversarial attacks for categorical sequences can also be harmful. Successful attacks for inputs in the form of categorical sequences should address the following challenges: **(1)** non-differentiability of the target function, **(2)** constraints on transformations of initial sequences, and **(3)** diversity of possible problems. We handle these challenges using two black-box adversarial attacks. The first approach adopts a Monte-Carlo method and allows usage in any scenario, the second approach uses a continuous relaxation of models and target metrics, and thus allows a usage of state-of-the-art methods for adversarial attacks with little additional effort. Results for money transactions, medical fraud, and NLP datasets suggest that the proposed methods generate reasonable adversarial sequences that are close to original ones, but fool machine learning models.

Keywords: Adversarial attack · Discrete sequential data · Natural language processing

1 Introduction

The deep learning revolution has led to the usage of deep neural network-based models across all sectors in the industry: from self-driving cars to oil and gas. However, the reliability of these solutions are questionable due to the vulnerability of almost all of the deep learning models to adversarial attacks [1] in computer vision [2,3], NLP [4,5], and graphs [6]. The idea of an adversarial attack is to modify an initial object, so the difference is undetectable to a human eye, but fools a target model: a model misclassifies the generated object, whilst for a human it is obvious that the class of the object remains the same [7].

© Springer Nature Switzerland AG 2021
W. M. P. van der Aalst et al. (Eds.): AIST 2020, LNCS 12602, pp. 356–368, 2021.
https://doi.org/10.1007/978-3-030-72610-2_27

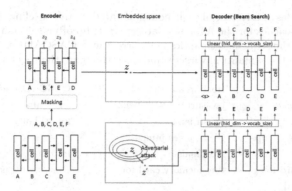

Fig. 1. Top figure: learning of our seq2seq model with the masking of tokens in an initial sequence. We also use beam search and an attention mechanism. Bottom figure: our adversarial attack, modification of a sequence $\mathbf{z}$ in the embedded state to be sure that the decoding of the adversarial sequence $D(\mathbf{z}')$ is close to the decoding $D(\mathbf{z})$, whilst the classifier score is significantly different.

For images we can calculate derivatives of the class probabilities with respect to the colour of pixels in an input image. Thus, moving along this direction we can apply slight alterations to a few pixels, and get a misclassified image, whilst keeping the image almost the same. For different problem statements attacks can be different, but in general a continuous space of images is rich enough for providing adversarial images.

The situation is different for sequential categorical data due to its discrete nature and thus absence of partial derivatives with respect to the input. The space of possible modifications is also limited. For certain problems a malicious user can not modify an object arbitrarily. For example, whilst trying to increase a credit score we can not remove a transaction from the history available to the bank; we only add another transaction. Both of these difficulties impose additional challenges for creation of adversarial attacks for categorical sequential data.

A survey on adversarial attacks for sequences [4,5] presents a list of possible options to overcome these difficulties. With respect to white-box attacks, there are two main research directions. Many approaches work with the initial space of tokens as input attempting to modify these sequences of tokens using operations like addition or replacement [8–10]. Another idea is to move into an embedded space and leverage on gradients-based approaches in this space [11]. We also note that most of these works focus on text sequence data.

We propose two approaches that can alleviate the aforementioned problems with differentiability and a limited space of modification actions, and work in the space of embedded sequences. The first approach is based on a Monte-Carlo search procedure in an embedded space, treating as the energy the weighted sum of the distance between the initial sequence and the generated one and the difference between the probability scores for them. The first term keeps

Table 1. Examples of adversarial sequences generated by the baseline HotFlip and our CASCADA approaches for the AG news dataset. HotFlip often selects the same strong word corrupting the sequence semantics and correctness. CASCADA is more ingenious and tries to keep the semantics, whilst sometimes changing the sequence too much.

Initial sequence x	HotFlip adversarial	CASCADA adversarial
jayasuriya hits back for sri lanka	jayasuriya arafat back for sri lanka	snow hits over back for sri lanka
determined jones jumps into finals	arafat jones jumps into finals	ibm music jumps into match
tiny memory card for mobiles launched	tiny memory card for economy economy	artificial memory card for hewitt pistons
nokia plots enterprise move	nokia plots economy economy	nokia steers enterprise move
sony shrinking the ps	sony shrinking economy ps	sony blames the indies cross
sackhappy d bags bills	google d bags bills	textile d bags bills
tunisian president ben ali reelected	nba president ben ali reelected	bayern hat toshiba got reelected

two sequences close to each other, whilst the second term identifies our intention to fool the classifier and generate a similar but misclassified example for a particular object. This approach is universal, as it does not require derivatives for the first and second terms whilst traversing the embedded space. The number of hyperparameters remains small, and each hyperparameter is interpretable with respect to the problem statement. The second approach illustrates adopts differentiable versions of sequential distance metrics. We use a trained differentiable version of the Levenshtein distance [12] and a surrogate classifier defined on embeddings of sequences. In this case our loss is differentiable, and we can adopt any gradient-based adversarial attack. The two approaches, which we name MCMC and CASCADA attacks, are summarised in Fig. 1. Examples of generated sequences for the AG News dataset are presented in Table 1.

The generative model for adversarial attacks is a seq2seq model with masking [13]. So, the constructed RNN model can be reused for generating adversarial attacks based on these two approaches and creating adversarial attacks with a target direction as well as training embeddings for sequences. The validation of our approaches includes testing on diverse datasets from NLP, bank transactions, and medical insurance domains.

To sum up, we consider the problem of adversarial attack generation for categorical sequential data. The main contributions of this work are the following.

- Our first approach is based on an adaptation of Markov Chain Monte Carlo methods.
- Our second approach uses a continuous relaxation of the initial problem. This makes it possible to perform a classic gradient-based adversarial attack after applying a few new tricks.
- We construct seq2seq models to generate adversarial attacks using an attention mechanism and a beam search, and test the performance for attacking

models based on different principles, e.g. logistic regression for TF-IDF features from a diverse set of domains.
- Our adversarial attacks outperform the relevant baseline attacks; thus it is possible to construct effective attacks for categorical sequential data.

2 Related Work

There exist adversarial attacks for different types of data. The most popular targets for adversarial attacks are images [14, 15], although some work has also been done in areas such as graph data [16] and sequences [17].

It seems that one of the first articles on the generation of adversarial attacks for discrete sequences is [17]. The authors correctly identify the main challenges for adversarial attacks for discrete sequence models: a discrete space of possible objects and a complex definition of a semantically coherent sequence. Their approach considers a white-box adversarial attack with a binary classification problem. We focus on black-box adversarial attacks for sequences. This problem statement was considered in [9, 18, 19].

Extensive search among the space of possible sequences is computationally challenging [20], especially if the inference time for a neural network is significant. Authors of [18] identify certain pairs of tokens and then permute their positions within these pairs, thus working directly on a token level. Another black-box approach from [9] also performs a search at the token level.

It is also possible to use gradients for embeddings [11]. However, the authors of [11] limit directions of perturbations by moving towards another word in an embedded space, and the authors of [11, 21] traverse the embedding space, whilst achieving limited success due to the outdated or complex categorical sequence models. Also, they consider only general perturbations and only NLP problems, whilst it is important to consider more general types of sequences.

As we see from the current state of the art, there is still a need to identify an effective end2end way to explore the space of categorical sequences for the problem of adversarial attacks generation. Moreover, as most of the applications focus on NLP-related tasks, there is still a room for improvement by widening the scope of application domains for adversarial attacks on categorical sequences. Among the methods presented in the literature we highlight HotFlip [10] as the most justified option, so we use compare it with our embeddings-based methods.

3 Methods

We start this section with the description of the general sequence-to-sequence model that we use to generate adversarial sequences, with some necessary details on model training and structure. We then describe the classifier model that we fool using our adversarial model. Next, we describe, how our seq2seq model is used to generate adversarial examples and present our MCMC and CASCADA adversarial attacks. Finally, we provide a description of how to obtain a differentiable version of the Levenshtein distance.

3.1   Models

Sequence-to-sequence Models. Seq2seq models achieve remarkable results in various NLP problems, e.g. machine translation [22], text summarisation [23], and question answering [24]. These models have an encoder-decoder architecture: it maps an initial sequence $\mathbf{x}$ to dense representation using an encoder $\mathbf{z} = E(\mathbf{x})$ and then decodes it using a decoder $\mathbf{x}' = D(\mathbf{z})$ back to a sequence.

Following the ideas from CopyNet [25], we use a seq2seq model with an attention mechanism [22] for the copying problem and train an encoder and a decoder such that $\mathbf{x}' \approx \mathbf{x}$. The final network is not limited to copying the original sequence, but also discovers the nature of the data providing a language model. As the encoder $E(\mathbf{x})$ we use a bi-directional LSTM [26], and as the decoder $D(\mathbf{x})$ we use a uni-directional LSTM with Beam Search [27].

To train the model we mask some tokens from an input sequence, whilst trying to recover a complete output sequence, adopting ideas from MASS [28] and training a CopyNet [25] with the task to reconstruct an initial sequence. Masking techniques include swap of two random tokens, random deletion, random replacement by any other token, and random insertion. The objective for training the model is cross-entropy [29]. As we do not need any labelling, this unsupervised problem is easy to define and train.

In addition, we input a set of possible masking operations $\mathbf{m} = \{m_1, \ldots, m_s\}$. An example of such a set is $\mathbf{m} = \{AddToken, Replace, Delete\}$. We provide $\mathbf{m}$ to the model in addition to input sequence $\mathbf{x}$. As another example, for bank transactions, we can only use the addition of new tokens and $\mathbf{m} = \{AddToken\}$.

Classification Models. As a classifier $C(\mathbf{x})$ we use a one-layer bi-directional LSTM with one fully-connected layer over the concatenation of the mean $\frac{1}{d}\sum_{i=1}^{d} z_i$ and $\max(\mathbf{z})$ of a hidden state $\mathbf{z} = \{z_1, \ldots, z_d\}$ or a logistic regression with TF-IDF features. A classifier takes a sequence $\mathbf{x}$ as input and outputs class probabilities (a classifier score) $C(\mathbf{x}) \in [0, 1]^k$, where k is the number of classes or a class label $c(\mathbf{x})$ on the base of class probability scores $C(\mathbf{x})$.

3.2   Generation of Adversarial Sequences

We generate adversarial sequences for a sequence $\mathbf{x}$ by a targeted modification of a hidden representation $\mathbf{z} = E(\mathbf{x})$ given by encoder $E(\cdot)$ in such a way that the decoder generates an adversarial sequence $A(\mathbf{x})$ that is (1) similar to the original sequence and (2) have a lower probability of a targeted label.

The general attack scheme is presented in Algorithm 1. This attack works under the black-box settings: an attacker has no access to the targeted model. The algorithm uses an encoder, a decoder, word error rate WER between a generated and the initial sequences and a classifier that outputs class probability $C(\mathbf{x})$, and a class label $c(\mathbf{x})$. Slightly abusing the notation we refer to $C = C(\mathbf{x})$ as the classifier score for a class we want to attack in case of multiclass classification. CASCADA attack also uses a surrogate classifier and a surrogate word error rate distance.

The attack algorithm generates a set $\{z_1, \ldots, z_N\}$ of adversarial candidates via consecutive steps $z_i := G(z_{i-1})$ in the embedded space starting at z and selects the best one from the set. The difference between algorithms is in which function $G(z)$ we use.

Input: Number of steps N
Data: Original sequence x and true label c_x
Result: Adversarial sequence $x^* = A(x)$

$z_0 = E(x)$;
for $i \leftarrow 1$ **to** N **do**
    % attack generator step;
    $z_i := G(z_{i-1})$;
    $C_i := C(D(z))$ % score;
    generate class label c_i from score C_i;
    $w_i = WER(D(z_i), x)$;
end
if $\exists i$ s.t. $c_i \neq c_x$ **then**
    $x^* = x_i$ s.t.
    $i = \arg\min_{i:c_i \neq c_x} w_i$;
else
    $x^* = x_i$ s.t. $i = \arg\min_i C_i$;
end
Algorithm 1: The general attack scheme

Input: Embedding z, proposal variance σ^2, energy temperatures σ_{wer}, σ_{class}, initial class label c_0
Result: Attacked embedding $z' = G(z)$

$\varepsilon \sim \mathcal{N}(0, \sigma^2 I)$;
$z' := z + \varepsilon$;
$x' := D(z')$;
$C := C(x')$;
generate class label c from score C;
$w = WER(x', x)$;
$\alpha = \exp\left(\frac{-w}{\sigma_{wer}} + \frac{-[c_0 = c]}{\sigma_{class}}\right)$;
$u \sim \mathcal{U}([0, 1])$;
if $\alpha < u$ **then**
    $z' := z$;
end
Algorithm 2: The MCMC attack defines a generator step $z_i := G(z_{i-1})$, $[\cdot]$ is the indicator function

Naïve Random Walk Attack. The natural approach for generating a new sequence x^* in an embedded space is a random jump to a point z^* in that embedded space from the embedding of an initial sequence $z = E(x)$. An adversarial candidate is a decoder output $x^* = D(z^*)$. As we have a total budget N, we make up to N steps until we find a sufficiently good sequence. Whilst this algorithm seems to be quite simple, it can provide a good baseline against more sophisticated approaches, and can work well enough for an adequate embedding space.

Formally, for this variation of Algorithm 1 we use $z' = G(z) = z + \varepsilon, \varepsilon \sim \mathcal{N}(0, \sigma^2 I)$ with σ^2 being a hyperparameter of our algorithm. Note that in the case of a random walk we defer from the general attack scheme, and each time use the same initial sequence $z_0 = E(x)$ instead of z_{i-1} to get a new sequence z_i.

MCMC Walk. Markov chain Monte Carlo (MCMC) can lead to a more effective approach. We generate a new point using Algorithm 1 with $G(\cdot)$ defined in Algorithm 2 by an MCMC walk. This walk takes into account the similarity between the initial and the generated sequences and the adversity of the target sequence, so we can generate point $z_i := G(z_{i-1})$ at each step more effectively. Similar to the naïve random walk, the MCMC uses the noise variance for embedded space σ. In addition, the MCMC walk approach has temperature parameters σ_{wer} and σ_{class} that identify the scale of the energy we are seeking, and what is the trade-off between the distance among sequences and the drop in the classification score.

The MCMC random walk is designed to make smarter steps and traverses through the embedded space.

CASCADA Attack. Naïve and MCMC attacks can be inefficient. Both of these approaches are computationally expensive for deep seq2seq architectures.

The CASCADA (CAtegorical Sequences Continuous ADversarial Attack) attack is an end-to-end approach, which computes the WER metric and runs a beam search only once.

In the CASCADA approach we use Deep Levenshtein model $WER_{deep}(z, z')$ [12] and a surrogate classification model $C_s(z)$ on top of a seq2seq CopyNet. Both of these models act in the embeddings space. Therefore, we can evaluate derivatives with respect to arguments of $WER_{deep}(z_0, z)$ and $C_s(z)$ inside the target function, thus making it possible to run a gradient-based optimisation that tries to select the adversarial sequence with the best score.

We search for a minimum of a function $C_s(z) + \lambda WER_{deep}(z, z_0)$ with respect to z. The hyperparameter λ identifies a trade-off between trying to get a lower score for a classifier and minimising the distance between z and the initial sequence z_0. So, the attack z' is a solution of the optimisation problem:

$$z' = \arg\min_z C_s(z) + \lambda WER_{deep}(z, z_0).$$

After the generation of a set of candidates during the gradient descent optimisation $z_1, \ldots, z_N$, we apply the decoder to each candidate, obtaining $x_1 = D(z_1), \ldots, x_N = D(z_N)$ as a set of adversarial candidates.

Deep Levenshtein. To make gradient-based updates to an embedded state, we use a differentiable version of the Levenshtein distance function [30]. We use the Deep Levenshtein distance proposed by [12] and considered also in [30]. In our case, WER is used instead of the Levenshtein distance, since we work on the word level instead of the character level for NLP tasks, and for non-textual tasks there are simply no levels other than "token" level.

To collect the training data for each dataset we generate about 2 million pairs. For each pair we apply masks similar to CopyNet, obtaining an original sequence and a close but different sequence. We have also added pairs composed of different sequences from the training data for a better coverage of distant sequences. Our target is $WER_{norm}(\mathbf{x}, \mathbf{y}) = \frac{WER(\mathbf{x},\mathbf{y})}{\max(|\mathbf{x}|,|\mathbf{y}|)}$. We train a model $M(z)$ with the objective $\|\frac{1}{2}(\cos(M(E(\mathbf{x})), M(E(\mathbf{y}))) + 1) - WER_{norm}(\mathbf{x}, \mathbf{y})\|$. The mean absolute error for the learned Deep Levenstein distance $WER_{deep}(z, z') = \frac{1}{2}(\cos(M(z), M(z')) + 1)$ is 0.15 for all considered datasets.

4 Experiments

In this section we describe our experiments. The datasets and the source code are published online[1].

4.1 Datasets

To test the proposed approaches we use NLP, bank transactions, and medical sequence datasets.

We use **NLP dataset** AG news [31] dedicated to topic identification. The four largest classes from the corpus constitute our dataset. The number of training samples for each class is $30,000$ and the number of test samples is $1,900$. We also use **a transactions dataset**, aimed at predicting gender[2]. We use sequences of transactions codes (gas station, art gallery, etc.) and transaction amounts as an input. We also supplement these datasets with another dataset from the **medical insurance** [20] domain. The goal is to detect frauds based on a history of visits of patients to a doctor. Each sequence consists of visits with information about a drug code and amount of money spent for each visit.

For the attacked logistic regression model with TF-IDF features as inputs, the macro-average ROC AUC scores for Transcations-GENDER, Healthcare Insurance and AG News datasets are 0.70, 0.74, 0.88, and 0.96 correspondingly.

Preprocessing of the Datasets. For AG news we use a standard preprocessing procedure. For the healthcare insurance dataset each sequence of tokens consists of medical codes or the procedure assigned after the next visit to a clinic, and a label if the entire sequence for a patient is a fraud or not, with the percentage of frauds in the available dataset being 1.5% and total number of patients being $381,013$.

For the transactions datasets the preprocessing is more complex, so we describe it separately. For the gender prediction dataset we compose each token from the transaction type, the Merchant Category Code (MCC), and the transaction amount bin. We split all amounts into decile bins and then sort them, so index 0 corresponds to the cheapest purchases and index 9 corresponds to the most expensive purchases. An example encoding of a token from a sequence of transactions is 4814_1030_3 with 4814 being the MCC code, 1030 being the transaction type and 3 the index of the decile amount bin. Each sequence corresponds to transactions during the last three days with the mean sequence length being 10.25.

4.2 Metrics

The two types of metrics for the evaluation of the quality of adversarial attacks on sequences are the difference in the classifier score between an initial and a generated adversarial sequences and the distance between these sequences.

---

[1] The code is available at
https://github.com/fursovia/dilma/tree/master. The data is available at https://www.dropbox.com/s/axu26guw2a0mwos/adat_datasets.zip?dl=0.

[2] https://www.kaggle.com/c/python-and-analyze-data-final-project/data.

To measure the performance of the proposed approaches we use three metrics that identify the accuracy drop after adversarial attacks: the ROC AUC drop, the accuracy drop, and the mean classifier score drop. To measure the difference for the new adversarial sequences we use the word error rate (WER) between the initial and generated adversarial sequences.

We also propose a new metric for evaluating adversarial attacks on classifiers for categorical sequences, which combines distance-based and score-based approaches. To get a more realistic metric we perform a normalisation using $WERs$ between the initial and adversarial sequences, which we call the normalised accuracy drop $\mathrm{NAD}(A) = \frac{1}{|Z|} \sum_{i \in Z} 1\{c(\mathbf{x}_i) \neq c(A(\mathbf{x}_i))\} \left(\frac{L_i - \mathrm{WER}(A(\mathbf{x}_i), \mathbf{x}_i)}{L_i - 1} \right)$, where $c(\mathbf{x})$ outputs class labels instead of probabilities $C(\mathbf{x})$, $Z = \{i | c(\mathbf{x}_i) = y_i\}$, and L_i is the maximum length of $\mathbf{x}_i$ and the adversarial sequence $\mathbf{x}_i' = A(\mathbf{x}_i)$ generated by the adversarial attack A.

4.3   Main Experiment for Adversarial Attacks

We compare our approach with the current state of the art, HotFlip [10]. HotFlip at each step selects the best token to change, given an approximation of partial derivatives for all tokens and all elements of the dictionary. To complete the HotFlip attack in our setting we generate N sequences with beam search and then follow our general selection procedure described in Algorithm 1.

We run experiments to keep WER similar for the four considered approaches: HotFlip, random walk attack, MCMC walk attack, and CASCADA. We select hyperparameters to get approximately similar WER scores for different approaches. We generate $N = 100$ sequences for each of the four approaches and select the best one according to the criterion described above.

In Table 2 we present results for the proposed approaches, whilst attacking an independent logistic regression model with TF-IDF features and using LSTM model as a surrogate classifier. We see that embedding-based approaches provide decent performance and are a better way to generated more adversarial examples, while NAD metric puts too significant emphasis on WER values when comparing different approaches.

4.4   Constrained Adversarial Attack

We compare the performance of general and constrained adversarial attacks. In the first case the attack applies all possible modifications to sequences. In the second case only certain perturbations are allowed, e.g. an addition of a token or swapping two tokens. The comparison of performances for various attacks is presented in Table 3: all types of attacks have comparable performances for our CASCADA approach.

4.5   Reliability Study

The selection of hyperparameters often affects the performance of an adversarial attack. We run 599 different hyperparameters configurations for training seq2seq

Table 2. Fooling logistic regression with TF-IDF representations as inputs by running the considered attacks on the four diverse datasets. We maximise metrics with the ↑ signs and minimise metrics with the ↓ signs. Embedding-based methods work better when looking both at perplexity and accuracy drops.

Transactions Gender	ROC AUC drop ↑	Accuracy drop ↑	Probability drop ↑	Normalised WER ↓	Log perplexity ↓	NAD ↑
Random walk	0.539	0.40	0.189	0.561	4.29	0.334
HotFlip	0.243	0.26	0.091	**0.100**	5.15	**0.623**
MCMC walk	**0.640**	**0.55**	**0.245**	0.719	**4.28**	0.333
CASCADA	0.361	0.32	0.121	0.198	4.49	0.426
AG News						
Random walk	0.406	0.66	0.487	0.704	5.21	0.274
HotFlip	0.342	0.67	0.477	**0.218**	6.76	**0.723**
MCMC walk	**0.452**	**0.72**	**0.525**	0.757	**5.16**	0.270
CASCADA	0.422	0.62	0.492	0.385	6.29	0.494
Healthcare insurance						
Random walk	0.566	0.47	0.094	0.725	4.90	0.258
HotFlip	**0.778**	**0.92**	**0.294**	**0.464**	6.75	**0.371**
MCMC walk	0.364	0.29	0.062	0.695	4.50	0.194
CASCADA	0.131	0.26	0.045	0.492	**4.28**	0.106

Table 3. Constrained adversarial attacks on logistic regression with TF-IDF using various masking tokens for the AG news dataset. Log perplexity is almost similar for all approaches.

Masker	Accuracy drop ↑	Normalised WER ↓	NAD ↑
No constraints	**0.62**	**0.39**	**0.492**
Add	**0.62**	0.51	0.382
Replace	0.59	0.50	0.366
Swap	0.61	0.52	0.333

models and the CASCADA adversarial attack based on these models. The results are presented in Fig. 2. We observe that by varying hyperparameters, we select a trade-off between the similarity of initial sequence and an adversarial one and corresponding classifier probability drop. Moreover, varying of hyperparameters for a selected trade-off we observe robust results without significant drop of quality for particular runs or configurations.

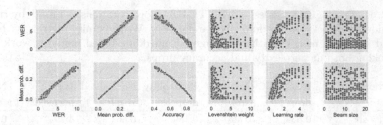

Fig. 2. Mean WER and accuracy drops for various configurations of hyperparameters for the Transactions Gender dataset: the learning rate, the Deep Levenshtein weight, and the beam number. Mean WER and accuracy drop are inversely related as expected, whilst the seq2seq model is robust against changes of hyperparameter values.

5 Conclusion

A construction of an adversarial attack for a categorical sequence is a challenging problem. We consider two approaches to solve this problem: directed random modifications and two differentiable surrogates, for a distance between sequences and for a classifier, that act from an embedded space. The first approach is based on the application of MCMC to generated sequences, and the second approach uses surrogates for constructing gradient attacks. At the core of our approaches lies a modern seq2seq architecture, which demonstrates an adequate performance. To improve results we adopt recent ideas from the NLP world, including masked training and the attention mechanism.

For considered applications, which include NLP, bank card transactions, and healthcare, our approaches show a reasonable performance with respect to common metrics for adversarial attacks and sequence distances. Moreover, we can limit the space of possible modifications, e.g. use only addition operations during an adversarial sequence generation.

Acknowledgments. The work presented in Sect. 3 by Alexey Zaytsev was supported by RSF grant 20–71-10135. The work presented in Sect. 4 by Evgeny Burnaev was supported by RFBR grant 20-01-00203.

References

1. Yuan, X., He, P., Zhu, Q., Li, X.: Adversarial examples: attacks and defenses for deep learning. IEEE Trans. Neural Netw. Learn. Syst. **30**(9), 2805–2824 (2019)
2. Akhtar, N., Mian, A.: Threat of adversarial attacks on deep learning in computer vision: a survey. IEEE Access **6**, 14410–14430 (2018)
3. Khrulkov, V., Oseledets, I.: Art of singular vectors and universal adversarial perturbations. In: IEEE CVPR, pp. 8562–8570 (2018)
4. Zhang, W.E., Sheng, Q.Z., Alhazmi, A., Li, C.: Adversarial attacks on deep-learning models in natural language processing: a survey. ACM Trans. Intell. Syst. Technol. (TIST) **11**(3), 1–41 (2020)

5. Wang, W., Tang, B., Wang, R., Wang, L., Ye, A.: A survey on adversarial attacks and defenses in text. arXiv:1902.07285 preprint (2019)
6. Sun, L., Wang, J., Yu, P.S., Li, B.: Adversarial attack and defense on graph data: a survey. arXiv:1812.10528 preprint (2018)
7. Kurakin, A., Goodfellow, I.J., Bengio, S.: Adversarial machine learning at scale. In: ICLR (2017)
8. Samanta, S., Mehta, S.: Towards crafting text adversarial samples. arXiv:1707.02812 preprint (2017)
9. Liang, B., Li, H., Su, M., Bian, P., Li, X., Shi, W.: Deep text classification can be fooled. In: IJCAI (2017)
10. Ebrahimi, J., Rao, A., Lowd, D., Dou, D.: Hotflip: white-box adversarial examples for text classification. In: Annual Meeting of ACL, pp. 31–36 (2018)
11. Sato, M., Suzuki, H., Shindo, J., Matsumoto, Y.: Interpretable adversarial perturbation in input embedding space for text. In: IJCAI (2018)
12. Moon, S., Neves, L., Carvalho, V.: Multimodal named entity recognition for short social media posts. In: Conference of the North American Chapter ACL: Human Language Technologies, pp. 852–860 (2018)
13. Bowman, S., Vilnis, L., Vinyals, O., Dai, A., Jozefowicz, R., Bengio, S.: Generating sentences from a continuous space. In: SIGNLL CoNNL, pp. 10–21 (2016)
14. Szegedy, C., et al.: Intriguing properties of neural networks. In: ICLR (2014)
15. Goodfellow, I.J., Shlens, J., Szegedy, C.: Explaining and harnessing adversarial examples. In: ICLR (2014)
16. Zügner, D., Akbarnejad, A., Günnemann, S.: Adversarial attacks on neural networks for graph data. In: ACM SIGKDD, pp. 2847–2856 (2018)
17. Papernot, N., McDaniel, P., Swami, A., Harang, R.: Crafting adversarial input sequences for recurrent neural networks. In: IEEE MILCOM, pp. 49–54 (2016)
18. Gao, J., Lanchantin, J., Soffa, M.L., Qi, Y.: Black-box generation of adversarial text sequences to evade deep learning classifiers. In: IEEE Security and Privacy Workshops, pp. 50–56. IEEE (2018)
19. Jin, D., Jin, Z., Zhou, J.T., Szolovits, P.: Is bert really robust? a strong baseline for natural language attack on text classification and entailment. In: AAAI (2020)
20. Fursov, I., Zaytsev, A., Khasyanov, R., Spindler, M., Burnaev, E.: Sequence embeddings help to identify fraudulent cases in healthcare insurance. arXiv:1910.03072 preprint (2019)
21. Ren, Y., et al: Generating natural language adversarial examples on a large scale with generative models. arXiv preprint arXiv:2003.10388 (2020)
22. Bahdanau, D., Cho, K., Bengio, Y.: Neural machine translation by jointly learning to align and translate. In: ICLR (2015)
23. Li, P., Lam, W., Bing, L., Wang, Z.: Deep recurrent generative decoder for abstractive text summarization. In: EMNLP, pp. 2091–2100 (2017)
24. Hu, R., Andreas, J., Rohrbach, M.: Learning to reason: end-to-end module networks for visual question answering. In: IEEE ICCV, pp. 804–813 (2017)
25. Gu, J., Lu, Z., Li, H., Li, V.O.: Incorporating copying mechanism in sequence-to-sequence learning. In: Annual Meeting of ACL, pp. 1631–1640 (2016)
26. Gers, F.A., Schmidhuber, J., Cummins, F.: Learning to forget: continual prediction with LSTM (1999)
27. Graves, A.: Sequence transduction with recurrent neural networks. arXiv:1211.3711 preprint (2012)
28. Song, K., Tan, X., Qin, T., Lu, J., Liu, T.-Y.: Mass: masked sequence to sequence pre-training for language generation. In: ICML, pp. 5926–5936 (2019)

29. Papineni, K., Roukos, S., Ward, T., Zhu, W.-J.: BLEU: a method for automatic evaluation of machine translation. In: Annual Meeting of ACL, pp. 311–318 (2002)
30. Fursov, I., Zaytsev, A., et al.: Differentiable language model adversarial attacks on categorical sequence classifiers. arXiv:2006.11078 preprint (2020)
31. Zhang, X., Zhao, J., LeCun, Y.: Character-level convolutional networks for text classification. In: NeurIPS, pp. 649–657 (2015)

Russia on the Global Artificial Intelligence Scene

Dmitry Kochetkov[1]([✉]) [iD], Aliaksandr Birukou[2,3] [iD], and Anna Ermolayeva[2] [iD]

[1] Higher School of Economics, Moscow, Russia
[2] Peoples' Friendship University of Russia (RUDN University), Moscow, Russia
[3] Springer Nature, Heidelberg, Germany

Abstract. Artificial Intelligence (AI) is a very active research area with a number of applications in business, communication, healthcare, etc. AI attracts great attention in Russia and in late 2019 was included in the country strategy till 2030. In this paper we analyze Russia's position in the space of international scholarly publications on AI. While AI research is interdisciplinary, its core fits the computer science field, where most of the original results are published in conference proceedings and not in journals. Therefore, we consider the list of key AI conferences based on the Australian CORE and Microsoft Academic rankings. We conduct a comparative analysis Russia vs World based on the Scopus data. The metrics used in the analysis allow us to tackle both quantity and quality dimensions. The results of the study are essential for developing measures for the further implementation of the National Strategy for the Development of Artificial Intelligence in Russia and better understanding its position in the international scientific community.

Keywords: Artificial intelligence · Conference proceedings · Ranking · Russia

1 Introduction

Modern Artificial Intelligence appeared in the fifties of the last century and aimed at simulating human intelligence. Before the advent of AI, the only way to study intelligence was by observing living organisms. The research on Artificial Intelligence had three main objectives: 1) creating algorithms that can affect intelligence; 2) creating programs that can demonstrate the effect of human intelligence, using similar processes of human thought activity and 3) creating programs that could supplement or replace human intelligence to perform certain jobs [18]. Now Artificial Intelligence is woven into all areas of our life and its development is paramount. Being a very active research area, AI finds application

Supported by the RUDN University Strategic Academic Leadership Program (recipients A. Birukou and A. Ermolayeva) and the Russian Foundation for Basic Research, project 18-00-01040 KOMFI (recipient D. Kochetkov).

in commerce [12], medicine linguistics [7], robotics [2] and in many others areas. Although the use of AI could possibly eliminate some jobs, at the same time, AI provides many advantages. For example, Artificial Intelligence systems have a great impact on those who work in the commercial sphere, since AI-based systems allow you to access a large amount and quality of information in real time [12]. Another example of using AI is the creation of a neural network that processes students' critical reviews of pseudoscientific works, as well as identifies psychological personality types that are used to determine students who have the ability to work as a scientist [19].

At the end of 2019, Russian government created the Department for Artificial Intelligence and Innovation. In October of the same year, the President of the Russian Federation approved the national strategy for the development of AI until 2030 and proposed the introduction of AI in public administration. Main priorities for the development of the AI strategy in Russia[1] include support of AI research activities, development of AI-based software, availability of more high-quality data that enables AI tools, attracting best AI talent to Russia. Recently, Moscow started a five-year project to improve conditions for the development and implementation of AI. This project will simplify the procedure for working with personal data in AI. Earlier, Moscow launched a database of the projects in the field of Artificial Intelligence in order to collect the best Russian and international AI practices. Also in 2019, AI training programs were launched in 100 Russian universities.

The long-term development of AI is impossible without a strong scientific foundation. While the research on AI is interdisciplinary [3], most of the funding and research are in the field of computer science. According to [13], more than 60% of scientific research conducted in computer science are published in the proceedings of conferences. The same statement is true for AI research where most of the original results are published in conference proceedings.

In the course of our research, top AI conferences were taken, but if you look at the context of communities [15], everything is much more complicated. For example, in [1], it is shown that scientists with a high rating can publish the results of their research in both high-rated and low-rated publications. Also, [21] shows the importance of considering communities for evaluating conferences.

In this paper, we analyze Russia's position in the international research on AI published in conference proceedings. We look at the top AI conferences defined by the Australian CORE and Microsoft Academic conference rankings and use the citation information from Scopus. Section 2 describes the materials and outlines the methodology of the analysis. Section 3 presents the results, while Sect. 4 discusses the implications for the implementation of the National Strategy for the Development of Artificial Intelligence in Russia.

---

[1] https://www.economy.gov.ru/material/departments/d01/razvitie_iskusstvennogp_intellekta/.

2   Data and Methods

In the first step, we have identified a list of top conferences in the field of Artificial Intelligence. We used two data sources: the Australian CORE 2018 conference ranking [4] and the Microsoft Academic field conference rating [14]. CORE provides assessments of major conferences in the computing disciplines and includes 1626 conferences. We filtered conferences by the 'Artificial Intelligence' subject area (FoR code 0801, 364 conferences) and A* rank (19 conferences), the latter denoting the top conferences. We also took the top 20 conferences on Artificial Intelligence from the Microsoft Academic database, which lists 4472 conferences, out of which 1601 are on AI. As both lists overlap, as a result, we received a list of 31 conferences (Table 1).

At the second step we computed the number of citation received by the papers published in the proceedings of those conferences. This was based on the Scopus data [17], time frame 2010–2019. Data were taken by year (2010, 2011,...,2019) as cited articles from the same conference for this time period (2010–2019). We used Scopus because it covered all the conferences in our analysis except for IEEE InfoVis (Fig. 1).

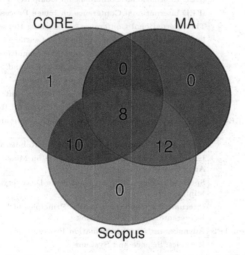

Fig. 1. Coverage.

Eight conferences are indexed in all three databases, and we can consider them as a core:

1. National Conference of the American Association for Artificial Intelligence (AAAI)
2. Meeting of the Association of Computational Linguistics (ACL)
3. IEEE Conference on Computer Vision and Pattern Recognition (CVPR)
4. IEEE International Conference on Computer Vision (ICCV)

Table 1. The list of top conferences in Artificial Intelligence included in the CORE or Microsoft Academic AI conference rankings (alphabetical order). **Bold acronyms** highlight the Core 8 conferences

	Abbreviation	Full
1	**AAAI**	National Conference of the American Association for Artificial Intelligence
2	AAMAS	International Joint Conference on Autonomous Agents and Multiagent Systems
3	**ACL**	Meeting of the Association of Computational Linguistics
4	COLT	Annual Conference on Computational Learning Theory
5	**CVPR**	IEEE Conference on Computer Vision and Pattern Recognition
6	EC	ACM Conference on Economics and Computation
7	ECCV	The European Conference on Computer Vision
8	EMBC	The IEEE Engineering in Medicine and Biology Society
9	EMNLP	The Conference on Empirical Methods in Natural Language Processing
10	FOGA	Foundations of Genetic Algorithms
11	ICAPS	International Conference on Automated Planning and Scheduling
12	ICASSP	The International Conference on Acoustics, Speech, and Signal Processing
13	**ICCV**	IEEE International Conference on Computer Vision
14	ICIP	IEEE International Conference on Image Processing
15	ICLR	The International Conference on Learning Representations
16	**ICML**	International Conference on Machine Learning
17	ICPR	International Conference on Pattern Recognition
18	ICRA	The International Conference on Robotics and Automation
19	IEEE InfoVis	IEEE Information Visualization Conference
20	**IJCAI**	International Joint Conference on Artificial Intelligence
21	IJCAR	International Joint Conference on Automated Reasoning
22	INTERSPEECH	Conference of the International Speech Communication Association
23	IROS	International Conference on Intelligent Robots and Systems
24	ISMAR	IEEE/ACM International Symposium on Mixed and Augmented Reality
25	KDD	Special Interest Group on Knowledge Discovery and Data Mining
26	KR	International Conference on the Principles of Knowledge Representation and Reasoning
27	**NIPS/NeurIPS**	Advances in Neural Information Processing Systems
28	RSS	Robotics: Science and Systems
29	**SIGGRAPH**	ACM SIG International Conference on Computer Graphics and Interactive Techniques
30	SMC	IEEE International Conference on Systems, Man, and Cybernetics
31	UAI	Conference in Uncertainty in Artificial Intelligence

5. International Conference on Machine Learning (ICML)
6. International Joint Conference on Artificial Intelligence (IJCAI)
7. Advances in Neural Information Processing Systems (NIPS/NeurIPS)
8. ACM SIG International Conference on Computer Graphics and Interactive Techniques (SIGGRAPH)

Table 2. Metrics

	Metric	Definition
1	Total output	Total number of publications
2	Average output per year	Total number of publications divided by the number of years with the number of publications higher than zero
3	Total citation score (TCS)	Total number of citations
4	Citations per paper (CPP)	Total citation score divided by total output
5	Mean normalized citation	Average number of citations per a publication score (MNCS) normalized by publication year, title, and affiliation country

During the next step, we defined the metrics for the citation analysis, listed in Table 2). The citations per paper (CPP) metric can be interpreted as the expected citation rate (ECR). This concept is related to a number of widely used metrics; e.g., the impact factor [6]. Finally, we calculated the citation rates of publications by Russian authors in the conference proceedings listed above. A publication was considered from Russia if at least one author had Russian affiliation. This gave us the actual citation rates for the authors from Russia. The metrics used in the analysis allow us to tackle both quantity (publication output) and quality (citations) dimensions.

MNCS is a size-independent item-oriented citation indicator. It can be defined as in [20]:

$$MNCS = \frac{1}{n} \sum_{i=1}^{n} \frac{c_i}{e_i} \qquad (1)$$

Where n is the number of publications, c_i is the actual citation rate, and e_i is the expected citation rate (in our discussion it is also referred to as CPP). In other words, it helps to identify publications which outperformed expectations.

Normalized citation metrics are widely used in analytical tools, for example, Field-Weighted Citation Impact (FWCI) in Elsevier SciVal [5] and Category Normalized Citation Impact (CNCI) in Clarivate Analytics InCites [16] (for more examples see [10]). In both cases, citation rates are normalized by publication year and source type. In our case, we were dealing with the sources of the same type, so the normalization was based on publication year, title, and affiliation country.

3 Results

Proceedings of IEEE Information Visualization Conference have not been indexed in Scopus, that is why we had to exclude it. Table 3 represents the results of calculations for the remaining 30 conferences.

First of all, the extremely small number of publications by Russian scientists at the analyzed conference proceedings is really striking. In almost all cases,

Table 3. Citation metrics for the AI conferences (alphabetical order). **Bold acronyms** highlight the Core 8 conferences and * denotes conferences with MNCS (RU) > 1

Conferences	Total output	Output (RU)	Average output per year	MNCS (RU)	TCS	TCS (RU)	CPP	CPP (RU)
AAAI	3958	8	439.78	0.73	34172	39	8.63	4.88
AAMAS	3400	8	340.00	0.59	20170	22	5.93	2.75
ACL*	3842	21	384.20	1.98	74351	473	19.35	22.52
COLT	402	2	57.43	0.83	4429	17	11.02	8.50
CVPR	6250	21	694.44	0.56	415899	981	66.54	46.71
EC	394	2	65.67	0.00	2231	0	5.66	0.00
ECCV*	3107	17	443.86	1.86	68784	786	22.14	46.24
EMBC	12953	32	1619.13	0.95	48334	121	3.73	3.78
EMNLP	1813	4	201.44	0.31	56109	29	30.95	7.25
FOGA	90	1	18.00	0.00	635	0	7.06	0.00
ICAPS*	626	1	62.60	2.39	5150	8	8.23	8.00
ICASSP	14625	20	1462.50	0.83	116017	94	7.93	4.70
ICCV*	3329	12	665.80	1.57	144935	757	43.54	63.08
ICIP*	8678	6	867.80	1.52	45414	53	5.23	8.83
ICLR	1696	15	282.67	0.20	41690	72	24.58	4.80
ICML*	3653	26	365.30	1.59	82020	807	22.45	31.04
ICPR	5580	21	558.00	0.50	30606	46	5.48	2.19
ICRA*	8974	12	897.40	2.07	126521	131	14.10	10.92
IJCAI	5281	25	586.78	0.49	45140	111	8.55	4.44
IJCAR	252	3	42.00	0.55	2057	4	8.16	1.33
INTERSPEECH*	7861	41	786.10	2.07	54804	379	6.97	9.54
IROS	9023	20	902.30	0.35	87510	53	9.70	2.65
ISMAR*	1105	2	110.50	2.42	8891	9	8.05	4.50
KDD	2416	3	241.60	0.13	71748	13	29.70	4.33
KR*	417	1	83.40	1.97	3290	23	7.89	23.00
NIPS/NeurIPS	3466	19	433.25	0.58	199271	473	57.49	24.89
RSS*	363	1	51.86	3.42	4177	23	11.51	23.00
SIGGRAPH	10789	22	1078.90	0.74	83862	167	7.77	7.59
SMC*	6842	22	760.22	1.14	21185	86	3.10	3.91
UAI*	1459	5	145.90	1.53	8456	6	5.80	1.20

it is less than 1% of the total number of publications. There may be several explanations: a) the growth of the publication activity in Russia started with the launch of the 5–100 academic excellence project in 2013. Effectively, one can only see the results of this project since 2014 - i.e., only in the second half of the period we analyzed. b) the publication culture in most of the disciplines (except for computer science) is based on journals. For instance, the system for research evaluation in Russia, introduced in 2019, i.e., the Comprehensive Methodology for Evaluating Publication Performance (CMEPP) is based on the quartile of the journal as defined by its Impact Factor (IF) in the Web of Science: a journal article published in the first quartile journal gets 20 points, the second quartile - 10 points, the third quartile - 5 points, and the fourth quartile - 2.5 points. All other publications, including conference papers, book chapters, journal articles, indexed only in Scopus or Russian Science Citation

Index, receive 1 point. This scale applies to the natural sciences, engineering, and life sciences. Such a system significantly reduces the incentives for publications in conference proceedings for Russian scientists and contradicts the international practice of publishing in computer science. Previous research [9] shows that if we consider Scimago Journal Ranking (SJR), then for top conferences, it will be higher than that of many first quartile journals (e.g., Proceedings of the IEEE International Conference on Computer Vision [8].

As for citation, in 13 cases out of 30 (including the 3 core conferences), the mean citations per paper for Russian documents is greater or equal to average for this conference, i.e. actual citation rate of Russian authors is greater than or equal to the expected citation rate. However, it is challenging to estimate citation rates based on such a small number of publications. Correlation analysis showed a relatively high rate of Pearson's correlation between the mean citation rate of conference proceedings in general and the citation rate of publications by Russian authors (Fig. 2). Thus, the expected citation rate is the theoretical probability and the actual citation rate is experimental. Theoretical probability is probability that is determined on the basis of reasoning. Experimental probability is probability that is determined on the basis of the results of an experiment repeated many times. With an increase in the number of observations (i.e. publications), the actual citation rate should approach the expected rate. It is also interesting to observe that the average citations per paper for "core" conferences is twice that for the entire set of publications (we compared publications of Russian authors). Therefore, focusing on these eight conferences seems to be a rather promising strategy for increasing one's research impact. Another observation is that for several conferences with more than 10 publications from Russia

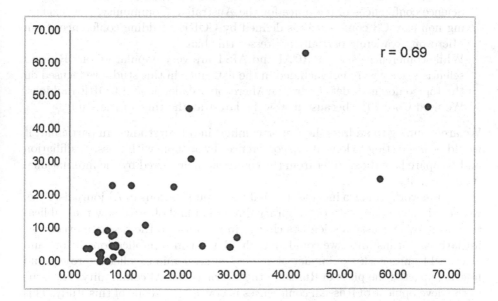

Fig. 2. Correlation analysis of CPP (total) and CPP of papers by Russian authors.

$MNCS(RU) > 1$: ACL, ECCV, ICCV, ICML, ICRA, INTERSPEECH, SMC. This might suggest that Russian AI school is particularly strong in those areas, even though this claim requires further investigation.

4   Conclusion and Future Work

The results presented in this paper constitute an initial attempt to compare the publication practices of the AI researchers from Russia with the international ones. While computer science conferences play the major role in disseminating research results internationally, they do not fancy many publications by authors with the Russian affiliation. From the point of view of maximising the citations it might be advisable to focus on the conferences which have above-average citation per paper rate. However, the bigger picture should also include the analysis of the strengths of the Russian AI community. An interesting research question is how those strengths could be efficiently used to maximise the presence and impact of the top Russian AI schools in the international AI community.

While we aimed at creating an objective list of top AI conferences, where the selection of conferences is guided by clear criteria, there are a couple of shortcomings:

- The ICDM conference is missing from the list due to the error on the CORE website: ICDM is present on the website, but not in the spreadsheet available for the download. As we used the spreadsheet, ICDM was not included in our analysis.
- As the AI is a very popular topic now, our selection is by no means comprehensive and subject to biases. For instance, CORE only lists Computer Science conferences of relevance for the Australian Community, thus excluding non core CS conference (as defined by CORE). Adding conferences from Microsoft Academic partially addresses this bias.
- While conferences such as RCAI and AIST are very popular among Russian scientists, they were not included in the list, since in this study we focused on the top conferences defined by the Microsoft Academic and CORE rankings.
- We used Core 2018 because it was the latest at the time of the study.

We are planning to address these shortcomings in future studies. In particular, it would be interesting to look at top conferences by authors with Russian affiliation and compare how those differ from the conferences preferred by the international AI community.

Future work will also include the analysis of publications in AI journals. Also, we would like to explore the temporary dimension and observe how the publication strategy or Russian scientists change in response to the research excellence initiatives. For instance, we could split the dataset into publications before and after 2014 and check the impact of the 5–100 academic excellence project and how it impacted the place of Russia in the international AI community. The issue of the development of Russian conferences is beyond the scope of this study. This issue has been addressed in a number of studies, for example, [11].

Acknowledgements. This paper has been supported by the RUDN University Strategic Academic Leadership Program (recipients A. Birukou and A. Ermolayeva). The study was also supported by the Russian Foundation for Basic Research, project 18-00-01040 KOMFI "The Impact of Emerging Technologies on Urban Environment and the Quality of Life of Urban Communities" (recipient D. Kochetkov).

References

1. Biryukov, M., Dong, C.: Analysis of computer science communities based on DBLP. In: Lalmas, M., Jose, J., Rauber, A., Sebastiani, F., Frommholz, I. (eds.) ECDL 2010. LNCS, vol. 6273, pp. 228–235. Springer, Heidelberg (2010). https://doi.org/10.1007/978-3-642-15464-5_24

2. Campos-Garduño, E., Tang, Y., Du, J.: Control of an autonomous mobile robot in cluttered environments guided by a single CCD camera. In: Proceedings of the 2019 2nd International Conference on Algorithms, Computing and Artificial Intelligence, pp. 212–218 (2019)

3. Chen, Y.: IoT, cloud, big data and AI in interdisciplinary domains. Simulation Modelling Practice and Theory **102**, (2020). https://doi.org/10.1016/j.simpat.2020.102070

4. Core rankings portal, July 2020. https://www.core.edu.au/conference-portal

5. What is field-weighted citation impact (FWCI)?, July 2020. https://service.elsevier.com/app/answers/detail/a_id/14894/supporthub/scopus//what-is-field-weighted-citation-impact-/

6. Garfield, E., et al.: The impact factor. Curr. Contents **25**(20), 3–7 (1994)

7. Hämäläinen, M., Alnajjar, K.: A template based approach for training NMT for low-resource Uralic languages-a pilot with Finnish. In: Proceedings of the 2019 2nd International Conference on Algorithms, Computing and Artificial Intelligence, pp. 520–525 (2019)

8. Proceedings of the IEEE International Conference on Computer Vision, July 2020. https://www.scimagojr.com/journalsearch.php?q=110561&tip=sid&clean=0

9. Kochetkov, D., Birukou, A., Ermolayeva, A.: The importance of conference proceedings in research evaluation: a methodology for assessing conference impact. J. Scientometrics. arXiv:2010.01540 preprint (2020)

10. Kochetkov, D.M.: A correlation analysis of normalized indicators of citation. Publications **6**(3), 39 (2018)

11. Kutuzov, A., Nikishina, I.: Double-blind peer-reviewing and inclusiveness in Russian NLP conferences. In: van der Aalst, W.M.P., et al. (eds.) AIST 2019. LNCS, vol. 11832, pp. 3–8. Springer, Cham (2019). https://doi.org/10.1007/978-3-030-37334-4_1

12. Lopes da Costa, R., Dias, Á., Pereira, L., António, N., Capelo, A.: The impact of artificial intelligence on commercial management. Impact Artif. Intell. Commer Manag. **4**, 441–452 (2019)

13. Meho, L.I.: Using scopus's citescore for assessing the quality of computer science conferences. J. Informetrics **13**(1), 419–433 (2019)

14. Microsoft academic, July 2020. https://academic.microsoft.com/home

15. Mussi, A., Casati, F., Birukou, A., Cernuzzi, L.: Discovering scientific communities using conference network. In: Proceedings of XXX Sunbelt Social Networks Conference (Sunbelt XXX) (2010)

16. Category normalized citation impact, July 2020. http://help.prod-incites.com/inCites2Live/indicatorsGroup/aboutHandbook/usingCitationIndicatorsWisely/normalizedCitationImpact.html

17. Scopus, July 2020. https://www.scopus.com/

18. Simon, H.A.: Artificial intelligence: an empirical science. Artif. Intell. **77**(1), 95–127 (1995)

19. Ibragim, S., Akhat, B., Dinara, M., Anastasiya, G., Mariya, K., Grigoriy, M.: Example of the use of artificial neural network in the educational process. In: Arai, K., Kapoor, S., Bhatia, R. (eds.) FICC 2020. AISC, vol. 1129, pp. 420–430. Springer, Cham (2020). https://doi.org/10.1007/978-3-030-39445-5_31

20. Waltman, L., van Eck, N.J., van Leeuwen, T.N., Visser, M.S., van Raan, A.F.: Towards a new crown indicator: an empirical analysis. Scientometrics **87**(3), 467–481 (2011)

21. Yavorskiy, R., Voznesenskaya, T., Rudakov, K.: Visualization of data science community in Russia. In: van der Aalst, W.M.P., et al. (eds.) AIST 2018. LNCS, vol. 11179, pp. 3–9. Springer, Cham (2018). https://doi.org/10.1007/978-3-030-11027-7_1

New Properties of the Data Distillation Method When Working with Tabular Data

Dmitry Medvedev[✉] and Alexander D'yakonov

Lomonosov Moscow State University, Moscow, Russia

Abstract. Data distillation is the problem of reducing the volume of training data while keeping only the necessary information. With this paper, we deeper explore the new data distillation algorithm, previously designed for image data. Our experiments with tabular data show that the model trained on distilled samples can outperform the model trained on the original dataset. One of the problems of the considered algorithm is that produced data has poor generalization on models with different hyperparameters. We show that using multiple architectures during distillation can help overcome this problem.

Keywords: Dataset distillation · Knowledge distillation · Neural networks · Synthetic data · Gradient descent · Tabular data

1 Introduction

Data distillation is an aggregation of all possible information from the original training dataset to reduce its volume. The algorithm proposed in [1] tries to produce a small synthetic dataset, which can be used to train models reaching the same quality as with training on the original dataset. In addition, the algorithm also reduces the number of optimization steps needed for training on new data, limiting this number in the objective.

Besides pure scientific interest, research in this new area can be very helpful in practice. For example, often the solution to one problem requires many different models to be trained on the same dataset. The creation of a new synthetic dataset that allows to simultaneously reduce training time for a large number of models with different architectures and hyperparameters would be very helpful. However, the mentioned algorithm has drawbacks. Distilled data is poorly generalized for models not involved in the distillation process. In this paper, we examine the work of the algorithm on tabular data trying to address this problem.

The contribution of this article is the discovery of new properties of the distillation of tabular data. The first one is the generalization increase when distilling with different model architectures simultaneously. The second one is the ability of a model trained on distilled samples to outperform a model trained on the original dataset.

W. M. P. van der Aalst et al. (Eds.): AIST 2020, LNCS 12602, pp. 379–390, 2021.
https://doi.org/10.1007/978-3-030-72610-2_29

The rest of the work is divided into 7 sections. In Sect. 2, we do a short overview of related work. A detailed description of the data distillation algorithm with its complexity analysis is located in Sect. 3. Section 4 consists of descriptions of architectures and the tabular dataset used in the research. In Sect. 5, we show results of experiments and examine the properties of synthetic tabular data. In Sect. 6 we examine the possibility of training models with different architectures on one synthetic dataset. Finally, we present our conclusions in Sect. 7. For more details, you can check out our code.[1]

2   Related Work

The basis of the data distillation algorithm is the optimization of synthetic data and learning rates with backpropagation through the training iterations. The application of backpropagation [6] for optimization of hyperparameters was proposed in [9] and [10]. Backpropagation through L-BFGS [11] and SGD with momentum [12] was presented in [7], and a more memory-efficient algorithm was proposed in [8]. In addition, [8] conducted experiments with data optimization.

The algorithm examined in our work was developed in [1], where successful distillation of the MNIST dataset [4] was shown. Leaving only 10 examples (one for each class), and thus reducing the dataset volume by 600 times, they were able to train the LeNet model [5] to quality close to the quality of training on the original dataset. Also, they proposed to use fixed distribution for network initialization to increase distilled data generalization, but still couldn't reach quality obtained with fixed initialization. It is important to note that authors of the considering algorithm [1] were inspired by network distillation [2], that is, the transfer of knowledge from an ensemble of well-trained models into a single compact one.

The way to distill both objects and their labels was shown in [3]. Authors showed that such distillation increases accuracy for several image classification tasks and allows distilled datasets to consist of fewer samples than the number of classes. Also, they showed the possibility to distill text data.

3   Distillation Algorithm

The simplest version of the algorithm developed in [1] is the one-step version. Let θ_0 be the initial model's weights vector sampled from a fixed distribution $p(\theta_0)$, x be the original data, $\tilde{x}$ be the synthetic data (randomly initialized vectors), $\tilde{\eta}$ be a synthetic learning rate (positive scalar needed in the gradient descent method), and $l(x, \theta)$ be a loss function. If we do gradient descent step and get updated weights θ_1 (index corresponds to step number) then the distillation problem can be written in the following form:

$$\tilde{x}^*, \tilde{\eta}^* = \underset{\tilde{x}, \tilde{\eta}}{\operatorname{argmin}} \mathbb{E}_{\theta_0 \sim p(\theta_0)} l(x, \theta_1) = \underset{\tilde{x}, \tilde{\eta}}{\operatorname{argmin}} \mathbb{E}_{\theta_0 \sim p(\theta_0)} l(x, \theta_0 - \tilde{\eta} \nabla_{\theta_0} l(\tilde{x}, \theta_0)). \quad (1)$$

---

[1] https://github.com/dm-medvedev/dataset-distillation.

To launch the gradient descent method and find the optimum of this problem we have to calculate second-order derivatives, which is possible for the majority of loss functions and model's architectures. In general, when we want to train a model on distilled data for a few steps or even for a few epochs, the algorithm looks a bit more complicated. To describe it we introduce the concepts of external and internal steps and epochs, and the concept of internal models. So, at each internal step of each internal epoch, several internal models are trained on synthetic data. In [1], internal models have the same architecture and different initializations $\theta_0^{(j)} \sim p(\theta_0)$. At each external step of each external epoch, the loss function of these trained models is evaluated on the original data. After this, we can calculate the direction to make the descent step and optimize the synthetic data and learning rates. As a result, the general-case algorithm at each external step solves the following optimization problem:

$$\theta_0^{(j)} \sim p(\theta_0); \quad j = 1, ..., m;$$

$$\theta_{k+1}^{(j)} = \theta_k^{(j)} - \tilde{\eta}_k \nabla_\theta l(\tilde{x}_{i(k)}, \theta_k^{(j)}); \quad k = 0, ..., n-1; \quad i(k) = k \bmod s; \quad (2)$$

$$\mathcal{L} = \frac{1}{m} \sum_{j=1}^{m} l(x, \theta_n^{(j)}) \to \min_{\tilde{x}, \tilde{\eta}}.$$

In (2) s is the number of internal steps of one internal epoch; n is the total number of steps of the internal loop and m is the number of internal models. Optimization requires an estimation of gradients $\nabla_{\tilde{x}}\mathcal{L}$ and $\nabla_{\tilde{\eta}}\mathcal{L}$:

$$d\mathcal{L} = \sum_{j=1}^{m} \frac{\partial \mathcal{L}}{\partial \theta_n^{(j)}} d\theta_n^{(j)} = \sum_{j=1}^{m} \frac{\partial \mathcal{L}}{\partial \theta_n^{(j)}} d\big(\theta_{n-1}^{(j)} - \tilde{\eta}_{n-1} \nabla_\theta l(\tilde{x}_{i(n-1)}, \theta_{n-1}^{(j)})\big)$$

$$= \{g_{n-1}^{(j)} := \tilde{\eta}_{n-1} \nabla_\theta l(\tilde{x}_{i(n-1)}, \theta_{n-1}^{(j)})\} = \sum_{j=1}^{m} \left[\left(\frac{\partial \mathcal{L}}{\partial \theta_n^{(j)}} - \frac{\partial \mathcal{L}}{\partial \theta_n^{(j)}} \frac{\partial g_{n-1}^{(j)}}{\partial \theta_{n-1}^{(j)}} \right) d\theta_{n-1}^{(j)} \right.$$

$$(3)$$

$$\left. - \left(\frac{\partial \mathcal{L}}{\partial \theta_n^{(j)}} \frac{\partial g_{n-1}^{(j)}}{\partial \tilde{\eta}_{n-1}} \right) d\tilde{\eta}_{n-1} - \left(\frac{\partial \mathcal{L}}{\partial \theta_n^{(j)}} \frac{\partial g_{n-1}^{(j)}}{\partial \tilde{x}_{i(n-1)}} \right) d\tilde{x}_{i(n-1)} \right].$$

It is clear that if we continue to express $\theta_{n-1}^{(j)}$ through $\theta_{n-2}^{(j)}$ and so on we will get an expression with $\theta_0^{(j)}$. After summing up all the necessary terms, we obtain the following formulas for gradients:

$$\nabla_{\tilde{\eta}_k}\mathcal{L} = \frac{\partial \mathcal{L}}{\partial \theta_{k+1}} \cdot \frac{\partial g_k}{\partial \tilde{\eta}_k}; \qquad \nabla_{\tilde{x}_j}\mathcal{L} = \sum_{k=0}^{n-1} I[j = i(k)] \cdot \frac{\partial \mathcal{L}}{\partial \theta_{k+1}} \cdot \frac{\partial g_k}{\partial \tilde{x}_i(k)}. \quad (4)$$

Algorithms 1, 2 and 3 show the implementation of the method in pseudo-code. It's clear from the algorithm description that memory and time complexity is high, since we need to store n copies of the internal model and to perform backward and forward passes through all these n copies. This limitation negatively affects the performance, since the increment of n significantly increases the quality of the model trained using the distilled dataset.

Algorithm 1. Main Cycle

1: **Input:** $p(\theta_0)$; m; n; s; T — number of external steps; batch sizes.
2: **Initialization** $\tilde{x}, \tilde{\eta}$
3: **loop** for each $t = 1, ..., T$:
4:      $\nabla_{\tilde{x}}\mathcal{L} = \mathbf{0}, \nabla_{\tilde{\eta}}\mathcal{L} = \mathbf{0}$ ▷ accumulated values of opt. directions.
5:      Get a minibatch of real training data x
6:      **loop** for each model $j = 1, ..., m$:
7:          Model initialization $\theta_0^{(j)} \sim p(\theta_0)$
8:          Res $\leftarrow$ **Forward**   ▷ see Algorithm 2
9:          $\nabla_{\tilde{x}}\mathcal{L}, \nabla_{\tilde{\eta}}\mathcal{L} \leftarrow$ **Backward**   ▷ see Algorithm 3
        Update $\tilde{x}, \tilde{\eta}$

Algorithm 2. Forward Pass

1: **Input:** $\theta_0^{(j)}, \tilde{x}, \tilde{\eta}, x$.
2: Res $\leftarrow \theta_0^{(j)}$
3: **loop** for each internal step $k = 0, ...n - 1$:
4:      $g_k = \tilde{\eta}_k \nabla_\theta l(\tilde{x}_{i(k)}, \theta_k^{(j)})$ ▷ see equation (3)
5:      $\theta_{k+1}^{(j)} = \theta_k^{(j)} - g_k$
6:      Res $\leftarrow g_k, \theta_{k+1}^{(j)}$ ▷ remember computational graph and model
7: $\frac{\partial\mathcal{L}}{\partial\theta_n^{(j)}} = \frac{1}{m}\frac{\partial l(x, \theta_n^{(j)})}{\partial\theta_n^{(j)}}$
8: Res $\leftarrow \frac{\partial\mathcal{L}}{\partial\theta_n^{(j)}}$
9: **Output:** Res

Algorithm 3. Backward Pass

1: **Input:** $\nabla_{\tilde{x}}\mathcal{L}, \nabla_{\tilde{\eta}}\mathcal{L}$, Res   ▷ Res — computational graphs and weights of models.
2: $\frac{\partial\mathcal{L}}{\partial\theta_n^{(j)}} \leftarrow$ Res
3: **loop** for each internal step $k = n - 1, ...0$:
4:      $g_k, \theta_k^{(j)} \leftarrow Res$
5:      $\nabla_{\tilde{x}_{i(k)}}\mathcal{L} \leftarrow \nabla_{\tilde{x}_{i(k)}}\mathcal{L} + \frac{\partial\mathcal{L}}{\partial\theta_{k+1}^{(j)}}\frac{\partial g_k}{\partial\tilde{x}_{i(k)}}$
6:      $\nabla_{\tilde{\eta}_k}\mathcal{L} \leftarrow \nabla_{\tilde{\eta}_k}\mathcal{L} + \frac{\partial\mathcal{L}}{\partial\theta_{k+1}^{(j)}}\frac{\partial g_k}{\partial\tilde{\eta}_k}$
7:      $\frac{\partial\mathcal{L}}{\partial\theta_k^{(j)}} = \frac{\partial\mathcal{L}}{\partial\theta_{k+1}^{(j)}}\left(1 - \frac{\partial g_k}{\partial\theta_k^{(j)}}\right)$
8: **Output:** $\nabla_{\tilde{x}}\mathcal{L}, \nabla_{\tilde{\eta}}\mathcal{L}$

---

4   Data and Models

We consider a simple two-dimensional binary classification problem, which dataset volume is 1,500 objects (see Fig. 1a). The dataset is divided into training and test parts in a 2:1 ratio. We distill the training part and use the test part for quality estimation. We suppose that using such a small amount of data and producing less extreme distillation can help us to explore new properties of the algorithm. Note that it is not always possible with big visual datasets and large models due to the complexity of the distillation. For experiments we use three fully connected architectures: 1-layer, 2-layers and 4-layers. Figure 1(b, c) schematically shows non-linear architectures.

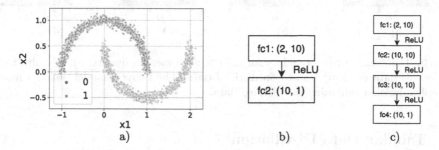

Fig. 1. Used data and non-linear architectures: a) whole original dataset, b) 2-layers model, c) 4-layers model.

First, we estimate the quality of training on the whole dataset. We train models 25 times with random initialization from Xavier distribution. Each training takes 500 epochs. Note that 500 epochs are more than enough for convergence for all three architectures (see Fig. 2a, b), and it seems that the training procedure could be stopped after 200 epochs. Hereinafter, such figures show the boundaries of the 95% confidence interval estimated with the bootstrap procedure.

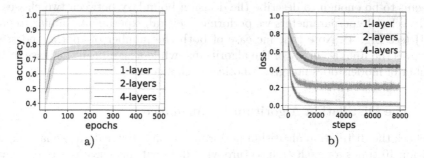

Fig. 2. The quality of training on the whole dataset: a) the convergence of accuracy, b) the convergence of loss.

In addition Table 1 shows mean and standard deviation of achieved accuracy; the largest model reached the highest quality.

Figure 3(a, c) shows the boundaries of the decision rules. For each architecture, it shows the median-quality model. Note that the 2-layers model was not able to significantly outperform the linear solution, despite the significant number of made gradient descent steps and use of the whole training dataset.

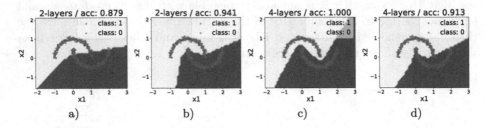

Fig. 3. The decision rule boundary for models with median quality: a), c) trained on the whole dataset; b, d) trained on distilled data. The 2-layers model builds a more complex decision rule using the distilled data.

5   Tabular Data Distillation

5.1   Examining Hyperparameters

The distillation algorithm has several hyperparameters: the number of internal epochs, internal steps and internal models, as they significantly affect complexity, it is important to choose the most appropriate ones. The number of internal models affects the total time of distillation. The number of steps and epochs influences n and thus affects the total time of distillation and the size of needed memory. Note that the number of steps needed to train any model on distilled data is fixed. It means that there is a risk that models may not converge in the preselected number of steps. Another parameter is the number of synthetic objects taking part in each inner step. We choose this parameter to be 4 since it seems to be enough to describe the decision boundary between two classes.

To select hyperparameters we performed several experiments similar to ones in [1] (see Fig. 4). Note that increase of both the number of epochs and steps improves the final accuracy of the algorithms, while a significant increase in the number of models does not give a noticable change.

5.2   The Distillation Algorithm Performance

To check the distillation algorithm performance on tabular data, we launch distillation 10 times for each architecture with different initializations of θ_0, $\tilde{x}$ and $\tilde{\eta}$). Each launch takes 50 outer epochs which is equal to 800 outer iterations since the batch size of 64. For this experiment, we choose the number of internal steps to be 40, the number of internal epochs to be 5 and the number of internal models to be 3. This set of hyperparameters leads to the total number of synthetic objects equals 320, thus the volume of data is reduced by more than 3 times. Note that during training there are $3 \times 800 \times 5 \times 40$ forward passes through the model, and for each gradient descent step we need to store 5×40 model's copies. As a result, total distillation time for all three architectures and 10 restarts reaches 4 h, while the usual training procedure of models on the whole original data lasts only a few minutes. To compare the standard training procedure described in Sect. 4 with training on distilled data we present similar plots.

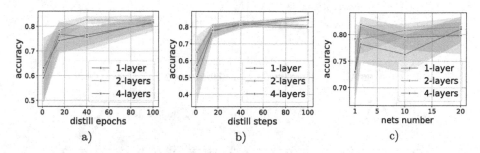

Fig. 4. The dependence of the test quality on: a) the number of internal epochs (1 internal model and 1 internal step); b) the number of internal steps (1 internal model and 1 internal epoch); c) the number of internal models (10 internal steps and 1 internal epoch).

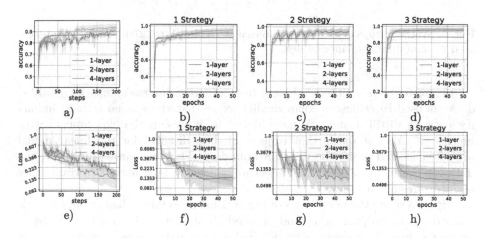

Fig. 5. The convergence of the accuracy and logarithmic loss functions on the test part for various strategies of increasing the number of training steps.

Table 1 shows mean and standard deviation of accuracy. Comparing them we can see that the quality of 1-layer and 2-layers architecture has grown.

We suggest that the distilled data has become additional parameters and allowed the network to better solve the problem. Note that at the same time, the quality of 4-layers models has decreased a bit.

Figures 2 and 5(a, e) show the convergence of both training procedures. Note that the number of steps is not enough to train the 2-layers and 4-layers models on the distilled data. Therefore, we assume that quality will increase if the number of iterations is greater.

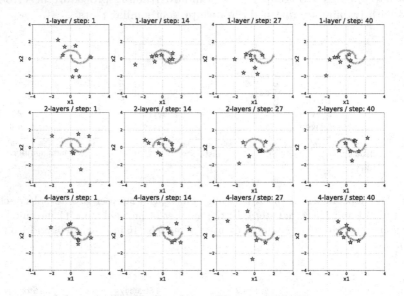

Fig. 6. Synthetic objects of different internal steps (1st, 14th, 27th and last) that were used to train median-quality models.

Figure 6 shows synthetic objects from different internal steps for distilled datasets used to achieve median quality. Note that there is no strong similarity between distilled and original objects, as it was in the experiments with images [1]. Also, it seems that objects of data distilled for the linear model are often located along the decision boundary. Thus we can assume that the data can overfit for a specific architecture which can be a barrier to train other architectures on such data.

Figure 3 shows the boundaries of the decision rules for median-quality models trained on the original (a, c) and distilled (b, d) data. The progress of the 2-layers architecture is clear: using distilled data models can build more complex decision rules, which bring them closer to bigger models.

5.3   The Problem of a Small Number of Epochs

Due to memory and time complexity, it is undesirable to use a large number of internal epochs in the distillation procedure. Fixation of a small number of steps causes the aforementioned problem: it doesn't allow training procedure on distilled data to converge. To overcome this problem, we explore three strategies for artificially increasing the number of epochs when training new models.

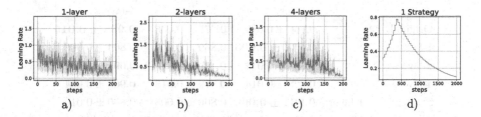

Fig. 7. Learning rates: a-c) obtained by distillation and averaged for 10 different initializations; d) used in the first strategy.

Figure 7(a, b, c) shows synthetic learning rates. Note that there are strong fluctuations from iteration to iteration, so it seems difficult to just replace such a complex scheme with an universal one. Nevertheless, we show that a standard strategy (see Fig. 7d) can improve the quality of models trained on distilled data (see Table 1). This strategy increases step sizes in 1.1 times at the beginning of each new epoch and then similarly decreases them in 0.95 times. Note that the new strategy allowed non-linear models to converge (see Fig. 5b, f) even without using synthetic learning rate.

Other strategies use synthetic learning rates. The second strategy repeats synthetic epochs multiplied by a coefficient. Thus the synthetic learning rate stays unchanged for the first 5 epochs. For the next 5 epochs, each learning rate is multiplied by 0.98 and repeated again. Then, to get 5 more epochs, we again multiply each learning rate but by $0.98 \cdot 0.98 = 0.9604$, and so on we repeat this 10 times. Note that in contrast to the smooth convergence of the previous strategy (see Fig. 5b, f), the new one has strong fluctuations during training (see Fig. 5c, g), but outperforms the previous one (see Table 1).

The third strategy attempts to correct the inaccurate connection of the epochs of the previous strategy. Instead of cyclically repeating all five epochs, only the last one is repeated. Note that the training procedure convergence has indeed become smoother (see Fig. 5d, h) and models reach the higher quality (see Table 1).

6 Data Generalization to Different Architectures

One of the possible practical applications of the data distillation is the fast training of a large number of different architectures with different initializations to acceptable quality. Therefore, it is important for the synthetic data to be well generalized to all variations of networks. To examine this issue, we tried to train models of different architectures on each distilled dataset.

Table 1. Mean and standard deviation of accuracy on the test part for different sets of synthetic data and models. Bold font indicates the biggest value in the column per strategy.

Data models	Test models		
	1-layer	2-layers	4-layers
Original	0.766 ± 0.089	0.877 ± 0.005	**0.995 ± 0.015**
1-layer	**0.871 ± 0.003**	0.869 ± 0.004	0.864 ± 0.006
2-layers	0.808 ± 0.014	**0.941 ± 0.043**	0.691 ± 0.182
4-layers	0.825 ± 0.014	0.879 ± 0.013	0.906 ± 0.054
Strategy1 + 1-layer	**0.863 ± 0.006**	0.860 ± 0.008	0.860 ± 0.010
Strategy1 + 2-layers	0.808 ± 0.010	**0.956 ± 0.047**	**0.985 ± 0.015**
Strategy1 + 4-layers	0.818 ± 0.012	0.911 ± 0.059	0.916 ± 0.062
Strategy2 + 1-layer	**0.869 ± 0.004**	0.867 ± 0.006	0.865 ± 0.005
Strategy2 + 2-layers	0.804 ± 0.012	**0.954 ± 0.065**	0.672 ± 0.229
Strategy2 + 4-layers	0.827 ± 0.017	0.937 ± 0.035	**0.941 ± 0.055**
Strategy3 + 1-layer	**0.870 ± 0.003**	0.866 ± 0.006	0.863 ± 0.008
Strategy3 + 2-layers	0.807 ± 0.011	**0.961 ± 0.041**	0.834 ± 0.210
Strategy3 + 4-layers	0.835 ± 0.016	0.910 ± 0.041	**0.955 ± 0.039**

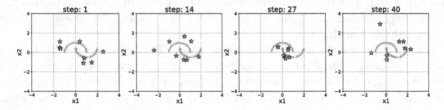

Fig. 8. Synthetic objects of different internal steps (1st, 14th, 27th and last) that were used to train median-quality models. The data distilled using all three architectures.

Table 1 (first subtable) depicts the results. Note that the synthetic data generally acts acceptable for architectures smaller than used in distillation procedure. The worst result was obtained when training a 4-layers model on the data distilled for a 2-layers model. Since the data distilled for 2-layers models doesn't seem much different from the data distilled for 4-layers models, we assume that the problem can be caused by synthetic learning rates. So it makes sense to try each of the three strategies described in the previous section. Next subtables in Table 1 show the result: it seems that for all three cases there are improvements in the quality. The most significant changes touched the worst case. Note that using the strategy without distilled learning rates helped to get much better quality, even higher than when training on data distilled specifically for this architecture.

The natural assumption is that data distillation using all three architectures can lead to better results. To do so we select $m = 3$, but instead of using three internal models with the same architecture, we use three different ones: 1-layer, 2-layers and 4-layers. Table 2 shows the results: for all non-linear architectures, we were able to reach higher accuracy using undistilled learning rates. Note that new synthetic data (see Fig. 8) looks similar to the data from previous experiments (see Fig. 6).

Table 2. Mean and standard deviation of accuracy on the test part for different sets of synthetic data and models. Bold font indicates the biggest value in the column. Data distilled using all three architectures.

Data models	Test models		
	1-layer	2-layers	4-layers
Raw steps	0.859 ± 0.005	0.881 ± 0.004	0.867 ± 0.122
Strategy 1	0.851 ± 0.007	$\mathbf{0.970 \pm 0.028}$	$\mathbf{0.986 \pm 0.014}$
Strategy 2	$\mathbf{0.862 \pm 0.007}$	0.941 ± 0.027	0.984 ± 0.017
Strategy 3	0.858 ± 0.006	0.897 ± 0.014	0.965 ± 0.045

7 Conclusion

In this work, we examined the distillation of tabular data using the algorithm proposed in [1]. We observed that models trained using distilled data can outperform models trained on the whole original data. We show that synthetic objects have generalizability and can be successfully used in the training of different architectures. In addition, we found that it is sometimes better not to use synthetic learning rates, and explored strategies to increase the number of training steps. As future work, we plan to change the memory complexity according to the work [8]. In addition, we want to reduce time complexity and experiment with much more diverse datasets and architectures. Also, we want to improve the distilled data generalizing ability using stochastic depth networks [13]. Finally, we would like to bring the distribution of synthetic objects closer to the original one.

Acknowledgments. We thank anonymous reviewers for many useful recomendations. In particular, we thank Sergey Ivanov for detailed feedback on the initial version of this paper. This research was performed at the Center for Big Data Storage and Analysis of Lomonosov Moscow State University and was supported by the National Technology Initiative Foundation (13/1251/2018 of December 11, 2018).

References

1. Wang, T., Zhu, J., Torralba, A., Efros, A.A.: Dataset distillation. CoRR abs/1811.10959 (2018)
2. Hinton, G., Vinyals, O., Dean, J.: Distilling the knowledge in a neural network. In: NIPS Deep Learning and Representation Learning Workshop (2015)
3. Sucholutsky, I., Schonlau, M.: Soft-label dataset distillation and text dataset distillation. CoRR abs/1910.02551 (2019)
4. MNIST Handwritten Digit Database. http://yann.lecun.com/exdb/mnist/. Accessed 24 June 2020
5. Lecun, Y., Bottou, L., Bengio, Y., Haffner, P.: Gradient-based learning applied to document recognition. Proc. IEEE **86**, 2278–2324 (1998)
6. LeCun, Y., et al.: Backpropagation applied to handwritten zip code recognition. Neural Comput. **1**(4), 541–551 (1989)
7. Domke, J.: Generic methods for optimization-based modeling. In: Proceedings of the Fifteenth International Conference on Artificial Intelligence and Statistics, pp. 318–326. PMLR (2012)
8. Maclaurin, D., Duvenaud, D., Adams, R.: Gradient-based hyperparameter optimization through reversible learning. CoRR abs/1502.03492 (2015)
9. Bengio, Y.: Gradient-based optimization of hyperparameters. Neural Comput. **12**(8), 1889–1900 (2000)
10. Baydin, A., Pearlmutter, B.: Automatic differentiation of algorithms for machine learning. In: Proceedings of the AutoML Workshop at the International Conference on Machine Learning (ICML), Beijing, China, 21–26 June 2014 (2014)
11. Liu, D.C., Nocedal, J.: On the limited memory BFGS method for large scale optimization. Math. Program. **45**, 503–528 (1989). https://doi.org/10.1007/BF01589116
12. Polyak, B.: Some methods of speeding up the convergence of iteration methods. USSR Comput. Math. Math. Phys. **4**, 1–17 (1964)
13. Huang, G., Sun, Y., Liu, Z., Sedra, D., Weinberger, K.: Deep networks with stochastic depth. CoRR abs/1603.09382 (2016)

Unsupervised Anomaly Detection
for Discrete Sequence Healthcare Data

Victoria Snorovikhina$^{(\boxtimes)}$ and Alexey Zaytsev

Skolkovo Institute of Science and Technology, Moscow 121205, Russia

Abstract. Fraud in healthcare is widespread, as doctors could prescribe unnecessary treatments to increase bills. Insurance companies want to detect these anomalous fraudulent bills and reduce their losses. Traditional fraud detection methods use expert rules and manual data processing. Recently, machine learning techniques automate this process, but hand-labeled data is extremely costly and usually out of date. We propose a machine learning model that automates fraud detection in an unsupervised way. Two deep learning approaches include LSTM neural network for prediction next patient visit and a seq2seq model. For normalization of produced anomaly scores, we propose Empirical Distribution Function (EDF) approach. So, the algorithm works with high class imbalance problems.

We use real data on sequences of patients' visits data from Allianz company for the validation. The models provide state-of-the-art results for unsupervised anomaly detection for fraud detection in healthcare. Our EDF approach further improves the quality of LSTM model.

Keywords: Unsupervised anomaly detection · Deep learning · Discrete sequence data

1 Introduction

Healthcare is an essential part of modern society, and the modern medical system is one of the main achievements of humankind. However, both private healthcare companies and government healthcare systems face fraudulent cases every day, and this number keeps increasing every year. Clinics as service providers prescribe unnecessary expensive medications and procedures. Moreover, a patient and a doctor can falsify a patient's diagnosis to get money for medical services. Insurance companies have to cover such excessive bills and want to detect these fraudulent expenses.

Traditionally detection of such frauds was a manual routine for expensive subject area experts [2], but now since machine learning techniques and deep learning tools become a natural part of business processes, automatic fraud detection systems were built [1].

A machine learning guided fraud detection is faster and requires lower human involvement, but the solution is not ideal, as the problem itself is hard.

© Springer Nature Switzerland AG 2021
W. M. P. van der Aalst et al. (Eds.): AIST 2020, LNCS 12602, pp. 391–403, 2021.
https://doi.org/10.1007/978-3-030-72610-2_30

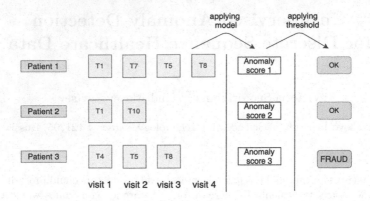

Fig. 1. Pipeline of a proposed solution. For each patient, we have information on a code of prescribed treatment for each visit. We recover these treatments with a generative model and get anomaly scores on the base of errors of our generative model. Then if the anomaly score if higher than a selected threshold, we signal about a fraud. Our generative model can deal with sequences of various length and treatments from a dictionary of a large size.

The typical approach is to hire experts to obtain labeled data [10] and then construct a classification model from an available imbalanced dataset with many fair and a few fraudulent records. Due to a large amount of data and complicated fraud patterns, only experienced auditors are able to detect fraudulent cases; thus, data collection is expensive. Also, machine learning models are able to catch only identified types of frauds. Moreover, as the resulting dataset is imbalanced, we have to carefully construct machine learning involving methods aimed at the solution of imbalanced classification problem [3, 14].

Unsupervised fraud detection systems can successfully deal with these issues. For example, the authors in [10] identify if a particular doctor conducts fraud or not using an open dataset. They assume that doctors with common specialties behave in the same way with similar average bills and medicine price rate [1]. So, these types of models help to detect the doctor, who is prone to fraud. The work [2] detect frauds at the patient level using the private dataset, as there are no open data for this problem. The model takes general information about a patient as input: the number of medical procedures that were provided, the average procedure bill, and so on. So, the existing approaches in healthcare exist but utilize only hand-crafted features [2], thus not being able to detect frauds on the base of complex semi-structured data.

In other areas, anomaly and fraud detection methods are wider, and can roughly be divided into four directions. The first direction considers both classic and neural network supervised [1] and unsupervised Machine Learning algorithms [17]. The second direction considers various probabilistic approaches [18] and relies on an approximation of the generative distribution of the observed data. The third direction adopts autoencoder models [25] and learn data representations using sequence to sequence (seq2seq) architectures. The hybrid models also exist [27]. Recently, the state of the art approaches for anomaly detection based on sequence data are autoencoder models [26] and recurrent neural networks (RNN) [12].

This work advances unsupervised anomaly detection for healthcare semi-structured data. We deal with discrete sequence variables and modify existing anomaly detection approach to handle complex data from large dictionaries. The main idea is to reconstruct a sequence of treatment using a generative model and compare it to the initial one, treating reconstruction error values as anomaly scores. To signal about an anomaly, we apply a threshold to these scores.

The raw data consist of semi-structured sequences of treatments for different patients, plus additional features like patient's age, sex, etc. Treatments are coded, so it could be interpreted as a set of pre-defined tokens. The pipeline of the proposed solution could be seen in Fig. 1.

To sum up, our contributions are the following:

- We apply unsupervised anomaly detection to fraud detection based on healthcare records.
- We adopt a classic anomaly detection approach for regression to a classification problem with a generative model for sequences of treatments. We consider the local LSTM model for the prediction of a single token and a sequence to sequence model based on LSTM to recover the whole sequence.
- For the first model, we provide a new normalization procedure to handle a large dictionary size about 2000 and thus an imbalanced classification problem with a large number of classes.

The paper is organised as follows: In Sect. 2, related work on anomaly detection in sequences (especially in a healthcare) is summarised. Section 3 describes the proposed solution, models and approaches of errors definition. Approach to handle sequence imbalance is also in this section. Section 4 is devoted to machine experiments with real data. Finally, Sect. 5 concludes the paper.

2   Related Works

There are two types of works related to the problem at hand: fraud detection in healthcare and anomaly detection in general, especially anomaly detection for semi-structured sequences. For a general review of anomaly detection in industry and healthcare look at [9], for a recent survey on applications of deep learning to unsupervised anomaly detection, [5,15] and [13] can be useful. Here we present the part of the research that we believe is the most relevant to our studies.

In healthcare, there is a widely-used open dataset *Medicare claims data* [6], which includes aggregated information by doctors and patients. The data have labeling of doctors: do they fraud or not? In [10] authors used Logistic Regression and Random Forest to work with this dataset.

In [2], the authors focus on supervised detection of upcoding fraud, when doctors replace code for an actual service with a code of a more expensive one.

For example, a procedure that lasted fifteen minutes can be coded as a more expensive thirty minutes visit. Used data consisted of a sequence of coded visits.

The paper [1] considers unsupervised approaches to healthcare fraud detection. In particular, the authors investigate the applicability of k-nearest Neighbors, Mahalanobis distance, an autoencoder, and a hybrid approach based on a pre-trained autoencoder without labeled data as input for supervised classifiers. In [19], authors use Generative Adversarial Network to detect anomalies for healthcare providers.

Applications of supervised deep learning models also attract attention in deep learning models [7,8]. The authors in [8] used embedding techniques and both classical Gradient Boosting and deep learning approaches.

For unsupervised anomaly detection in general, there are clustering techniques [17]: authors used the Isolation Forest algorithm, which is considered as one of the most popular and easy in the usage of anomaly detection algorithms. See also usage of probabilistic approaches in [18], usage of sequence to sequence architectures in [25], and usage of hybrid models in [27].

In [21], authors have investigated a problem of human trafficking, which requires detection and interpretability, so they applied used Formal Concept Analysis. In [4], authors also have investigated this algorithm, but for specific sequences. In [20], authors used Hidden Markov models in a healthcare field, which provided better results than FCA.

In [12], the authors used the LSTM to predict each subsequent measurement of the spacecraft. They proposed an automatic threshold selection to determine an anomaly, indeed due to the mean and standard deviation of LSTM errors. It is worth to mention that this domain of study is using raw data as input to a neural network, which provides better accuracy compared to a model based on processed data.

We see that no one proceeds semi-structured medical insurance data to detect fraud in an unsupervised manner. Moreover, general machine learning literature lacks methods that can deal with a moderate token dictionary size for the anomaly detection problem and, in particular, automatically select multiple thresholds for a base anomaly score for each considered class label for a dictionary.

3   Methods

3.1   General Scheme

We have a set of size n of patients. Every patient i is represented as a sequence of observations $X_i = \{\mathbf{x}_{1i}, \mathbf{x}_{2i}, \ldots, \mathbf{x}_{T_i i}\}$, $t \in \overline{1, T_i}$. The total number of visits for a patient is T_i. Each vector $\mathbf{x}_{ji}$ is the description of a particular j-th visit of i-th patient. $\mathbf{x}_{ji}$ consists of treatment type from a dictionary of size d_t, cost type from a dictionary of size d_c and benefit type from a dictionary of size d_b. We also pass the general information $\mathbf{g}_i$ about the patient at each sequential step.

The pipeline of a proposed solution is in Fig. 2. Below we provide more details of each step from this pipeline.

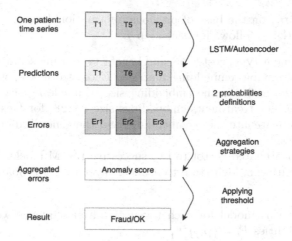

Fig. 2. General scheme of the proposed solution. We predict a sequence using a generation model and obtain errors for each token. Then we get a general anomaly score on the base of aggregation of these errors. Applying a selected threshold, we can say if a particular sequence of tokens is a fraud or not and identify fraudulent tokens as tokens with highest errors.

In order to detect whether a particular sequence has fraud visits or not, we will measure the likelihood of the sequence $p(X_i)$ using a seq2seq approach that we call the Autoencoder model and a token by a token approach which we call LSTM model. To do this, we either pass X_i through the seq2seq model and get probabilities for each token in output $\mathbf{p}_{ij}$ or predict a token using all previous tokes for LSTM model to get another vector $\mathbf{p}_{ij}$. Then we calculate the likelihood of a particular token j using the following formula: $e_{x_{ij}} = 1 - p_{x_{ij}}$, if x_{ij} is a true label token or $e_{x_{ij}} = p_{x_{ij}}$, if x_{ij} is a false label token, where $e_{x_{ij}}$ is an error. We recover through an autoencoder only a part of sequence related to treatments or treatment types. Thus, we provide results of experiments for large (treatments) and small (treatment types) number of classes (tokens).

To get an estimate of a sequence likelihood we, first, built either vector of errors or a matrix of errors. Vector corresponds to the case, where errors on only true token labels are taking into account, the matrix is consist of errors both on true and false token labels. Secondly, we use sum/max pooling to get a single anomaly score for a sequence. The final prediction is , we compare the obtained score with a threshold. We select a threshold to get the recall 0.8. This is the only number we calibrate using fraud labels, we state that our approach is unsupervised.

3.2   Sequence Models

LSTM Model   [11] is the most widely used type of Recurrent neural networks (RNNs) [23]. We use LSTM architecture to get the probabilities for the next treatment $\mathbf{p}_{ij}$ on the base of previous information $\{\mathbf{x}_{1i}, \ldots, \mathbf{x}_{(j-1)i}\}$. The architecture works as follows for each step j:

1. Embed treatment type, cost type, and benefit type using separate embedding layers with trainable embedding matrices E of size $d \times e$, where d is the dictionary size, and e is the embedding size. Embedding size is one for all feature types. For treatments embeddings size is 128, for treatment type it equals 32. Concatenate these embeddings and the general information about a customer $\mathbf{g}_i$.
2. Pass this concatenated vector to two successive LSTM blocks.
3. Pass the resulting hidden state to a linear layer to get probabilities of each token $\mathbf{p}_{ij}$

By applying this model for each token of initial sequence we get a set of vector of probabilities $P_i = \{\mathbf{p}_{ij}\}_{j=1}^{T_i}$

Autoencoder Model is a sequence-to-sequence architecture [24]. We learn the model to copy a sequence, such that the generated sequence is as close to initial as possible. The intuition is that the network learns the representation of sequences structure, so it would be difficult to recover fraudulent sequences with unexpected treatments inside.

The model consists of an encoder and a decoder. The encoder constructs a representation of an input sequence $\mathbf{r}_i = E(X_i)$ that equals to the hidden state of the last recurrent block. The decoder tries to generate the initial sequence for the representation: $X_i' = D(\mathbf{r}_i) \approx X_i$. As a result, the model outputs the probability distribution for every element of a sequence.

We used a bi-directional LSTM network [22] as encoder, unidirectional LSTM network as a decoder. Both had two layers and embedding sizes 128. We also used a context attention vector [16]. Every decoder hidden state is passed through a dense layer by applying a soft-max function, and we obtain probability distribution for the next treatment.

Table 1. Features for the description of each visit of a patient

Feature	Description
Treatment	2204 unique values
Treatment type	17 unique values (aggregated treatments)
Treatment number	Prescribed number of each treatment
Factor	Factor of the treatments' amount
Cost	Cost of a particular visit
Cost type	11 cost categories
Benefit type	24 treatments' combinations types

3.3   Anomaly Score

Given this probability distribution $p(X_i)$ either from LSTM or Autoencoder models, let us define anomaly score.

For every patient the output of a model is a set of vectors of probabilities $P_i = \{\mathbf{p}_{ij}\}_{j=1}^{T_i}$. Length of vectors $\mathbf{p}_{ij}$ equals to the size of the dictionary of treatment d.

Given true labels, we define error as one minus the probability of a true label: $e_{ijx_{ij}} = 1 - p_{ijx_{ij}}$, where $p_{ijx_{ij}}$ is x_{ij}-th element of probability vector that corresponds to the index of the true token label x_{ij}. We calculate errors that correspond to high probabilities of false labels as $e_{ijk} = p_{ijk}$, $k \neq x_{ij}$.

If we concatenate all errors for true classes $e_{ijx_{ij}}$ we get a vector of errors $\mathbf{e}_i = \{e_{ijx_{ij}}\}_{j=1}^{T_i}$. If we concatenate all errors for all classes we have a matrix of errors $E_i = \{e_{ijk}\}_{j=\overline{1,T_i};k=\overline{1,d}}$, where d is the dictionary size.

To get a single anomaly score a_i from a vector or a matrix, we aggregate them using pooling. We consider sum pooling and max pooling, as in our experiments, mean pooling worked worse. For the vector aggregation we get:

$$a_i^{\text{sum}} = \sum_{j=1}^{T_i} e_{ijx_{ij}}, \ a_i^{\text{max}} = \max_{j=\overline{1,T_i}} e_{ijx_{ij}}.$$

For the matrix aggregation, we use E_i instead of $\mathbf{e}$ to extract sum or maximum.

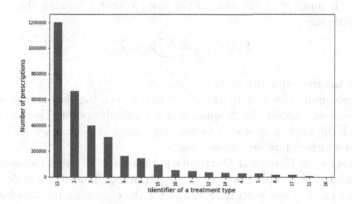

Fig. 3. Distribution of treatment types

3.4   EDF Approach

To normalize the probability scores and get meaningful aggregation, we transform the error scores based on their empirical distribution function (EDF). EDF is an approximation of a theoretical distribution function, based on an observed sample for a random variable. Assume, there is a sample of n independent real

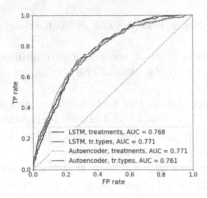 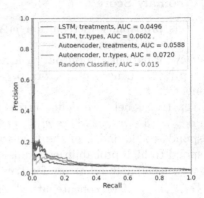

(a) ROC curves of unsupervised fraud detection models with respect to large number of classes (treatments) and small number of classes (treatment types). Best ROC AUC is 0.771.

(b) PR curves of unsupervised fraud detection models with respect to large number of classes (treatments) and small number of classes (treatment types). Best PR AUC is 0.0720.

Fig. 4. Comparison of performance curve for presented algorithms

values with common distribution function $\mathbf{e} = \{e_i\}_{i=1}^n$, then EDF value for a particular e is number of elements in the sample that is smaller than e divided by the sample size n:

$$\mathrm{EDF}(e) = \frac{1}{n} \sum_{i=1}^n [e_i < e],$$

where $[\cdot]$ is the indicator function.

Thus, as cumulative distribution function defines the probability of a variable, EDF defines the relative frequency of a particular point. Therefore, having calculated EDF for errors of every class separately, it provides a better understanding of which points are anomalous.

We construct d Empirical Distribution Functions for each element using a separate validation sample not used during training. Then we transform errors by replacing error values with EDF values for the corresponding label and get a normalized vector $\hat{\mathbf{e}}_i$ or a matrix $\hat{E}_i$. We aggregate these errors in the next step in a similar way, replacing errors with EDF-normalized errors.

4   Experiments

We compare our anomaly detectors based on recovery of treatments and treatment types to each other and a baseline for a considered applied problem from healthcare insurance.

Table 2. Quality comparison of unsupervised fraud detection for LSTM and Autoencoder models. Precision is given for the corresponding recall 0.8. For LSTM additional normalization provided by EDF is useful, while Autoencoder can capture all information without EDF

Model	Treatments			Treatment types		
	ROC AUC	PR AUC	Precision	ROC AUC	PR AUC	Precision
LSTM	0.743	0.0425	0.0164	0.761	0.0681	0.0170 .
LSTM + EDF	0.768	0.0499	**0.0333**	**0.771**	0.0601	**0.0325**
Autoencoder	**0.771**	**0.0588**	0.0331	0.761	**0.0720**	0.0319
Autoencoder + EDF	0.750	0.0483	0.0317	0.760	0.0654	0.0319

4.1 Data

The data for the current research was provided by a major insurance company [8]. The dataset consists of 350 thousand records with anonymous patient's IDs and target labels (fraud or not) for patients. About 1.5% records are fraudulent.

For each patient, we have general features age, sex, insurance type, and total invoice amount and visit-specific features given in Table 1. In our model, visits are coded either as treatments or treatment types. In Fig. 3, we provide a distribution of treatment types concerning its prescribed frequently: the histogram demonstrates a strong class imbalance.

4.2 Results

Metrics. The problem at hand is an imbalanced binary classification, so we use traditional metrics like ROC AUC and area under precision-recall curve PR AUC, where positive samples are the fraudulent ones. We also use ROC and PR curves, as well as precision and recall.

Training Process. We conduct experiments with sequences of treatments and treatment types independently.

Patients have a different number of visits; thus, all sequences were padded with zeros to the closest power of two to an initial sequence length. For the padded elements, a network returns zeros.

The training sample includes 95% of the data, and the test sample includes the remaining 5% of the data. We use 5% of the training sample as a validation set to compare model performance and calculate the EDF function. The test consists of 17000 for patients with 300 fraudulent cases. Distribution of classes in validation, training, and test datasets are the same.

The training process consists of 100 epochs for the LSTM model and 70 epochs for the Autoencoder model. We use the Adam optimization algorithm and a cross-entropy loss. An initial learning rate is 0.001 for LSTM model; 3×10^{-6} and 10^{-6} for treatments and treatment types respectively for autoencoder model. Exponential learning rate decay with a coefficient 0.95 is used. We train the

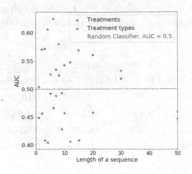 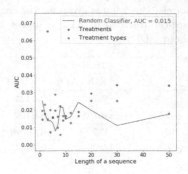

(a) ROC AUC values with respect to visit sequence length.

(b) PR AUC values with respect to visit sequence length.

Fig. 5. Performance of models based on treatments and treatment types for different lengths of sequences. The models work better than a random classifier most of the time.

Table 3. Comparison of the proposed models and a baseline. We present result for our models LSTM and Autoencoder and for a baseline Isolation Forest. For Isolation Forest we can't reach Recall 0.8, so we present precision for the maximum possible value of recall. The best combination of precision and recall is marked in bold. Our models are better, than a baseline

	LSTM		Autoencoder		Isolation Forest (Baseline)	
	Recall	Precision	Recall	Precision	Recall	Precision
Treatments	**0.80**	**0.0333**	0.80	0.0331	0.07	0.08
Treatment types	0.80	0.0325	0.80	0.0319	0.06	0.06

LSTM network in the end-to-end fashion. Parameters at the first iteration are initialized randomly.

The used best hyperparameters come from cross-validation for training data and are given below. Embedding size is one for all feature types, for treatments, it is 128, and for treatment types, it is 32. Batch sizes are 128 for Autoencoder and 256 for the LSTM model. For treatment models, we use sum aggregation for a matrix of errors with EDF for the LSTM model and sum aggregations for a matrix of errors for the Autoencoder model. For treatment types, we use sum aggregation for the matrix of errors with EDF for the LSTM model and sum aggregations for a matrix of errors for the Autoencoder model.

Results. A comparison of quality for both models with and without the EDF approach for the best parameters and aggregation strategies are in Table 2. Precision is calculated with the expected recall 80%. Corresponding ROC and PR curves are in Figs. 4, PR curve. LSTM and Autoencoder models provide similar quality of anomaly detection with ROC AUC 0.77. For a large number of classes in LSTM model, the difference in precision are $\sim 1.7\%$; for a small number of classes is $\sim 1.5\%$. There is no difference in the Autoencoder model either with large or small number of classes. We also examine dependence of ROC AUC, PR AUC on the lengths of visits sequence. They are in Figs. 5a and 5b respectively.

In Table 3 we present a comparison with the Isolation Forest Algorithm based on Word2Vec embeddings of tokens for treatments sequences. Since the algorithm returns class labels, we compare recall and precision values.

5   Conclusion

We have investigated the unsupervised anomaly detection problem. The applied problem is from healthcare insurance, and the data is semi-structured sequences.

We present unsupervised anomaly detection algorithms for semi-structured data never used before in the healthcare industry and compared them. Moreover, we propose an approach to natural normalization of errors based on the Empirical Distribution Function for better handling class imbalance within tokens. On top of these errors, we examine various aggregation strategies to provide a single anomaly score for a sequence.

The overall quality of anomaly detection is similar for various LSTM and Autoencoder models and a various number of classes. Both models outperformed reasonable baselines, and thus provide a new baseline. The usage of normalization further increases the quality of the LSTM model.

Acknowledgments. We thank Martin Spindler for providing the data and Ivan Fursov for providing code for data processing. This work was supported by the federal program "Research and development in priority areas for the development of the scientific and technological complex of Russia for 2014–2020" via grant RFMEFI60619X0008.

References

1. Bauder, R.A., Khoshgoftaar, T.M.: Medicare fraud detection using machine learning methods, pp. 858–865 (2017)
2. Bauder, R., Khoshgoftaar, T., Seliya, N.: A survey on the state of healthcare upcoding fraud analysis and detection. Health Serv. Outcomes Res. Method. **17**, 07 (2016)
3. Branco, P., Torgo, L., Ribeiro, R.P.: A survey of predictive modeling on imbalanced domains. ACM Comput. Surv. (CSUR) **49**(2), 1–50 (2016)
4. Buzmakov, A., Egho, E., Jay, N., Kuznetsov, S., Napoli, A., Raïssi, C.: On mining complex sequential data by means of FCA and pattern structures. Int. J. Gen. Syst. **45** 135–159 (2016)
5. R. Chalapathy and S. Chawla. Deep learning for anomaly detection: A survey. arXiv:1901.03407 (2019)
6. Christopher, I.L.: National cancer institute's surveillance epidemiology and end results (seer) data analysis from nine population-based us cancer registries. JAMA **289**, 1421–1424 (2003)

7. Farbmacher, H., Löw, L., Spindler, M.: An explainable attention network for fraud detection in claims management. Technical report, Technical Report, University of Hamburg (2019)
8. Fursov, I., Zaytsev, A., Khasyanov, R., Spindler, M., Burnaev, E.: Sequence embeddings help to identify fraudulent cases in healthcare insurance. ArXiv, abs/1910.03072 (2019)
9. Habeeb, R.A.A., Nasaruddin, F., Gani, A., Hashem, I.A.T., Ahmed, E., Imran, M.: Real-time big data processing for anomaly detection: a survey. Int. J. Inf. Manage. **45**, 289–307 (2019)
10. Herland, M., Bauder, R.A., Khoshgoftaar, T.M.: Approaches for identifying U.S. medicare fraud in provider claims data. Health Care Manage. Sci. **23**(1), 2–19 (2018). https://doi.org/10.1007/s10729-018-9460-8
11. Hochreiter, S., Schmidhuber, J.: Long short-term memory. Neural Comput. **9**(8), 1735–1780 (1997)
12. Hundman, K., Constantinou, V., Laporte, C., Colwell, I., Söderström, T.: Detecting spacecraft anomalies using LSTMs and nonparametric dynamic thresholding. In: Proceedings of the 24th ACM SIGKDD International Conference on Knowledge Discovery & Data Mining (2018)
13. Kiran, B.R., Thomas, D.M., Parakkal, R.: An overview of deep learning based methods for unsupervised and semi-supervised anomaly detection in videos. J. Imaging **4**(2), 36 (2018)
14. Kozlovskaia, N., Zaytsev, A.: Deep ensembles for imbalanced classification. In: IEEE ICMLA, pp. 908–913. IEEE (2017)
15. Kwon, D., Kim, H., Kim, J., Suh, S.C., Kim, I., Kim, K.J.: A survey of deep learning-based network anomaly detection. Cluster Comput. 1–13 (2019)
16. Qiuxia L., Wenguan W., Khan, S., Shen, J., Sun, H., Shao, L.: Human vs machine attention in neural networks: A comparative study. ArXiv, abs/1906.08764, (2019)
17. Liu, F.T., Ting, K., Zhou, Z.: Isolation forest, pp. 413–422 (2009)
18. Miljković, D.: Review of novelty detection methods. In: Proceedings of the 33rd International Convention (MIPRO), pp. 593–598 (2010)
19. Krishnan, N., Vukosi, N.M.: Unsupervised anomaly detection of healthcare providers using generative adversarial networks. Responsible Design, Implementation and Use of Information and Communication Technology, p. 12066 (2020)
20. Poelmans, J., Dedene, G., Verheyden, G., Van der Mussele, H., Viaene, S., Peters, E.: Combining business process and data discovery techniques for analyzing and improving integrated care pathways. In: Perner, P. (ed.) ICDM 2010. LNCS (LNAI), vol. 6171, pp. 505–517. Springer, Heidelberg (2010). https://doi.org/10.1007/978-3-642-14400-4_39
21. Poelmans, J., Elzinga, P., Ignatov, D., Kuznetsov, S.: Semi-automated knowledge discovery: identifying and profiling human trafficking. Int. J. Gen. Syst. **41**, 11 (2012)
22. Schuster, M., Paliwal, K.K.: Bidirectional recurrent neural networks. Trans. Sig. Proc. **45**(11), 2673–2681 (1997)
23. Sherstinsky, A.: Fundamentals of recurrent neural network and long short-term memory network. Phys. D Nonlinear Phenomena **404**, 132306 (2020)
24. Sutskever, I., Vinyals, O., Le, Q.V.: Sequence to sequence learning with neural networks. CoRR, abs/1409.3215, 2014
25. Wiewel, F., Yang, B.: Continual learning for anomaly detection with variational autoencoder. In: IEEE ICASSP, pp. 3837–3841. IEEE (2019)

26. Zimmerer, D., Kohl, S., Petersen, J., Isensee, F., Maier-Hein, K.: Context-encoding variational autoencoder for unsupervised anomaly detection. arXiv preprint arXiv:1812.05941 (2018)
27. Zong, B., et al.: Deep autoencoding gaussian mixture model for unsupervised anomaly detection (2018)

Theoretical Machine Learning
and Optimization

Lower Bound Polynomial Fast Procedure for the Resource-Constrained Project Scheduling Problem Tested on PSPLIB Instances

E. Kh. Gimadi[1,2] , E. N. Goncharov[1] , and A. A. Shtepa[2(✉)]

[1] Sobolev Institute of Mathematics, Prosp. Akad. Koptyuga, 4, Novosibirsk, Russia
{gimadi,gon}@math.nsc.ru
[2] Novosibirsk State University, Pirogova, 1, Novosibirsk, Russia
http://www.math.nsc.ru/, http://www.nsu.ru/

Abstract. We consider the Resource-Constrained Project Scheduling Problem (RCPSP) with respect to the makespan minimization criterion. The problem accounts for technological constraints of activities precedence together with resource constraints. No activity preemption is allowed. We consider relaxation of RCPSP with special types of non-renewable resources to get a lower bound of the problem. We present new lower bound algorithm for the RCPSP with time complexity depending on the number of activities n as $O(n \log n)$, and we test it on PSPLIB instances. Numerical experiments demonstrate that the proposed algorithm produces on some series of instances lower bounds very close to the best existing lower bounds published in the PSPLIB, while their calculation time is a fraction of a second. We get especially good marks for large-sized instances.

Keywords: Project management · Resource-constrained project scheduling problem · Renewable resources · Cumulative resources · PSPLIB · Lower bound

1 Introduction

We consider the Resource-Constrained Project Scheduling Problem with respect to the makespan minimization criterion. The problem accounts for technological constraints of activities precedence together with resource constraints. No activity preemption is allowed. This problem is denoted as $m, 1|cpm|C_{\max}$ according to the classification scheme proposed in [10]. According to the classification proposed in [3], this problem is denoted as $PS \mid prec \mid C_{\max}$.

Supported by the program of fundamental scientific researches of the SB RAS No. I.5.1., project No. 0314-2019-0014, and by the Russian Foundation for Basic Research, project No. 20-31-90091.

The RCPSP can be defined as a combinatorial optimization problem. A partial order on the set of activities is defined with a directed acyclic graph. For every activity we know duration and the set of consumed resources and their amounts.

The literature (see review [3]) traditionally considers renewable and non-renewable limited resources. In the case of non-renewable resource constraints restrict any quantity of resource allocated for unit time period $t \in \mathbf{Z}^+$ and non-utilized in that period can be consumed at any subsequent period $t' > t$ unlike it happens with renewable resources, where any part of resource non-consumed at time period t is just wasted. Typical representatives of renewable resources are production facilities, equipment, human resources. Non-renewable resources are limited for the entire project as a whole, allowing, for example, to simulate the project budget.

In Russian works [5–8] [14, 21] initially some other classification was used. Limited resources were divided into cumulative (or stored) and non-cumulative (non-stored) ones. The concept of a non-cumulative resource exactly matches the concept of a renewable resource described above. Resource constraints of the cumulative type are also formulated as a balance between the consumed and allocated amount of the resource, while another mechanism for calculating the consumed and allocated resources is used. For each activity j and resource k, we set the intensity $r_{jk}(t)$ of the consumption of the resource at the moment t from the start of its execution. To calculate the amount of resource k required to perform the activity j for a time τ, it is necessary to integrate the function $r_{jk}(t)$ over t in the segment $[0, \tau - s_j]$, where s_j is the starting time of activity j. Similarly, in order to calculate the resource k allocated at time moment τ, it is necessary to integrate the function $R_k(t)$ of the intensity of resource k allocation in the segment $[0, \tau]$. For the schedule to be acceptable with respect to the cumulative resource k, it is necessary that for each moment of time $t \geq 0$, the total amount of the resource consumed by the time t, does not exceed the amount of the resource allocated at the time t. An example of cumulative resources is materials with a long shelf life.

It is easy to see that cumulative resources cannot be taken into account with a combination of renewable and non-renewable resources. Moreover, the introduced concept of partially renewable resources [2], which is a generalization of renewable and non-renewable resources, also does not allow us to describe the restrictions on cumulative resources. In turn, a non-renewable resource can be considered as being cumulative, the entire volume of which is available from the beginning of the project. In this case, the resource is, as it were, allocated with non-zero intensity until the project begins, and with zero intensity after this moment. Cumulative resources are interesting for at least two reasons. Firstly, they more adequately take into account the specifics of some resources with the property of cumulativeness. Secondly, the concept of cumulative resources allows expanding the possibilities for an approximate solution of the "traditional" scheduling problems, since cumulative resources are a weakening of renewable ones. From the results obtained, in particular, it follows that in polynomial time it is possible to obtain lower bounds for RCPSP with renewable resources.

In [21], for the RCPSP with restrictions on cumulative resources, an algorithm was described without justifying the optimality of the obtained solution and analysis of time complexity. In [14], the basic properties of schedules with stored resources were studied and on their basis an algorithm was constructed, from the description of which one can draw a conclusion about its pseudopolynomial time complexity. In [5–8] [14,21], an asymptotically optimal algorithm was proposed for solving a problem with durations of activities from $\mathbf{R}^+$, deadlines, and restrictions on cumulative resources. Its time complexity depends on the number of arcs u in a partial order reduction graph as a function of order $u \log u$, and the absolute error tends to zero with increasing dimension of the problem. In work [8], a discrete analogue of the RCPSP was considered, when all activities have integer durations, and all functions defining resource limitations (such as: the amount of renewable resource consumed by each activity, the intensity of consumption of the cumulative resource, and also the functions of resource availability) are piecewise constant, and all intervals of constancy of these functions are integer. A polynomial time algorithm for solving this problem is given in the particular case, when the model has no renewable resource constraints. (We denote this problem by PS^σ, by analogy with the problem PS^π considered in [2].)

The algorithmic solution to the PS^σ problem was initiated by the need to develop mathematical software for planning large-scale projects related to the construction of the Baikal-Amur Railway in Soviet Union, the development of territorial-industrial complexes in Siberia and the Far East [5], and, further, with the implementation of the West Siberian oil and gas complex. The approach to solving the PS^σ problem in these works is based on the use of so-called T–late schedules.

As noted above, the solution of the RCPSP, weakened through the complete replacement of renewable resources by cumulative ones, makes it possible to obtain a lower bound for the schedule in polynomial time.

There is a large number of publications devoted to the discussion on lower bounds for RCPSP. Recent comprehensive reviews can be found in Neron et al. [16] and Knust [11]. The analysis of the computational complexity of some algorithms and the quality of the obtained bounds are discussed in Gafarov, Lazarev, and Werner [4]. The majority of efficient algorithms can be referred to as "destructive" methods. Such an algorithm starts with a defined project deadline and tries to find a feasible schedule for it. If a feasible solution does not exist, the deadline is increased (usually by incrementing the deadline with 1 unit of time) and the calculation procedure restarts. The calculation continues until the algorithm cannot reveal any contradiction with the defined deadline or until the end of the allocated calculation time. Similarly, constraint programming methods can be applied as for example in Schutt, Feydy, Stuckey, and Wallace [18] and Laborie [15]. Neumann and Schwindt [17] introduced cumulative resources and their practical applications.

In the study of Baptiste and Pape [1], each renewable resource is considered as a system of several identical processors with the number of processors equal to the capacity of the resource. The problem is formulated using mixed integer programming and heuristics are used to solve it.

In this paper we present new lower bound algorithm for the RCPSP with time complexity depending on the number of activities n as $O(n \log n)$. We consider relaxation of RCPSP with special type of non-renewable resources to get a lower bound of the problem. We conduct numerical experiments on the datasets of instances from the PSPLIB electronic library. The results of the computational experiments suggest that the proposed algorithm is a very competitive and on some series of instances yields results close to the best one (this best lower bounds published in PSPLIB), while their CPU time is dramatically small, it is equal to a fraction of a second. We get especially good marks for large-sized instances.

2   Problem Setting

The RCPSP problem can be formulated as follows. A project is taken as a directed acyclic graph $G = (N, A)$. We denote by $N = \{1, ..., n\} \cup \{0, n + 1\}$ the set of activities in the project, where activities 0 and $n + 1$ are dummy. The last activities define the start and the completion of the project, respectively. The precedence relation on the set N is defined with a set of pairs $A = \{(i, j) \mid i \ precedes \ j\}$. If $(i, j) \in A$, then activity j cannot start before activity i has been completed. The set A contains all pairs $(0, j)$ and $(j, n + 1)$, $j = 1, ..., n$.

We have two types of resources: renewable and cumulative one. A set of renewable resources is denoted as $\mathcal{K}^\rho$, and a set of cumulative resources as $\mathcal{K}^\nu$. For each renewable resource $k \in \mathcal{K}^\rho$, $R_k^\rho(t)$ units are available at time t. For each cumulative resource $k \in \mathcal{K}^\nu$, $R_k^\nu(t)$ units of resource of type k arrive at time t and can be consumed at any time $t_1 \geq t$. An activity j has deterministic duration $p_j \in Z^+$ and requires $r_{jk}(t) \geq 0$ units of resource of type k, $k \in \mathcal{K}^\rho \bigcup \mathcal{K}^\nu$ at time $t = 1, ..., p_j$. We assume that $r_{jk}(t) \leq R_k^\rho(t)$, $j \in N$, $k \in \mathcal{K}^\rho$, $t \in \mathbf{Z}^+$ (else problem has no feasible solution). The durations of dummy activities 0 and $n + 1$ are zero, while other activities have non-zero durations. Moreover, dummy activities have zero resource consumption.

Now, we introduce the problem variables. We denote by $s_j \in \mathbf{Z}^+$ the starting time of activity $j \in N$. Since each activity is executed without preemption, the completion time of activity j is equal to $c_j = s_j + p_j$. We define schedule S as $(n + 2)$–vector $(s_0, ..., s_{n+1})$. A makespan of a project $C_{\max}(S)$ corresponds to the moment, when the last activity $n + 1$ is completed, i.e. $C_{\max}(S) = c_{n+1}$. We also introduce the notation $J(t) = \{j \in N \mid s_j < t \leq c_j\}$, $t \in Z^+$, is a set of activities, which are processed during interval $[t - 1; t)$ in schedule S. The problem is to find a feasible schedule $S = \{s_j\}$ respecting to the resource and precedence constraints so that the completion time of the project is minimized. It can be formalized as follows: minimize the makespan of the project

$$C_{\max}(S) = \max_{j \in N}(s_j + p_j) \tag{1}$$

s.t.

$$s_i + p_i \le s_j, \quad \forall (i,j) \in A; \tag{2}$$

$$\sum_{j \in J(t)} r_{jk}(t - s_j) \le R_k^\rho(t), \ k \in \mathcal{K}^\rho, \ t \in \mathbf{Z}^+; \tag{3}$$

$$\sum_{t'=1}^{t} \sum_{j \in J(t')} r_{jk}(t' - s_j) \le \sum_{t'=1}^{t} R_k^\nu(t'), \ k \in \mathcal{K}^\nu, \ t \in \mathbf{Z}^+; \tag{4}$$

$$s_j \in \mathbf{Z}^+, \quad j \in N; \tag{5}$$

Inequalities (2) determine a precedence relation of activities. Conditions (3)–(4) ensure compliance with constraints on renewable and cumulative resources, respectively, i.e. the total amount of consumed resources during interval $[t - 1; t)$ can not exceed the quantity of corresponding available resource during this interval. The last set of inequalities (5) determines the restrictions on problem variables.

The problem (1)–(5) is known to be NP-hard for $\mathcal{K}^\rho \ne \emptyset$.

3 Problem with Renewable Resources

Let us consider special case of the problem (1)–(5), where $\mathcal{K}^\nu = \emptyset$ and the problem isn't dynamic for resource consumption and availability, this problem we denote as $RCPSP^\rho$. We have K different renewable resources, i.e. $K = |\mathcal{K}^\rho|$. Also R_k is the constant amount of available renewable resource k in each unit interval of time, $J(t) = \{j \in N \mid s_j < t \le c_j\}$ is a set of activities, which are processed during interval $[t - 1; t)$. Each activity has deterministic duration p_j and constant intensity of resource consumption r_{jk} of resource k by activity $j \in N$. We assume, that $r_{jk} \le R_k, j \in N, k = 1, ..., K$.

The problem is to find a feasible schedule $S = \{s_j\}$ that minimizes the completion time of the project $C_{max}^\rho(S)$, and can be formalized as follows: minimize the makespan of the project

$$C_{max}^\rho(S) = \max_{j \in N}(s_j + p_j) \tag{6}$$

s.t.

$$s_i + p_i \le s_j, \quad \forall (i,j) \in A; \tag{7}$$

$$\sum_{j \in J(t)} r_{jk} \le R_k, \ k = 1, ..., K, \ t \in \mathbf{Z}^+; \tag{8}$$

$$s_j \in \mathbf{Z}^+, \quad j \in N. \tag{9}$$

Inequalities (7) define precedence relation on a set of activities N. Conditions (8) determine resource constraints. And the last type of inequalities (9) restricts variables of the problem.

4 Problem with Cumulative Resources

Now we consider another special case of the problem (1)–(5) with condition $\mathcal{K}^\rho = \emptyset$ and the problem is static for resource consumption and availability. We denote this problem as $RCPSP^\nu$. Analogically we determine parameters of the problem: graph of precedence relation $G = (N, A)$, K is the number of different cumulative resources, R_k is constant amount of available cumulative resource k in each unit interval of time, $J(t) = \{j \in N \mid s_j < t \le c_j\}$ is a set of activities, which are processed during interval $[t-1; t)$. For arbitrary activity j we have deterministic duration p_j and constant intensity of resource consumption r_{jk} of resource k by activity $j \in N$. So we want to find feasible schedule $S = \{s_j\}$ minimizing makespan of whole project $C^\nu_{max}(S)$ with precedence and resource restrictions:

$$C^\nu_{max}(S) = \max_{j \in N}(s_j + p_j) \tag{10}$$

s.t.

$$s_i + p_i \le s_j, \quad \forall (i, j) \in A; \tag{11}$$

$$\sum_{t'=1}^{t} \sum_{j \in J(t')} r_{jk} \le R_k \cdot t, \ k = 1, ..., K, \ t \in \mathbf{Z}^+; \tag{12}$$

$$s_j \in \mathbf{Z}^+, \quad j \in N. \tag{13}$$

The set of inequalities (11) describes precedence relation of activities. Conditions (12) define constraints on cumulative resources. Inequalities (13) are for restrictions on the variables of the problem.

Lemma 1. *Let the problems $RCPSP^\rho$ and $RCPSP^\nu$ have identical parameters, besides, in the first problem the set of parameters connect to renewable resources and in the second one to cumulative resources. Then solution of the $RCPSP^\nu$ is a lower bound for solution of the $RCPSP^\rho$.*

Proof. Let us consider problems $RCPSP^\rho$ and $RCPSP^\nu$. As we can see all parameters, objective functions and constraints are identical, except the constraints (8) and (12). It is clear that the inequalities (12) are implications of the corresponding conditions (8). So we have broader feasible set for (12) versus (8). Thus, conversion from $RCPSP^\rho$ to $RCPSP^\nu$ enriches the feasible set of the problem, and the minimum of objective function $C^\nu_{max}(S)$ less or equal to $C^\rho_{max}(S)$. Finally, solution of the problem $RCPSP^\nu$ gives us a lower bound for the problem $RCPSP^\rho$.

So we advisedly introduce identical notation (K is a number of resources, R_k is an amount of available resource k in each unit interval of time, r_{jk} is an intensity of consumption of resource k and etc.), because we want to solve $RCPSP^\nu$ to obtain lower bound for $RCPSP^\rho$. Henceforth, when we refer to resources, we do not specify the type of resource: renewable or cumulative, implying that we solve the problem with cumulative resources in order to obtain lower bound for the problem with renewable resources.

5   Fast Exact Algorithm for the $RCPSP^\nu$

In [8], the problem was considered in a more general setting in comparison with (1)–(5), which assumes the possibility of taking into account the deadlines, as well as more detailed description of the intensity of resource consumption functions by each activity, and availability of resources. An exact algorithm was built for this problem, its time complexity is

$$O(\widehat{D}(u + \widetilde{I} + I \log_2 f|K|) + |N_{\text{dir}}| \log_2 N_d),$$

where
$\widehat{D}$ is the length of the binary notation of the maximum deadline;
u is the number of arcs of the reduction graph G;
$\widetilde{I}$ is the total (over all resources $k \in \mathcal{K}^\nu$) number of constancy intervals of functions $R_k^\nu(t)$;
I is the total (over all resources and all activities) number of constancy intervals of functions $r_{jk}(t)$;
f is the width of the partial order;
$|K|$ is the number of constrained resources;
$|N_{\text{dir}}|$ is the number of activities with given deadlines;
N_d is the number of different deadlines.

   In this paper, we consider the problem $RCPSP^\nu$ (10)–(12), in which there are no restrictions on the deadlines, and the functions of the intensity of consumption and resource availability are constant in given time intervals. For this case, we propose the following algorithm.

5.1   Algorithm A

1. We calculate the critical time T_{cr}.
2. We find T_{cr}–late schedule $\{t_j\}$ of activities $j \in N$, sorted by the starting time
   $$0 = t_0 \le t_1 \le \dots \le t_{n+1} = T_{cr}.$$
3. We find a set of different starting or completion times T_h of activities in T_{cr}–late schedule: $0 = T_0 < T_1 < \dots < T_H = T_{cr}$, where H is the number of different starting or completion times of activities. We denote $\mathcal{H} = \{0, 1, \dots, H\}$.
4. We find a set of activities J^+ ordered by non-decreasing of their starting times in T_{cr}–late schedule, and compose a list of A_h^+, $h \in \mathcal{H}$, corresponding to this set, a list of the initial addresses (numbers) of the activities with the starting time in T_{cr}–late schedule equal to T_h.
5. We find a set of activities J^- sorted by non-decreasing of their completion times in T_{cr}–late schedule, and compile a list of A_h^-, $h \in \mathcal{H}$, corresponding to this set, a list of the initial addresses (numbers) of the activities with the completion time in T_{cr}–late schedule equal to T_h.
6. We calculate the dynamics of resource consumption for each type of resource by time layers $(T_h, T_{h+1}]$, $h = 0, \dots, H-1$.
   For this, in the loop on resources $k = 1, \dots, K$ and on layers $h = 0, \dots, H-1$

the procedure $\mathcal{P}(k,h)$ is performed. This procedure calculates the total intensity $r(k,h)$ of resource consumption k in the layer h and the total amount Q_h^k of resources of the type k consumed in the layer h.

Description of Procedure $\mathcal{P}(k,h)$

Remark. We believe that for $h > 0$, the total intensity $r(k,h-1)$ of resource consumption k in the layer $(h-1)$ has already been calculated. For $h = 0$, this value is zero.

(a) We calculate the total intensity of activities with starting times T_h;
(b) We calculate the total intensity of activities with finish times equal to T_h;
(c) We get the total intensity $r(k,h)$ of resource consumption k in the layer h, summing up the total intensity $r(k,h-1)$ of resource consumption k in the layer $(h-1)$ with the result of the item (a), and subtracting the result of the item (b);
(d) We calculate the total amount Q_h^k of resource of the type k consumed in the layer h: $Q_h^k = r(k,h) \cdot (T_{h+1} - T_h)$.

7. We calculate for each k the integral resource functions $\widetilde{R}_h^k = R_k \cdot T_h$ (availability) and $\widetilde{Q}_h^k$ (consumption) by layers: $\widetilde{Q}_0^k = 0$, $\widetilde{Q}_h^k := Q_h^k + \widetilde{Q}_{h-1}^k$, $h = 1, ..., H$.

8. We calculate for each resource k (with availability R_k) the minimum shift of the schedule, at which it is valid for this resource: $\Delta_k = \max\limits_h \left\lceil \frac{\widetilde{Q}_h^k}{R_k} - T_h \right\rceil$.

9. We calculate the maximum shift of the schedule: $\Delta = \max\limits_{1 \le k \le K} \Delta_k$.

10. We calculate the length of the optimal schedule of the problem $RCPSP^\nu$, which gives a lower bound for the problem $RCPSP^\rho$: $T^* = T_{cr} + \Delta$.

11. The optimal schedule of the $RCPSP^\nu$ problem: $s_j^* = t_j + \Delta$, $j \in N$.

Description of Algorithm A is complete.

5.2   Time Complexity of Algorithm A

Theorem 1. *Algorithm A for finding the optimal schedule in the problem* $RCPSP^\nu$ *with limited cumulative resources has time complexity* $\mathcal{O}\Big(n(\log_2 n + d_G + K)\Big)$, *where d_G is the arithmetic average number of immediate predecessors in the reduction graph G.*

Proof. We give estimates of the time complexity for the items of Algorithm A. The Items 1–2 are realized in the $\mathcal{O}(nd_G)$ and $\mathcal{O}\Big(n(\log_2 n + d_G)\Big)$ time complexity, respectively.

The Item 3 is fulfilled in a linear time complexity of n.

The Items 4–5 are performed in the $\mathcal{O}(n \log_2 n)$ time complexity.

The Item 6 is realized in time complexity $\mathcal{O}(nK)$, using available data on intensity of resource consumption, as well as the lists $\{T_h\}$, $\{A_h^+\}$, $\{A_h^-\}$, $h \in \mathcal{H}$.

The Items 7–8 are fulfilled in $\mathcal{O}(nK)$–time, taking into account $H \le n$.

The Items 9–11 are performed in the time $\mathcal{O}(n + K)$.

As a result, we get an estimate of the announced time complexity.
The theorem is proved.

5.3   The Correctness of the Calculation of the Shift Value Δ

The following considerations are important for calculating Δ. For each resource k, we calculate the value Δ_k of the minimum shift of the graph of the integral resource consumption function to the right, at which the (shifted) graph is completely placed under the graph of the integral function $\tilde{R}_h^k$ of resource availability. The maximum over all k from the numbers Δ_k is selected as the value of the Δ.

We consider two cases for some T_{cr}–late schedule.

Case 1: If $\tilde{Q}_h^k \leq R_k \cdot T_h$, $h \in \mathcal{H}$, then this schedule is optimal.

Case 2: If the set $\mathcal{H}^k \subset \mathcal{H}$ of layers h is distinguished with the inverse inequality $\tilde{Q}_h^k > R_k \cdot T_h$. Let consider some layer $h \in \mathcal{H}^k$. Obviously, there is a minimal integer shift-addition to the T_h, which changes the sign of the inequality: $\tilde{Q}_h^k \leq R_k \cdot (T_h + \delta_h^k)$. From the last inequality we get this quantity: $\delta_h^k = \left\lceil \frac{\tilde{Q}_h^k}{R_k} - T_h \right\rceil$. The largest over all $h \in \mathcal{H}^k$ of these quantities is equal to the shift Δ_k, making $(T_{cr} + \Delta_k)$–late schedule is optimal in terms of resources type k. Finally, the shift $\Delta = \max_{k \in K} \Delta_k$ makes $(T_{cr} + \Delta)$–late schedule an optimal for all resources.

6   Computational Experiments

This section presents results of the computational experiments done with the algorithm proposed in this paper. The algorithm proposed in this article was coded in C++ in the Visual Studio system and run on a 2.3 GHz CPU and 6 Gb RAM computer under the operating system Windows 10.

In order to evaluate the performance of the proposed algorithm, we use the standard set presented in Kolisch and Sprecher [12] referred as j30, j60, j90 and j120. These instances are available in the project scheduling library PSPLIB along with their either optimal or best-known solutions, and lower bound values, that have been obtained by various authors over the years.

The datasets j30, j60 and j90 consist of 480 instances (48 series, 10 instances in each series) with 30, 60 and 90 non-dummy activities, respectively. The dataset j120 contains 60 series of instances with 120 non-dummy activities, 10 instances in each series, 600 instances in total. Each instance considers four types of resources. Three parameters: network complexity (NC), resource factor (RF) and resource strength (RS) are combined together to define the full factorial experimental design. The NC reflects the average number of the immediate successors of an activity. The RF sets the average percent of various resource type demand by activities. The RS measures scarcity of the resources. Zero value of the RS factor corresponds to the minimum need for each resource type to execute all activities while the RS value of one corresponds to the required amount of each resource type obtained from the early start time schedule. The parameter values used to built up these instances for the set j30, j60 and j90 are: $NC \in \{1.5, 1.8, 2.1\}$, $RF \in \{0.25, 0.5, 0.75, 1\}$ and $RS \in \{0.1, 0.2, 0.3, 0.4, 0.5\}$. For the instance sets j120 the parameter RS can take the values $RS \in \{0.1, 0.2, 0.3, 0.4, 0.5\}$. It is known [20] that values of the parameters $RF = 1$, $RS = 0.2$ match hard enough series for the datasets j30, j60 and j90 ($RF = 1$, $RS = 0.1$ for the dataset j120). Several authors (e.g. [9, 19]) have pointed out that the parameter RS is a very influential parameter as far as solution quality of exact and heuristic algorithms is concern. The average value of RS is much lower for the set j120, then for the others and the instances with lower values of RS are the most difficult to solve, as several researches have indicated. For more details about the PSPLIB test problems, readers are referred to Kolisch and Sprecher [12, 13].

Table 1. APD Lower Bounds from the best known heuristic Lower Bounds for the dataset j90.

Series	APD optimal solutions from critical path, %	APD LB from best known LB, %	Time, ms	Series	APD optimal solutions from critical path, %	APD LB from best known LB, %	Time, ms
45	67,66	3,07	4,3	7	0	0	5,4
13	56,98	2,15	4,1	8	0	0	6,4
29	56,98	3,29	3,8	10	0	0	10,4
41	52,31	5,71	4,2	11	0	0	7,1
25	47,92	4,39	4	12	0	0	5,6
9	46,25	2,47	5,4	14	0	0	4,1
37	32,09	16,01	5,1	15	0	0	3,8
21	28,98	15,89	4	16	0	0	4
5	24,72	6,40	6,4	18	0	0	4,3
1	9,73	8,00	5	19	0	0	4,3
17	9,69	8,55	4,2	20	0	0	4,1
33	8,38	7,36	4,4	23	0	0	4,3
46	3,82	1,36	4,2	24	0	0	4,5
42	2,28	2,12	4,1	27	0	0	4,1
30	1,46	0,33	4,4	28	0	0	4,5
34	1,42	1,35	4,3	31	0	0	4,3
38	0,72	0,71	4,4	32	0	0	3,5
26	0,47	0,10	4,4	36	0	0	3,8
6	0,42	0,41	5	39	0	0	4,8
22	0,41	0,40	4,3	40	0	0	4,9
35	0,10	0,10	4,5	43	0	0	4,5
2	0	0	5,8	44	0	0	4,1
3	0	0	7,3	47	0	0	4,4
4	0	0	5,8	48	0	0	4,5

Table 2. APD Lower Bounds from the best known heuristic Lower Bounds for the dataset j120.

Series	APD optimal solutions from critical path, %	APD LB from best known LB,%	time, ms	Series	APD optimal solutions from critical path, %	APD LB from best known LB, %	Time, ms
56	152, 97	5, 14	17	22	11,40	9,82	9,6
16	142, 37	1, 23	8,6	42	11,31	8,62	16,8
36	128, 01	1, 73	9,1	39	8,82	1,32	15,3
51	122, 72	4, 33	16,6	19	8,55	1,41	10,5
31	108, 93	1, 88	14	54	8,19	3,93	16,3
11	105, 57	1, 64	12,3	8	6,90	4,22	10,6
6	74, 41	4, 63	8,1	28	5,69	4,23	10,4
46	71, 79	8, 45	16	34	5,45	2,41	10,7
26	67, 68	7, 02	14,6	43	4,68	4,33	15,2
37	64, 75	1, 98	10,7	49	4,19	3,79	15,3
57	64, 46	1, 46	15,4	60	3,93	1,46	18,7
17	59, 30	1, 51	11,4	14	3,65	1,37	11,2
52	54, 02	3, 09	17,4	29	2,74	1,53	12,5
12	53, 25	2, 78	12,7	44	1,91	1,79	15,8
32	39, 92	2, 61	9,6	4	1,60	1,52	9,9
47	32, 83	6, 32	16,4	23	1,19	1,15	10
27	31, 73	4, 63	11,6	24	1,17	1,11	11,5
58	30, 72	2, 41	15,9	9	0,75	0,48	10,5
7	29, 80	5, 21	14,7	25	0,75	0,72	11,5
18	25, 78	2, 56	10,6	50	0,64	0,60	15,8
1	25, 12	14, 71	11	3	0,50	0,48	10
41	24, 67	16, 54	16,1	20	0,41	0,13	10,3
38	24, 30	2, 10	12,2	40	0,38	0,13	14,9
21	21, 07	14, 05	12,2	55	0,31	0	15,4
53	19, 06	2, 83	15,6	30	0,25	0,24	12,8
33	18, 62	2, 84	11,3	45	0,18	0,17	16,5
13	13, 43	4, 21	9,6	5	0	0	15,3
48	12, 74	6, 91	14,9	10	0	0	9,4
2	12, 74	10, 34	13,2	15	0	0	10,10
59	11, 80	3, 31	16,6	35	0	0	12,10

The instances can be found in Kolisch and Sprecher [12] and are downloadable at http://www.om-db.wi.tum.de/psplib/. The measure of the solution quality is the average percent deviation (APD) obtained lower bounds for problem RCPSP from the best lower bounds (BLB) that were taken from the PSPLIB website as of June 2020.

In the tables below, a series of instances are arranged in non-decreasing order APD of the best heuristic solutions (for a sets of j90 and j120) from the lower bounds obtained by the critical path algorithm. The best heuristic solutions for a sets of j90 and j120 were taken from the PSPLIB website as of June 2020 as well. Tables 1, 2 show the APD of the obtained lower bounds from the best known lower bounds and the CPU time (in milliseconds) taken to get them for the datasets j90 and j120, respectively.

As we can see from Tables 1–2, the quality of the lower bounds increases with the dimension of the instances. We have got the lower bounds very close to the best known lower bounds for the hard series from the dataset j120 (see Table 2) with the largest gap between the best solutions found and the length of the critical path.

The average CPU time of algorithm on a regular desktop personal computer on all 480 instances from the j60 dataset was 0,0034 s, on all instances from the j90 dataset was 0,0048 s, and on all 600 instances from the j120 dataset was 0,0134 s.

7   Conclusion

We have proposed a new lower bound algorithm for the Resource-Constrained Project Scheduling Problem with respect to the makespan minimization criterion. This algorithm has polynomial time complexity depending on the number of activities n as $O(n \log n)$. We have considered relaxation of RCPSP with special types of non-renewable resources to get a lower bound of the problem. We have conducted numerical experiments on the datasets of instances from the PSPLIB electronic library. The results of the computational experiments suggest that the proposed algorithm is a very competitive and on some series of instances yields results close to the best one (this best lower bounds published in PSPLIB library), while their CPU time is dramatically small, it is equal to a fraction of a second. We get especially good marks for large-sized instances. Electronic library PSPLIB allowed testing examples of problems with 120 activities. The authors would be glad, if it were possible to carry out the same tests on instances with the dimension of the problem by orders of magnitude greater. In addition, it would be interesting to test a similar algorithm using examples that take into account deadlines and a more detailed description of resource functions.

References

1. Baptiste, P., Pape, C.L.: Constraint propagation and decomposition techniques for highly disjunctive and highly cumulative project scheduling problems. Constraints **5**, 119–139 (2000)
2. Bottcher, J., Drexl, A., Salewski, F.: Project scheduling under partially renewable resource constraints. Manage. Sci. **45**(4), 543–559 (1999)
3. Brucker, P., Drexl, A., Möhring, R., et al.: Resource-constrained project scheduling: notation, classification, models, and methods. Eur. J. Oper. Res. **112**(1), 3–41 (1999)
4. Gafarov, E., Lazarev, A., Werner, F.: On Lower and Upper Bounds for the Resource-Constrained Project Scheduling Problem. Technical report. Otto-von-Guericke Universitaet (2010)
5. Gimadi, E.Kh.: On Some Mathematical Models and Methods for Planning Large-Scale Projects. Models and Optimization Methods. In: Proceedings of AN USSR Sib. Branch, Mathematics Institution, Novosibirsk: Nauka, vol. 10, pp. 89–115 (1988) (in Russian)
6. Gimadi, E., Sevastianov, S.V.: On solvability of the project scheduling problem with accumulative resources of an arbitrary sign. In: Leopold-Wildburger U., Rendl F., Wäscher G. (eds.) Operations Research Proceedings, pp. 241–246. Springer, Berlin (2003) https://doi.org/10.1007/978-3-642-55537-4_39
7. Gimadi, E.K., Sevastianov, S.V., Zalyubovsky, V.V.: On the Project Scheduling Problem under Stored Resource Constraints. In: Proceedings of 8th IEEE International Conference on Emerging Technologies and Factory Automation, Antibes - Juan les Pins, France, pp. 703–706 (2001)
8. Gimadi, E.K., Zalyubovskii, V.V., Sevastyanov, S.V.: Polynomial solvability of scheduling problems with storable resources and deadlines. Diskretnyi Analiz i Issledovanie Operazii, Ser. 2, **7**(1), 9–34 (2000) (in Russian)
9. Hartmann, S., Kolisch, R.: Experimental evaluation of state-of-the-art heuristics for the resource-constrained project scheduling problem. Eur. J. Oper. Res. **127**(2), 394–407 (2000)
10. Herroelen, W, Demeulemeester, E., De Reyck, B.: A Classification Scheme for Project Scheduling. In: Weglarz J. (ed.). Project Scheduling-Recent Models, Algorithms and Applications, International Series in Operations Research and Management Science, Kluwer Academy Publish., Dordrecht, vol. 14, pp. 77–106 (1998)
11. Knust, S.: Lower bounds on the minimum project duration. In: Schwindt, C., Zimmermann, J. (eds.) Handbook on Project Management and Scheduling Vol.1. IHIS, pp. 43–55. Springer, Cham (2015). https://doi.org/10.1007/978-3-319-05443-8_3
12. Kolisch, R., Sprecher, A.: PSPLIB - a Project Scheduling Problem Library. Eur. J. Oper. Res. **96** 205–216 (1996) http://www.om-db.wi.tum.de/psplib/
13. Kolisch, R., Sprecher, A., Drexl, A.: Characterization and generation of a general class of resource-constrained project scheduling problems. Manage. Sci. **41**, 1693–1703 (1995)
14. Kozlov, M.K., Shafranskii, V.V.: Scheduling the implementation of work packages for a given dynamics of the receipt of stored resources. Izv. USSR Acad. Sci. Tech. Cybern. **4**, 75–81 (1977) (in Russian)
15. Laborie, P.: Algorithms for propagation of resource constraints in AI planning and scheduling: existing approaches and new results. Artif. Intell. **143**, 151–188 (2003)
16. Neron, E., et al.: Lower bounds for resource constrained project scheduling problem. In: Jozefowska, J., Weglarz, J. (eds.), Perspectives in Modern Project Scheduling, Chap. 7, pp. 167–204. Springer, Boston (2006)

17. Neumann, K., Schwindt, C.: Project scheduling with inventory constraints. Math. Methods Oper. Res. **56**, 513–533 (2002)
18. Schutt, A., Feydy, T., Stuckey, P., Wallace, M.: Explaining the cumulative propagator. Constraints **16**(3), 250–282 (2011)
19. Valls, V., Ballestin, F., Quintanilla, S.: A hybrid genetic algorithm for the resource-consrtained project scheduling problem. Eur. J. Oper. Res. **185**(2), 495–508 (2008)
20. Weglarz, J.: Project Scheduling. Recent Models, Algorithms and Applications. Kluwer Academy Publication, Boston(1999)
21. Zukhovickii, S.I., Radchik, I.A.: Mathematical Network Planning Methods. Nauka, Moscow (in Russian) (1965)

Fast Approximation Algorithms for Stabbing Special Families of Line Segments with Equal Disks

Konstantin Kobylkin[1,2]([envelope]) [iD]

[1] Krasovsky Institute of Mathematics and Mechanics, Ural Branch of RAS,
Sophya Kovalevskaya Street 16, 620108 Ekaterinburg, Russia
[2] Ural Federal University, Mira Street 19, 620002 Ekaterinburg, Russia

Abstract. An NP-hard problem is considered to stab a given set of n straight line segments on the plane with the smallest size subset of disks of fixed radii $r > 0$, where the set of segments forms a straight line drawing $G = (V, E)$ of a planar graph without proper edge crossings. To the best of our knowledge, only 100-approximation $O(n^4 \log n)$-time algorithm is known (Kobylkin, 2018) for this problem. Moreover, when segments of E are axis-parallel, 8-approximation is proposed (Dash et al., 2012), working in $O(n \log n)$ time. In this work another special setting is considered of the problem where G belongs to classes of special plane graphs, which are of interest in network applications. Namely, three fast $O(n^{3/2} \log^2 n)$-expected time algorithms are proposed: a 10-approximate algorithm for the problem, considered on edge sets of minimum Euclidean spanning trees, a 12-approximate algorithm for edge sets of relative neighborhood graphs and 14-approximate algorithm for edge sets of Gabriel graphs. The paper extends recent work (Kobylkin et al. 2019) where $O(n^2)$-time approximation algorithms are proposed with the same constant approximation factors for the problem on those three classes of sets of segments.

Keywords: Operations research · Computational geometry · Approximation algorithms · Straight line segment · Hippodrome

1 Introduction

Facility location problem represents an important application area for many combinatorial optimization problems. Usually, in facility location problems there

The work is carried out within the research, conducted at the Ural Mathematical Center. It is also supported by the Russian Foundation for Basic Research, project №-19-07-01243. The paper is a substantially extended version of the short paper Kobylkin, K., Dryakhlova, I.: Practical approximation algorithms for stabbing special families of line segments with equal disks. In: Kotsireas, I., Pardalos, P. (eds.) Learning and Intelligent Optimization, LION 2020. Lecture Notes in Computer Science, vol. 12096. Springer, Cham.

W. M. P. van der Aalst et al. (Eds.): AIST 2020, LNCS 12602, pp. 421–432, 2021.
https://doi.org/10.1007/978-3-030-72610-2_32

are objects of interest, e.g. customers, roads, offices or production units, facility objects, say, inventories, stores, markets, cellular base stations, petrol or charging stations, and the problem is to place facilities at the vicinity of objects of interest. Both types of objects are often implied to be geographically distributed. Here optimization is usually done over locations of facilities to achieve the minimum average (or maximal) distance from the placed facilities to the objects of interest. Alternatively, total number is minimized of the located facilities while providing the necessary degree of coverage of all objects of interest. The latter class of problems is called a class of coverage problems.

Both types of facility location problems can adequately be modeled by optimization problems from computational geometry. Here objects of interest are encoded by some simple geometric structures, e.g., points, straight line (or curvilinear) segments and rectangles etc. Besides, facility objects are modeled by translates of fixed objects like points, identical disks, axis-parallel squares or rectangles. The corresponding optimization problem from the class of coverage problems can look as follows: given a set $\mathcal{K}$ of geometric objects on the plane, the smallest cardinality set $\mathcal{C}$ of objects is to be found, chosen from a class $\mathcal{F}$ of simply shaped objects, such that each object from $\mathcal{K}$ is intersected by an object from $\mathcal{C}$ in some prescribed way.

In this paper, subquadratic time small constant factor approximation algorithms are designed for the following problem in which $\mathcal{F}$ is a set of radius r disks and $\mathcal{K}$ coincides with a finite set E of straight line segments on the plane. INTERSECTING PLANE GRAPH WITH DISKS (IPGD): Given a straight line drawing (or a plane graph) $G = (V, E)$ of an arbitrary simple planar graph without proper edge crossings and a constant $r > 0$, find the smallest cardinality set $\mathcal{C}$ of disks of radius r such that $e \cap \bigcup_{C \in \mathcal{C}} C \neq \varnothing$ for each edge $e \in E$. Here each isolated vertex $v \in V$ is treated as a zero-length segment $e_v \in E$. Moreover, the vertex set V assumed to be in general position, i.e. no triple of points of V lies on any straight line.

Below the term "plane graph" is used to denote any straight line embedding of a planar graph whose (straight line) edges intersect at most at their endpoints.

The IPGD problem finds its applications in sensor network deployment and facility location problems related to optimal coverage of objects of some infrastructure with sensors or facilities.

Suppose one needs to provide a certain degree of coverage of a given road network with facility stations, which could be campings, petrol or charging stations, police precincts etc. Geometrically, the network roads can be modeled by piecewise linear arcs on the plane. One can split these arcs into chains of elementary straight line segments such that any two of the resulting elementary segments intersect at most at their endpoints. To cover the road network with facility stations to some extent, it might be reasonable to place the minimum number of stations such that each piece of every road (represented by an elementary segment) is within a given distance from some of the placed stations. This modeling approach gives a geometric combinatorial optimization model, which coincides with the IPGD problem.

The IPGD problem has close connections with the classical geometric HIT-TING SET problem on the plane. To describe a HITTING SET formulation of the IPGD problem, some notation is given below. Suppose $N_r(e) = \{x \in \mathbb{R}^2 : d(x, e) \leq r\}$, $\mathcal{N}_r(E) = \{N_r(e) : e \in E\}$ and $d(x, e)$ is Euclidean distance between a point $x \in \mathbb{R}^2$ and a segment $e \in E$, i.e. Euclidean distance between x and its projection on e; for a zero-length segment $x \in \mathbb{R}^2$ $N_r(x)$ denotes a radius r disk centered at x. Each object from $\mathcal{N}_r(E)$ is a Euclidean r-neighborhood of some segment of E also called r-$hippodrome$ or r-$offset$ in the literature [4].

Thus, the IPGD problem can equivalently be formulated as follows: given a set $\mathcal{N}_r(E)$ of r-hippodromes on the plane whose underlying straight line segments form an edge set of some plane graph $G = (V, E)$, find the minimum cardinality point set C such that $C \cap N \neq \varnothing$ for every $N \in \mathcal{N}_r(E)$. In fact, C represents a set of centers of radius r disks, forming a solution to the IPGD problem. In the sequel, a set $C_0 \subset \mathbb{R}^2$ is called a $piercing$ set for $\mathcal{N}_r(E)$ when $C_0 \cap N \neq \varnothing$ for all $N \in \mathcal{N}_r(E)$.

1.1 Related Work and Our Results

Settings close to the IPGD problem are originally considered in [4]. Motivated by applications from sensor monitoring for urban road networks, they explore the case in which $\mathcal{F}$ contains equal disks and E consists of (generally properly overlapping) axis-parallel segments. Their algorithms can easily be extended to the case of sets E of straight line segments with bounded number of distinct orientations.

In [11] constant factor approximation algorithms are first proposed for the IPGD problem. Namely, a 100-approximate $O(n^4 \log n)$-time algorithm is given for the problem in its general setting where E is formed by an edge set of an arbitrary plane graph. Moreover, due to applications, 68- and 54-approximate algorithms are given in [12] for special cases where E is an edge set of a generalized outerplane graph and a Delaunay triangulation respectively as well as a 23-approximation algorithm is proposed under the assumption that all pairs of non-overlapping segments from E are at the distance more than r from each other.

Let us give some definitions. Let V be a finite point set in general position on the plane. Assuming that no 4 points of V lie on any circle, a plane graph $G = (V, E)$ is called a $Gabriel$ graph [14] when $[u, v] \in E$ iff intersection of V is empty with interior of the disk with diameter $[u, v]$. Under the same assumption a plane graph $G = (V, E)$ is called a $relative$ $neighborhood$ graph [6] when $[u, v] \in E$ iff $\max\{d(u, w), d(v, w)\} \geq d(u, v)$ for any $w \in V \backslash \{u, v\}$. Both types of plane graphs defined above appear in a variety of network applications. They represent convenient network topologies, simplifying routing and control in geographical (e.g. wireless) networks. They can also be applied when approximating complex networks.

In [13] faster $O(n^2)$-time 10-, 12- and 14-approximate algorithms are designed for the NP-hard [10] IPGD problem when E is being an edge set of a minimum Euclidean spanning tree, a relative neighborhood graph and a Gabriel graph respectively. This paper extends this latter work by presenting faster $O(n^{3/2} \log^2 n)$-expected time 10-, 12- and 14-approximation algorithms for the IPGD problem for classes of minimum Euclidean spanning trees, relative neighborhood graphs and Gabriel graphs respectively.

2   Our Approximation Algorithms

2.1   Some Preliminaries

Our subquadratic $O(1)$-approximation algorithms are improved versions of the $O(n^2)$-time $O(1)$-approximation algorithms, given in [13]. The latter algorithms operate on two concepts whose definitions are given below.

Definition 1. *A subset* $\mathcal{I} \subseteq \mathcal{N}_r(E)$ *is called a maximal (with respect to inclusion) independent set in* $\mathcal{N}_r(E)$, *if* $I \cap I' = \varnothing$ *for any* $I, I' \in \mathcal{I}$, *and for any* $N \in \mathcal{N}_r(E)$ *there is some* $I \in \mathcal{I}$ *with* $N \cap I \neq \varnothing$.

Given $N \in \mathcal{N}_r(E)$, let $e(N)$ be a straight line segment such that $N_r(e(N)) = N$. Let also $\mathcal{N}_{r,e}(E) = \{N \in \mathcal{N}_r(E) : N \cap N_r(e) \neq \varnothing\}$ for $e \in E$. Of course, each maximal independent set $\mathcal{I}$ in $\mathcal{N}_r(E)$ defines a (possibly non-unique) partition of the form $\mathcal{N}_r(E) = \bigcup_{I \in \mathcal{I}} \mathcal{N}_I$, where $\mathcal{N}_I \subseteq \mathcal{N}_{r,e(I)}(E)$, $I \in \mathcal{I}$; moreover, families $\mathcal{N}_I$ and $\mathcal{N}_{I'}$ are non-intersecting for distinct $I, I' \in \mathcal{I}$. It is easy to see that each maximal independent set in $\mathcal{N}_r(E)$ also induces the corresponding partition of E.

Definition 2. *Let* $G = (V, E)$ *be a plane graph and* $\alpha > 0$ *be some* (r-*independent) absolute constant. An edge* $e \in E$ *is called* α-*coverable with respect to* E, *if for any constant* $\rho > 0$ *one can construct at most* α-*point piercing set* $U(\rho, e, E) \subset \mathbb{R}^2$ *for* $\mathcal{N}_{\rho,e}(E)$ *in polynomial time with respect to* $|\mathcal{N}_{\rho,e}(E)|$.

It turns out (see Lemmas 1 and 5 in [13] for proof) that any edge of every Gabriel and relative neighborhood graph is α-coverable for some suitable positive integer α.

Lemma 1. *Any edge* $e \in E$ *is 12-coverable of an arbitrary subgraph* $G = (V, E)$ *of a relative neighborhood graph. More precisely, for any* $\rho > 0$ *respective piercing set* $U(\rho, e, E)$ *for* $\mathcal{N}_{\rho,e}(E)$ *can be found in* $O(1)$ *time.*

Lemma 2. *Any edge* $e \in E$ *is 14-coverable of an arbitrary subgraph* $G = (V, E)$ *of a Gabriel graph. Namely, for any* $\rho > 0$ *respective piercing set* $U(\rho, e, E)$ *for* $\mathcal{N}_{\rho,e}(E)$ *can be found in* $O(1)$ *time.*

Let $G = (V, E)$ be a Euclidean minimum spanning tree with a root $v_0 \in V$, $\mathrm{depth}(u) = \mathrm{depth}(u|v_0, G)$ be the (graph-theoretic) distance in G from v_0 to an arbitrary $u \in V$ and $V(u|v_0)$ be the subset of those vertices $w \in V$ such that the shortest path in G from w to v_0 (with respect to the number of its edges) passes through u. If an edge $e = [u_1, u_2] \in E$ is such that $\mathrm{depth}(u_1) = \mathrm{depth}(u_2) - 1$, then int $N_{2\Delta}(u_2) \cap (V \backslash V(u_2|v_0)) = \varnothing$, where $\Delta = d(u_1, u_2)/2$ and int N denotes interior of a set $N \subset \mathbb{R}^2$.

It can also be proved (see the lemma 3 from [13]) that in any subgraph of a minimum Euclidean spanning tree there always exists a 10-coverable edge.

Lemma 3. *Let $G_0 = (V_0, E_0)$ be a subgraph without isolated vertices of a Euclidean minimum spanning tree $G = (V, E)$. Let depth$(\cdot|v_0)$ be a distance function on V with respect to a chosen $v_0 \in V$ as defined above. Then an edge $e = [u_1, u_2] \in E_0$ is 10-coverable with respect to E_0, if $u_2 \in \text{Arg}\max_{u \in V_0} \text{depth}(u)$. Besides, for any constant $\rho > 0$ the corresponding piercing set $U(\rho, e, E_0)$ of size at most 10 can be found in $O(1)$ time.*

Roughly, $O(1)$-approximate algorithms from [13] compute a partition $\mathcal{N}_r(E) = \bigcup_{I \in \mathcal{I}} \mathcal{N}_I$, defined by some maximal independent set $\mathcal{I}$ for $\mathcal{N}_r(E)$. Moreover, on the way of constructing $\mathcal{I}$, a constant sized piercing set $U(I)$ is computed for each $\mathcal{N}_I \subseteq \mathcal{N}_{r,e(I)}(E)$, $I \in \mathcal{I}$, as described in proofs of lemmas 1, 3 and 5 from [13]. Finally, it turns out that $\left\{ N_r(u) : u \in \bigcup_{I \in \mathcal{I}} U(I) \right\}$ is an $O(1)$-approximate solution to the IPGD problem.

The key to the performance gain, achieved in our algorithms over the algorithms of [13], lies in the efficient way of constructing a partition of $\mathcal{N}_r(E)$, which is induced by $\mathcal{I}$. Here, our algorithms additionally maintain a special data structure. For a given set F of non-intersecting straight line segments and a point $x \in \mathbb{R}^2$, this data structure allows to efficiently compute the segment $f \in F$, which is the nearest to x with respect to Euclidean distance. In fact, it implicitly implements a Euclidean Voronoi diagram for F. Voronoi diagram for sets of pairwise non-overlapping straight line segments is a generalization of the well-known Euclidean Voronoi diagram for point sets on the plane.

Definition 3. *Let F be a set of pairwise non-intersecting straight line segments. A Voronoi diagram $\mathcal{V}(F)$ for F is a partition of the plane into a set of open regions and their boundaries where each region represents a locus of those points, which are closer to a particular segment $f \in F$ than to any other segment of $F \backslash \{f\}$. Boundary of each region is composed of curvilinear edges and vertices. Each (relatively open) edge is the locus of points, which are equidistant from two distinct segments $f, f' \in F$ while being at the larger distance from segments of $F \backslash \{f, f'\}$. Each vertex represents a endpoint of an edge of $\mathcal{V}(F)$, which is equidistant from more than two distinct segments of F.*

Open regions, edges and vertices of $\mathcal{V}(F)$ are called Voronoi cells, edges and vertices respectively.

An additional assumption is imposed below on sets of segments for simplicity.

Definition 4. *A set F of pairwise non-overlapping straight line segments is called to be in general position if:*

1. *no quadruple exists of segments from F which is touched by any single disk;*
2. *the set is in general position of endpoints of segments from F.*

Generality of position can be achieved by a small perturbation of endpoints of E.

2.2 Implementation of Algorithms

Let $G = (V, E)$ be a plane graph and $r > 0$ be a constant, forming an input of the IPGD problem. Work of our algorithms can be split into two stages. At their first stage a partition $\mathcal{N}_r(E) = \bigcup_{I \in \mathcal{I}} \mathcal{N}_I$ is efficiently extracted, which is induced by a maximal independent set $\mathcal{I}$ in $\mathcal{N}_r(E)$. Then, during the second stage, another pass is performed over the built set $\mathcal{I}$ to construct a piercing set $U(I)$ of $\mathcal{N}_I$ for each $I \in \mathcal{I}$. Here, $U(I)$ is built in the analogous way to that done in the algorithms from [13]. Merging those piercing sets together into a point set $C = \bigcup_{I \in \mathcal{I}} U(I) \subset \mathbb{R}^2$, a set $\mathcal{C} = \{N_r(c) : c \in C\}$ is yielded as an approximate solution to the IPGD problem instance, defined by G and r. In our algorithms below, the sought partition $\mathcal{N}_r(E) = \bigcup_{I \in \mathcal{I}} \mathcal{N}_I$ is found implicitly in the form of constructing the corresponding partition of E, induced by $E' = \{e(I) : I \in \mathcal{I}\}$.

Algorithmic work at the first stage is split into n_1 phases, where $n_1 \leq \sqrt{n}$. During the ith phase a pass is performed over some part of E to iteratively grow a subset $F \subseteq E$ by adding segments from E into F one by one such that:

1. $\mathcal{N}_r(F)$ contains pairwise non-overlapping r-hippodromes;
2. the upper bound $|F| \leq \sqrt{n}$ holds.

During ith phase a special incremental data structure is applied. It allows to do the following two operations:

1. query: given a straight line segment $e \notin F$, return a segment $f \in F$ such that $N_r(e) \cap N_r(f) \neq \varnothing$ or report that $N_r(e) \cap N_r(f) = \varnothing$ for all $f \in F$; namely, in the former case the segment f is returned along with the truth value of a special indicator variable named $flag$; otherwise, $flag = False$ is returned;
2. insertion: insert a segment $e \notin F$ into F.

Here the segment e is allowed to intersect segments of F at most at their endpoints. When $F = \varnothing$, $flag = False$ is returned.

Below a pseudo-code is presented of two of our algorithms. Their input is formed by a graph G which is either a Gabriel or a relative neighborhood graph. Moreover, they contain a constant parameter $\alpha > 0$, which is specific to the class of plane graphs from which G is chosen. Here $\alpha = 14$ for the case where G is a Gabriel graph whereas $\alpha = 12$, when G is a relative neighborhood graph.

COVERING SEGMENTS WITH r-DISKS.

Input: a constant $r > 0$ and a plane graph $G = (V, E)$;
Output: an α-approximate solution $\mathcal{C}$ of radius r disks for the IPGD problem instance, defined by G and r.

---

1. set $n = |E|$, $E' := \varnothing$, $E'' := \varnothing$, $E_0 := E$ and $C := \varnothing$; // *stage 1*
2. while $|E''| \leq \sqrt{n}$, process edges of E_0 one by one: do steps 3–5 //*phase begins*
3. choose $e \in E_0$ and perform a query to the data structure built for $F = E''$;
4. if $flag = True$ and a segment $g \in E''$ are returned, set $E_g := E_g \cup \{e\}$; otherwise, when $flag = False$ is returned, insert e into E'' and set $E_e := \varnothing$;
5. set $E_0 := E_0 \backslash \{e\}$;
6. if $E_0 = \varnothing$, go to step 9; otherwise, process edges of E_0 one by one: do steps 7–8
7. choose $e \in E_0$ and perform a query to the data structure built for $F = E''$;
8. if $flag = True$ and $g \in E''$ are returned, set $E_g := E_g \cup \{e\}$ and $E_0 := E_0 \backslash \{e\}$;
9. set $E' := E' \cup E''$ and $E'' := \varnothing$;
10. if $E_0 \neq \varnothing$, go to step 2; //*phase finishes*
11. for each $e' \in E'$ repeat steps 12–13 // *stage 2*
12. construct a piercing set $U(e')$ of at most α points for $\mathcal{N}_r(E_{e'})$, applying e.g. the corresponding $O(1)$-time procedure, mentioned in lemmas 1 and 2 above;
13. set $C := C \cup U(e')$;
14. return $\mathcal{C} := \{N_r(c) : c \in C\}$ as an α-approximate solution.

As the pseudo-code above shows, in accordance with the basic algorithm of [13], the COVERING SEGMENTS WITH r-DISKS algorithm computes a maximal independent set in $\mathcal{N}_r(E)$ by iteratively growing a subset $E' \subseteq E$ such that r-hippodromes of $\mathcal{N}_r(E')$ are pairwise non-overlapping. The algorithm grows the set E' by chunks E'' of size $\sqrt{n}$ except the last chunk, where each chunk is computed in a single phase. While $|E''| \leq \sqrt{n}$, each chunk E'' is grown by performing query and insertion operations of the data structure, built on top of E'', for the remaining unprocessed segments of E_0, which are tried one by one. When $|E''| > \sqrt{n}$, those remaining segments $e \in E_0$ are removed from E_0 for which query operation returns $flag = True$. As shown below, the used randomized data structure allows to perform query operation in $O\left(\log^2 |E''|\right)$ expected time. Besides, the maintained bound $|E''| \leq \sqrt{n}$ helps keep total expected time complexity of insertion operations as low as $O(n \log n)$ in each phase.

The described organization of processing of segments of E and favourable query times of the used data structure allow to formulate the following

Theorem 1. *Suppose $G = (V, E)$ is a plane graph whose edge set is in general position. The* COVERING SEGMENTS WITH r-DISKS *algorithm is*

1. *12-approximate when G is a subgraph of a relative neighborhood graph;*
2. *14-approximate if G is a subgraph of a Gabriel graph.*

It admits an implementation, which works in $O\left((n + \text{OPT}\sqrt{n}) \log^2 \text{OPT}\right)$ expected time and $O(n)$ expected space, where OPT *is the problem optimum.*

Theorem 2. *The* COVERING SEGMENTS WITH r-DISKS *algorithm can be slightly modified to become a 10-approximate $O\left((n + \text{OPT}\sqrt{n}) \log^2 n\right)$-expected time and $O(n)$-expected space algorithm when G is a subgraph of a minimum Euclidean spanning tree whose edge set is in general position.*

Proofs of Theorems 1 and 2 are given in Sect. 4.

It can be seen that expected complexity of the COVERING SEGMENTS WITH r-DISKS algorithm depends on OPT. For example, if OPT $< \sqrt{n}$, then this algorithm is of almost linear expected time complexity. Due to an obvious bound OPT $\leq n$, the algorithm has at most $O\left(n^{3/2} \log^2 n\right)$ expected time complexity.

To compare performance of the COVERING SEGMENTS WITH r-DISKS algorithm with the basic algorithm from [13] it is enough to observe that the former algorithm would coincide with the latter one if the restriction $|E''| = 1$ was imposed on all chunks E''. This restriction leads to the $O(n\text{OPT})$ expected and worst case time complexity: here one has at most OPT singleton chunks as well as constant query and insertion times.

Key to the achieved performance gain in our algorithms over the related work lies in efficient implementation of query operations of the data structure, built on top of E'', within the COVERING SEGMENTS WITH r-DISKS algorithm. This core data structure is described in the section below.

3   Description of the Core Data Structure

3.1   Implementing the Query Operation with a Few Nearest Neighbor Queries

Suppose a set F is given of pairwise non-intersecting straight line segments on the plane. Besides, let e be a straight line segment such that its intersection with each segment of F (if it is nonempty) can only be at the common endpoint. The query operation for e on F can in theory be implemented by computing the segment $f \in F$, being the closest to e, and checking if $N_r(e) \cap N_r(f) \neq \varnothing$. It means that the query operation admits its implementation by performing a segment nearest neighbor query operation on F.

Unfortunately, this approach fails to efficiently work within the COVERING SEGMENTS WITH r-DISKS algorithm. In fact, there is a variety of incremental algorithms to maintain efficient point nearest neighbor queries for sets of point sites. The most efficient known data structure [2] provides $O(\log^2 |F|)$ expected query time and $O(\log^4 |F|)$ expected insertion time. However, there is a lack of available incremental data structures to maintain segment nearest neighbor queries. The only available incremental randomized data structure to work with straight line segment sites and segment queries, which we are able to find, implicitly maintains their segment Voronoi diagram [7][1].

In this data structure segment nearest neighbor queries cost $O(\log^2 |F| + t)$ expected time, where t is the number of segments of F, sharing a Voronoi edge with e in the Voronoi diagram $\mathcal{V}(F \cup \{e\})$. Thus, t can be $\Omega(|F|)$ on average in general. Therefore with this idea one can not guarantee for the query to have sublinear expected time complexity and, as a consequence, for the COVERING

---

[1] Its C++ implementation is built in the CGAL library (see https://www.cgal.org/), providing robust geometric computations.

SEGMENTS WITH r-DISKS algorithm to have subquadratic expected time complexity.

Luckily, query operations, performed in the COVERING SEGMENTS WITH r-DISKS algorithm, can efficiently be implemented using a constant number of nearest neighbor queries of the special type in which sites are straight line segments whereas query objects are points.

Nearest Neighbor Query. Given a point $x \in \mathbb{R}^2$, find a segment $f \in F$, being the closest to x among segments of F with respect to Euclidean distance between points and segments.

Lemma 4. *Let e be a query straight line segment in a query operation for a set F of pairwise non-intersecting striaght line segments on the plane, being in general position. Suppose the following assumptions are hold:*

1. *$F \cup \{e\}$ is a subset of edges of a Gabriel graph;*
2. *$N_r(f) \cap N_r(g) = \varnothing$ for any distinct $f, g \in F$.*

Then, the query operation can be implemented using at most 14 nearest neighbor query operations and at most 14 operations to check if two r-hippodromes intersect.

Proof. Let $\Delta = d(x, y)/2$, where x and y are endpoints of e. First, nearest neighbor query operation is applied for both x and y, returning segments f_x and f_y of F respectively. If either $N_r(f_x) \cap N_r(e) \neq \varnothing$ or $N_r(f_y) \cap N_r(e) \neq \varnothing$, it is done. Otherwise, consider three cases.

CASE 1. Let $\Delta \leq r$. Let also z_1 and z_2 be points at the intersection $\mathrm{bd}\, N_{2r}(x) \cap \mathrm{bd}\, N_{2r}(y)$, where $\mathrm{bd}\, N$ denotes boundary of a set $N \subset \mathbb{R}^2$. Perform another two nearest neighbor queries for z_1 and z_2. Let $f_{z_i} \in F$ be closest to z_i. If $N_r(e) \cap N_r(f_{z_{i_0}}) \neq \varnothing$ for some $i_0 \in \{1, 2\}$, then it is done. If not, it can be proved that $N_r(e) \cap N_r(f) = \varnothing$ for all $f \in F$.

Indeed, suppose, in contrary, that $f_0 \in F$ exists such that $N_r(e) \cap N_r(f_0) \neq \varnothing$ or, equivalently, f_0 intersects $N_{2r}(e)$. Obviously, the case $f_0 \cap \bigcup_{u \in \{x,y\}} N_{2r}(u) \neq \varnothing$ is impossible. Therefore f_0 must intersect $N_{2r}(e) \setminus \bigcup_{u \in \{x,y\}} N_{2r}(u)$. Of course, if f_0 intersects $N_{2r}(e)$ inside the same half-plane H_{i_0}, bounded by the straight line through e, as that in which the point z_{i_0} lies for some $i_0 \in \{1, 2\}$, then both f_0 and $f_{z_{i_0}}$ must intersect $N_{\sqrt{8r^2 - 4r\sqrt{4r^2 - \Delta^2}}}(z_{i_0})$. As $\Delta \leq r$, it implies that $H_{i_0} \cap N_{2r}(e) \setminus \bigcup_{u \in \{x,y\}} N_{2r}(u)$ is covered by $N_{2r}(f_{z_{i_0}})$ and, therefore $N_r(f_{z_{i_0}}) \cap N_r(f_0) \neq \varnothing$, a contradiction with the assumption 2. Thus, at most 4 nearest neighbor query operations are enough in the considered case.

CASE 2. If $r < \Delta \leq 2r$, split e into two subsegments of length Δ and apply the same technique for each subsegment as in the previous case. Here at most 7 nearest neighbor query operations are enough.

CASE 3. Suppose that $\Delta > 2r$. Let z be the midpoint of e. Recall that $N_\Delta(z)$ does not contain endpoints of segments of $F \cup \{e\}$ in its interior according to the assumption 1. Let z_1 and z_2 be projections of 4 points from $\operatorname{bd} N_\Delta(z) \cap \operatorname{bd} N_{2r}(e)$ onto e, where $z_1 \in [x, z]$ and $z_2 \in [y, z]$. Below it is proved that length of $[z_1, x]$ is less than $2r$. Indeed, $d(x, z_1) = \Delta - \sqrt{\Delta^2 - 4r^2} \leq 2r$. Moreover, $d(x, z_1) \leq r$ when $\Delta > \frac{5r}{2}$.

From [13] (see Sect. 2 and the Lemma 5 therein) it follows that if some $f_0 \in F$ intersects $N_{2r}(e)$, then f_0 must intersect $N_{2r}(e) \backslash N_\Delta(z)$. Therefore at most 14 nearest neighbor query operations are enough for $2r < \Delta \leq \frac{5r}{2}$ and 8 nearest neighbor query operations are sufficient when $\Delta > \frac{5r}{2}$.

Any relative neighborhood graph and minimum Euclidean spanning tree is a subgraph of a Gabriel graph. Therefore the following corollary holds.

Corollary 1. *Let e be a query straight line segment. Suppose the assumptions are hold:*

1. *$F \cup \{e\}$ is a subset of edges of either a relative neighborhood graph or a minimum Euclidean spanning tree;*
2. *$N_r(f) \cap N_r(g) = \varnothing$ for any distinct $f, g \in F$.*

Then, the query operation for e on F can be implemented using at most 14 nearest neighbor query operations and at most 14 operations to check if two r-hippodromes intersect.

Due to the Lemma 4 and the Corollary 1, the data structure from [7] can be used to implement query and insertion operations, performed within the Covering segments with r-disks algorithm. To the best of our knowledge, this data structure has the most efficient implementation of point nearest neighbor queries for segment sites. Being incorporated into our algorithm, it implicitly stores a Voronoi diagram $\mathcal{V}(E'')$ of the set E''. Its performance is summarized in the following lemma (see works [1,7–9,15] for proofs).

Lemma 5. *Let F be a set of pairwise non-intersecting straight line segments in general position on the plane. A randomized data structure can be built incrementally in $O(|F|^2 \log |F|)$ expected time and $O(|F|)$ expected space cost such that:*

1. *given a point $x \in \mathbb{R}^2$, nearest neighbor query for x and F can be performed in $O(\log^2 |F|)$ expected time;*
2. *given a segment $e \notin F$ such that $\mathcal{N}_r(F \cup \{e\})$ contains only pairwise non-overlapping r-hippodromes, insertion of e into F can be done in $O(|F| \log |F|)$ expected time.*

In distinction to the data structure from [7] only a single type of randomization is applied in the implementation of the used data structure, which is related to generating a random hierarchy of nested subsets of F.

In [7] another type of randomization is also used implied by a random order of insertion of segments into F : i.e. it is assumed that the order of insertion of segments into F is a random permutation on F and all insertion orders are equally likely. Applying both randomization types allows to reduce expected time complexity of insertion operations to $O(1)$. In the COVERING SEGMENTS WITH r-DISKS algorithm this favourable symmetry of insertion of segments into F can not be guaranteed even if E is preliminarily randomly shuffled at the algorithm step 1.

4   Proofs of Theorems 1 and 2

4.1   Proof of the Theorem 1

Proof. In each phase (steps 2–10 of the COVERING SEGMENTS WITH r-DISKS algorithm) query operations are performed for at most n segments of E. Therefore running these operations takes $O(n \log^2 \text{OPT})$ expected time in each phase due to the Lemma 4, the Corollary 1 and the Lemma 5. As $|E''| \leq \min\{\sqrt{n}, \text{OPT}\}$, at most $\min\{\sqrt{n}, \text{OPT}\}$ insertions are done in each phase. Thus, insertion operations take $O(\min\{n, \text{OPT}^2\} \log \text{OPT})$ expected time, again, due to the Lemma 5. There are at most $\frac{\text{OPT}}{\sqrt{n}} + 1$ phases in the algorithm as $\text{OPT} \leq n$. Steps 11–13 require $O(\text{OPT})$ time in view of Lemmas 1 and 2. Thus, total expected complexity of the algorithm is of the order

$$O\left(n\left(\frac{\text{OPT}}{\sqrt{n}} + 1\right) \log^2 \text{OPT}\right),$$

or of the order $O\left((n + \text{OPT}\sqrt{n}) \log^2 \text{OPT}\right)$. Its expected space cost is $O(n)$.

4.2   Proof of the Theorem 2

Proof. One can maintain an $O(n)$ sized search tree (see e.g. the Chap. 13 of [3]) to report an edge of E_0 in $O(\log n)$ time at the algorithm step 3, which is incident to a vertex, being the most distant from some fixed vertex of G. This search tree can be preliminary created in $O(n \log n)$ time by performing a breadth-first search over G. Each time when a segment is removed from E_0 at steps 5 and 8 of the COVERING SEGMENTS WITH r-DISKS algorithm, the corresponding node is removed from the tree in $O(\log n)$ time. Thus, first, n insertions in a row are performed into the tree; second, n consecutive deletions are done.

5   Conclusion

The paper presents randomized subquadratic small constant factor approximation algorithms for three special cases of the problem of intersecting a given set of straight line segments on the plane with the least number of identical disks of potential interest in facility location. When $\text{OPT} = O(\sqrt{n})$, these algorithms have almost linear expected time complexity of $O(n \log^2 n)$, where OPT is the problem optimum. In the general case their expected complexity is $O\left((n + \text{OPT}\sqrt{n}) \log^2 n\right)$, i.e. with $\frac{\text{OPT}}{\sqrt{n}}$ superlinear multiplicative overhead. Approximation factors of the proposed algorithms are still prohibitively high to be practical as being only theoretically guaranteed upper bounds on approximation ratios of the algorithms in the worst case. In the follow-up paper their actual approximation factors will be explored for real-world facility location problems.

References

1. Boissonnat, J.D., Wormser, C., Yvinec, M.: Curved Voronoi diagrams. In: Boissonnat, J.D., Teillaud, M. (eds.) Effective Computational Geometry for Curves and Surfaces, pp. 67–116. Springer, Heidelberg (2006). https://doi.org/10.1007/978-3-540-33259-6_2
2. Chan, T.M.: Dynamic geometric data structures via shallow cuttings. In: Proceedings of the 35th International Symposium on Computational Geometry, SoCG 2019, pp. 24:1–24:13 (2019)
3. Cormen, T.H., Leiserson, C.E., Rivest, R.L., Stein, C.: Introduction to Algorithms, 3rd edn. MIT Press, Cambridge (2009)
4. Dash, D., Bishnu, A., Gupta, A., Nandy, S.: Approximation algorithms for deployment of sensors for line segment coverage in wireless sensor networks. Wirel. Netw. **19**(5), 857–870 (2012)
5. Devillers, O.: The Delaunay hierarchy. Int. J. Found. Comput. Sci. **13**, 163–180 (2002)
6. Jaromczyk, J., Toussaint, G.: Relative neighborhood graphs and their relatives. Proc. IEEE **80**(9), 1502–1517 (1992)
7. Karavelas, M.I.: A robust and efficient implementation for the segment Voronoi diagram. In: Proceedings of the 1st International Symposium on Voronoi Diagrams in Science and Engineering, Tokyo, pp. 51–62 (2004)
8. Karavelas, M., Yvinec, M.: The Voronoi Diagram of Convex Objects in the Plane. Research report. RR-5023, INRIA. 27 p. (2003)
9. Klein, R., Mehlhorn, K., Meiser, S.: Randomized incremental construction of abstract Voronoi diagrams. Comput. Geom. **3**(3), 157–184 (1993)
10. Kobylkin, K.: Stabbing line segments with disks: complexity and approximation algorithms. In: van der Aalst, W.M.P., et al. (eds.) AIST 2017. LNCS, vol. 10716, pp. 356–367. Springer, Cham (2018). https://doi.org/10.1007/978-3-319-73013-4_33
11. Kobylkin, K.: Constant factor approximation for intersecting line segments with disks. In: Battiti, R., Brunato, M., Kotsireas, I., Pardalos, P.M. (eds.) LION 12 2018. LNCS, vol. 11353, pp. 447–454. Springer, Cham (2019). https://doi.org/10.1007/978-3-030-05348-2_39
12. Kobylkin, K.: Efficient constant factor approximation algorithms for stabbing line segments with equal disks, CoRR abs/1803.08341, 31 p. (2018). https://arxiv.org/pdf/1803.08341.pdf
13. Kobylkin, K., Dryakhlova, I.: Approximation algorithms for piercing special families of hippodromes: an extended abstract. In: Khachay, M., Kochetov, Y., Pardalos, P. (eds.) MOTOR 2019. LNCS, vol. 11548, pp. 565–580. Springer, Cham (2019). https://doi.org/10.1007/978-3-030-22629-9_40
14. Matula, D., Sokal, R.: Properties of Gabriel graphs relevant to geographic variation research and the clustering of points in the plane. Geogr. Anal. **12**(3), 205–222 (1980)
15. Motwani, R., Raghavan, P.: Randomized Algorithms, 476 p. Cambridge University Press, Cambridge (1995)

Process Mining

Checking Conformance Between Colored Petri Nets and Event Logs

Julio C. Carrasquel[1(✉)], Khalil Mecheraoui[1,2], and Irina A. Lomazova[1]

[1] National Research University Higher School of Economics, Myasnitskaya ul. 20, 101000 Moscow, Russia
{jcarrasquel,ilomazova}@hse.ru, k_mecheraoui@esi.dz
[2] University of Constantine 2—Abdelhamid Mehri, Nouvelle ville Ali Mendjeli BP: 67A, 25000 Constantine, Algeria

Abstract. Event logs of information systems consist of recorded traces, describing executed activities and involved resources (e.g., users, data objects). Conformance checking is a family of process mining techniques that leverage such logs to detect whether observed traces deviate w.r.t some specification model (e.g., a Petri net). In this paper, we present a conformance checking method using colored Petri nets (CPNs) and event logs. CPN models allow not only to specify a causal ordering between system activities, but also they allow to describe how resources must be processed upon activity executions. By replaying each trace of an event log on top of a CPN, we present how this method detects: (1) control-flow deviations due to unavailable resources, (2) rule violations, and (3) differences between modeled and real produced resources. We illustrate in detail our method using the study case of trading systems, where orders from traders must be correctly processed by a platform. We describe experimental evaluations of our method to showcase its practical value.

Keywords: Process mining · Conformance checking · Petri Nets · Colored Petri nets · Trading systems · Order books

1 Introduction

Conformance checking is a family of process mining techniques to diagnose whether or not a system process is being executed as described by its specification model [1,3]. Two main inputs are considered in such methods: *event logs* and *process models*. On the one hand, an event log describes *real behavior* of a process. It consists of recorded traces, each of them consisting of executed activities and *resources* involved in such executions. Resources may be users or data objects processed by a system. On the other hand, a process model allows to describe *expected behavior* of a system process, based on its specification. Regarding the model notation, conformance checking methods consider Petri nets—a formalism for modeling and analysis of concurrent distributed systems [15].

This work is supported by the Basic Research Program at the National Research University Higher School of Economics.

W. M. P. van der Aalst et al. (Eds.): AIST 2020, LNCS 12602, pp. 435–452, 2021.
https://doi.org/10.1007/978-3-030-72610-2_33

In particular, Petri nets allow to specify the *control-flow* of a system, that is, a causal ordering between system activities (e.g., activity a must be followed by b). Thus, conformance checking methods use Petri nets and event logs to determine, for instance, to which degree the modeled control-flow is being complied by the real system, as observed in the recorded traces. For example, "a loan approval was executed, but it was not inspected before, and this must not happen according to the model". This is why conformance checking has become a research subject of interest in several application domains, i.e., for auditing business processes [20].

Nonetheless, most of the conformance checking methods merely focus on the control-flow aspect (i.e., only considering event activities), thereby neglecting other valuable information recorded in event logs, for example, processed resources. This imposes severe limitations in study cases where the system's correct execution can be only determined by checking involved resources in events (i.e., "a trade can be executed if a buy order and a sell order are available").

To address such limitation, certain conformance methods propose the use of enriched models, such as in [13], where Petri nets with data (DPN) are employed. In DPNs, data variables are attached to transitions (representing activities), and thus this model allows to specify data constraints on activity executions (for instance, "a loan is rejected if the requested amount is higher than a threshold"). In DPNs, however, the system's control-flow is still defined separately and data objects play a minor role, being statically attached to transitions. In consequence, a conformance checking method with DPNs does not allow to clearly validate whether dynamic resources are evolving as expected, while they are processed by the system, nor how new resource states may affect the overall system execution.

In this paper, we present a conformance checking method between event logs and colored Petri nets (CPN) [11]—a Petri net extension resembling the object-oriented paradigm. In CPNs, tokens carry values, representing object instances of some classes (called *colors*). Besides, arc expressions adjacent to transitions allow to specify how objects are transformed upon activity executions. Our conformance method is based on replaying each trace of an event log on top of a CPN model. When replaying each trace, the distinct observed resources (object instances) are injected as tokens in the model. Then, for each event of a trace, we try to fire a transition associated to the activity executed in the event, and selecting as input tokens the ones which represent the real resources observed in the event. Following such scheme, we explain in this work how our method can detect three kinds of deviations: (1) control-flow deviations caused by the absence of resources (i.e., it is not possible to fire a transition with the resources indicated in the event); (2) rule violations (e.g., according to priority rules on transitions, some resources must be served first); and (3) differences between modeled and real resources regarding their evolution along a trace (for instance, after a transition firing, the resulting values of produced tokens must be equal to their corresponding resources observed in an event).

For this method, we consider a specific class of CPNs with certain restrictions. For instance, all tokens in a model must be unique (e.g., using identifiers). Also, each token involved in a transition firing must be of a different class. These restrictions come to be natural in various information systems where, for example, objects can be distinguished. In the next sections, we explain in detail these restrictions, describing how they guarantee a correct and efficient replay.

We illustrate our method throughout the paper with the study case of trading systems [9]. These systems receive buy/sell orders from agents to trade securities (company shares), placing them in lists called *order books*. Then, orders in a same book are matched to produce trades. In a system, there can be as many order books as securities are traded (e.g., an order to buy 3 stocks of the company yandex is placed in the order book "yandex"). Event logs of these systems consist of traces, each of them related to a trading session in an order book (see Fig. 1).

trace	timestamp	activity	buy order				sell order			
			id	tsub	price	qty	id	tsub	price	qty
001	09:13:07.536	submit buy order	Wpl	09:13:07.536	22.00	3				
001	09:13:07.537	new buy order	Wpl	09:13:07.536	22.00	3				
001	09:13:07.544	submit sell order					Wpm	09:13:07.544	19.00	1
001	09:13:07.545	new sell order					Wpm	09:13:07.544	19.00	1
001	09:13:07.565	submit sell order					Wpn	09:13:07.565	21.00	3
001	09:13:07.566	new sell order					Wpn	09:13:07.565	21.00	3
001	09:13:07.581	trade 2	Wpl	09:13:07.536	22.00	2	Wpm	09:13:07.544	19.00	0
001	09:13:07.582	trade 3	Wpl	09:13:07.536	22.00	0	Wpn	09:13:07.565	21.00	1
001	09:13:11.236	discard sell order					Wpn	09:13:07.565	21.00	0

Fig. 1. Log trace of a trading system. Each row shows an activity fired and orders involved, with attributes id, arrival time (tsub), price and quantity (qty). (Color figure online)

Given a CPN modeling a trading session, and an event log of real sessions (i.e., see Fig. 2), our method detects the following deviations: (1) control-flow errors due to absent resources, e.g., trades occurred with unavailable orders, (2) violation of priority rules when serving orders, and (3) differences between modeled and real produced resources (e.g., an order attribute was incorrectly modified).

The remainder of this paper is structured as follows. In Sect. 2, we introduce CPNs, its formal definition and execution semantics. In Sect. 3, we describe event

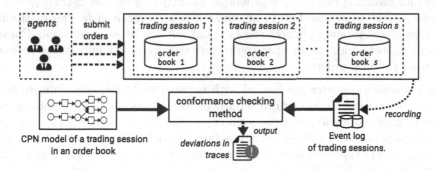

Fig. 2. Validating trading sessions via conformance checking with CPNs. (Color figure online)

logs. In Sects. 4 and 5, we describe our conformance checking method, its implementation and experimental validation. In Sect. 6, we conclude our paper with a discussion on the novelty of our contribution. Also, we briefly mention how our method compares to other conformance proposals, as well as methods within the sphere of data science.

2   Colored Petri Nets

In this section, we present colored Petri nets (CPN), using as an example the model of a trading session in an order book. Then, we introduce the formal definition of CPNs and their execution semantics, as well as we consider some model restrictions. In general, Petri nets consist of two kinds of nodes: *transitions* modeling activities, and *places* storing *tokens*, which model buffers with resources. Pictorially, transitions and places are drawn as boxes and circles respectively. Directed arcs connect input places to transitions, and transitions to output places. Activity executions and resource processing are modeled by *transition firings*, consuming and producing tokens in input and output places respectively.

Colors and Tokens. CPNs are an extension of Petri nets, where tokens carry values of some types. Formally, a token is a tuple $(d_1, ..., d_n) \in D_1 \times ... \times D_n$, s.t. $\{D_1, ..., D_k\} \subseteq \mathfrak{D}$ are *data types* from a *data type domain* of interest $\mathfrak{D}$. A cartesian product $D_1 \times ... \times D_n$ between any combination of data types from $\mathfrak{D}$ is a *color*. We denote by Σ the set of all colors that can be obtained from $\mathfrak{D}$. Resembling the object-oriented paradigm, colors denote object classes, whereas tokens are object instances. For example, we define buy and sell order classes with colors $OB = O_B \times \mathbb{N} \times \mathbb{R}^+ \times \mathbb{N}$ and $OS = O_S \times \mathbb{N} \times \mathbb{R}^+ \times \mathbb{N}$, where O_B and O_S are sets of order identifiers, $\mathbb{N}$ is the set of natural numbers, and $\mathbb{R}^+$ is the set of positive real numbers. In Fig. 3, tokens stored in p_1 and p_2 represent buy and sell orders, e.g., the token $(b1, 1, 22.0, 3)$ in place p_1 models a buy order with identifier $b1$, submitted in time 1, to buy 3 stocks at a price per unit of 22.0.

Places. Places store tokens of a specific color. We define a function `color`, mapping each place to a color in Σ. In Fig. 3, places p_1 and p_2 are the initial places for incoming buy and sell orders, so $\text{color}(p_1) = OB$ and $\text{color}(p_2) = OS$. Places p_3 and p_4 denote buffers of buy/sell orders, received by the platform, whereas places p_5 and p_6 model the buy and sell side of an order book. Places p_7 and p_8 store filled orders that traded successfully, and finally places p_9 and p_{10} store canceled orders.

Arc Expressions. Arcs are labeled with expressions to formally indicate how tokens are processed upon transition firings. We consider a language of expressions $\mathcal{L}$. Each expression is of the form $(e_1, ..., e_n)$ s.t., for each $i \in \{1, ..., n\}$, e_i is either a constant, a variable, or a function. We define a function $\mathcal{E}$ that maps each arc to an expression from $\mathcal{L}$. Let us consider some examples in Fig. 3. The expressions $\mathcal{E}(p_1, t_1) = \mathcal{E}(t_1, p_3) = (o, ts, pr, q)$ in arcs (p_1, t_1) and (t_1, p_3) specify that, when transition t_1 fires, one token in p_1 shall be consumed from place p_1 and transferred (without modifications) to place p_3. This is how we model processing of resources. In such firing, variables in the expression are binded to token values, e.g., $(o, ts, pr, q) = (b1, 1, 22.0, 3)$. As another example, let us consider transition t_6. It specifies a trade where a buy order is partially filled (q2 out of q stocks were bought), so the order should return with its remainder to the buy side (place p_5). The expression $\mathcal{E}(t_6, p_5) = (o, ts, pr, q - q2)$ in arc (t_6, p_5) makes such modification, decrementing the buy order's stock quantity by q2, s.t. q2 is the stock quantity of the sell order binded from place p_6.

Transitions and Activity Labels. We consider a function Λ, mapping each transition to a label from a finite set $\mathcal{A}$ of activity labels. Thus, as shown in Fig. 3, each transition represents an activity in a trading session. Transitions t_1, t_2

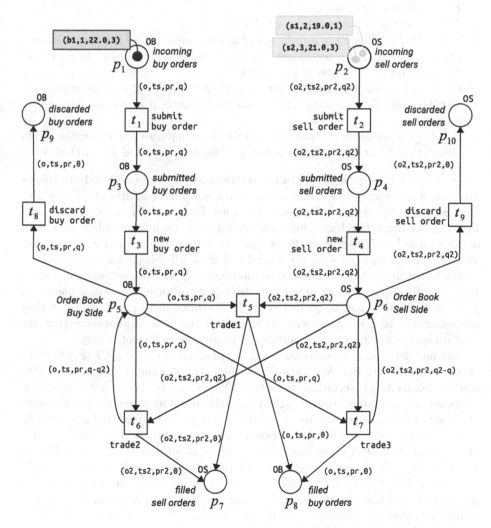

Fig. 3. CPN modeling a trading session in an order book. (Color figure online)

model submission of incoming orders from participants. Transitions t_3, t_4 model insertion of submitted orders in an order book side. Then, a trade may occur between a buy order and a sell order. In particular, transition t_5 (activity `trade1`) models a trade where both orders were filled (all their stocks were bought/sold). Transitions t_6, t_7 (activities `trade2` and `trade3`) model the situation where only one of the orders is filled, whereas the second one is partially filled (returning to the order book). Finally, transitions t_8 and t_9 represent activities to discard orders from the order book.

Definition 1. *[Colored Petri net] A colored Petri net is a 6-tuple* $CP = (P, T, F, \mathtt{color}, \mathcal{E}, \Lambda)$, *where:*

- *P is a finite set of places;*
- *T is a finite set of transitions, s.t. $P \cap T = \emptyset$;*
- *$F \subseteq (P \times T) \cup (T \times P)$ is a finite set of directed arcs (called the flow relation);*
- *$\mathtt{color} : P \to \Sigma$ is a place-coloring function, mapping each place to a color in Σ, such that Σ is a finite set of colors;*
- *$\mathcal{E} : F \to \mathcal{L}$ is an arc-labeling function, mapping each arc r to an expression of a language $\mathcal{L}$, s.t. $\mathtt{color}(\mathcal{E}(r)) = \mathtt{color}(p)$ where p is a place adjacent to an arc r;*
- *$\Lambda : T \to \mathcal{A}$ is an activity-labeling function, mapping each transition to an element in $\mathcal{A}$, s.t. $\mathcal{A}$ is a finite set of activity labels, $\forall t, t' \in T : \Lambda(t) \neq \Lambda(t')$.*

We consider CPNs with the following restrictions. On the one hand, all tokens are unique, so each place stores a set of tokens (not a multiset). Notice also that only one token can be consumed at once from each input place (e.g., see Fig. 3). On the other hand, for each transition t, each input place of t is of a different color. These restrictions come to be natural in many information systems. As exemplified with the model in Fig. 3, all tokens are unique having distinct identifiers. Also, orders can be modified (e.g., to update their stock size), but cannot disappear when processing them (e.g., all orders in initial places p_1, p_2 must arrive to places p_7, p_8 if they trade all stocks, or to places p_9, p_{10} if they are canceled). In Sect. 4, we explain how these restrictions guarantee that the conformance checking method performs a correct and efficient replay.

We now define execution semantics of our model. Let $CP = (P, T, F, \mathtt{color}, \mathcal{E}, \Lambda)$ be a colored Petri net. A *marking* M is a function, mapping each place $p \in P$ to a set of tokens $M(p)$, according to its color. We denote by M_0 an initial marking. Markings model system states, e.g., in Fig. 3, the initial marking of the net models the start of a simple trading session, with orders yet not submitted and with an empty order book. A *binding* b of a transition $t \in T$ is a function, assigning a value $b(\mathtt{v})$ to each variable $\mathtt{v}$ occurring in arc expressions adjacent to t. Let be $^\bullet t$ be a set of input places of a transition $t \in T$. Transition t is *enabled* in marking M w.r.t. a binding b iff $\forall p \in {}^\bullet t : b(\mathcal{E}(p, t)) \in M(p)$, that is, each input place of t has at least one token to be consumed. The *firing* of an enabled transition t in a marking M w.r.t. to a binding b yields a new marking M' such that $\forall p \in P : M' = M(p) - \{b(\mathcal{E}(p, t))\} \cup \{b(\mathcal{E}(t, p))\}$.

3   Event Logs

In this section, we now introduce *event logs*, describing how they are structured.

Definition 2. *[Event Log] An event log of is a finite set of traces $L = \{\sigma_1, ..., \sigma_s\}$ where, for each $i \in \{1, ..., s\}$, a trace $\sigma_i = \langle e_1, ..., e_m \rangle$ is a finite sequence of events, s.t. $m = |\sigma_i|$ is the trace length.*

Each event e in a trace is a tuple of the form $(a, \{r_1, ..., r_k\})$ where $a \in \mathcal{A}$ is an activity label, s.t. $\mathcal{A}$ is a finite set of activity labels, and for each $j \in \{1, ..., k\}$, we say that r_j is a resource involved in the execution of activity a.

Table 1. Example of a trace σ in an event log L of a simple trading session.

Event (e)	Activity (a)	Resources ($R(e)$)
e_1	submit buy order	(b1, 1, 22.0, 3)
e_2	new buy order	(b1, 1, 22.0, 3)
e_3	submit sell order	(s1, 2, 19.0, 1)
e_4	new sell order	(s1, 2, 19.0, 1)
e_5	submit sell order	(s2, 3, 21.0, 3)
e_6	new sell order	(s2, 3, 21.0, 3)
e_7	trade 2	(b1, 1, 22.0, 2), (s1, 2, 19.0, 0)
e_8	trade 3	(b1, 1, 22.0, 0), (s2, 2, 21.0, 1)
e_9	discard sell order	(s2, 2, 21.0, 0)

As an example, Table 1 shows a trace σ of an event log L. Each event e in σ indicates which activity was executed and a set of involved resources, e.g., in event e_2, activity **new buy order** was executed, placing order (**b1**, 1, 22.0, 3) in the order book. We introduce function $R(e)$ to return the set of resources involved in an event e. As introduced in Sect. 1, since our conformance method aims to associate observed resources in an event with tokens in a CPN model, we assume that each resource $r \in R(e)$ is a tuple belonging to some color in Σ, s.t. Σ is the set of all possible colors in a CPN model. With slight abuse of notation, we use color(r) to denote the color of a resource r. For example, (**b1**, 1, 22.0, 3) in event e_1 is a buy order, so color((**b1**, 1, 22.0, 3)) = OB, where OB defines the structure of buy orders, as we exemplified in Sect. 2.

For each resource $r = (r^{(1)}, ..., r^{(n)})$ in an event $e = (a, R(e))$, its tuple components $r^{(1)}, ..., r^{(n)}$ represent the state of the resource after the execution of activity a. In other words, after executing a, some attributes of r could have been modified. However, we shall assume that the first component of r, i.e., $r^{(1)}$, is the *resource identifier*, which cannot be modified by any activity. For compactness, we denote by id(r) = $r^{(1)}$ the identifier of $r = (r^{(1)}, ..., r^{(n)})$, e.g., id($r_1$) = **b1** for $r_1 = (\mathbf{b1}, 1, 22.0, 3)$.

By using identifiers, we consider that resources can be distinguished (as we assumed with tokens in a model). This allows us to identify the distinct objects involved in a trace, e.g., in Table 1 we identify three distinct resources: one buy order **b1**, and two sell orders **s1** and **s2**. Also, this allows us to track how a resource is modified. For example, let us consider the order **s2**, which initially had 3 stocks in event e_5. In event e_8 its stock size was reduced to 1 after executing a trade, and then in event e_9 its stock size went to 0 after the order was discarded.

Let $r = (r^{(1)}, ..., r^{(n)})$ be a resource. For $j \in \{1, ..., n\}$, we have that each resource attribute $r^{(j)}$ can be accessed using an attribute name. Also, all resources of the same color share the same set of attribute names. For instance, for the color of buy orders **OB**, we consider the attribute names $\{\text{id}, \text{tsub}, \text{price}, \text{qty}\}$. We define a *member access function* **#**, such that given a resource $r = (r^{(1)}, ..., r^{(n)})$ and the name of the jth-component, it returns the value of $r^{(j)}$, i.e., $\#(r, \text{name}_j) = r^{(j)}$. For simplicity, we use notation $\text{name}_j(r)$ instead of $\#(r, \text{name}_j)$. For example, for $r = (\text{b1}, 1, 22.0, 3)$, we have that $\text{tsub}(r) = 1$, $\text{price}(r) = 22.0$, and $\text{qty}(r) = 3$.

4   Conformance Checking Using Colored Petri Nets and Event Logs

In this section, we present a conformance checking method for CPNs (cf. Definition 1) and event logs (cf. Definition 2). Before describing our method, we first explain the restrictions that CPNs must satisfy to guarantee a correct and efficient replay.

Model Restrictions. As described in Sect. 2, tokens (in all model markings) must be unique (to have distinct identifiers), and for each transition t, all input places of t are of different colors. We illustrate the need of such restrictions with the next example. Consider the replay of trace $\sigma = \langle e_1, e_2 \rangle = \langle (\text{a}, \{\text{green}, \text{red}\}), (\text{b}, \{\text{red}\}) \rangle$ on the CPN of Fig. 4(a). All places and variables x, y in arcs are of a same color A. The CPN breaks the restrictions: there are clones with identifiers **green** and **red**, and t has input places of the same color.

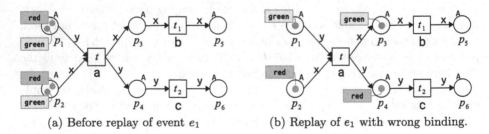

(a) Before replay of event e_1 (b) Replay of e_1 with wrong binding.

Fig. 4. Replay of trace σ on a CPN model without restrictions (a). After replaying event e_1 with wrong binding b_1, event $e_2 = (\text{b}, \{\text{red}\})$ cannot be replayed (b). (Color figure online)

To replay event $e_1 = (\mathtt{a}, \{\mathtt{green}, \mathtt{red}\})$, we may fire t ($\Lambda(t) = \mathtt{a}$) with binding $b_1 = \langle \mathtt{x} = \mathtt{green}, \mathtt{y} = \mathtt{red} \rangle$ or $b_2 = \langle \mathtt{x} = \mathtt{red}, \mathtt{y} = \mathtt{green} \rangle$. Consider to fire t with binding b_1, yielding the marking of Fig. 4(b). Now, event $e_2 = (\mathtt{b}, \{\mathtt{red}\})$ cannot be replayed. The event does not showcase a real deviation, but a wrong binding selection: if t fires with binding b_2, then e_2 can be replayed. Backtracking (return to previous events and to try other bindings) is needed in such situations to assert if a deviation has been found, or instead previous firings with wrong bindings blocked the replay. Backtracking may be computationally expensive. Instead, let us consider now the model of Fig. 5(a) where restrictions are complied.

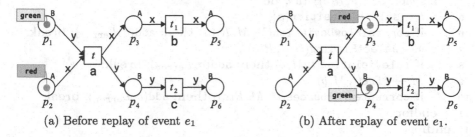

(a) Before replay of event e_1 (b) After replay of event e_1.

Fig. 5. CPN model with the restrictions: the $\mathtt{green}$ value now is of a color B; after replaying e_1 (with the only allowed binding), $e_2 = (\mathtt{b}, \{\mathtt{red}\})$ can be replayed. (Color figure online)

Now, to replay $e_1 = (\mathtt{a}, \{\mathtt{green}, \mathtt{red}\})$, there is only one binding associated to the observed resources in event e_1, that is, $b = \langle \mathtt{x} = \mathtt{red}, \mathtt{y} = \mathtt{green} \rangle$. No other binding may be selected as $\mathtt{green}$ and $\mathtt{red}$ values are now forced to come from separate sources. After firing t, $e_2 = (\mathtt{b}, \{\mathtt{red}\})$ is guaranteed to be replayed, so backtracking is avoided. In this way, our method associates each observed resource to a token from a specific input place (such token cannot appear in other place), and thus only one binding can satisfy the event replay. Thus, with these restrictions, the method guarantees a correct and less expensive replay.

Conformance Checking Method. We proceed now to explain our conformance checking method, based on individual replay of each trace on top of a CPN. As introduced before, the method seeks to fire transitions, labeled with activities indicated in the events, and selecting input tokens according to resources observed in the events. If the latter is not possible or, as we will present, other deviations are found, the trace replay is stopped. As output, this method returns a list of non-fitting traces (not completely replayed), and a fitness metric. This metric indicates a degree of conformance between a CPN and an event log.

Algorithm 1: Conformance Checking using CPNs

Input: $CP = (P, T, F, \mathsf{color}, \mathcal{E}, \Lambda)$, a CPN with an empty initial marking;
   $P_0 \subseteq P$, a non-empty set of initial places;
   L, an event log (finite set of traces);
Output: L_{error} - set of non-fitting traces;
   $\mathtt{fitness}$ - degree of conformance;

1 $L_{\mathsf{error}} \leftarrow \emptyset;$ $\mathtt{fitness} \leftarrow 0;$
2 **foreach** $\sigma \in L$ **do**
3     $M \leftarrow \emptyset;$ $M \leftarrow \mathtt{populateInitialPlaces}(P_0, R(\sigma));$
4   **foreach** $e = (a, R(e))$ **in** σ **do**
5        $t \leftarrow \mathtt{selectTransition}(a);$
6   **if** $\mathtt{controlFlowDeviation}({}^\bullet t, M, R(e))$ **then:** $\mathtt{add}(\sigma, L_{\mathsf{error}});$ **break;**
7        $b \leftarrow \mathtt{selectBinding}(t, M, R(e));$
8   **if** $\mathtt{ruleViolation}(t, M, b)$ **then:** $\mathtt{add}(\sigma, L_{\mathsf{error}});$ **break;**
9        $M \leftarrow \mathtt{fire}(t, M, b);$
10   **if** $\mathtt{corruptedResources}(t^\bullet, M, R(e))$ **then:** $\mathtt{add}(\sigma, L_{\mathsf{error}});$ **break;**
11   **endfor**
12 **endfor**
13 $\mathtt{fitness} \leftarrow 1 - (|L_{\mathsf{error}}| \,/\, |L|);$
14 **return** $(L_{\mathsf{error}}, \mathtt{fitness});$

Initial Setting. Algorithm 1 presents our method whose input is a CPN with an empty initial marking, an event log L, and a set of initial places $P_0 \subseteq P$. At the start of each trace replay, each place in P_0 is populated with the distinct resources in σ, according to its color (function $\mathtt{populateInitialPlaces}$).

Let us consider the replay of σ in Table 1 on the CPN of Fig. 3. The CPN shows a marking after the distinct resources in σ were placed in the initial places: buy order b1 in p_1, and sell orders s1 and s2 in p_2. For each resource to insert as a token in an initial place, we set its token values according to its first occurrence in a trace, e.g., b1 is placed with values $(\mathtt{b1}, 1, 22.0, 3)$ as shown in event e_1.

Control-Flow Deviation Due to the Absence of Input Resources. In a trace σ, after setting the model marking according to the distinct resources in σ, we start to replay σ on the CPN. For each event $e = (a, R(e))$ in σ, we try to fire a transition t, s.t. $\Lambda(t) = a$. To fire, we check if, in a current marking M, each resource involved in e is contained in an input place of t. Let ${}^\bullet t$ be the set of input places of t. To this aim, we check the truth value of the next formula:

$$\forall p \in {}^\bullet t \;:\; \exists! r \in R(e) \; \exists(d_1, ..., d_n) \in M(p) \;:\; \mathtt{id}(r) = d_1 \wedge \mathtt{color}(r) = \mathtt{color}(p)$$

The function $\mathtt{controlFlowDeviation}$ checks if the previous formula evaluates to false. If so, the replay is stopped, e.g., some resource in $R(e)$ is not available in an input place. Otherwise, a binding is selected (function $\mathtt{selectBinding}$), s.t. a token $(d_1, ..., d_n)$ will be consumed from each input place p, i.e., $b(\mathcal{E}(p, t)) = (d_1, ..., d_n)$, and each token corresponds to a resource r in $R(e)$, i.e., $\mathtt{id}(r) = d_1$.

For example, let us consider the replay of a trace sigma σ'_1 on the CPN of Fig. 3. Let us assume that σ'_1 consists of the first six events of Table 1 (submission of buy order b1 and sell orders s1,s2) plus the two events shown below.

e_7	discard sell order	(s1, 2, 19.0, 0)
e_8	trade 2	(b1, 1, 22.0, 2), (s1, 2, 19.0, 0)

Event e_8 (trade 2 between b1 and s1) will not be replayed as s1 was discarded in e_7 (moved to place p_{10}). Hence, s1 is not anymore available in the sell side (place p_6). Clearly, a trade cannot be executed with a canceled order. In this way, we can detect deviating events with unavailable resources.

Rule Violations. Let b be the selected binding to fire t according to the resources in $R(e)$. For each input place p of t, we want to check if the token to consume is the one that should be selected (among all possible ones in p) according to some rule. For example, in trading systems, it is mandatory to know if a priority rule is being complied, e.g., a buy order with highest price must trade before other buy orders. Thus, we define a *marking dependent rule* $\Phi(t)$ as follows:

$$\Phi(t) \equiv \bigwedge_{\forall p \in \bullet t} \phi_p(M(p), b(\mathcal{E}(p,t)))$$

where each $\phi_p(M(p), b(\mathcal{E}(p,t)))$ is a local rule in place p. In Algorithm 1, we set ruleViolation$(t, M, b) \equiv \neg\Phi(t)$. Before firing t, if a rule $\phi_p(M(p), \mathcal{E}(p,t))$ is violated, the trace replay is stopped. Otherwise, if all rules are complied or no rule was defined for transition t, then t fires with selected binding b. For the CPN of Fig. 3, let us assign the rule below to transitions t_5, t_6, t_7 (trade activities).

$$\Phi(t) \equiv \phi_{\text{BUY}}(M(p_5), r_1) \wedge \phi_{\text{SELL}}(M(p_6), r_2).$$

$$\phi_{\text{BUY}}(M(p_5), r_1) \equiv \forall_{(o,ts,pr,q) \in M(p_5)} \text{ id}(r_1) \neq o : (\text{price}(r_1) > \text{pr})$$
$$\vee (\text{price}(r_1) = \text{pr} \wedge \text{tsub}(r_1) < \text{ts})$$

$$\phi_{\text{SELL}}(M(p_6), r_2) \equiv \forall_{(o,ts,pr,q) \in M(p_6)} \text{ id}(r_2) \neq o : (\text{price}(r_2) < \text{pr})$$
$$\vee (\text{price}(r_2) = \text{pr} \wedge \text{tsub}(r_2) < \text{ts})$$

where r_1 and r_2 are the buy and sell orders to consume. The local rule ϕ_{BUY} for place p_5 states that r_1 must be the order with highest price (or with earlier submitted time than other order with same highest price). The local rule ϕ_{SELL} for place p_6 is defined similarly, but stating that r_2 must be the order with lowest price. Notably, $\phi_{\text{BUY}}(M(p_5), r_1) \wedge \phi_{\text{SELL}}(M(p_6), r_2)$ is a price-time priority rule, that trading sessions must comply. For instance, let us consider the replay of the trace in Table 1 on CPN of Fig. 3. It can be observed that this rule is being complied when executing trades, e.g., sell order s1 is served before order s2. For other trace, if the rule is not complied, the replay of that trace is stopped.

Checking Differences Between Modeled and Real Produced Resources. We aim to exploit the fact that each event $e = (a, R(e))$ has information about the new state of each resource after executing a. Recall that each resource in $R(e)$ could have been modified by a. Let us consider the firing of a transition t w.r.t a selected binding b, and yielding a new marking M. Then, we proceed to check whether each resource in real life, after executing a, was modified as performed in the model, after firing t. Let $t^\bullet$ be the set of output places of t. Then, we verify if the following formula is satisfied:

$$\forall p \in t^\bullet : \exists! r \in R(e) \, \exists (d_1, ..., d_n) \in M(p) : r = (d_1, ..., d_n) = b(\mathcal{E}(t, p))$$

In Algorithm 1, the function `corruptedResources` checks if the previous formula evaluates to false. If so, the replay is stopped, i.e., some resource in $R(e)$ was not modified as indicated by the output arc expressions of t. For example, let us consider the replay of a trace σ_2' on the CPN of Fig. 3. Let us assume that σ_2' consists of the first six events of Table 1 plus the event shown below.

| e_7 | trade 2 | (b1, 1, 22.0, 1), (s1, 2, 19.0, 0) |

Before the execution of event e_7, we recall that orders b1 and s1 have the following states: (b1, 1, 22.0, 3) and (s1, 2, 19.0, 1). After replaying event e_7 in the CPN, the token b1 is transformed to (b1, 1, 22.0, 2). As specified by the arc expression $\mathcal{E}(t_6, p_5)$, the stock size of b1, which is 3, was decremented by the stock size s1, which is 1. Thus, the resulting stock size of b1 is 2. However, event e_7 states that after trading the stock size of order b1 is 1. Evidently, the stock size of b1 was corrupted when trading. For such a case, the trace replay is stopped.

The output of Algorithm 1 is a set of non-fitting traces L_{error}, e.g., traces with any kind of the deviations explained before. Also, the algorithm computes a *fitness metric* (a ratio of completely replayed traces) as follows: `fitness` $= 1 - (|L_{\text{error}}| / |L|)$ where $L_{\text{error}} \subseteq L$. Since $|L| \geq |L_{\text{error}}| \rightarrow$ `fitness` $\in [0, 1]$.

We close this section with a brief analysis on the time complexity of our method. Whilst we do not carry out a deeper study, for example, using asymptotic notation, we do identify the crucial parameters that mainly influence the time performance of our algorithm. Let $CP = (P, T, F, \text{color}, \mathcal{E}, \Lambda)$ be a colored Petri net and $L = \{\sigma_1, ..., \sigma_s\}$ be an event log, as described in Definitions 1 and 2. Now, let us examine the required operations to replay a single event $e = (a, R(e))$ of a trace σ (i.e., see lines 4–10 of Algorithm 1). On the one hand, these operations seek to fire a transition t s.t. $\Lambda(t) = a$, which it may require up to $|T|$ transitions to visit. On the other hand, a common term of these functions is to compare each resource in $R(e)$ against its corresponding token in an input place of t and, after firing, in an output place of t (e.g., when checking availability of resources). Both, the sets of input and output places of t are bounded by $|P|$. Thus, it can be inferred that the number of steps required to replay a single event is in the order of magnitude $|T| + (|R(e)| \cdot |P|)$. With this term, it is clear to identify that the computing time required for replaying an event is mainly affected by the number of places $|P|$ in a CPN and the number of observed resources $|R(e)|$ in an event. Afterwards, it is easy to see that this event replay routine will be repeated at most $|\sigma| \cdot |L|$ times, s.t. $\forall \sigma' \in L : |\sigma| \geq |\sigma'|$, that is, $|\sigma|$ is the maximum number of events per trace, whereas $|L|$ is the number of traces in the event log.

5   Implementation and Experimental Validation

In this section, we describe the implementation and evaluation of our conformance method using colored Petri nets. We developed this implementation in Python programming language. We have carried out experimental works with both real and artificially generated event logs, which allow us to show the practical value of our method for detecting system deviations. The implementation and all material of our experiments are available in our project repository [7].

Our solution is supported by a Python library called SNAKES [19]. This library facilitates the prototyping of high-level classes of Petri nets, including CPNs. This allows us to instantiate CPN models as Python objects, which can be used as input to our method. Figure 6 illustrates the organization of our prototypical implementation. Users of our solution simply need to invoke a program called the "conformance checker". This program receives three parameters: an option indicating the conformance method to use (e.g., replay with CPNs), an event log stored in a log repository, and a Petri net model stored in a model repository. This generic organization allows us to seamlessly extend our solution, incorporating other conformance methods and models of our research.

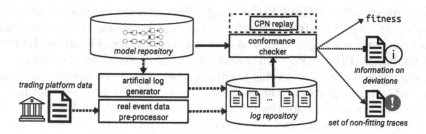

Fig. 6. Organization of our prototypical implementation. (Color figure online)

Upon execution of our method, the program provides command-line messages, indicating resulting metrics, i.e., fitness. Besides, it generates two files:

```
=============== CONFORMANCE RESULTS ================
Total number of traces: 4
Non-fitting traces: 3

Fitness : 0.2500
Control-flow deviations detected: 1
Rule violations deviations detected: 1

Resource corruptions detected: 1

Deviations written in file: deviations_14081.csv
Non-fitting traces cloned to file: nf_traces_14081.csv
===================================================
```

Fig. 7. Output summary resulting after the execution of our method. (Color figure online)

a file consisting of a set of non-fitting traces (i.e., traces that were not completely replayed) and a file with specific information of trace deviations: in which event a trace replay was stopped, which kind of deviation occurred, and comments that may help engineers to localize failures. For example, Fig. 7 shows a resulting message fragment after replaying an event log, where three traces suffered from the kinds of deviations explained in the previous section (e.g., control-flow deviation, rule violation, or resource corruption). In addition, Fig. 8 shows a fragment of an output file specifying the deviating events in each non-fitting trace.

TRACE	EVENT OCCURRED AT	ACTIVITY	DEVIATION	ADDITIONAL INFORMATION
1111037	18-02-2019T09:13:07.581	trade2	CONTROL-FLOW	resource with id: s1 is not available.
1111038	18-02-2019T09:13:07.581	trade2	RULE-VIOLATION	resource with id: s2 does not have priority over other resources in the same place
1111039	18-02-2019T09:13:07.581	trade2	RESOURCE-CORRUPTED	resource with id: b1 has observed state: ('b1',1,22.0,3) but expected state was: ('b1',1,22.0,2)

Fig. 8. Fragment of specific deviation diagnostics generated by our method. (Color figure online)

Experiment with a Real Event Log. We conducted an experimental work using an event log from a trading platform. This log was obtained by pre-processing data from a real trading system. The data is a recorded set of Financial Information Exchange (FIX) protocol messages [8]. These messages were exchanged by participants and a platform during trading sessions, so they encapsulate activities executed by both agents and the platform. We developed a pre-processor in Java which extracts such event log from this set of messages (also available via [7]). The event log consists of individual traces, each of them related to a trading session in an order book. In particular, the expected behavior in each of these trading sessions is specified by the CPN model shown in Fig. 3.

Table 2 shows characteristics of the event log. Also, it presents conformance results, resulting from the execution of our method using the mentioned log and the CPN of Fig. 3. The table shows the number and kinds of deviations detected.

Table 2. Event log characteristics and obtained conformance results.

Event log characteristics		Conformance results	
Number of pre-processed FIX Messages	552935	Number of non-fitting traces	8
Number of traces (trading sessions)	73	Fitness	0.890
Total number of events in the log	2259	Rule violations detected	1
Average number of events per trace	30	Resource corruptions detected	7

The fact that the majority of traces were completely replayed evidences that most of the trading sessions comply with the model. Regarding the non-fitting traces, the obtained file with information about deviations indicate that the rule violations and resource corruptions originated in trade activities. The obtained information may support experts to confirm whether a failure is occurring in those activities, or instead the CPN model should be slightly refined.

Experiments with Artificial Event Logs. In addition to the previous experiment, we considered to test our method with slightly more stressed scenarios, where system runs may be hampered by sporadic (yet critical) deviations. More precisely, we aimed to evaluate the impact of each of the kinds of deviations previously introduced when a system is managing a certain number of resources (e.g., buy or sell orders). Thus, we considered an experiment where we slightly modified the correct specification model (the CPN of Fig. 3), obtaining three "incorrect" variants (see Table 3). These variants can be seen as instances of a correct trading platform, but sporadically suffering from one kind of deviation.

Table 3. Description of "incorrect" variants to generate artificial event logs.

Model variant	Description of the deviation that may occur
System A	*Control-flow deviation*: with 5% of probability, activities discard buy order or discard sell order do not cancel orders, so orders keep in the order book (places p_5 and p_6 in Fig. 3), and may continue to trade.
System B	*Rule violation*: with 2% of probability, and upon execution of activities trade1, trade2, or trade3, this variant does not respect the priority rule, i.e., buy/sell orders with highest/lowest prices are not served first.
System C	*Resource corruption*: With 5% of probability, stock quantities of buy/sell orders change to 0 when placing them in the order book (misbehavior of activities new buy order or new sell order of Fig. 3)

We built models of these variants (using SNAKES) to generate artificial event logs, thereby representing observed behavior of the "incorrect" instances. Recall that the fitness metric considered in this work relates to the number of completely replayed traces. We used this metric to assess the probability that a system run can be jeopardized by the occurrence of a deviation. Table 4 presents results of this experiment. Each cell indicates the average fitness value after executing our method, between the correct model (Fig. 3) and ten artificial event logs from an 'incorrect' variant, and with a certain number of resources per class (e.g., number of buy orders and sell orders). All event logs consist of 500 traces.

Table 4. Fitness between the CPN (Fig. 3) and logs of each incorrect variant.

Model variant	5 resources per class	25 resources per class
System A	0.6436	0.05854
System B	0.9816	0.8569
System C	0.6196	0.05852

Let us consider variants A and C. Albeit the probability of a deviation is expected to be low (i.e., 5%), the average ratio of correct traces is only above 60%. Also, when the number of resources processed by a system inscreases, the fitness value notably decreases. This is an expected pattern since when considering more resources during a system run, the length of traces are enlarged (i.e., more activities processing resources), and the probability of one deviation is increased.

6   Discussion and Conclusion

Conformance checking methods detect deviations in system processes using event logs and process models. These methods use replay [18], as seen in this paper, or alignments [2], which relate traces with model executions. A limitation of these methods is that they only focus on control-flow, i.e., whether system's activities comply a causal ordering. Whilst certain proposals tackle this issue with slightly enriched models (e.g., data Petri nets) [12,13], they use notations whose backbone does not allow to describe transformation of objects. Tackling such problem, we presented in this paper a conformance method using colored Petri nets—an extension where tokens carry data of some classes (colors). Arc expressions specify how tokens are transformed upon transition firings. The method replays events on a CPN, firing transitions labeled with an event's activity, and choosing as input tokens the ones related to observed resources in the event. We showed deviations that can be detected: errors due to absent resources, rule violations or resource corruptions. We provided a prototype and experiments [7].

To make feasible the use of CPNs, we considered restrictions, e.g., tokens must be unique and cannot be destroyed. Also, all input places of a transition must be of different colors. We described how the latter allows us to avoid backtracking when replaying a trace. Interestingly, by looking the CPN in Fig. 3, this restricted model resembles a union of workflow nets [1], a Petri net class used in process mining. Each "lane" processing a resource class can be seen as a workflow net, and some transitions allow interaction between these workflows.

Through the paper, we illustrated our method using trading systems [9]. In this regard, a direction of our research focuses on the definition of formal models to analyze different aspects of these systems [4–6]. However, it is easy to see that the method provided in this work can be easily applied in other domains.

There are similar approaches for checking system's compliance in the broader field of data science, e.g., passive analysis [10] or action rules [17]. Also, event logs with resources are similar to (multi-dimensional) sequence databases in data mining [16]. However, such methods do not use formal models such as Petri nets. Instead, we showed how rules to comply can be exhaustively described into a single model, so that traces can be systematically compared against such model.

For future work, we want to enhance our method, replaying traces after one deviation is found. When a transition cannot fire due to an absent resource, we cannot consider injection of "missing" tokens (as proposed in [1]) since we would violate resource uniqueness in a model. Instead, we plan to study an approach based on "moving" tokens. Also, we plan to research how this method may relate to another work, where we check conformance of system-agent interactions [14].

References

1. van der Aalst, W.: Process Mining: Data Science in Action. Springer (2016)
2. Adriansyah, A., Munoz-Gama, J., Carmona, J., van Dongen, B., van der Alst, W.: Measuring Precision of Modeled Behavior. IseB **13**(1), 37–67 (2015)
3. Carmona, J., van Dongen, B., Solti, A., Weidlich, M.: Conformance Checking: Relating Processes and Models. Springer, Cham (2018)
4. Carrasquel, J.C., Lomazova, I.A.: Modelling and Validation of Trading and Multi-Agent Systems: An Approach Based on Process Mining and Petri Nets. In: van Dongen, B., Claes, J. (eds.) Proceedings of the ICPM Doctoral Consortium. CEUR, vol. 2432 (2019)
5. Carrasquel, J.C., Lomazova, I.A., Itkin, I.L.: Towards a Formal Modelling of Order-driven Trading Systems using Petri Nets: A Multi-Agent Approach. In: Lomazova, I.A., Kalenkova, A., Yavorsky, R. (eds.) Modeling and Analysis of Complex Systems and Processes (MACSPro). CEUR, vol. 2478 (2019)
6. Carrasquel, J.C., Lomazova, I.A., Rivkin, A.: Modeling trading systems using petri net extensions. In: Köhler-Bussmeier, M., Kindler, E., Rölke, H. (eds.) International Workshop on Petri Nets and Software Engineering (PNSE). CEUR, vol. 2651 (2020)
7. Conformance Checking with Colored Petri Nets: Project Repository (Github). https://github.com/jcarrasquel/hse-uamc-conformance-checking
8. FIX Community - Standards. https://www.fixtrading.org/standards/
9. Harris, L.: Trading and Exchanges: Market Microstructure for Practitioners. Oxford University Press (2003)
10. Itkin, I., Yavorskiy, R.: Overview of applications of passive testing techniques. In: Lomazova, I., Kalenkova, A., Yavorsky, R. (eds.) Modeling and Analysis of Complex Systems and Processes (MACSPro). CEUR, vol. 2478 (2019)
11. Jensen, K., Kristensen, L.M.: Coloured Petri Nets: Modelling and Validation of Concurrent Systems. Springer, Cham (2009)
12. de Leoni, M., Munoz-Gama, J., Carmona, J., van der Aalst, W.: Decomposing alignment-based conformance checking of data-aware process models. In: Meersman, R., et al. (eds.) On the Move to Meaningful Internet Systems (OTM), pp. 3–20. Springer, LNCS (2014)
13. Mannhardt, F., de Leoni, M., Reijers, H.A., van der Aalst, W.: Balanced multi-perspective checking of process conformance. Computing **98**(4), 407–437 (2016)
14. Mecheraoui, K., Carrasquel, J.C., Lomazova, I.A.: Compositional Conformance Checking of Nested Petri Nets and Event Logs of Multi-Agent Systems. CoRR abs/2003.07291 (2020)

452   J. C. Carrasquel et al.

15. Murata, T.: Petri nets: Properties, analysis and applications. Proc. IEEE **77**(4), 541–580 (1989)
16. Pinto, H., Han, J., Pei, J., Wang, K., Chen, Q., Dayal, U.: Multi-dimensional sequential pattern mining. In: International Conference on Information and Knowledge Management, pp. 81–88. ACM (2001)
17. Ras, Z.W., Wyrzykowska, E., Tsay, L.: Action rules mining. In: Encyclopedia of Data Warehousing and Mining, pp. 1–5. IGI Global (2009)
18. Rozinat, A., van der Alst, W.: Conformance Checking of Processes Based on Monitoring Real Behavior. Inf. Syst. **33**(1), 64–95 (2008)
19. SNAKES - Petri net library: https://snakes.ibisc.univ-evry.fr/
20. van der Aalst, W., van Hee, K., van der Werf, J., Verdonk, M.: Auditing 2.0: using process mining to support tomorrow's auditor. Computer **43**(3), 90–93 (2010)

Data and Reference Semantic-Based Simulator of DB-Nets with the Use of Renew Tool

Anton Rigin[✉][iD] and Sergey Shershakov[iD]

Faculty of Computer Science, National Research University Higher School
of Economics, 20 Myasnitskaya Street, 101000 Moscow, Russia
amrigin@edu.hse.ru, sshershakov@hse.ru
https://www.hse.ru/

Abstract. Complexity of software systems is constantly growing, which
is even more aggravated by concurrency of processes in systems, so modeling and validating such systems is necessary for detecting and eliminating failures. One of the most well-known formalisms for solving this problem is Petri nets and their extensions such as colored Petri nets, reference
nets, and similar models. Many software systems use databases for storing persistent data. However, Petri nets and their mentioned extensions
are not designed for modeling persistent data manipulation since these
formalisms aim at modeling a control flow of the considered systems.
DB-nets, as a novel formalism, aim to solve this problem by providing
three following layers: (1) the control layer represented by a colored Petri
net with extensions, (2) the data logic layer, which allows to retrieve and
update the persistent data, and (3) the persistence layer representing a
relational database for storing the persistent data. To date, there are
no publicly available software tools that implement simulation of db-net
models with reference semantics support. The paper presents a novel
software tool for the db-nets simulation. The simulator is developed as
a pure plugin for the Renew (Reference Net Workshop) software tool
without modifying existing Renew source code. Such an approach for
development of the simulator allows to reinforce existing Renew reference
semantics. The SQLite embeddable relational DBMS is used as a base
tool to implement the db-net persistence layer. In the paper, theoretical
foundations and architecture of the developed simulator are described.
The results of the work can be used in research projects that involve modeling complex software systems with persistent data for both academic
and industry-oriented applications.

Keywords: DB-nets · Petri nets · Reference nets · Renew tool ·
Relational DBMS · Persistent data · Software modeling · Software
validation

1 Introduction

In the current days, the complexity of software systems is constantly growing.
One of the most important constituents of such complexity is concurrency of

© Springer Nature Switzerland AG 2021
W. M. P. van der Aalst et al. (Eds.): AIST 2020, LNCS 12602, pp. 453–465, 2021.
https://doi.org/10.1007/978-3-030-72610-2_34

processes supported by such systems since concurrency brings a lot of uncertainty and non-deterministic behavior.

One of the most popular modeling formalisms solving this problem is Petri nets [4]. It allows to represent a concurrent system as a directed bipartite graph. Resources of the system are modeled by a *marking* that describes availability of a modeled action to run and represents the current system state [4,5]. There are several well-known extensions of Petri nets. *Colored Petri nets* support *data types* called "colors", *arc expressions* and *guard expressions* [1,2]. *Reference Petri nets* or simply *reference nets* is another formalism based on Petri nets. Reference nets allow to use *references* to objects in marking, including references to other nets [6]. One of the well-known software tools implementing the semantics of reference nets is Renew (the Reference Net Workshop) [6].

Many software systems use databases for storing persistent data. Despite that Petri nets and their well-known extensions can model the complex concurrent software systems behavior, they still cannot easily model data manipulation in a persistent database. To overcome the aforementioned problem, a solution called db-nets was recently proposed [3]. A db-net is a formalism that consists of the control layer (a colored Petri net with extensions) with two additional layers for handling data in a connected database. Working together, these three layers combine a model workflow with data manipulation and, hence, solve the problem of data of classical Petri net-based solutions [3]. Moreover, the db-nets can be built with usage of the reference semantics since any reference data type can be potentially considered as an ordinary data type (color) in a colored Petri net. It allows to use complex data types with heavy payload as tokens in the net.

Although the db-nets can offer new possibilities of modeling the concurrent complex software systems and their validation, especially those which use the persistent data, there is only one their software implementation which is available publicly [7]. This A. Sinha's implementation is done as a plugin for the CPN Tools which is a software tool for manipulating colored Petri nets. In our work the reference semantics is also supported because of using the Renew tool. This tool is designed as a collection of plugins written in Java, and it is an open-source solution. In this work we extend this tool in order to support a simulation of a db-net run. This paper reveals aspects of this simulator development. The simulator provides the ability to validate possible behavior of a designed complex concurrent software system even if we need to consider the persistent data used by the system. This can be used for modeling and validating the behavior of the real safety-critical software systems, for example, smart contracts on a blockchain virtual engine, different financial software systems with high price of failure, medical and nuclear plant control software systems, as well as for further research of possibilities of db-nets.

The rest of the paper is organized as follows. Section 2 presents the theoretical foundations of the work. In Sect. 3 the design and implementation of the developed software simulator are described, and the running example is presented. In Sect. 4 the conclusion is given as well as the further ways of the work continuation.

2   Theoretical Foundations

The db-net formalism is based on the Petri net and colored Petri net formalisms. In the paper the reference semantics that was initially introduced for the reference nets [6] is also considered.

2.1   Petri Nets and Their Well-Known Extensions

The Petri net is a powerful tool for modeling the processes and behavior of concurrent software systems invented by Petri [4]. The Petri nets are used by different researchers for a long time since they provide wide range of the concurrent software modeling abilities. Many other models are built based on them.

Petri nets can be represented as the directed bipartite graph where the vertices (nodes) of the graph are places and transitions of the net, and the edges of the graph are arcs of the net connecting the places and transitions of the net.

Formally, the Petri net N is a triple $N = (P, T, F)$ [5], where:

1. P is the finite set of places of the net N;
2. T is the finite set of transitions of the net N;
3. $P \cap T = \emptyset$;
4. $F \subseteq (P \times T) \cup (T \times P)$ is the set of arcs of the net N.

Petri nets support modeling behavior using tokens which form markings. Conceptually, marking is a distribution of tokens across places. Formally, a marking M is a function $M : P \mapsto \mathbb{N}_0$. This function maps each place into the number of tokens which are contained by this place. If there are no tokens in the place at the given moment, then this function value is equal to 0 at the corresponding moment. The token represents the resource of the system. The whole marking represents the state of the system modeled by the Petri net [5].

Petri nets allow to model the concurrent systems avoiding need to depict all possible states which may constitute extremely large set or even infinite set which is difficult or impossible for people to comprehend.

There are several extensions of the Petri nets for solving the specific problems.

The colored Petri net is the extension of the Petri net which allows to define and use the typed ("colored") tokens. Each color represents a data type. In this case arcs may have type constraints in their arc expressions in order to prevent the transmission of token of unsupported type. These nets were described by Jensen and Kristensen [1,2].

The reference Petri net is the extension of the Petri net which have reference semantics rather than value semantics. The tokens in such nets are references to objects, for example, to other Petri nets – this concept called hierarchical Petri nets, nets within nets or nested nets [6].

Despite the mentioned Petri net extensions are useful in many cases, they do not allow to model the persistent data manipulation in the database. The possible solution for this problem is the db-net formalism which is discussed as the main topic of this work further.

2.2 DB-Nets

The db-net is a formalism which consists of three layers: (1) the control layer represented by a colored Petri net with extensions, (2) the data logic layer represented by queries and actions which allow the control layer to retrieve and manipulate the data in the database and (3) the persistence layer represented by a relational database schema together with constraints which declare data consistency rules. The last two layers and the extensions of a colored Petri net used in the control layer allow to model the data manipulation in a persistent database while the control layer allows to model the control flow of the system as well as the local (non-persistent) data manipulation. Therefore, the db-nets solve the problem of the persistent data manipulation in the model. The db-nets were proposed recently by Montali and Rivkin [3]. The db-net layers structure is shown in Fig. 1.

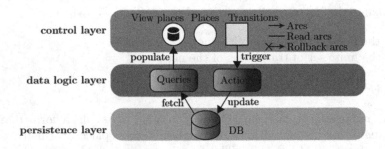

Fig. 1. The db-net structure [3]. (Color figure online)

Persistence Layer. The persistence layer of the db-net formalism is a relational database schema together with constraints which declare the data consistency rules. The database schema consists of relational table declarations – names of relational tables and types of their fields. The consistency rules are presented as boolean queries which may be evaluated to be equal to "true" or "false". If any of the consistency rules are evaluated as equal to "false" after performed update action, then the database is inconsistent, and the rollback of the last update action is performed [3]. In the db-net software implementation the persistence layer can be represented as a database in one of the relational database management systems (RDBMSs).

Data Logic Layer. The data logic layer of the db-net formalism is a set of queries which may be used on the persistence layer for retrieving the persistent data together with a set of parameterized actions which may be performed on the persistence layer for updating (modifying) the persistent data [3].

The query in the data logic layer is a relational query for retrieving the data from the RDBMS. The action in the data logic layer is represented by sets of added and deleted facts (table rows). During performing an action, rows from the set of added facts are inserted into the corresponding database tables and rows from the set of deleted facts are removed from them. Each added or deleted fact consists of the table name and the parameters which are the column variables for the inserted or removed row. The variable may be a simple variable (in this case it is replaced by a real value in the db-net control layer), an external input variable (in this case it is replaced by the data inputted by a user) or a randomly generated fresh variable (in this case it is randomly generated during the action performing) [3].

Control Layer. The control layer of the db-net formalism is a colored Petri net with the following extensions [3]: (1) in addition to the traditional Petri net places there view places may exist which allow to retrieve the persistent data from the db-net persistence layer using the data logic layer queries; (2) actions from the data logic layer may be assigned to transitions for being performed during the transition firing (execution); (3) in addition to the traditional Petri net arcs there are also read arcs, for connecting view places with transitions, and rollback arcs, for handling the case of the action rollback because of emerged data inconsistency in the db-net persistence layer.

The example of the db-net control layer is shown in Fig. 2. In this example the simple information system for task tracking is modelled. This system allows to create (register) new tasks (tickets), assign created tickets to employees, release employees from their tickets, log resolved tickets and control that for each employee not more than one ticket is assigned at the moment.

The left two circles in the example depict the view places while other circles represent the traditional Petri net places. The text near the view places is the corresponding query assignment. The lines connected to the view places show read arcs. The rectangles depict the transitions, the text pieces on these rectangles are the corresponding actions assignments. The arrows are the traditional Petri net arcs except of the red arrow from the *"Awake"* transition to the *"Stalled tickets"* place which is the rollback arc.

The db-nets are used as the basic formalism for the software tool developed in the current work.

3   Design and Implementation Details

The Renew architecture allows to extend its functionality by developing plugins. The simulator for db-net formalism is developed as a Renew plugin.

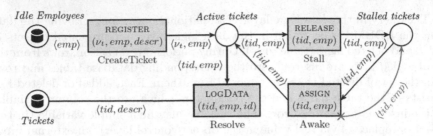

Fig. 2. The example of the db-net control layer [3]. (Color figure online)

3.1 Renew Software Tool

Renew (the Reference Net Workshop) is an open-source software tool for modeling the reference and colored Petri nets. Renew is a plugin-based tool which means that each its component and each element of this tool source code belongs to one of its plugins. All the Renew plugins are written in the Java programming language and use Apache Ant as a build system [6].

The main Renew plugin is the `Loader` plugin which loads all other plugins in the corresponding Renew build during the Renew tool running start. Other base Renew plugins include but not limited to the `Simulator` plugin, for the Petri net model simulation, the `Formalism` plugin, for parsing and compiling the Petri net and its constituents as formalisms, number of the GUI plugins and others [6].

Adding new plugin is possible through implementing the subclass of the `de.renew.plugin.PluginAdapter` class from the `Loader` plugin and adding several build and configuration files as well as adding the information about the new plugin into the main Apache Ant `build.xml` build file [6].

3.2 Developed DB-Nets Renew Plugin Overall Structure

In this work the *DB-nets Renew Plugin* is developed in order to add into the Renew tool the ability to work with the db-nets together with the existing Renew reference semantics (ability to use reference data types for tokens).

The *DB-nets Renew Plugin* is developed without modifying any existing Renew source code. Modeling the db-net in the Renew GUI, as well as its simulation, saving and opening are implemented.

The *DB-nets Renew Plugin* Java classes extend and use the classes of the existing Renew plugins. The *DB-nets Renew Plugin* UML class diagram[1] is available in the project GitHub repository[2]. Only classes implemented/interfaces declared inside the plugin are shown on the UML class diagram together with classes/interfaces which are inherited/implemented by them (the latter are

---

[1] Link: https://github.com/Glost/db_nets_renew_plugin/tree/master/root/docs/other/RenewDBNetsPluginFormalismsClassDiagram.

[2] Link: https://github.com/Glost/db_nets_renew_plugin.

shown without class members) respectively. The plugin classes which override only constructor, the GUI and utility classes as well as the autogenerated classes are not shown in order to save the visibility.

To parse the db-net text inscriptions (queries, actions and their calls, variables declarations and usages, relational database schema and data consistency rules) the JavaCC-based parser is used which is automatically translated into the Java class. The parser is based on the Renew parser for the reference nets implementation.

For implementing the db-net persistence layer, the SQLite RDBMS [8] together with the Java Database Connectivity (JDBC) interface is used.

3.3 DB-Net Control Layer Implementation

In the current work, the existing functionality of the Renew tool was reused and extended to allow handling the db-net control layer elements. The functionality linked with view places is based on the Renew places implementation and all the classes connected with the view places extend the corresponding Renew classes where it is possible. The same is true for the read and rollback arcs and for the db-net transitions. The elements of the implementation of the colored Petri net elements behavior which are not overridden by the implementation of the corresponding db-net elements are executed in the same way as for the corresponding elements of the implementation of the counterparts in colored Petri nets. The overridden elements of the implementation are executed based on the db-net semantics. Since the Renew uses the reference semantics for the tokens, the developed plugin allows to use the data and reference semantics together.

The implementation of the db-net control layer elements is as follows.

View Places and Read Arcs Behavior Implementation. A view place can be only input place for the transition, and it should be connected with the transition by read arc. A view place does not contain any tokens but simulates their existence by retrieving the data from the persistence layer database using the data logic layer query which is bound to the place.

In the *DB-nets Renew Plugin*, the *read arc binder*, which extends the Renew *input arc binder*, executes the related view place query call (the usage of the query in the particular view place) on the persistence layer database. The retrieved result is assigned to the read arc variables using the fired transition variables mapper, after which it can be used by the fired transition for performing modifying action and/or moving tokens to other places through the transition output arcs.

DB-Nets Transitions Behavior Implementation. The transition in the db-net can be bound with the action for modifying the data in the db-net persistence layer relational database. Each parameter, which is a value for inserting to and/or deleting from the database, gets its value in one of the following three ways depending on the transition action call declaration during the db-net designing: (1) from a value of the input arc variable, (2) as a constant in a model (for example, the model (Fig. 3) has the "test_descr" string parameter which is defined manually in the register action call on the *"Create Ticket"* transition) or (3) as a generated fresh value from the database sequence.

In the first case, the standard *input arc binders* as well as the *read arc binders* provide all the necessary variable values. In the second case, the value is assigned to the parameter variable by the *action call values binder* based on the parsed constant in the transition action call declaration. In the third case, the parameter value in the action call should be string and have the form of the dbn_autoincrement_tableName where the tableName is the name of the persistence layer database table whose sequence should be used for generating the fresh value. Only database sequence can be used for generating the fresh value in our current implementation. After generating, the corresponding sequence is updated in the database using the SQL/DML update query.

Because of the mentioned sequence update, execution of *action call values binder* is not guaranteed to be read-only. In conjunction with the concurrent Renew binders execution it leads to the problem of the race condition including the fact that one generated value may be retrieved several times from the database before the sequence update. This can issue errors because of the unique constraints that may be violated during performing the actions. To avoid this, the semaphore-based synchronization is used. The *action call values binder* acquires its transition instance semaphore permit. As a result, only one thread may execute the *action call values binder* code as well as the action performing code at one moment since the sequence update is committed with other transaction operations in the end of the action performing in order to allow the full rollback in the case if the action performing failed. When transition firing ends, the semaphore permit is released by the transition instance. It allows to eliminate the described problem.

When all the variables are bound with their values, the action is performed through the execution of the prepared statement series – one prepared statement for each action edited (added/deleted) fact (the database table row). The deleting statements are executed before the adding statements execution, but all they are executed in one common transaction. If an error occurs during executing of any prepared statement, for example, because of violating some of the database constraints, the transaction is rollbacked and the rollback arc usage is requested.

Rollback Arc Implementation. In the case when the execution of an action failed, for example, because of violating some of the database constraints, the transaction is rollbacked and the rollback arc usage is requested instead of the simple output arcs. If the fired transition has a rollback arc, then it will be used as the output arc instead of the simple output arcs. Otherwise, the simple output arcs will be used.

For implementing this logic, the wrapper was developed for the *output arc executable* (the latter is responsible for the tokens moving). This wrapper checks if the transition has a rollback arc *and* the rollback was requested during the current transition firing. If it was, then the simple *output arc executable* performing is blocked. In this case, the *rollback arc executable* will execute. Otherwise, the rollback arc will not be used, and the wrapper will allow the output arc to be used as usually.

3.4   DB-Net Data Logic Layer Implementation

The db-net data logic layer consists of the queries for retrieving the persistent data and the actions for modifying this data as it was described in Sect. 2.2.

Action Implementation. The actions are declared in the Renew declaration node for the whole db-net and used in the particular transitions as the action calls which also should be declared in the db-net model. Each action declaration contains the lists of the action parameters, added facts (inserted database table rows) and deleted facts (deleted database table rows). Each edited (added or deleted) fact consists of the database relation (table) name, names of the columns and their values which should be inserted/deleted. Such value can be either the action parameter (in this case it is bound with a real value on the start of the corresponding transition firing) or the constant (in this case it is used as declared). The action is performed through executing a series of the prepared statements as described in Sect. 3.3.

Query Implementation. As well as the actions, the queries are also declared in the Renew declaration node for the whole db-net and used in the particular view places as the query calls which also should be declared in the db-net model. The query should be declared as the SQL string using the SQLite dialect. The query may contain the parameters in the form `${variableName}` where the `variableName` is the name of variable, whose value should be used for replacing this string. This variable should be known to the fired transition, which means that its value should be achievable through one of three ways of getting parameter values described in Sect. 3.3. For example, if the query is `SELECT id, description FROM ticket WHERE id = ${ticket_id};` and the `ticket_id` variable value equals 42, then `${ticket_id}` will be replaced with 42, and the resulting SQL query will be `SELECT id, description FROM ticket WHERE id = 42;`. After binding all the parameters, the query is executed by the *read arc binder* as described in Sect. 3.3.

Parameterized query implementation is different from the original formalism where for the similar case the database tables *join* operation is supposed to be used [3]. It is done in order to optimize the querying process in the implementation.

3.5   DB-Net Persistence Layer Implementation

The persistence layer in the *DB-nets Renew Plugin* is represented by the SQLite RDBMS. It is connected to the Renew using the JDBC interface.

For creating the database connection, the JDBC URL is declared in the db-net declaration node as well as the database schema in the series of the SQL/DDL and DML queries. When the db-net simulation starts and the db-net control layer instance is created, it tries to create the connection based on the given JDBC URL. If it is successful, then the SQL/DDL and DML queries are executed to create the necessary database schema. If there are some problems, then simulation is stopped, and the standard Renew tool error message is outputted for the user.

The created database connection instance is stored in the db-net control layer instance and therefore can be accessed by any db-net control layer elements *instances, occurrences* and *executables* for performing the queries and actions.

When the simulation stops, the database connection is closed either by executing the db-net control layer instance `finalize()` method by the Java Virtual Machine (JVM) garbage collector or by the next db-net control layer instance with the same JDBC URL. For this goal, the static connections map (the connections mapped by their JDBC URLs) is used.

This described technical solution has led to the software product presented in the next Section.

3.6   Developed Simulator Running Example

The developed *DB-nets Renew Plugin* allows to model (design) the db-net through the standard Renew GUI. The example of the modeled db-net is shown on the screenshot in Fig. 3. It is the db-net model which was shown in [3] and was presented in the current paper in Fig. 2.

The multiline text in the top part of the screen is the db-net declaration node. It contains the JDBC URL declaration for creating the database connection (in this example the "`dbn_jdbc_url <{ jdbc:sqlite:emp.db }>`" string (without quotes) is such a declaration; if such a file does not exist, but it is possible to create it, it will be created automatically), the database schema declaration via the SQL/DDL and DML queries (in the "`dbn_ddl <{ ... }>`" expression), the queries' and actions' declarations.

In the given example, one of the declared queries is the tickets query – "`dbn_query tickets <{ SELECT id, description FROM ticket WHERE id = ${ticket_id}; }>`" . The "`${ticket_id}`" expression is replaced with `ticket_id` variable value in the read arc of the view place where the corresponding query call is used. This query retrieves the data from the `tickets` table by the given *id*.

One of the declared actions in the example is the register action – `dbn_action register { params = <ticket_id, emp_name, ticket_description>, add = { ticket(id: ticket_id, description: ticket_description), resp (emp_name: emp_name, ticket_id: ticket_id) }, del = { } }}`. The action parameter names here are `ticket_id`, `emp_name` and `ticket_description`. They are

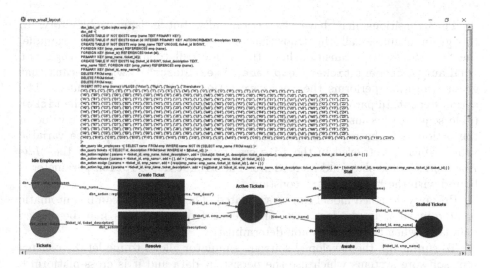

Fig. 3. The example of the db-net for the employees and tickets model. (Color figure online)

replaced by real values in the transition where the corresponding action call is used. This action adds to the `ticket` relation (database table) the row where the `id` column value is equal to the `ticket_id` action parameter value in the corresponding action call and the `description` column value is equal to the `ticket_description` action parameter value in the corresponding action call. This row is one of two added facts of this action. Also, the action adds the row into the `resp` table where the `emp_name` column value is equal to the `emp_name` action parameter value in the corresponding action call and the `ticket_id` column value is equal to the `ticket_id` action parameter value in the corresponding action call. This action does not have deleted facts.

The yellow-brown circles (have the *"Idle Employees"* and *"Tickets"* names in the given example) are the view places. The lines which start/end in them are the read arcs. Other circles are obvious Petri net places and other arcs are obvious Petri net arcs except of the red arc (from the *"Awake"* transition to the *"Stalled Tickets"* place) which is the rollback arc. The rectangles represent the transitions. Each transition should be the db-net transition, the obvious Petri net transitions are not allowed in the developed plugin.

Each view place should have the inscription with the query call. In the given example, the *"Idle Employees"* view place have the query call "`dbn_query : idle_employees`" which means that the declared `idle_employees` query is executed to retrieve the persistent data. The *"Tickets"* view place have the query call "`dbn_query : tickets`" which means that the declared `tickets` query is executed to retrieve the data. The "`${ticket_id}`" expression in this query will be replaced by the `ticket_id` variable value.

The transitions may have the action calls, but it is not obligatory. In the given example, all the transitions have the action calls. For example, the "*Create Ticket*" transition have the action call "dbn_action : register (dbn_autoincrement_ticket, emp_name, "test_descr")" which means that given transition performs the register action when it fires (executes). The action ticket_id parameter is bound with the generated value from the ticket table sequence because of the dbn_autoincrement_ticket parameter value in the action call, the emp_name parameter is bound with the emp_name variable value, which is retrieved in the idle_employees query call through the "*Idle Employees*" view place and its read arc. The ticket_description parameter is bound with the "test_descr" constant.

Properly designed model can be simulated. The simulation can be automatic or step-by-step. In the last case, fired (executed) transitions may be chosen manually, when the choice is non-deterministic.

To sum it up, the developed software simulator allows to model the concurrent software systems which use the persistent data and it is cross-platform – tested on the Windows 10 and Ubuntu 18.04.4 LTS.

4   Conclusion

In the current work the software simulator supporting the db-net formalism together with the existing Renew reference semantics is developed in the form of the Renew plugin. The developed tool allows to model the db-nets in the Renew GUI and to simulate them with usage of the SQLite RDBMS in the db-net persistence layer implementation.

This tool is the first software implementation of the db-net formalism with reference semantics support available in public. This simulator can be used in the research tasks as well as in the industrial tasks when the complex concurrent software system which uses the persistent data is under consideration. Also, this simulator can be used for further research and modifying the db-nets as well as for developing new formalisms which consider the persistent data.

The possible ways of the further development of the plugin are adding ability for user to input the external data during the simulation by the Renew request, increasing the usability by adding more understandable error output and other improvements.

References

1. Jensen, K.: A brief introduction to coloured Petri Nets. In: Brinksma, E. (ed.) TACAS 1997. LNCS, vol. 1217, pp. 203–208. Springer, Heidelberg (1997). https://doi.org/10.1007/BFb0035389
2. Jensen, K., Kristensen, L.M.: Coloured Petri Nets: Modelling and Validation of Concurrent Systems. Springer, Heidelberg (2009)

3. Montali, M., Rivkin, A.: DB-Nets: on the marriage of colored Petri nets and relational databases. In: Koutny, M., Kleijn, J., Penczek, W. (eds.) Transactions on Petri Nets and Other Models of Concurrency XII. LNCS, vol. 10470, pp. 91–118. Springer, Heidelberg (2017). https://doi.org/10.1007/978-3-662-55862-1_5
4. Petri, C.A., Reisig, W.: Petri net. Scholarpedia http://www.scholarpedia.org/article/Petri_net. Accessed 25 May 2020
5. Reisig, W.: Understanding Petri Nets: Modeling Techniques, Analysis Methods. Case Studies. Springer, Heidelberg (2013)
6. Renew - The Reference Net Workshop. http://www.renew.de/. Accessed 25 May 2020
7. Sinha, A.: DB-Nets-Extension: a CPN tools extension in order to facilitate DB-Nets - GitHub. https://github.com/sinhaaman/DB-Nets-Extension. Accessed 4 Oct 2020
8. SQLite Home Page. https://www.sqlite.org/. Accessed 25 May 2020

Author Index

Printed in the United States
by Baker & Taylor Publisher

Printed in the United States
by Baker & Taylor Publisher Services